SAP Fiori® Apps for SAP S/4HANA®

SAP PRESS

SAP PRESS is a joint initiative of SAP and Rheinwerk Publishing. The know-how offered by SAP specialists combined with the expertise of Rheinwerk Publishing offers the reader expert books in the field. SAP PRESS features first-hand information and expert advice, and provides useful skills for professional decision-making.

SAP PRESS offers a variety of books on technical and business-related topics for the SAP user. For further information, please visit our website: *www.sap-press.com*.

Wolfgang Fitznar, Dennis Fitznar
Using SAP S/4HANA: An Introduction for Business Users
2022, 420 pages, paperback and e-book
www.sap-press.com/5065

Krishnamoorthy, Murray, Reynolds, Teesdale
SAP Transaction Codes: Your Quick Reference to
Transactions in SAP ERP (2nd Edition)
2016, 648 pages, paperback and e-book
www.sap-press.com/4070

Bardhan, Baumgartl, Dascalescu, Dudgeon, Górecki,
Lahiri, Maund, Meijerink, Worsley-Tonks
SAP S/4HANA: An Introduction (5th Edition)
2025, 744 pages, hardcover and e-book
www.sap-press.com/5973

Janet Salmon, Stefan Walz
Controlling with SAP S/4HANA: Business User Guide (2nd Edition)
2025, 697 pages, hardcover and e-book
www.sap-press.com/6063

Tritschler, Walz, Rupp, Mucka
Financial Accounting with SAP S/4HANA: Business User Guide (2nd Edition)
2023, 571 pages, hardcover and e-book
www.sap-press.com/5698

Anand Seetharaju, Jon Simmonds, Tinotenda F. Chiraudi

SAP Fiori® Apps for SAP S/4HANA®

The Quick Reference Guide

 Rheinwerk
Publishing

Editor Megan Fuerst
Acquisitions Editor Emily Nicholls
Copyeditor Melinda Rankin
Cover Design Graham Geary
Layout Design Graham Geary
Production Graham Geary
Typesetting SatzPro, Germany
Printed and bound in the United States of America, on paper from sustainable sources

ISBN 978-1-4932-2723-5
1st edition 2026

© 2026 by:
Rheinwerk Publishing, Inc.
2 Heritage Drive, Suite 305
Quincy, MA 02171
USA
info@rheinwerk-publishing.com
+1.781.228.5070

Represented in the E.U. by:
Rheinwerk Verlag GmbH
Rheinwerkallee 4
53227 Bonn
Germany
service@rheinwerk-verlag.de
+49 (0) 228 42150-0

Library of Congress Cataloging-in-Publication Control Number: 2025034927

Contents

Preface

If you're familiar with the classic SAP Business Suite, you've likely encountered some of the 16,000 transaction codes that exist in the system. These transaction codes formed the bedrock of a robust and long-standing enterprise resource planning (ERP) system, which is still used by thousands of customers around the globe today. However, since the advent of SAP S/4HANA in 2015, a new user experience has been introduced: SAP Fiori.

What exactly is SAP Fiori, and how does it differ from the traditional SAP GUI transaction codes that have served SAP so well for so long?

SAP Fiori is both a design system and a technology platform. This book deals with SAP Fiori as a technology platform, but it is worth spending a few short sentences on what SAP Fiori offers as a design system.

SAP Fiori as a design system is built for three key experiences: the web, mobile devices, and a conversational user experience (UX). The system is a suite of tools and guidelines that can be harnessed to design SAP Fiori apps that meet a consistent and SAP-approved standard. SAP offers various tools for achieving this, including SAP Screen Personas, SAPUI5, SAP Build, SAP Build Work Zone, and Joule. The documentation available in the SAP Fiori suite of products also includes guidelines for working with mobile development kit (MDK) and software development kit (SDK) tools, Apple's iOS, Android, Bot Builder, Slack, and Alexa.

This book serves as a reference for the most useful and commonly used SAP Fiori apps in SAP S/4HANA. The entries list the apps in alphabetical order, include the SAP Fiori app ID, and give a small amount of detail about the purpose of each app and how it can be used in a business context. In addition, we will highlight tips wherever applicable to help you maximize the potential use of these apps. The focus of this book is on the private cloud and on-premise offerings of SAP S/4HANA and the associated SAP Fiori apps. This is not to say that these apps are not available for SAP Cloud ERP (the public cloud offering), but the focus remains on the more widely used SAP Cloud ERP Private and the on-premise solutions.

SAP Fiori has over 15,000 apps available as part of its private cloud offering. This book does not attempt to detail all 15,000; that would require far too many pages and would likely not be at all intuitive. Instead, we have focused entirely on usefulness, relying heavily upon our extensive experience in SAP S/4HANA projects to focus on the SAP Fiori apps that are likely to give the most value.

The apps contained within the book are native to SAP S/4HANA 2023, which is the latest version at the time of writing. The SAP Fiori version used for detailing these apps is SAP Fiori 3.

Some of the apps also have an associated SAP transaction code; where these exist, we have listed them to help those who are familiar with SAP GUI map their workflow to SAP Fiori. It is not, however, the intention of this book to detail all the SAP transaction codes that have also been rendered, unchanged, as SAP Fiori apps. For that purpose, consider consulting another SAP PRESS book: *SAP Transaction Codes: Your Quick Reference to Transactions in SAP ERP* (Krishnamoorthy et al. 2016).

Target Audience

This book is written for professionals who are using SAP Fiori as part of their day-to-day work. This includes SAP consultants, SAP architects, SAP project managers, SAP developers, super users, business users, configuration specialists, business process specialists, architects, and administrators. Maybe you need to understand SAP Fiori apps and how they can help you in an upcoming SAP S/4HANA implementation or SAP migration, or maybe you simply want to delve a little deeper into your chosen area of SAP.

Regardless of your specific role, if you interact with SAP Fiori apps, this is the book for you.

Navigating This Book

The apps detailed in this book are organized based upon the functionalities within which they are used. There are several apps, such as My Inbox, that cross all SAP S/4HANA functionalities, and these types of apps are detailed in Chapter 11.

The apps are listed alphabetically, but they also contain the SAP Fiori app ID. This allows you to use the book's extensive index to find the app you are looking for. You can search by one of three methods:

- Search by app name. This is the most common method of searching.
- Search by SAP Fiori app ID.
- Search by keyword. If you are not aware of the app ID or the exact name of the app, then you can search by keywords that relate to the specific function of the app.

In SAP Fiori, as with SAP GUI transactions, functionality can be achieved via a variety of different apps. In these circumstances, we mention all the various SAP Fiori apps.

Note that in some cases, apps may appear in multiple chapters when they're relevant to more than one functional area. For example, there is a potential overlap between materials management and warehouse management. If apps are truly cross functional, they appear in Chapter 11.

This book contains the following chapters:

- **Chapter 1** explains the SAP Fiori apps for the financial accounting functionality, including the following subfunctions: asset management, general ledger, accounts receivable, accounts payable, and cash and funds management. This chapter does not include many treasury or financial supply chain management transactions. The apps included in this chapter are intended to meet an organization's external and legal reporting requirements and control the consistent and structured process for recording financial transactions.

- **Chapter 2** explains the SAP Fiori apps in the controlling functionality, including the following subfunctions: cost and profit center accounting, internal order and product cost planning, and cost object controlling and profitability analysis. The apps in this chapter are intended to meet an organization's business requirements for internal and management reporting. The controlling subareas and associated SAP Fiori apps facilitate controlling master data, resource planning, recording actual business transactions, and flexible reporting.

- **Chapter 3** explains the SAP Fiori apps in the sales and distribution functionality, including the following subfunctions: master data for business partners and pricing, sales management (including inquiries, quotations, contracts, scheduling agreements, and sales orders), shipping, and billing. These apps are intended to meet the business requirements for the sale and distribution of materials and services as a part of the wider supply chain process. For example, the process may start with an inquiry from a potential customer, leading to a quotation and then to a sales order. The operations of this sales order include credit checks and availability checks (including advanced available-to-promise [ATP]) according to the production or purchasing lead times. Follow-on processes include issuing delivery documents, picking/packing and shipments, goods issue with associated material movements, and finally billing, transfer to accounting, and revenue recognition.

- **Chapter 4** explains the SAP Fiori apps for the warehouse management and inventory management functionalities. These necessarily include extended warehouse management (EWM) and stock room management, but they also include the extensive suite of apps aligned to transportation management (TM). Processes covered by the SAP Fiori apps within this section include inbound movement of goods, outbound movement of goods, and stock transfers between locations.

- **Chapter 5** explains the SAP Fiori apps for the production planning and manufacturing functionalities. This chapter covers a wide range of functionalities to ensure efficient and effective production planning and execution. The chapter includes process areas such as production engineering and operations (PEO), predictive material and resource planning (pMRP), demand management, MRP, process and production order execution, capacity planning and leveling, Kanban, reporting, and more.

- **Chapter 6** explains the SAP Fiori apps for the plant maintenance functionality. This chapter covers plant maintenance apps that support the preventive, corrective, and predictive maintenance processes. These apps cover key features such as maintenance requests, maintenance planning, maintenance notification handling, cost analysis, work permit management, job scheduling, reporting, and more.

- **Chapter 7** explains the SAP Fiori apps for the materials management functionality. This chapter covers apps used in procurement, inventory management, and logistics, ensuring efficient material flow within an organization. In this chapter, we cover SAP Fiori apps for vital processes such as central procurement, supplier evaluations, sourcing projects, purchasing, good receipts and issues, and more. This area is central to many functions in SAP, so the processes necessarily flow onward to and from other areas such as sales and distribution, production planning, financial accounting, quality management, plant maintenance, and warehouse management.

- **Chapter 8** explains the SAP Fiori apps for the quality management functionality. Quality management encompasses quality planning, quality inspection, quality improvement, and quality analytics. This chapter details a wide range of applications that are useful for organizations that manage quality control, quality assurance, and compliance with industry standards and regulations. The chapter covers apps for process areas that include but are not limited to master inspection characteristics, defect management, quality tasks, inspection lots and plans, nonconformance, failure mode and effects analysis (FMEA), quality levels, reporting, workflows, and more.

- **Chapter 9** provides a comprehensive overview of all SAP Fiori apps for the Project System component. The chapter enables you to quickly locate the relevant apps for structuring projects, scheduling tasks, tracking costs and revenues, managing procurement, and monitoring progress from initial planning to execution within an enterprise.

- **Chapter 10** explores all SAP Fiori apps for the flexible real estate management component. The chapter lets you quickly locate and utilize apps to perform critical tasks such as analyzing real estate objects, tracking valuations, managing contracts, and overseeing property assignments.

- **Chapter 11** explains the SAP Fiori apps that touch upon all the various functional areas, known as *cross-functional apps*. This includes apps such as My Inbox that are useful in their own right in SAP Fiori without having to be tied to a specific function. You will also find apps here that are covered elsewhere. This is because some apps related to areas such as business partners in SAP are also used across many functions, and they play an important role in functions such as financial accounting, sales and distribution, and materials management. This chapter offers an overview for administrators, configuration experts, and business process specialists of the SAP Fiori apps that are relevant to their roles.

- **Chapter 12** concludes the book, with a short summary followed by recommendations for further reading on the topic of SAP Fiori.

SAP Fiori App IDs

An SAP Fiori app ID can be found in the **About** information, which can be accessed from the **Me** section of the user menu in SAP Fiori, as shown in Figure 1.

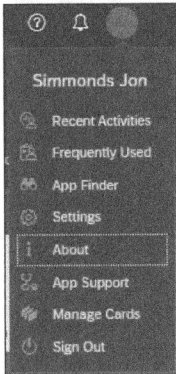

Figure 1 Navigating the Me Section Within an SAP Fiori App

Once you click the **About** option, the app ID is one of several data options that appears on the screen, as shown in Figure 2.

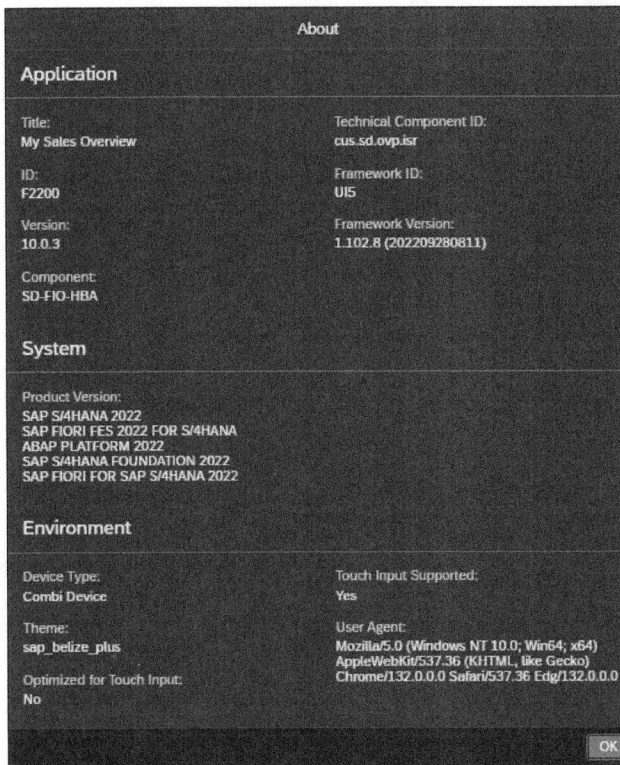

Figure 2 App ID F2200 Shown for Selected App

Additional data points available here include the title of the app as well as technical information such as the version of the app and the technology the app is built on (e.g., **UI5**). The information under **System** shows the SAP version being used, along with the SAP Fiori and ABAP platforms. The **Environment** section shows the type of device upon which the SAP Fiori launchpad is currently being rendered, whether the device supports touch input, and which SAP Fiori theme is being used.

Acknowledgments

We the authors would like to thank our family, friends, and colleagues for their support in writing this book. Our individual acknowledgments can be found in the following sections.

Anand Seetharaju

First and foremost, I extend my deepest gratitude to my family for their unwavering encouragement and support throughout the writing of this book. Your willingness to sacrifice weekends and share in the journey has been invaluable, allowing me the time and space to complete this work.

I am profoundly grateful to Emily Nicholls for providing me with this incredible opportunity and for her insightful guidance, which helped shape the book's initial structure. Emily, your confidence in my abilities has inspired me to push beyond my limits and strive for excellence.

I would also like to express my heartfelt thanks to Megan Fuerst for her timely guidance and encouragement throughout the writing process. Your professionalism and constructive criticism have significantly enhanced the quality of this book, and I truly appreciate your support.

A special acknowledgement goes to my coauthors, Jon Simmonds and Tinotenda Chiraudi. Your support and encouragement made the writing process not only productive but also enjoyable. Jon, the foundation you laid at the beginning and your exceptional project management skills were crucial to bringing this book to fruition.

Lastly, I want to thank my friends and colleagues for their invaluable feedback and support. Your insights have enriched my work and motivated me to keep pushing forward.

Jon Simmonds

As with any SAP project (and SAP PRESS projects are no different), it takes a small village to make everything tick. This village is made up of friends, family, and colleagues. It goes without saying that without support from each of these three groups, the book would not have been possible.

This project came about after a lengthy discussion with Emily Nicholls, who I would like to thank for her support and enthusiasm in getting it off the ground. Without Emily, this book would not be in your hands (or on your screen).

Similarly, a big thank you goes to Megan Fuerst for her wisdom and thoughtful guidance to all of us throughout the process of writing this book. Thank you, Megan!

My coauthors, Anand and Tinotenda, have been unswerving in their commitment and made the writing of this book a pleasure rather than a chore. Their knowledge and flexibility have been greatly appreciated. I've thoroughly enjoyed being part of the author team with them.

I must also mention Matthew Phu, a mentor and boss, and someone who not only has granted me the freedom to pursue these SAP PRESS adventures but also provided the SAP systems and SAP Fiori launchpad, and access to the associated experts as required, for those adventures to bear fruit.

The biggest thanks must go to my family: Susan, my wife, as well as my two (not so young now) children, Ella and Charlie. They never complained when I disappeared for evenings and weekends to complete chapters.

Tinotenda F. Chiraudi

Writing this book has been a journey filled with challenges, learning, and growth. It would not have been possible without the unwavering support of the people who stood by me and believed in me even when I was miles away, working on projects around the world.

First and foremost, I want to express my deepest gratitude to my family. To my parents, thank you for instilling in me the values of perseverance, curiosity, and hard work. Your constant encouragement and sacrifices have been the foundation of everything I have achieved. To my siblings, thank you for being my cheerleaders and for always reminding me of the importance of balance and laughter, even during the busiest of times.

I also want to extend heartfelt thanks to Anand and Jon—true luminaries in the industry and two individuals I deeply admire. Your work has been a constant source of inspiration, and your dedication to excellence has set a standard I strive to emulate. Your insights and contributions to the SAP field have not only shaped my professional journey but also motivated me to push boundaries and aim higher.

My thanks to Megan Fuerst and Emily Nicholls, whose editorial guidance was instrumental throughout the journey of bringing this book to life. Their keen eye for detail, thoughtful suggestions, and unwavering support helped shape this manuscript into its final form. I am deeply thankful for their patience, encouragement, and belief in my work and that of my colleagues, Jon and Anand.

To my colleagues, mentors, and friends who have supported me throughout this journey, thank you for your guidance and encouragement.

Finally, to the readers of this book, thank you for allowing me to share my experiences and insights with you. I hope this work inspires you, challenges you, and contributes to your own journey of growth and success.

Feedback

The construction and writing of this book has been a lengthy process, and sometimes it feels like we are working in our own little world, ploughing through page counts and chapter targets. However, what makes this book valuable for readers is the usability of its content—and therefore, we need to know how useful it has been. As with all SAP PRESS books, we welcome feedback from our readers. The SAP Fiori user experience is no longer a new area, but it is still (sadly) underutilized in the SAP arena. Many SAP consultants and business users are all too familiar with the old SAP GUI, and old habits die hard: We understand that. Hopefully this book will give our readers a little more confidence to dip their toes into the SAP Fiori waters, where you will find the temperature pleasant and inviting.

Please let us know your thoughts and experiences.

Introduction

If you're picking up this book as a reference guide, it's a safe bet that you are already reasonably familiar with SAP as an enterprise resource planning (ERP) system. It's another reasonable bet that you have experience in SAP ERP 6.0 and maybe even SAP S/4HANA, but your experience of SAP Fiori may be limited or even nonexistent. Fear not! SAP Fiori may seem bewildering in its scope, but there are many standardized principles that, once mastered, will allow you to navigate the user experience of SAP Fiori with confidence.

In this introductory chapter, which we promise to keep as short as possible, we will go over an introduction to the user interface of SAP and how it has evolved through the years, from the SAP R/1 days to SAP ERP 6.0 and on to SAP S/4HANA and SAP Fiori.

We will also explore what is meant when we say *SAP Fiori* (hint: it's not just a user interface technology platform!), what the different types of SAP Fiori apps are, and how the platform is structured in terms of security and navigation.

The SAP S/4HANA Era

The era of SAP S/4HANA has taken decades to build. SAP started its life in 1972 as the brainchild of five former IBM employees. Initially, SAP used the IBM DOS operating system to develop its first solution under the name SAP R/1.

The full integration of modules such as finance, materials management, and sales and distribution was groundbreaking for its time, and further integration of these modules led to the first real user interface (UI) in the SAP world in the era of SAP R/2. The look and feel of the interface relied upon text, with no icons or graphics at all and navigation through the keyboard only; although it would appear extremely dated by today's standards, it was modern for its time.

The definition and development of SAP, including the introduction of the ABAP programming language, were major milestones in the march toward the new graphical user interface (GUI), branded SAP R/3, which was finally released in 1991 to extremely positive reviews. The SAP R/3 GUI introduced for the first time a modern, full-color GUI that could be navigated via a mouse as well as a keyboard—revolutionary stuff for the early 1990s, when home personal computers (PCs) were just coming into fashion. SAP rode that trend incredibly successfully, and SAP R/3 rapidly became the benchmark for ERP GUIs.

The SAP R/3 GUI was a long way ahead of its time; as a result, it stuck around for many years. Even now, deep in the SAP S/4HANA era, the SAP R/3 GUI is familiar to many.

Tweaks were made to the GUI, especially after the dot-com boom, with the integration of a new and colorful interface under the banner *mysap.com*.

By the middle of the 2000s, the SAP GUI needed some freshening up, so SAP launched SAP ERP Central Component (SAP ECC), as SAP ECC 5.0, soon to be upgraded to SAP ECC 6.0, the GUI for which is shown in Figure 3.

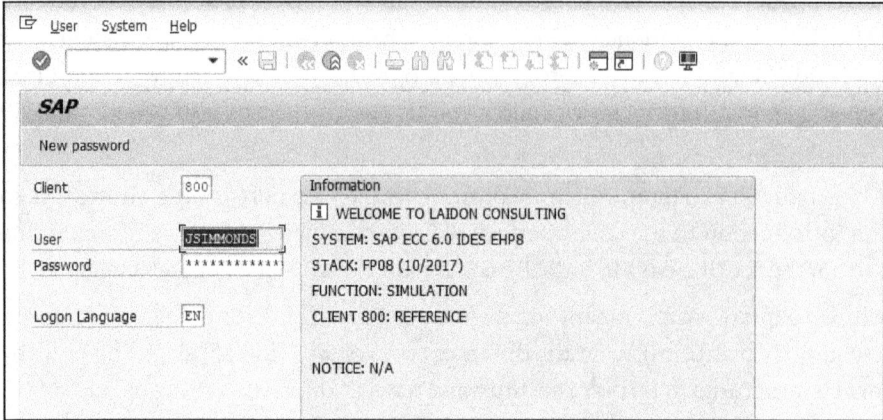

Figure 3 SAP ECC 6.0 GUI

SAP ECC 6.0, more commonly known as SAP ERP, has been the most successful and widely used ERP system in the world. However, the aggressive expansion of SAP into new markets and new acquisitions meant that a whole barrage of new solutions was being added to the SAP portfolio. These acquisitions included SuccessFactors, Ariba, Fieldglass, and Concur, all of which came with their own UIs.

A consolidated approach to the UI was required. Enter SAP S/4HANA in 2015 and, along with it, SAP Fiori, the first incarnation of which came in 2013, assigned to SAP ERP 6.0.

SAP Fiori 1.0 was the first SAP UI that was device agnostic and built on SAPUI5 technology. This platform first introduced analytical and fact sheet apps (more information on these later).

SAP Fiori also used tiles for the first time. These tiles could summarize data held within the individual apps on the front of the tiles, with color-coded numbers and basic charts.

SAP Fiori was branded as a *user experience* (UX) rather than a *user interface*. So, what's the difference? In SAP's own words, the term *user experience* is used rather than user interface because the overall experience of the user is considered, rather than just the look of the screen. As SAP S/4HANA progressed, and the UX took center stage more and more, SAP began to evolve the SAP Fiori approach, with the next iteration—SAP Fiori 2.0—delivered in 2015 and shipped as standard with SAP S/4HANA 1610.

The SAP Fiori overhaul as part of the launch of SAP Fiori 2.0 moved the offering much closer to the look and feel of today's SAP Fiori experience. The screen was separated into three areas: the Me Area to the left, which is accessed by clicking your user icon; the Workspace in the middle, with all your business apps; and finally, the Notifications space to the right.

The reception to the new SAP Fiori experience was largely positive, with the design winning the Red Dot Design Award in 2016, an annual international design competition for industry, brand, and communication design. However, SAP was still facing criticism that its portfolio of products was becoming unwieldy, with different UX setups for different products. SAP addressed this criticism with the launch of SAP Fiori 3, which first shipped with SAP S/4HANA Cloud 1908 and on-premise SAP S/4HANA 1909.

For the first time, SAP Fiori was the face of all SAP's applications—marking an important step toward a unified SAP UX. The top of the screen was rebranded as the shell bar and, next to the **Home** button, it contained the launchpad to access any SAP applications. Further changes also were introduced, such as the Me Area being moved to the user actions menu on the right of the screen.

Further innovations were delivered in 2020 and shipped with SAP S/4HANA 2020 (for private cloud). These included the important restructuring of the launchpad to encompass spaces and pages instead of groups. Previously in SAP Fiori, apps were organized into catalogs and then presented in the launchpad in *groups*, flat structures to show the available apps. With spaces and pages, as shown in Figure 4, *spaces* (such as **Internal Sales**) allow a hierarchical experience, while *pages* (such as **Sales Processing**) allow a more detailed opportunity to categorize apps into logical areas.

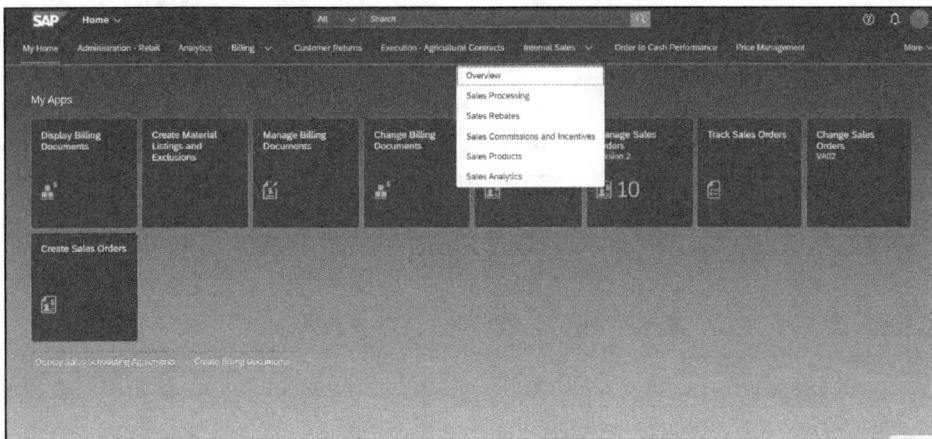

Figure 4 Spaces and Pages in SAP Fiori

Furthermore, each page, such as **Sales Products**, could be subdivided into *sections*, such as **Master Data** and **Product Proposals**, as shown in Figure 5.

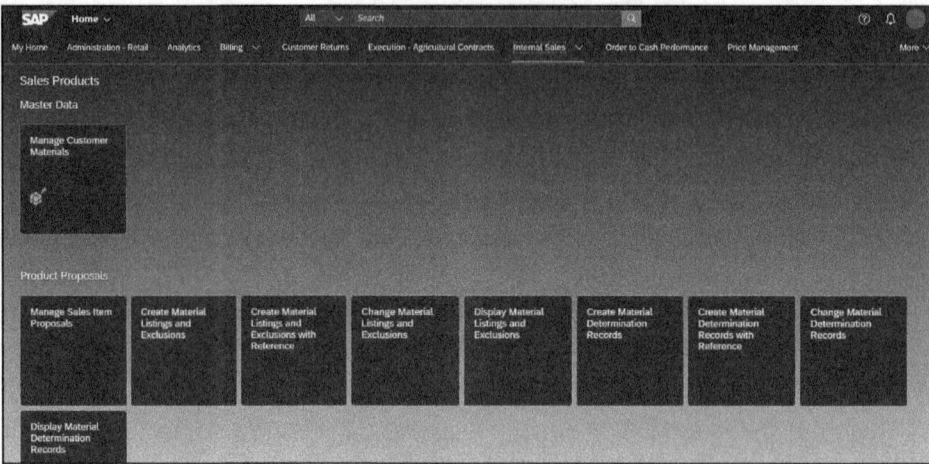

Figure 5 SAP Fiori Sections Within Spaces and Pages

So, what's next for the SAP Fiori UX? It has been six years since SAP Fiori 3 was launched. Many of its innovations are now embedded into SAP application releases, such as SAP S/4HANA 2023, and new updates are being delivered as part of SAP Fiori 3, meaning that there may not be an SAP Fiori 4.

In addition, it is important to note that all new innovations are being delivered to SAP Cloud ERP (specifically, the SAP S/4HANA Cloud Public Edition software) first, before being rolled out to SAP Cloud ERP Private (specifically, the SAP S/4HANA Cloud Private Edition software). On-premise customers are still getting access to updates, but not as frequently as cloud customers. This underpins SAP's *cloud-first* strategy. An example of this can be seen in SAP's sustainability and AI innovations, which are only available to public and private cloud customers.

Additional SAP Fiori themes also are constantly being delivered and upgraded. The most innovative of these is the Horizon theme, in which the structure of the SAP Fiori screen has been redefined to offer a fresher and more modern look, with additional options for customer branding.

The SAP Fiori User Experience

We have already mentioned that SAP Fiori is a UX rather than simply a UI, but let's delve a little deeper into what makes it so.

The structure and standardization of SAP Fiori is designed in such a way as to make the user's professional life seamless and controlled, while also being delightful. One of the ways in which this is achieved is by using SAP Fiori elements.

SAP Fiori Elements

Have you ever noticed that standard SAP Fiori apps always seem to have the same styling and the same look and feel? SAP manages this by employing *SAP Fiori elements*, a toolbox for SAP Fiori that contains floorplans.

SAP Fiori elements *floorplans* are standardized templates for the visualization of SAP business data in SAP Fiori apps. Each floorplan offers a template for a type of app, in which the structure conforms to a standardized format: headers, filters, table format, quick actions, and so on. Figure 6 shows the difference between a traditional ABAP change in SAP GUI and an SAP Fiori change using SAP Fiori elements.

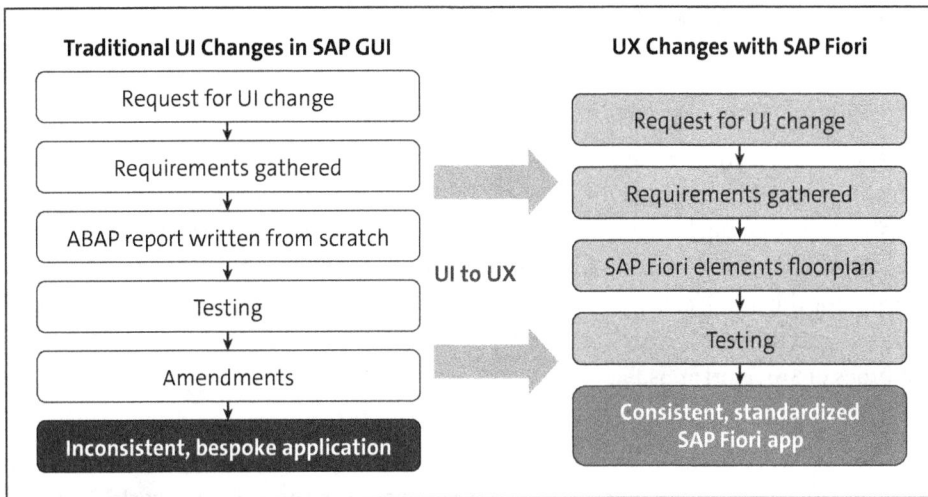

Figure 6 Simplification and Standardization Using SAP Fiori Elements

In traditional UI changes using ABAP, standardization was not enforced, leading to disparate solutions that were both difficult to upgrade and inconsistent. In contrast, SAP Fiori elements floorplans enforce a standardized look and feel in your SAP Fiori apps, which offers a consistent approach with built-in scalability.

There are five floorplans delivered as part of SAP Fiori elements:

- **List report**
  List reports are designed for large amounts of data for which no drilldown is necessarily required, but the user can still filter and switch between different views of the data.
- **Worklist**
  Like the list report, the worklist floorplan is designed for apps that present a list of items for action, but the list report has more flexibility to display items based upon a list of criteria.

- **Object page**
  An object page can be used to display, create, or edit data. This type of floorplan can be very useful for objects such as materials, for which an overview is required, but you also need the flexibility to make amendments.

- **Overview page**
  Designed as a hub for certain role definitions, the overview page floorplan can be used to show pertinent data all in one place, with multiple drilldowns available to related apps.

- **Analytical list page**
  One of the most-loved SAP Fiori floorplans, the analytical list page is designed to run an analysis and extract the root causes of issues.

Application Types

Floorplans represent the ways in which SAP delivers standard SAP Fiori apps. The apps themselves can be separated into three different types—transactional, analytical, and fact sheet—which can be designed using these floorplans. SAP Fiori covers visual representations of existing functionality and new functionality that has been designed specifically for the SAP Fiori UX, as well as revamped existing SAP GUI transactions that have been stylized within an SAP Fiori framework. Let's look at the details of the different types of SAP Fiori apps.

Transactional Apps

Transactional apps are the most common types of apps in the SAP Fiori world. These include web representations of SAP GUI transactions as well as revamped transactions in the form of apps. In addition, SAP Fiori has introduced many apps that are not available in SAP GUI.

In general, these types of apps are available on SAP S/4HANA and are available for SAP Fiori systems that are connected to SAP ERP 6.0 as well. Some, but not all, transactional apps require an SAP HANA database, especially when dealing with large volumes of data; therefore, their usability for SAP ERP 6.0 clients is limited unless an SAP Business Suite on SAP HANA landscape is deployed.

Typically, these transactional apps use the overview, list report, and worklist floorplans. Examples of such apps include Manage Sales Orders (List Report), shown in Figure 7, and Create Material (SAP GUI).

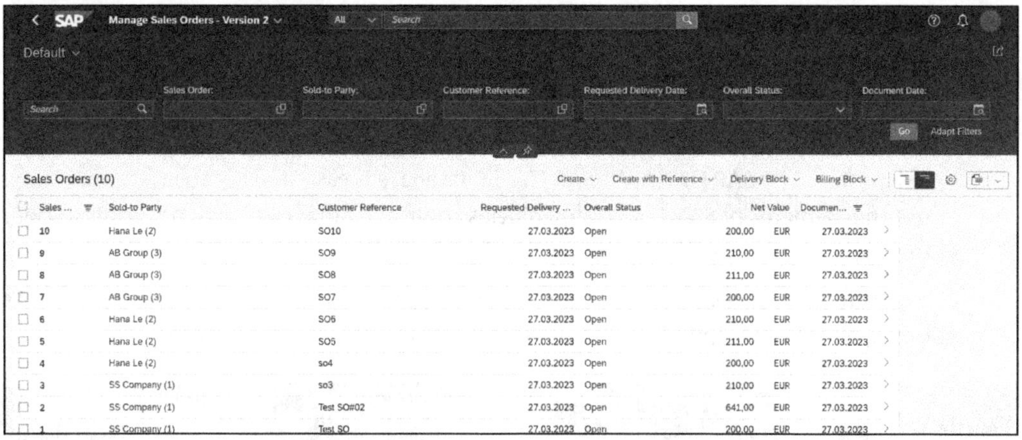

Figure 7 Transactional App Example: Manage Sales Orders

Analytical Apps

SAP Fiori users often find that analytical apps are the most useful. Certainly, when reviewing SAP's list of "lighthouse apps" (apps that offer the most business value in deployment), analytical apps tend to dominate.

These apps are usually restricted to SAP S/4HANA, although some are available on SAP ERP powered by SAP HANA. An SAP HANA database is essential because analytical apps use in-memory SAP HANA capabilities to provide business insights at high speed from large quantities of data.

Typically, these analytical apps use the overview page or analytical list page floorplans. Examples of analytical apps include My Sales Overview (overview) and Sales Order Fulfillment (analytical list), as shown in Figure 8.

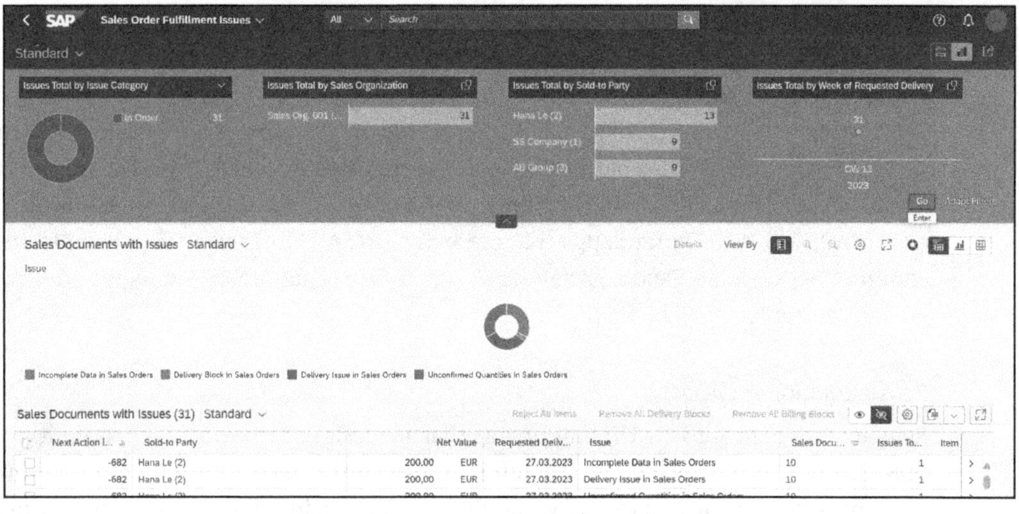

Figure 8 Analytical App Example: Sales Order Fulfillment

Fact Sheet Apps

Fact sheet apps are normally called from within other SAP Fiori apps and don't tend to have their own SAP Fiori tiles.

These apps are usually restricted to SAP S/4HANA, although some are available for SAP ERP powered by SAP HANA. An SAP HANA database is essential because fact sheet apps use the in-memory SAP HANA capabilities to provide business insights at high speed from large quantities of data.

Typically, these fact sheet apps use the overview page or analytical list page floorplan. An example fact sheet app is Customer 360° View (overview), as shown in Figure 9.

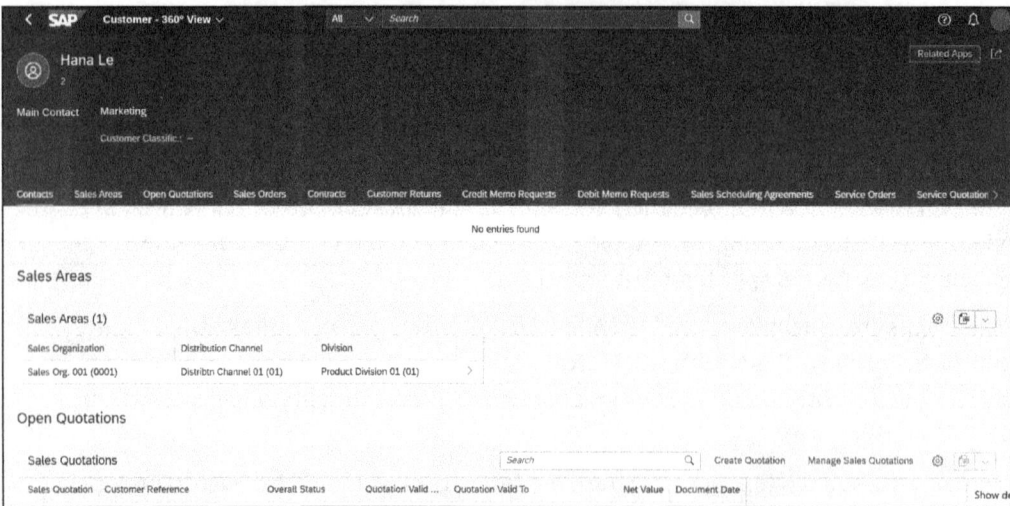

Figure 9 Fact Sheet App Example: Customer 360° View

Web Dynpro Apps

These apps are generally the same as SAP GUI transaction codes, but they use older SAP technology based upon the Web Dynpro framework, which includes both Java and ABAP. In essence, a Web Dynpro app is often a way of rendering an SAP transaction so that it looks and feels like a native SAP Fiori app. In these instances, because the app is a replication of the SAP GUI transaction, they are not covered here. There are, however, some SAP Fiori Web Dynpro apps that are native to SAP Fiori only—for example, Incoming Sales Orders – Flexible Analysis (F1249). These types of SAP Fiori apps are covered in the book.

Reuse Component Apps

Reuse component apps are UI components that are designed to be used across multiple different SAP Fiori applications. Therefore, reuse component apps are not SAP Fiori apps that can be navigated to, and we don't discuss them in this book as a general rule. However, critical transactional apps that are also reuse components will be covered.

Roles

SAP Fiori is a *role-based* application. This means that its structure is designed for a specific role, such as accounts payable administrator. To illustrate how this differs from traditional SAP GUI usage, imagine a user logging into the SAP GUI screen on SAP ERP 6.0: They would be confronted with the full SAP menu tree. This would mean that they must navigate their way through a whole series of menu paths that are not designed for the role that they play; that is, many of the transactions in those menu paths will never be employed by this specific user.

Now consider the same user working with SAP Fiori. Because SAP Fiori is a role-based application, the user will only see the applications that are relevant to their specific role.

To allow this to work, SAP uses a multilayered security structure based around the roles used in SAP Fiori. Let's first look at a traditional SAP GUI security structure, as illustrated in Figure 10.

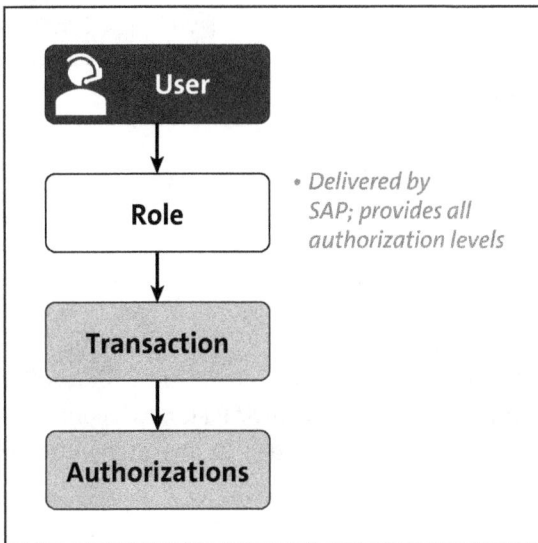

Figure 10 SAP GUI Security

SAP GUI security needs to be deployed from the bottom up; that is, the authorizations are provided first, then they are assigned to a transaction, then assigned to a role, then that role is assigned to a user. These authorizations still exist in SAP Fiori, but there are multiple additional layers on top, as illustrated in Figure 11.

SAP Fiori's security is top down, from the business role, instead of bottom up, from the business task. This is what makes SAP Fiori *role based*.

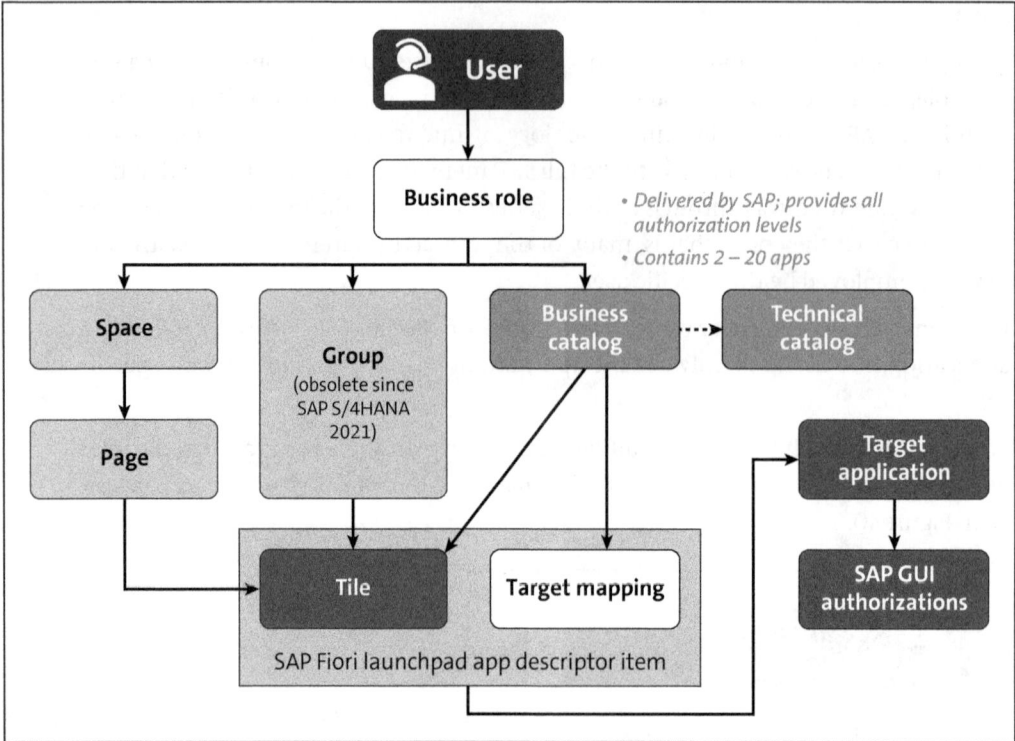

Figure 11 SAP Fiori Security

Summary

This brings us to the end of the introduction—and it sounds like it's time for a coffee break! You should now have a rudimentary understanding of SAP Fiori as a concept, a technology, and a design system, which will help you take a deeper dive into your area of interest in the following pages.

Chapter 1

Financial Accounting

SAP Fiori apps play a crucial role in modern financial accounting processes in SAP S/4HANA. These apps allow you to streamline operations, enhance your experience, and enable efficient data processing. They are designed to provide an intuitive interface to perform financial tasks in day-to-day events or period-end activities such as journal entry creation, account reconciliation, and reporting. In this chapter, we'll provide a comprehensive overview of SAP Fiori apps relevant to financial accounting, presenting them in alphabetical order for simplicity. Each app's entry will include an overview of the app's primary purpose, its key features, and how it integrates within the broader SAP financial landscape.

Account Balance Audit Trail (F1393)

Application Type: Analytical

This app is useful for external and internal auditors who need to compare and analyze the opening and closing balances of open item–managed general ledger, accounts receivable, and accounts payable accounts for a fiscal period or for a date range. In addition, the app allows you to review accounts to ensure the validity and legality of financial records and to group, sort, and filter results using various characteristics, such as company code, account, and posting date. The app's key features include displaying line items based on individual filters, comparing opening and closing balances, correlating open line items with related postings, and exporting results to Microsoft Excel. The app also supports creating and saving complex filter conditions; sorting, filtering, and grouping results lists; displaying detailed data; and saving a configuration as a home page tile. Also, if the app is integrated with SAP Jam, you can create and share links with current filter conditions via email or post them to SAP Jam.

Transaction Code

The corresponding SAP GUI transaction is Transaction S_ALR_87009850.

Account Determination (F1273)

Application Type: Transactional

This app is a configuration tool used to automate the process of assigning general ledger accounts to business transactions. It allows you to define rules and settings that determine which accounts should be used for automatic posting in various functionalities such as material management, sales and distribution, and financial accounting. This ensures that financial documents are posted correctly and efficiently, reducing manual effort and improving the accuracy of transactions. Since SAP S/4HANA 2022, the app includes enhancements such as intercompany margin accounts in the controlling area, clearing accounts for multiple valuation in the financial accounting area, new transaction keys **DPL** and **DRL** to use when defining accounts for tax clearing, and a new transaction group

(**WEC**) and transaction key (**BDS**) to use when defining accounts for bill of exchange transactions.

Transaction Code

The corresponding transaction in SAP GUI is Transaction OBYC.

Account Determination –
For Classic Contract Management
(FARR_ACCT_DETERMINATION_OVP)

Application Type: Web Dynpro

This app allows you to define account determination rules for contract-based revenue recognition in the revenue accounting and reporting (RAR) functionality in SAP S/4HANA. Account determination rules determine the actual accounts used for recognizing costs and revenue in RAR. The app allows account mapping based on business rules, ensures seamless integration with revenue accounting contracts for accurate financial postings, leverages Business Rule Framework plus (BRF+) for maintaining and validating determination rules, and includes consistency checks to verify compliance and accuracy before execution. BRF+ provides a structured framework for rule maintenance and storage, utilizing an integrated BRF+ decision table expression. You can check rules for consistency and errors, import and export decision tables into spreadsheets, and modify table settings by adding or removing condition columns or adjusting the column order.

Accounts Payable Overview (F2917)

Application Type: Analytical

This analytical overview app allows you to monitor important accounts payable indicators and access relevant accounts payable apps. As a result, you get a comprehensive and real-time overview of the accounts payable process, help-ing to streamline operations and ensure efficient handling of supplier invoices, payments, and other accounts payable–related tasks. The app allows you to filter data to view only the most relevant information. For example, you can search for accounts payable information using search parameters such as display currency, company code, supplier, accounting clerk, country key, reconciliation account, and item payment block. The app provides key features such as tracking parked invoices, blocked invoices, cash discount utilization, days payable outstanding (DPO), payable aging, suppliers with debit balances, invoices blocked in supplier master data, posted invoices in the current period, tasks and activities in the My Inbox app, due invoices free for payment, and invoice processing statistics.

Accounts Receivable Overview (F3242)

Application Type: Analytical

This analytical overview app lets you monitor key accounts receivable indicators and access related apps with ease in a comprehensive analytical dashboard. The app uses cards to display its analytical indicators. You can search for accounts receivable information using search parameters such as display currency, net due interval 1, net due interval 2, net due interval 3, company code, accounting clerk, reconciliation account, and posting key. The app includes features to display the following:

1. Accounts receivables aging analysis
2. Overdue and future receivables by age
3. Accounts receivable breakdown, highlighting the top 10 accounts receivable by total amount
4. Days sales outstanding (DSO), comparing the actual and best possible DSO
5. Cash collection tracker, tracking the total cash received and deviations from targets
6. Top 10 debtors, listing the total accounts receivable and top debtors by various criteria

In addition, default values can be maintained for filter criteria such as company code, customer, and display currency.

Accrual Postings – Simulated/Posted (W0184)

Application Type: Web Dynpro

This app provides an overview of accrual postings based on specified criteria related to the valid-to date and other optional filters. You can search for accrual postings and simulated postings with various other optional parameters, such as fiscal year, fiscal period, company code, component, accrual object, accrual transaction type, accrual item type, ledger, accrual deferral account, cost center, profit center, work breakdown structure (WBS), and lifecycle status. The app presents results data in the **Data Analysis** tab, including key information such as accrual objects, item types, and posted accrual amounts. In addition, the app allows you to compare simulated accrual amounts with actual posted values. Accrual postings can be reviewed for purchase orders, service entry sheets, and manual accruals. The app's features include running reports for actual postings, comparing simulated and actual values, using visual filters for analysis, drilling down into accrual objects for details, and navigating to related apps. The app utilizes the CDS view C_AccrEngnRptgSimlnActl-CostQry to simulate accrual postings.

As a prerequisite to use the app, you should run the Simulate Accrual Postings job via the Schedule Accruals Jobs app (F3778).

> **Tip**
>
> Although similarities exist between the Accrual Postings – With Simulation (W0183) and Accrual Postings – Simulated/Posted (W0184) apps, the former app displays posted, planned, and simulated amounts, while the latter app displays simulated and actual accrual postings for comparative purposes.

> **Transaction Code**
>
> The corresponding transaction in SAP GUI is Transaction ACEPOSTINGRUN.

Accrual Postings – With Simulation (W0183)

Application Type: Web Dynpro

This app provides an overview of accrual postings based on specified criteria related to the valid-to date and optional filters. You can search for accrual postings and simulated postings with various other optional parameters, such as fiscal year, fiscal period, company code, component, accrual object, accrual transaction type, accrual item type, ledger, accrual deferral account, cost center, profit center, work breakdown structure (WBS), and lifecycle status. When you click the **Go** button, the app displays all accrual objects, accrual subobjects, item types, and accrual amounts. The **Accrual Amount (TC)** column includes both posted and planned accruals, with unposted entries appearing as simulated amounts. In addition, you can review accrual postings for purchase orders, service entry sheets, and manual accruals. Also, you can add additional dimensions from the available fields to rows and columns. Thus, the app provides the ability to view actual and planned postings, use visual filters for analysis, drill down into accrual objects for details, and navigate to related apps. The app leverages the CDS view C_AccrEngnRptg-SmlteFutrPostgQ to simulate accrual postings.

> **Tip**
>
> Simulated amounts may change if an accrual journal entry is manually reversed, the periodic accrual amount is adjusted (such as modifying the total accrual amount or accrual periods), or the total deferral amount for purchase order deferrals is updated—for example, due to a new prepaid invoice. To view reports with the latest data, you must

schedule the Post Periodic Accruals and Simulate Accrual Postings jobs in the Schedule Accruals Jobs (F3778) app.

Transaction Code

The corresponding transaction in SAP GUI is Transaction ACEPOSTINGRUN.

Actual Cash Flow (F0513A)

Application Type: Analytical

This app provides an overview of daily cash flows for the past 90 days, helping cash managers and cash management specialists identify unusual inflows and outflows and take appropriate actions. With this app, you can display the aggregated cash flow, cash flow by liquidity item, and cash flow by company code; filter data by dimensions like calendar day and planning group; navigate detailed cash flow items; switch between charts and tables; and monitor past daily cash flows. The forecasted amount is displayed in the bank account currency and is derived from transaction data provided by memo records and the One Exposure from Operations hub. In addition, you can post comments on SAP Jam, send emails, and benefit from improved performance through caching. Data can be refreshed manually or updated automatically based on the cache duration settings.

Transaction Code

The equivalent transaction in SAP GUI is Transaction FF7A.

ADB Trial Balance (W0166)

Application Type: Web Dynpro

Average daily balance (ADB) refers to the average balance of an account over a defined period (e.g., for intramonth dates), calculated by summing daily balances and dividing by the number of days in the period. This updated app allows the display of average daily balances and spot balances across all company codes and ledgers. You can display balances using search parameters such as the ledger, company code, key date, and time stamp (date or time). The app presents data in a grid format, and you can customize the interface via drag and drop to add measures and dimensions using **Navigation Panel** functions and to organize data hierarchically. In addition, you can split balances by profit center, segment, function area, business area, supplier, customer, material, or asset. The app also supports hierarchical views, including financial statement versions, segments, and profit centers, which can be displayed by selecting the relevant rows and choosing the **Hierarchy** function. You can apply filters to quickly access balances, visualize results in a chart view or table view, create layout variants with predefined selection parameters, and export data to spreadsheets. The app relies on the standard view using the business intelligence (BI) query 2CCADBTRIALBAL.

This app replaces the deprecated Average Daily Balance Key Figure Report app (F3987).

Transaction Codes

The closest SAP GUI transactions are Transactions F.01, S_ALR_87012277, S_ALR_87012279, and FAGLB03H.

Aging Analysis (S/4HANA) (F1733)

Application Type: Analytical

With this app, you can get a comprehensive view of aging information across your organization, enabling you to identify negative trends in total payable, net due, and overdue amounts. This helps you to react promptly and have your team take appropriate actions to reverse these trends. The app offers features such as displaying payable amounts by aging, company code, general ledger account, currency, supplier group, and supplier; filtering payable amounts by various criteria; and specifying key dates for data analysis. Also, key dates can be managed using the Manage Date Functions app, and you must have the analytics specialist template role (SAP_BR_ANALYTICS_SPECIALIST) assigned to define key dates. In addition, the SAP_BR_AR_MANAGER role must be assigned to you if you want to load the app in SAP Smart Business, which helps you to make quick and informed decisions by showing KPIs. The app also supports SAP Jam integration for comments and emails. The app uses the CDS view C_APFLEXI-BLEAGING, and has caching enabled for improved performance.

Transaction Code

The corresponding transaction in SAP GUI is Transaction F.46.

Allocated Amount Explanation (F4424)

Application Type: Transactional

This app allows revenue accounting and reporting (RAR) users to verify key price allocation details for a revenue contract, including the allocation date, related attributes (e.g., standalone selling price, allocable amount, remaining standalone selling price), reallocation outcomes, and triggers for price allocation. By searching for revenue contracts, you can view contract change events relevant to price reallocation and review details such as the change type, allocable amount, and reallocation outcome for each performance obligation. The app also offers insights such as business change reasons and change logs. In addition, you can filter by attributes like revenue contract and effective date, with fuzzy search options for fields such as company code and business partner.

Allowance for Doubtful Accounts (F2686)

Application Type: Transactional, Analytical

With this app, accounts receivable users can manage doubtful accounts within an organization in SAP S/4HANA. This app provides insight into the management of allowance for doubtful accounts, allowing you to assess the adequacy of current allowance levels for possible nonpayment of overdue receivables. It offers a clear view of overdue receivables and their associated allowances, with the ability to drill down into individual customer accounts at the journal entry level. The app provides chart views of overdue receivables and allowances by customer or by country/region, details of individual accounts, separation of manually created from automatically generated allowances, and the total allowance for each customer account shown as a total and as a percentage of the overdue amount.

Transaction Codes

The closest transactions in SAP GUI are Transactions F103, FBL5N, and S_ALR_87012168.

Analyze Accrual Postings (F3732)

Application Type: Transactional

This app provides an overview of accrual postings, allowing general ledger accountants to filter these postings by various criteria and navigate specific postings within related apps for further investigation. The app's features include the use of visual filters for analyzing accrual postings, direct navigation to related apps, and the **Details** and **View By** functions for a comprehensive examination of postings.

Analyze Clearing Locks (Version 2) (F1653A)

Application Type: Transactional, Analytical

With this app, you can display valid clearing locks for receivables and payables. The app provides features such as indicating the snapshot's age with the KPI tag, restricting selected data with filters, displaying locks in a chart and table, grouping locks by expiration date or duration, and listing the number of items and total locked amounts per business partner. For example, you can analyze current clearing locks by searching via parameters such as display currency, variant, lock reason, start date, expiration date, total duration in days, and lock object category. In addition, you can create worklists for follow-up activities by selecting business partners from the table.

> **Tip**
>
> The app only selects locks that are currently valid; it does not display locks that were valid in the past or will become valid in the future.
>
> In addition, to generate worklists in the backend system, business function FICA_EHP7_B1 must be activated in the backend systems in which you wish to process worklists via Transaction FPCKWL.

Analyze Clearing Reasons (Version 2) (F2125A)

Application Type: Transactional, Analytical

This updated app is part of contract accounts receivable and payable, which is a subledger set up for companies with a high-end customer base and a need for processing mass data, used by industries such as utilities and telecommunications. With this app, you can follow the clearing reasons of receivables within a certain time frame. Using the available filters, you can restrict data by criteria like main transaction

and net due date. The selected data is displayed in charts and tables, with bars representing predefined groups of clearing reasons. In addition, you can combine multiple dimensions to adapt the display, such as grouping data by clearing reason and division. Also, in the app's settings, you can customize the dimensions displayed in the chart and override the delivered clearing reason groups in the **Grouping Clearing Reasons for Evaluations** configuration activity. The app is available for the accounts payable and receivable manager roles.

This app replaces the Analyze Clearing Reasons app (F2125), which has been deprecated.

> **Tip**
>
> SAP recommends using this app when business functions of a contract accounting industry are active in the system.

Analyze Collection Success (Version 2) (F2209A)

Application Type: Transactional, Analytical

With this updated app, you can see an overview of the percentage of successful collections within a certain time frame or related to specific collection strategies and steps for the display currency. The app analyzes the percentage of dunned open items paid as a probable result of the dunning notice, reflecting a success percentage rate assigned to the notice. The app provides features that include using filters to restrict data to meet your requirements, displaying data in charts and tables, and combining multiple dimensions for display. In addition, you can filter data by criteria such as company code, segment, contract account category, date of issue, collection strategy, and collection step, and you can customize the display by grouping data by collection strategy and week of issue.

This app replaces the Analyze Collection Success app (F2209), which has been deprecated.

Transaction Codes

The corresponding transaction codes in SAP GUI are Transactions FBL5N, UDM_SPECIAL-IST, UDM_SUPERVISOR, and S_ALR_87012168.

Transaction Codes

The corresponding transaction codes in SAP GUI are Transactions FBL5N, UDM_SPECIAL-IST, UDM_SUPERVISOR, and S_ALR_87012168.

Analyze Collection Volume (Version 2) (F2208A)

Application Type: Transactional, Analytical

With this updated app, you can display the number of items or the total amount based on the dunning procedure used in a selected display currency. In addition, you can restrict the selection of due items within a time frame based on the dunned collection strategy to specific company codes and contract account categories. The app's features include using filters to restrict data to match your requirements, displaying collection volumes in a chart and table, combining multiple dimensions for display, and customizing displayed dimensions. For example, you can apply filter criteria using parameters such as company code, contract account category, date of issue, collection strategy, and collection step. Also, you can view additional information about business partners and navigate to other apps for further analysis.

This app replaces the Analyze Collection Volume app (F2208), which has been deprecated.

Although both Analyze Collection Success (Version 2) (F2209A) and Analyze Collection Volume (Version 2) (F2208A) provide detailed analysis, there are a few important differences between these apps:

- **Analyze Collection Success (Version 2)**
  Focuses on analyzing the success rate of collections based on a certain time frame or based on specific collection strategies. Emphasizes already collected and overdue receivables.
- **Analyze Collection Volume (Version 2)**
  Focuses on analyzing the collection volume, thereby providing insights into the volume of collections based on dunning procedures. Includes both open and overdue receivables.

Analyze Credit Exposure (F2541)

Application Type: Analytical

With this app, credit managers, credit controllers, and credit risk analysts can analyze customers' credit exposure by various dimensions and measures, helping to support risk diversification, segmentation, credit, and payment term decision-making. The app provides features to support the analysis with features such as viewing the top three countries with the highest credit exposure; drilling down for more precise analysis by risk class, business partner, credit segment, and country key; and enabling coaching for better performance. You can refresh the data manually, or it will update automatically based on the cache duration in the tile configuration. It provides comprehensive insights into the total credit exposure.

Tip

To understand how the app calculates exposure, consult SAP Note 3056144, which provides additional guidance on critical utilization and calculation.

Transaction Code

The corresponding transaction in SAP GUI is Transaction UKM_MALUS_DSP.

Analyze Credit Loss Allowances (W0134)

Application Type: Web Dynpro

This app provides real-time visibility into expected credit losses and their allowances based on existing system data. You can perform analysis using search parameters such as the key figure layout, ledger, company code, fiscal year, to period, and currency type. In addition, you can define your own selection criteria to generate a snapshot of these values and track reconciliation from the opening to closing balance within a specified period. Loss allowances are displayed by risk class, which is assigned to a business partner based on their probability of payment failure. The app presents key financial figures in the **Data Analysis** tab, including the carryforward beginning balance, credit loss allowances in the change balance amount, the released balance amount, the utilized balance amount, the transfer balance amount, the foreign exchange (FX) balance amount, and the ending balance amount. The app also integrates data from contract accounting, reflecting adjustments posted to the general ledger.

> **Tip**
>
> If flat-rate value adjustments have not been performed and posted to the general ledger for a given period, then the app does not display data for that time frame.
>
> The app uses the **DEFAULT** value in **Key Figure Layout**. However, you can create your own layout based on business needs in the Customizing activity under **Financial Accounting • General Ledger Accounting • Periodic Processing • Advanced Valuation in General Ledger Accounting • Settings for Credit-Risk-Based Impairment • Settings for Analytics • Define Analytical Key Figures and Layouts**.

Analyze Dunning Locks (Version 2) (F1655A)

Application Type: Transactional, Analytical

With this updated app, you can display valid dunning locks for receivables for a selected display currency. To improve performance, the app evaluates a system snapshot of receivables and payables instead of the current live data, which

you can create using the Create Snapshot of Locked Open Items app (F6214) prior to running this app. You can search for dunning locks using search parameters such as the lock reason, start date, expiration date, created-on date, total duration in days, number of extensions, and lock object category. In addition, the app only selects valid current locks. The app provides features such as indicating the snapshot's age with the KPI tag, restricting selected data with filters, displaying locks in a chart and table, adapting the chart display, and listing the number of items and total locked amounts per business partner.

This app is the successor to the deprecated Analyze Dunning Locks app (F1655).

> **Tip**
>
> The app only selects locks that are currently valid; it does not display locks that were valid in the past or will become valid in the future.
>
> In addition, to generate worklists in the backend system, business function FICA_EHP7_B1 must be active in the backend systems in which you wish to process the worklists via Transaction FPCKWL.

Analyze Dunning Run Exceptions (Version 2) (F2596A)

Application Type: Transactional

With this enhanced app, you can view the number of dunning exceptions in a dunning run and the total amount associated with these exceptions. The app displays the selected collection volumes in both a chart and a table for easy analysis. In addition, you can create custom tiles, select dunning exceptions using filters, group exceptions by dunning run ID or date, choose a display currency, and filter by exception reason and groups of exception reasons. The app provides features such as allowing exception reasons to be placed into user-specific groups, modifying or deleting existing groups, analyzing business partners' overdue items, creating worklists, and adding insights cards for monitoring KPIs dynamically.

This app replaces the Analyze Dunning Run Exceptions app (F2596), which has been deprecated.

Analyze Dunning Success (Version 2) (F2123A)

Application Type: Transactional, Analytical

With this updated app, you can see an overview of the success percentage of dunning activities within a certain time frame or related to applied dunning procedures and levels for the selected display currency. In addition, you can search for dunning success data using parameters such as company code, segment, contract account category, date of issue, dunning procedure, and dunning level. The app analyzes the percentage of dunned open items that have been paid as a probable result of the dunning notice sent, reflecting a success rate assigned to the notice, including paid items. The app provides features that include using filters to restrict data, displaying selected data in charts and tables, combining multiple dimensions for display, and customizing displayed dimensions in the chart and table views.

This app replaces the Analyze Dunning Success app (F2123), which has been deprecated.

Transaction Codes

If you are transitioning from SAP GUI, this app can be used for the functions of dunning-related features available through the following transactions:

- Transaction F150 for dunning run processing and report generation

- Transaction F.26 for generating dunning lists based on parameters

- Transaction F110 for processing payment runs for overdue receivables

- Transaction FD10N for viewing open items and overdue balances in the customer line item display

Analyze Dunning Volume (Version 2) (F2122A)

Application Type: Transactional, Analytical

With this revised app, you can display the number of items or the total amount based on the dunning procedure that has been dunned for a selected display currency. You can restrict the selection of the due item's total amount within a time frame based on the dunning procedure to specific company codes and contract account categories. For example, you can search for dunning volume data using parameters such as company code, segment, contract account category, date of issue, dunning procedure, and dunning level. The app provides features such as filtering data by criteria like postal code and dunning level, displaying selected dunning volumes in charts and tables, and combining multiple dimensions for display (e.g., grouping by dunning level and week of issue). You also can navigate to business partners in the list for further information and customize the displayed dimensions under **Settings** for both **Chart View** and **Table View**.

This app replaces the Analyze Dunning Volume app (F2122), which has been deprecated.

Transaction Code

The corresponding transaction in SAP GUI is Transaction FPM3.

Analyze External Collections Volume (Version 2) (F2402A)

Application Type: Transactional, Analytical

With this revised app, you can display the number of items or the total amount submitted to collection agencies for outstanding payments for a selected display currency. You can restrict the selection of the total amount and submitted items to specific company codes and contract account categories and define whether open, paid, or noncollectible items are analyzed. You also can search for external collection volumes using a collection agency, date of submission, submission reason, submission status, and item type. The app comes with features such as filtering data by criteria like submission reason and postcode, displaying data in charts and tables, and combining multiple dimensions for display. In addition, you can generate worklists for selected business partners, trigger follow-up activities, and branch to other analysis apps. Besides the standard functionalities, the displayed dimensions for both the chart and the table can be customized in the settings.

This app replaces the Analyze External Collection Volume app (F2402), which has been deprecated.

Tip

To generate worklists in the backend system, business function FICA_EHP7_B1 must be active in the backend systems in which you wish to process the worklists via Transaction FPCKWL.

Analyze Incoming Payments (F5588)

Application Type: Transactional

This app helps monitor and analyze incoming payments to streamline clarification processes and optimize team workload in contract accounting. The app allows you to assess the number of payments clarified manually and validate the efficiency of automated processes—including machine learning, if SAP Cash Application, add-on for contract accounting is implemented. The app displays payment data per company code, with filter options such as posting date and lot type, and analyzed across payment, credit, and check lots for automatic and manual clearing. In addition, the app provides both chart and table views, offering flexible configurations to display data such as payment usage, clarification duration, and automation levels. Moreover, you can group payments by criteria such as house bank or lot origin and configure charts to display metrics such as payment amounts, payment lots, open items, and clarification percentages. With SAP Cash Application, add-on for contract accounting, the **Details** view provides additional insights, including proposal rates for machine learning processes, thus ensuring comprehensive analysis for process improvement.

Tip

A separate license for SAP Cash Application, add-on for contract accounting is required to use advanced features such as machine learning capabilities.

Transaction Code

The equivalent transaction in SAP GUI is Transaction FPE3.

Analyze Installment Plans (Version 2) (F2363A)

Application Type: Transactional, Analytical

With this updated app, accounts receivable specialists, finance teams, and credit managers can display the number of installment plans created and the total amount paid into the original receivables in contract accounting for a selected display currency. The app allows you to monitor and evaluate installment payment agreements made with customers. You can restrict the total amount and dunning exceptions to a specific company code and contract account category. In addition, you can search for installation plans using dimensions such as net due date, installment plan status, category, deactivation date, and deactivation reason. The app provides key features such as filtering installment plans by criteria such as the number of installments and fulfillment rate, displaying installment plans in a chart and table with corresponding status, and combining multiple dimensions for display. In addition, you can generate a worklist for selected business partners, trigger follow-up activities, and customize displayed dimensions for both chart and table views.

This app is the successor to the deprecated Analyze Installments Plans app (F2363).

Tip

For the system to generate corresponding worklists in the backend system, SAP recommends that business function FICA_EHP7_B1 be active in the clients of the backend systems in which you intend to process the worklists. In addition, you can then use Transaction FPCKWL to process and monitor these worklists in the backend system.

Transaction Code

The corresponding transaction in SAP GUI is Transaction FPRH.

Analyze Open Credits (Version 2) (F2364A)

Application Type: Transactional, Analytical

With this updated app, you can display the number of credit items posted and the total amount of open credits for a selected display currency. The app allows you to restrict the selection to specific company code, segment, document type, net due date, and contract account catego-

ries. The app also provides other features, including filtering credits by document type and due date, displaying credits in charts and tables with amounts for the selected period, and combining multiple dimensions for display. You also can group credits by business partner and week of maturity, view additional information by clicking the business partner, and create worklists for selected business partners. In addition, you can customize the displayed dimensions for both chart and table views.

This app is the successor to the deprecated Analyze Open Credits app (F2364).

Tip

To generate worklists in the backend system, SAP recommends that business function FICA_EHP7_B1 be active in the backend systems to process the worklist.

Analyze Overdue Items (Version 2) (F0860A)

Application Type: Transactional, Analytical

With this updated app, you can display the number and total amount of overdue items. In addition, you can filter by various criteria, such as the display currency, net due date, dunning procedure, doubtful items, company code, contract, and segment. The selected overdue items are shown in charts and tables, with amounts displayed by week, month, or quarter in the chosen currency. Multiple dimensions can be combined to customize the display. You can group items by the main transaction and net due date, adjust dimensions under **Settings**, and create worklists for one or more business partners to trigger follow-up activities. Moreover, the app displays overdue items in both chart and table formats, allowing you to visualize and analyze the data from different perspectives.

This app replaces the Analyze Overdue Items app (F0860), which has been deprecated.

Tip

SAP recommends activating business function `FICA_EHP7_B1` to generate worklists in backend systems. Transaction FPCKWL is used to process and monitor the worklists in the backend system.

Transaction Codes

The closest transaction codes in SAP GUI are Transactions FBL5N, S_ALR_87012078, and S_ALR_87012168.

Analyze Payment Locks (Version 2) (F6134)

Application Type: Transactional, Analytical

This app displays valid payment locks for receivables and payables in contract accounting. You can search payment locks with search criteria using fields such as company code, house bank, payment method, lock reason, created on, display currency, and lock object category. To optimize performance with large data volumes, the app evaluates system snapshots of receivables and payables, which are created using the Create Snapshot of Locked Open Items app (F6214). The app's KPI tag shows the snapshot's age, and authorized users can access the snapshot creation app directly from that tag. The app displays data in both a chart view and a table view. The chart view shows amounts by lock object category in the selected currency, supporting grouping by expiration date or total duration. In addition, you can adjust chart dimensions and measures here to display items or amounts per category. On the other hand, the table view lists locked items and their total amounts per business partner.

Tip

To generate backend worklists, SAP recommends that business function `FICA_EHP7_B1` be active, and Transaction FPCKWL is used to process and monitor these worklists. You can display a snapshot of the payment lock with Transaction FPLOCK_SNAP via SAP GUI.

Note that the app only displays active locks that are currently valid; it does not display locks that were valid in the past or that are set to become valid in the future.

Analyze Payment Run Exceptions (Version 2) (F2124A)

Application Type: Transactional, Analytical

With this updated app, you can display the number and total amount of payments that could not be processed successfully, along with the resulting exceptions for a selected display currency. You can restrict the selection of the total amount or exceptions to specific company codes and contract account categories and define whether the system considers successfully processed payments linked to exceptions. For example, you can search for payment run exceptions using the date of the payment run, payment run ID, exception reason, exception reason group, and whether the payment was successful. Exceptions from the selected payment runs are displayed in a chart. Key features include filtering exceptions by reasons or groups, assigning exception reasons to custom groups, grouping exceptions by payment run date, selecting the display currency, and displaying business partners and related amounts. You can create worklists for business partners, trigger follow-up activities, display additional information, create custom tiles, and manage exception reason groups in the app settings.

This app replaces the Analyze Payment Run Exceptions app (F2124), which has been deprecated.

Analyze Posting Locks (Version 2) (F1652A)

Application Type: Analytical

With this updated app, you can display existing posting/clearing locks for contract accounts receivable and payable. You can search for posting locks using parameters such as start date, expiration date, created-on date, duration, and number of extensions. As a result, this app helps prevent posting delays and ensures that financial processes run efficiently in contract accounting. The app displays currently locked journal entries and other financial transactions. In addition, it helps you analyze lock reasons, whether due to system processes, user actions, or errors, and it assists finance teams in troubleshooting posting issues by providing detailed insights. The app is delivered with options that allow you to filter locks by document type, company code, user, and other relevant criteria. In addition, the app lets you view the amounts related to the selected locks in different currencies. Also, to improve performance, this updated app evaluates a system snapshot of receivables and payables instead of the current live data, which is important given the high data volume in contract accounting.

This app is the successor to the deprecated Posting Locks app (F1652).

Analyze Returns (Version 2) (F2600A)

Application Type: Analytical

With this updated app, you can display the number of returns made and their total amount, with the option to restrict the selection by company code and contract account category for a selected display currency and variant. In addi-

tion, you can perform tasks such as displaying return charges separately from the return amount; filtering returns by various criteria such as posting date, value date, returns lot, returns type, and house bank; grouping returns by posting or value date; choosing the display currency; and displaying business partner details. Also, you can generate worklists, trigger follow-up activities, and create tiles with individual threshold values. The app provides a chart and table view for easy processing of data.

This app replaces the Analyze Returns app (F2600), which has been deprecated.

> **Tip**
>
> To generate worklists in the backend system, SAP recommends that business function FICA_EHP7_B1 be active in the backend systems in which you wish to process the worklists via Transaction FPCKWL.

Analyze Unbilled Items (F1427)

Application Type: Transactional

With this app, you can track items that have reached their billing date but were not billed, as well as items that are not yet due. As a result, the app allows businesses to analyze potential billing delays and optimize revenue recognition. The app displays the number or total amount of affected items and filters them by various criteria, such as bill starting from, billable item type, status of billable item, and subprocess. In addition, the app can display the total amount of unbilled items within a time limit, limit the selection to certain company codes, break down items by due date, select items by amount or number, group unbilled items by various categories, and choose the display currency. Also, you can create tiles with threshold values and share analyses via email or Microsoft Teams, with navigation options available depending on your roles and authorizations.

Transaction Code

The equivalent transaction in SAP GUI is Transaction FKKBIXBIT_MON.

Analyze Unrated Items (F1977)

Application Type: Transactional

This app allows you to track consumption items that have reached their rating date but weren't considered during rating or those that aren't yet due. The app enhances visibility and control over unrated consumption items. You can display the number of affected items or their total amount and can filter unrated items by various criteria, such as rating from, consumption item type, type of consumption item ID, and status of consumption item. This app is available for the invoicing manager role (convergent invoicing), and its key features include displaying the total number of unrated items within a time period, showing a breakdown by due date, selecting unrated items by number, and grouping unrated items by various attributes. In addition, the app supports creating custom tiles with threshold values and sharing analyses via email or Microsoft Teams, with navigation options depending on roles and authorizations.

Analyze Write-Offs (Version 2) (F0861A)

Application Type: Transactional, Analytical

With this updated app, you can display the number of write-offs and their amounts made in contract accounting. The app allows you to restrict the selection to a specific company code, contract account category, date of write off, write-off type, dunning procedure, and collection strategy. The app provides features that include using filters to restrict data, displaying selected write-offs in charts and tables, and adapting dimensions for display. In addition,

you can filter write-offs by reasons and dunning procedures, group write-offs by date and reason, and customize chart dimensions. Moreover, you can display specific write-off documents by selecting them in the table.

This app replaces the deprecated Analyze Write-Offs app (F0861).

Application Jobs – Activate Global Hierarchies (F7019)

Application Type: Transactional

This app allows you to efficiently manage hierarchies by viewing, creating, editing, activating, and deactivating them as needed. The app is essential for organizing and structuring data across different dimensions, such as organizational units, cost centers, or profit centers. The app supports quick hierarchy creation through importing or copying existing structures to accommodate reporting changes. In addition, you can filter hierarchies by hierarchy ID, type, status, and validity time frame (valid from/to); generate simulated versions; and manage activation for runtime use. The app's features include exporting, editing, and importing hierarchies via spreadsheets and integrating hierarchies created in compatible applications. Also, the app provides access to change logs and supports maintaining node texts in multiple languages for applicable hierarchy types, ensuring accurate and adaptable hierarchy management.

Application Jobs – Generate Flexible Hierarchies (F7168)

Application Type: Transactional

This app allows you to create, modify, and customize hierarchies for cost centers, profit centers, and company code reporting per your business requirements. For instance, you can generate hierarchies based on a custom sequence of master data attributes, such as country, city, and region. In addition, the app supports mass updates to existing hierarchies based on master data changes, along with bulk downloads of master data and attributes used in hierarchies. Moreover, you can extend hierarchies by adding custom fields and mass-uploading extended attributes. Thus, the app lets you manage flexible hierarchies efficiently.

Transaction Codes

SAP GUI transactions related to hierarchies include Transactions KCH1 and KSH1, among others.

Application Jobs – Import to Global Hierarchies (F5097)

Application Type: Transactional

This app allows you to view, create, edit, activate, and deactivate hierarchies efficiently while accommodating structure changes or evolving reporting needs through import or copy functionalities. In addition, you can filter hierarchies by type, hierarchy ID, status, or validity time frame, manage validity periods, generate simulations, and export or import hierarchies via spreadsheets. The app's additional features include checking change logs, importing hierarchies from compatible applications, and maintaining node texts in multiple languages for applicable hierarchy types.

Application Jobs – Mass Regenerate Flexible Hierarchies (F7169)

Application Type: Transactional

As its name implies, this app allows you to schedule mass regeneration jobs to make hierarchy adjustments in order to reflect the latest data without manual intervention based on business requirements. You can generate mass regeneration of hierarchies based on master data attributes such as country, city, and region. In addition, the app supports mass updates to hierarchies based on data changes, along with bulk downloads of relevant master data and attributes. You also can extend hierarchies by adding custom fields and mass-uploading extended attributes, facilitating efficient hierarchy management and streamlined reporting.

Application Jobs – Run Balance Validation (F7327)

Application Type: Transactional

This app allows you to schedule and execute balance validation runs for general ledger accounting tasks, supporting both immediate and recurring validations. You can define validation parameters, including company codes, profit center, cost center, rule group, ledger, currency type, and fiscal period. The app features template-based scheduling, allowing automated execution and status monitoring options that show the status **Finished**, **In Process**, or **Failed** in the job list. Upon completion, you can access validation results with the View Balance Validation Results app (F6387). Each row represents an entity validated against a specific **Rule Group**. In addition, the app integrates with SAP Advanced Financial Closing, enabling balance validation jobs to be incorporated into SAP Advanced Financial Closing task templates using predefined model **010035000** (**Balance Validation in Company Code Currency**) or **010035001** (**Balance Validation in Group Currency**). Thus, with this app, you can maintain financial accuracy across selected entities.

Tip

A separate license is required to use SAP Advanced Financial Closing.

Application Jobs – Upload Master Data in Flexible Hierarchies (F7167)

Application Type: Transactional

This app allows you to create, modify, and customize hierarchies for cost center, profit center, and company code reporting as per your business requirements. You can generate hierarchies based on a custom sequence of master data attributes such as country, city, and region. In addition, the app supports mass updates to hierarchies based on master data changes, as well as bulk downloads of master data and attributes used in hierarchies. Moreover, you can mass-upload extended attributes, thereby allowing for flexible hierarchy management and streamlined reporting.

Application Log – Automatic Adjustment (F5010)

Application Type: Transactional

With this app, general ledger users can select log category **FIN_ICA** (**Intercompany Matching and Reconciliation**) and subcategory **AA** (**Automatic Adjustment Posting**) to review automatic adjustment postings within a specified period. The results include a **Success** status for successful postings or an **Error** status if the posting failed, with root causes indicated, such as mismatched transaction currencies, incomplete manual assignments, missing valid posting rules or templates, or errors returned by the accounting posting API. If no variance exists between matching items, making adjustments unnecessary, then the system shows an **Information** status.

> **Tip**
>
> If you select **Reverse in Next Period** when defining the posting rule or document template, the system generates a reversal document with a posting date in the period immediately following the matching assignment.

Approve Bank Account Applications (F5859)

Application Type: Transactional

This app allows cash managers to manage received applications for new bank accounts by approving or rejecting them. Upon approval, the system automatically creates an inactive bank account based on the application details, and an email notification is sent to the applicant. To facilitate the approval process, approvers can search with various filter criteria, such as application status, contract type, currency, applicant, company code, account status, or bank country/region name. In addition, you also can confirm the withdrawal of a bank account application, which halts the creation process. If an inactive bank account was already created, its status will be updated to **Withdrawn**.

> **Tip**
>
> As a prerequisite, you must be assigned authorization object F_CLM_BAOR to process and control the access to bank account applications.

Approve Bank Account Changes – Two-Person Verification (F6264)

Application Type: Transactional

This app offers cash managers an overview of bank account change requests triggered in the two-person verification mode that are awaiting approval centrally. The app provides a list page and object page, which helps you review changes and decide whether to approve or reject them. To use the app, authorization to approve bank account change requests (with authorization object F_CLM_BAM, activity 63) is required, and the two-person verification mode must be enabled. The app's features include checking pending change requests, with only those from the two-person verification mode displayed. For those requests, you can manage the workflow mode using the My Inbox for Bank Accounts app (F2797). In addition, you can compare old and new values, view bank account object pages before and after proposed changes, and approve or reject requests. Approved requests activate revisions and create new bank account versions, while rejected requests are marked as **Canceled**.

Approve Bank Payments (Version 2) (F0673A)

Application Type: Transactional, Analytical

With this updated app, you can review and process payment batches, including approving, rejecting, or deferring individual payments or entire batches. The app offers features such as searching for payment batches by batch ID, company code, and house bank; editing due dates and instruction keys; submitting decisions for multiple batches simultaneously; and deferring payments to a future date. In addition, with this app, you can mass-approve payment batches either online or as a background job with additional configuration and security settings.

This app replaces the deprecated Approve Bank Payments app (F0673).

Tip

For external payment batches, SAP recommends reviewing the app features as there may be variations in functionality. In other words, the app may behave differently when handling external payment batches created outside of the standard payment proposal process—for example, via a third-party system or manual upload.

Transaction Code

The corresponding transaction in SAP GUI is Transaction BNK_APP.

Approve Reconciliation Close Requests – Inbox (F5000)

Application Type: Transactional, Reuse Component

This app allows you to manage reconciliation close requests efficiently as part of the intercompany group reporting process. With this app, approvers can receive requests in their inbox, view reconciliation details for specific trading unit pairs and periods, search and filter tasks, claim approval tasks to reserve them or forward them to other you, and approve or reject the requests seamlessly.

Asset Accounting Overview (F3096)

Application Type: Analytical

This app provides asset accountants with a comprehensive and centralized dashboard for managing and monitoring fixed asset processes in their organization. The app displays information using cards such as asset balance, asset master worklist, and asset transactions. In addition, the app offers options to navigate to purchase order items by account assignment and other apps via **Quick Links**. The app features a range of filters for focused analysis and integrated navigation into related apps, streamlining processes and enhancing efficiency. Some cards use semantic objects and actions, which are part of the business role assigned to a user.

Tip

Although the app offers a variety of cards related to asset accounting functions, not all are enabled by default. You will activate only the ones most relevant to your needs, and authorization objects are implemented to access these cards. SAP recommends activating only the cards required for your scope to minimize any performance bottlenecks.

Consult SAP Notes 2798214, 2798215, 3484638, and 2867794 for more information related to the app's performance.

Asset Balances (F1617)

Application Type: Web Dynpro

With this app, you can display balances of fixed assets based on criteria such as company code, depreciation area, segment, or profit center. You can filter asset balances by various filter criteria such as key figure group, company code, ledger, depreciation area, to period of the fiscal year, and currency type. The app presents results in the **Data Analysis**, **Graphical Display**, and **Query Information** tabs with the flexibility of adding additional dimensions and measures to rows and columns under the **Navigation Panel** section. The report helps you follow the net book value of fixed assets derived from acquisition and production costs and depreciation. To support the app, SAP provides preconfigured key figure groups for reporting, bundling key figure codes like acquisition and production costs, retirements, transfers, cumulative depreciation, and the net book value at the reporting date. In addition, you can display cumulative acquisition values, accumulated depreciation for each type, and planned book values. Also, you can specify search criteria using filters, display query results in a grid format, export results to Microsoft Excel, and modify the grid by adjusting the available characteristics. SAP recommends using the following key figure groups:

- Posted values: ABS_POSTED
- Planned values: ABS_DEF
- Total depreciation: DEPR_TOTAL
- Write-ups: ABS_DEF

Transaction Codes

The corresponding transactions in SAP GUI are Transactions AR01 or AW01N.

Asset History Sheet (F1615)

Application Type: Web Dynpro

With this app, you can call up the asset history sheet to show value changes to fixed asset balances in a fiscal year for a depreciation area. The asset history sheet lets you document and explain balances of fixed assets for various accounting principles, local regulations, and management purposes. You can search the asset history by key figure group, company code, currency type, fiscal year, general ledger account, and deactivation-on date. SAP provides preconfigured key figure groups for reporting, such as acquisitions, retirements, transfers, and depreciation. You can display asset data by using features such as standard SAP key figure groups, specifying search criteria with filters, displaying results in a grid format, exporting results to Microsoft Excel, and modifying the results grid. In addition, you can exclude deactivated assets using the **Deactivation On** filter, which shows only assets deactivated within the selected fiscal year. SAP recommends using standard SAP key figure groups:

- **Key figures without hierarchy**
  - Posted values: AHS
  - Planned values: AHS_PLAN
- **Key figures with hierarchy**
  - Posted values: AHS_HRY
  - Planned values: AHS_HRY_PL

This app replaces the deprecated Asset History Sheet (Design Studio) app (F1615A).

Transaction Code

The corresponding transaction in SAP GUI is Transaction AR02.

Asset Master Worklist (F1592)

Application Type: Transactional

With this app, you can display a worklist of fixed assets and quickly get an overview of their status. In addition, you can generate a list according to specific criteria, such as company code or asset class, and refine it to show only incomplete, capitalized, or retired assets. For example, you can search for asset master data using a company code, asset class, asset, cost center, location, and key date. The app comes with features such as displaying fixed assets based on the chosen selection criteria, viewing and customizing master data fields, filtering the list by asset status, and searching for particular assets using the filter bar.

> **Tip**
>
> The app is not supported when the universal parallel accounting business function (FINS_PARALLEL_ACCOUNTING_BF) is active. When this function is active, SAP recommends using the Manage Fixed Assets app (F3425) instead.

> **Transaction Code**
>
> The related transaction in SAP GUI is Transaction AR31.

Asset Transactions (F1614)

Application Type: Transactional

With this app, you can evaluate daily operations in asset accounting. You can search for asset transactions using important parameters such as fiscal year, company code, depreciation area, currency type, and asset value date. The app is delivered with standard SAP key figure groups, which allow you to select asset transactions by category, such as acquisitions, retirements, or transfers. In addition, the report displays all transactions within the selected groups and their respective key figures, such as retirement revenues or costs, gains, losses, retired APC, and net book value. There are some differences between the Asset Transactions and Asset History Sheet apps:

- **Asset Transactions**
  This app displays actual transactional values,

and is used in daily operations on an ad hoc basis. This is an SAPUI5 app, which displays data based on selection criteria in the app header.

- **Asset History Sheet**
  This app displays both planned and actual values, and runs periodically to support legal requirements. This is a Web Dynpro app, which displays data based on grid-based layout showing various dimensions.

> **Tip**
>
> When you select the **Retirement** view in this app, results for retirement gain and retirement loss are displayed only if the configuration permits postings of asset retirement gain and loss to the general ledger. However, SAP recommends not allowing such separate postings, preferring that you net gains and losses to reduce complexity.

> **Transaction Code**
>
> The corresponding transaction in SAP GUI is Transaction AR05.

Assign FS Item Mappings (F3334)

Application Type: Transactional

The app allows general ledger accountants of group reporting link existing financial statement (FS) item mappings to consolidation versions and effective periods, ensuring that the mappings between FS items in a consolidation chart of accounts and general ledger chart of accounts are applied within a specified time frame. The app offers key features such as viewing or deleting assignments; deleted assignments revert to earlier periods with matching consolidation versions and charts of accounts. In addition, you can create new assignments by specifying a start date and consolidation version, with fields like charts of accounts auto-filled based on the mapping revision. Also, you can edit assignments to update FS item mappings and revisions. However, if the FS item mapping contains any transactional data, SAP recommends that you require the deletion and recreation of the assignment for changes made prior to deletion.

Transaction Code

The corresponding transaction in SAP GUI is Transaction FINCS_FSI_MAPPG_UPLD.

Assign Open Items (F2626)

Application Type: Transactional

This app simplifies the management and clearing of open items related to customer accounts. The app offers a user-friendly interface for viewing, assigning, and clearing open items. You can search for open items by company code or customer. The app supports viewing all open items related to a customer, finding and assigning matching credit and debit items, and clearing these assigned items. In addition, the app generates a clearing document number for reference. If any items cannot be cleared automatically, you can note their journal entry numbers and then use the Clear Incoming Payments app (F0773) to clear them. Also, you can customize the app to meet your specific needs, for smooth handling of managing open items.

Tip

You must ensure that clearing accounts are configured correctly to avoid discrepancies in reporting. We recommend reviewing assignments routinely for various use cases, such as manual incoming clearing, credit and debit memos, and intercompany clearing.

Transaction Code

The corresponding transaction in SAP GUI is Transaction FB15.

Audit Journal (F0997)

Application Type: Transactional

With this app, you can verify the accuracy of journals. The app provides helpful features for auditing, such as viewing both compact and complete journals, displaying changes in journal entries, viewing the referenced supplier (account type **K**) or customer (account type **D**) documents, and checking for gaps in journal entry numbers to support a comprehensive auditing process. You can search for journals using parameters such as company code, fiscal year, posting date, ledger, general ledger account, and journal entry type. The app is useful for the general ledger accountant and external auditor roles to perform a comprehensive review of journals.

Balance Carryforward Status (F4683)

Application Type: Analytical

The app displays the status of the balance carryforward run and when it occurred. With this app, you can filter balance carryforward runs by various parameters via the fiscal year, company code, ledger, run date, and status fields. Once you run the app with your chosen selection criteria, the app displays the status of the run—either success or failure. In addition, the app lets you navigate to Application Logs app (F5002) to view log details.

Tip

As a prerequisite, ICF nodes for service FIN_ BCF_STATUS must be activated via Transaction SICF and metadata must be loaded using OData service C_BALANCECARRYFWDSTATUS_CDS via Transaction /N/IWFND/MAINT_SERVICES.

> **Transaction Code**
>
> The corresponding transaction in SAP GUI is Transaction FAGLGVTR.

Balance Sheet – By Consolidation Units (CCONS_FPM_OVP_BS_BY_CU (CCONS_FPM_OVP_BS_BY_CU))

Application Type: Web Dynpro

This Web Dynpro–based app provides a structured view of a balance sheet at the consolidation unit level in the financial consolidation process. The app offers the ability to analyze and reconcile balance sheets across different consolidation units in an enterprise. The app's features include categorizing consolidation units, supporting drilldown capabilities, and reconciling intercompany balances. In addition, the app integrates data from both SAP and non-SAP sources. Thus, the app lets you focus on legal consolidation and management reporting.

Balance Sheet FX Risk (W0123)

Application Type: Web Dynpro

This app allows analysis of foreign exchange (FX) risk by aggregating open items and balances from financial accounting (FX exposure), financial transactions from the Transaction Manager (hedging instruments), and operational data from the One Exposure from Operations hub. In addition, the app lets you refine FX risk information by adding attributes such as bank accounts, financial transactions, and general ledger postings. The app presents key financial accounting risk measures such as absolute exposures, hedges, net exposures, hedge amounts, and exposure amounts at the company code level in the display currency. The app also supports dynamic filtering by transaction currency, exchange rate type, and company codes. You can analyze key figures, which are calculated based on settings from the Define Key Figures

app (FXM_KF_DEF), helping derive exposure and hedge totals across currencies. The app also features **Graphical Display** and **Query Information** tabs for visualization and filter tracking, supports **Jump To** functionality for accessing related general ledger accounts, and allows you to export results to Microsoft Excel for further analysis.

> **Transaction Code**
>
> The related transaction in SAP GUI is Transaction TPM10.

Balance Sheet FX Risk – In Transaction Currency (W0124)

Application Type: Web Dynpro

This app allows you to analyze balance sheet foreign exchange (FX) risk by reading open items and foreign currency balances in general ledger accounts, financial transaction data in the transaction currency from the Transaction Manager, and operational data from the One Exposure from Operations hub. You can search for key figures such as **I/Co Exposure**, **Third Party Exposure**, **Liquidity Funds**, and **Hedge** using search dimensions such as company code, display currency, exchange rate type, transaction currency, and key date. After you call the app, select the selection criteria, and click the **Go** button, the app directly presents measures for FX risk in transaction currency across the selected criteria. The **Data Analysis** tab displays key figures, and additional attributes can be included for refined analysis. The **Graphical Display** tab provides a visual representation, and the **Query Information** tab shows selected filters and variables. The **Jump To** function allows navigation to related ledger apps, and results can be exported to Microsoft Excel for further processing. Thus, the app allows detailed risk analysis by incorporating attributes related to financial transactions, bank accounts, and ledger postings into the content area.

Tip

As a prerequisite to use the app, you must complete basic settings in the Define Key Figures app (FXM_KF_DEF) for the financial instruments that you will search for. This master data activity also be carried out from SAP Easy Access menu path **Accounting • Financial Supply Chain Management • Treasury and Risk Management • Hedge Management and Accounting • Hedge Management • Balance Sheet FX Risk • Define Settings – B/S FX Risk**.

Balance Sheet FX Risk Overview – Based on Snapshots (F4765)

Application Type: Analytical

This app offers treasury users an overview of hedge management for balance sheet foreign exchange (FX) risk using snapshot data. The app displays the total net open amount in transaction currencies and contributing key figure types, including **Net Exposure of Snapshot**, **Hedged Amount**, and **Open B/S Hedge Request**. The app comes with interactive chart and table views, which let you perform flexible analysis. The chart view displays net open amounts across currencies, calculated as **Net Exposure + Open Hedge Requests + Hedged Amount**. The chart view also lets you switch between different chart types, such as a bar chart or column chart. On the other hand, the table view provides detailed snapshot data, including columns for the snapshot ID, company code, currency, and net open amount. In addition, you can drill down into snapshots, jump to related apps like Process Snapshots – Balance Sheet FX Risk (F4763) or Process Hedge Requests – Balance Sheet FX Risk (F4764), and switch between chart types for an enhanced visualization.

Balance Sheet/Income Statement (F0708)

Application Type: Transactional

With this app, you can display balance sheets and profit and loss statements for your company using operational, local, and global charts of accounts, with reports generated on the fly by analyzing all relevant line items. In addition, you can access the line items of specific general ledger accounts, as well as customer and supplier line items. You can search for financial statements using the variant, ledger, financial statement version, key period, company code, key companion period, and currency fields. Depending on the authorizations assigned, the app can display full balance sheets and profit and loss statements for multiple company codes, specific nodes, general ledger accounts, profit centers, and segments. In addition, you can view different currencies, zero-balance accounts, alternative account numbers, and time-dependent hierarchies. Also, you can compare key periods with actual and plan data, search for specific items, and export statements to Microsoft Excel.

The app lets you navigate to apps like Display G/L Account Balances (F2141), Display Line Items in General Ledger (F2217), Manage Customer Line Items (F0711), and Manage Supplier Line Items (F0712), and it lets you drill down into the Display Financial Plan Data report.

Tip

SAP recommends using the Balance Sheet/Income Statement – Multidimensional app (W0161) to download results to Microsoft Excel as it provides other additional features.

Transaction Code

The equivalent classic credit management transaction in SAP GUI is Transaction F.01.

Balance Sheet/Income Statement – Multidimensional (W0161)

Application Type: Web Dynpro

This Web Dynpro–based app lets you generate and analyze balance sheets across operational, local, and global charts of accounts by evaluating relevant line items in real time. You can search for balance sheets and income statements for various search dimensions such as ledger, company code, hierarchy ID, from period, to period, comparison to period, comparison fiscal year, currency type, plan category, and general ledger account. In addition, the app supports viewing balance sheets for multiple company codes, specific nodes, and various financial statement types, including normal, opening, and movement. Also, you can compare fiscal years, ledgers, and actual (**Plan Category ACT01**) versus plan data while adjusting the grid layout using the **Navigation Panel** function for a tailored analysis. You can export the results to a Microsoft Excel file or a PDF for further evaluation.

The app leverages optimized hierarchy tables, requiring replication via the Replicate Runtime Hierarchy app (F1478), and it integrates with related apps such as Display G/L Account Balances (New) (F0707A), Display Line Items in General Ledger (F2217), and Display Financial Plan Data Report (W0163). The app also supports email functionality and utilizes the CDS view C_ FinStmntComparison.

Tip

This app displays the group account only if the **Group Account Number** indicator is enabled in Transaction OB58 or the Manage Global Hierarchies app (F2918) for the specified financial statement version (FSV). In addition, the FSV must be structured based on the corresponding group chart of accounts.

Transaction Codes

The corresponding transactions in SAP GUI are Transactions FAGLB03 or FAGLB03H.

Bank Account Applications (F5860)

Application Type: Transactional

This app allows cash specialists to perform tasks such as checking pending applications, checking approved applications, reviewing application information, and navigating to the Manage Bank Accounts app (F1366A). With this app, you can search for details using search criteria such as account type, currency, and other relevant information. Although this app and the Approve Bank Account Applications app (F5859) both serve to manage the bank account management (BAM) workflow, there are a few differences between them:

- **Bank Account Applications**
  Initiate or submit requests for creating new bank accounts. Offers additional search capabilities such as checking pending applications, checking approved applications, and navigating to the Manage Bank Accounts app.
- **Approve Bank Account Applications**
  Approvers review submitted bank account applications and decide whether to approve or reject requests. Offers additional features, such as the withdrawal of applications, which stops the account creation process.

Bank Account Change Requests – Two-Person Verification (F6265)

Application Type: Transactional

This app provides cash management users with an overview of bank account change requests triggered in the two-person verification mode that are awaiting approval with the app's list page and object page. Thus, the app enforces dual-control mechanisms for modifying sensitive bank account information, thereby mitigating the risk of fraud and errors. On the list page, you can search for bank account change requests with filter criteria such as the title of request, request ID, request status, requester, created on/at, account currency, bank key, bank country/region, company code, and contract type. You can review changes by comparing old and new values and can access the **Current Version** and **Target Version** links to view the bank account object page before and after the proposed updates. Only requests initiated in the two-person verification mode are displayed; those from the workflow mode must be handled via the My Inbox for Bank Accounts app (F2797). Approved requests activate the revision and create a new bank account version in the system, while rejected requests are marked as **Canceled**.

Bank Account Master Data (WDA_FCLM_BAM_ACC_MASTER (APP_CFG_FCLM_BAM_ACC_MASTER))

Application Type: Web Dynpro

This Web Dynpro–based app helps the management of bank account details within bank account management (BAM). The app serves as a central repository for maintaining accurate bank account data. You can create, update, and manage bank account information, including account numbers, IBANs, and bank control keys. In addition, the app supports integration with house banks, allows tracking of internal and external bank relationships, and allows account connectivity across systems. The app also provides functionalities for setting overdraft limits, maintaining multilingual descriptions, and attaching reference documents such as agreements or compliance records.

Bank Hierarchy Maintain (WDA_FCLM_BAM_HIER_BP)

Application Type: Web Dynpro

This app allows you to structure and manage bank account hierarchies within SAP's bank account management (BAM) framework. Using this app, you can define hierarchical relationships between bank accounts, create and modify groupings based on business needs, and configure custom views for varied financial management purposes. In addition, the app integrates with cash and liquidity management, supports bulk account management, and ensures audit compliance by recording changes for regulatory transparency. The app stores data in tables V_T038P and V_T038.

Bank Relationship Overview (F3775)

Application Type: Analytical

This app provides key insights into bank relationship management, offering a centralized view of recent payments, bank profiles, fees, and account statuses for cash managers. The app lets you customize the overview page by rearranging or hiding cards. In addition, you can display payment details by house bank, see the number of banks and accounts per bank group, and visualize bank fees. You also can view bank accounts by status (**Active** or **Closed**) and workflow requests (**Approved**, **To Be Approved**, etc.), as well as track bank accounts by revision status. Also, by clicking various chart sections, you can navigate to related apps for detailed information, helpful for managing bank statements, accounts, and fees, for example. You also can reset the view to the default settings if needed.

Bank Risk (S/4HANA) (F0515A)

Application Type: Analytical

This app allows cash managers and cash management specialists to analyze deposit distributions based on bank ratings, identifying deposits in high-risk accounts. With this app, you can display balances in low-rated banks (by low rating), show balances by banks, display balances by all ratings, view balances by bank and company, switch between charts and tables, and export results to a spreadsheet. In addition, the app supports sending emails with a URL to replicate your current selection criteria and saving these criteria as default settings in a tile.

Tip

Bank ratings can be sorted by their values, either high to low or low to high, rather than alphanumerically, with mappings maintained in the VN_TP06 view. You can include other threshold ratings by updating this view and incorporating your own rating system, with low ratings defined as values larger than 12. Refer to the Data Setup Guide for SAP S/4HANA to learn about the required setup for bank ratings: *http://s-prs.co/v612000*.

Transaction Code

The equivalent transaction in SAP GUI is Transaction FTE_BSM.

Bank (S/4HANA) (F1760)

Application Type: Fact Sheet

As part of bank account management (BAM), this app offers a comprehensive overview of bank data, including key information such as the bank name, bank key, and country/region. In addition, the app displays important business facts like the bank's address, SWIFT code, and bank number, making critical data easily accessible via a user-friendly interface. You can use the app as a starting point to navigate to additional relevant information, such as details about related business partners, master data, or documents. Also, the app supports features like sorting, grouping, and displaying subtotals, and it lets you trigger emails and export results to Microsoft Excel for further analysis and reporting. This approach ensures that you have all the necessary tools at hand to manage and analyze bank data effectively.

Tip

From a sequence point of view, the Bank (S/4HANA) app (F1760) is first used to create and store global bank master data. The House Bank app (F1758) then defines a house bank using the previously created bank master data and links it to a specific company code. Finally, the House Bank Account app (F1759) establishes accounts under the house bank, specifying which accounts are designated for payments and treasury activities.

Transaction Codes

The app combines features from Transactions FI01 (Create Bank Master Record), FI02 (Change Bank Master Record), FI03 (Display Bank Master Record), and BAUP (Upload Bank Master Data).

Bank Statement Monitor (F6388)

Application Type: Transactional

This app allows you to monitor the status of end-of-day bank statements for individual bank accounts, providing both single-day views and a 14-day overview. Based on your bank account settings, you can identify issues such as missing statement pages, discrepancies between bank statement balances and general ledger account balances, or unposted items. The statuses monitored for different bank accounts may vary. If a status indicator is not activated for a bank account, it is set with the **Not Activated** status.

For accounts with the **No Posting Processing** posting category, bank statements are not relevant for posting and are not processed beyond import. In such cases, the posting-related statuses such as **Difference Status** and **Reconciliation Status** are marked as **Not Applicable**. In addition, the app's **End of Day – Single Day** tab displays four status indicators, with detailed insights available via the **Object** page. On the other hand, the **End of Day – Last 14 Days** tab provides a historical status overview based on specified criteria.

Moreover, the app lets you navigate to related apps, including Manage Bank Statements (F6388) and Reprocess Bank Statement Items (F1520), thus allowing further investigation and follow-up actions. Let's compare the Bank Statement Monitor (F6388) and Bank Statement Monitor – Intraday (F3671) apps:

- **Bank Statement Monitor**
  Focuses on end-of-day bank statements. Daily reconciliation is performed to track statement statuses, identify discrepancies, and monitor missing pages or unposted items.
- **Bank Statement Monitor – Intraday**
  Focuses on transactions throughout the day. Facilitates monitoring cash flows and reconciling memo records on an intraday basis.

Bank Statement Monitor – Intraday (F3671)

Application Type: Analytical

This app allows you to track the import status of intraday bank statements for specific bank accounts. The app provides real-time visibility into intraday bank statements and automatically creates memo records to reflect anticipated transactions prior to final posting. The app assigns specific planning types to classify and manage these short-term flows, which generally expire within a day, while also facilitating quick detection of missing or inconsistent statements. To facilitate the app's functionality, you must

ensure the accounts are flagged for monitoring by selecting the **Upload of Intraday Statements** checkbox in the Manage Bank Accounts app (F1366A). Although bank accounts can be linked to multiple house banks via the **House Bank Account Connectivity** setting, only accounts linked to one house bank account are supported by this app.

> **Tip**
> Accurate rule configuration and timely recurring appointment generation from monitoring rules for a specific period are critical for the app's functionality. Consult SAP Note 2190119 for additional information on background job scheduling.

Business Partner Financial Overview (F2429)

Application Type: Transactional

This app provides a 360-degree centralized snapshot of a business partner's financial data to finance, accounts receivable, accounts payable, and credit controllers in real time. On the initial screen, you can filter the desired business partner based on selection criteria and sort out the results list. In addition, you can navigate to the financial overview to access balances of receivables and credits, master data, and contract accounts. The overview includes invoicing documents, payment documents, returns, dunning notices, clarification cases, installment plans, promises to pay, items from worklists, receivables submitted to collection agencies, and write-offs. The financial overview app also covers collections worklists and enforcement actions, if used. You also can access the list for generating worklists from the business partner list, which is available for certain apps with the accounts payable/receivable manager role template.

Business Reconciliation (F4830)

Application Type: Transactional

This app allows revenue accountants to display inconsistent data between sender components and revenue accounting and reporting (RAR), generated from the Run Business Reconciliation in Background app (F4841), as well as reconcile data online for a limited number of revenue contracts. In addition, you can filter reconciliation data by criteria such as company code, accounting principle, operational document, and revenue contract, depending on the chosen run method or mode. The app displays results that can highlight discrepancies in reconciliation values (e.g., price, cost, invoice amount, and quantity), marked in red for inconsistencies in the standard view; or it can show aggregated values for revenue recognition, cost, contract liability, and contract asset in the accounting view. Additional options include viewing data-validation errors, checking performance obligations, and examining condition types, while the reconciliation results are saved to table FARR_D_BIZ_RECON. You can restrict searches to inconsistent contracts or validation errors and navigate to detailed views for deeper analysis.

Tip

When reconciling sales order management and RAR, Transaction FARR_BIZ_RECON is applicable only for classic contract management (CCM) contracts and does not yield results for optimized contract management (OCM) contracts. For reconciling OCM contracts, SAP recommends using this app or using the Run Business Reconciliation in Background app (F4841), which is specifically designed for this purpose. Consult SAP Note 3588350 on this topic.

Calculate Contract Liabilities and Contract Assets (F3875)

Application Type: Transactional

This app streamlines revenue accounting processes by calculating contract liabilities and assets based on multiple accounting principles in revenue accounting and reporting (RAR). The app supports compliance with the revenue standards of IFRS 15 and ASC 606, ensuring accurate revenue recognition. With this app, you can now schedule application jobs for revenue processes with multiple accounting principles, including transferring revenue, calculating contract liabilities and assets, starting a revenue posting run, reversing revenue postings, and reposting revenue postings. The app has three sections to process its calculations: **General Information**, **Scheduling Options**, and **Parameters**. As a report, the app offers functionalities that streamline accounting processes and ensure efficient management of revenue-related tasks.

Tip

This app is delivered with a generic SAPUI5 job scheduling framework. To enable job templates for this app, you must complete the required configuration tasks in the backend system for job catalog SAP_FIN_FARR_CONTR_LIABILITY, as described in SAP Note 3532833.

Calculate Revenue Catch-Up (F5052)

Application Type: Transactional

For revenue contracts created using optimized contract management (OCM), a revenue accountant can use this app to manage revenue catch-up tasks in revenue accounting and reporting (RAR). Revenue catch-up occurs due to retrospective contract modifications classified as changes in estimates. When the unit price is adjusted in the current period, RAR recalculates the fulfilled revenue from prior periods based on the newly adjusted unit price. The app allows the calculation of quarterly and yearly revenue catch-up based on selection data and helps value disclosure. The results of this calculation can then be displayed using the Revenue Catch-Up app (W0169). In addition, you can analyze the revenue catch-up by period, quarter, and year. On the other hand, the Revenue Schedule app (F3882) allows you to make corrections to calculated revenue catch-ups. The app supports quarterly or yearly catch-up, which can be scheduled based on selection criteria such as company code, accounting principle, revenue accounting contract, fiscal year, and fiscal period.

Cash Collection Tracker – Accounts Receivable (F2925)

Application Type: Analytical

This analytical app lets you track customers' collection progress against due receivables for a selected period. With this app, you get a comprehensive view of the collection process, enabling informed decisions and improved cash flow management in real time. The app lets you choose a period or date, and it then shows the sums of all open invoices from previous periods, invoices posted earlier but due in the selected period, and new invoices both posted and due in the selected period, forming the total target. The app provides important features such as viewing the collected amount for the period, seeing it as a percentage of the total target, viewing the outstanding due amount, and using an **As of Date** filter option to limit results to invoices posted until that date.

Cash Collection Tracker – Collections Management (F3182)

Application Type: Analytical

With this analytical app, accounts receivable users can monitor collection progress against due receivables for a chosen period. As a result, they can improve the management of customer collections with real-time visibility into the status of outstanding receivables. After you select a date or period, the app displays the total of all open invoices due from previous periods, invoices posted earlier but due in the selected period, and new invoices both posted and due within the selected period, forming the total target amount.

Tip

The app uses collection information from the collection specialist master data in business partner role UDM000 with collection management. The business role maintains collections-specific data for business partners, enabling functionalities like managing collection profiles and worklists. You must ensure that the default collection specialist is maintained in the **Collection Profile** tab in Transactions UDM_BP or BP for the Cash Collection Tracker app to function correctly.

Cash Discount Forecast (S/4HANA) (F1735)

Application Type: Analytical

This app allows you to forecast short-term cash discounts, providing predictions for discounted amounts on blocked invoices for upcoming payment days. With this app, you can perform tasks such as displaying expiring cash discounts over a defined future period; analyzing discounts by company, payment terms, and payment day; and offering insights into available and expired discounts. To prevent missing out on expiring discounts, SAP recommends that invoices be paid on the current payment day. The app uses CDS view C_APCASHDISCOUNTFORECAST and has caching enabled for better performance. In addition, the app supports SAP Jam integration for comments and emails.

Cash Discount Utilization (S/4HANA) (F1736)

Application Type: Analytical

With this app, you can monitor real-time cash discount utilization across your responsible area and identify which company code or location needs to improve its cash discount use. The app provides features such as displaying the utilization rate for a defined past period, comparing current versus target utilization rates, and checking if the current rate is critically low, needs attention, or is acceptable. You can display the cash discount utilization by company code or by supplier. The app also lets you display either the utilization amount or the utilization rate of the cash discount via configuration. In addition, you can analyze utilization by business dimensions like company, country, supplier group, and payment terms, distinguishing between discounts claimed and those lost. The app uses CDS view C_APCSHDISCUTILIZATION and has technical features for better performance.

Cash Flow Analyzer (F2332)

Application Type: Analytical

With this app, you can get an overview of aggregated amounts and line-item details for cash positions, medium- and long-term liquidity forecasts, and actual cash flows. The app allows you to analyze cash flows for days, weeks, months, quarters, or years across all bank accounts and liquidity items using filter criteria such as company code, currency, time period, value date, bank account hierarchy ID, display currency, exchange rate type, and snapshot time. The app's key features include calculating amounts based on various currencies, selecting default reports, displaying cash flow lists, customizing calendar settings, and switching between balance and delta views. You also can display hierarchies and liquidity items; take snapshots, including unreleased cash flows; and navigate to related apps for further analysis and bank transfers, making the app a comprehensive tool for managing and reporting the cash flow status to management. Moreover, in the **Balance** view of the app, amount-based fields can be groupable by bank, bank account, bank country, currency, house bank, house bank account, and company code. In contrast, in the **Delta** view, all fields can be grouped. This ensures flexibility and customization in how you can organize and analyze your financial data, depending on the view and specific requirements.

Transaction Codes

The corresponding transactions in SAP GUI are Transactions FF7A, FF7B, S_ALR_87012271, and S_ALR_87012272.

Cash Flow Comparison – Actual/Forecast (W0128)

Application Type: Analytical, Web Dynpro

This app allows you to compare actual cash flows with historical forecasts and analyze forecast records across different snapshot dates. You can search for comparison results using various search criteria such as value date from, value date to, snapshot date from, snapshot date to, snapshot time, value date (hierarchy), display currency, exchange rate type, and certainty level. The **Data Analysis** tab displays comparison results for actual-to-forecast and forecast-to-forecast comparisons within a specified time range, leveraging various dimensions and filters for a personalized report via the **Navigation Panel**. In addition, cash flow figures can be displayed in hierarchical formats such as week/day, year/month/day, or year/quarter/month/day using the **Value Date (Hierarchy)** filter, which allows for aggregation and expansion of details as needed. Thus, the app improves the accuracy of future predictions.

Cash Flow Comparison – By Timestamp (W0137)

Application Type: Web Dynpro

This app allows you to compare forecasted cash flows from any two past snapshot time stamps and evaluate their accuracy against actual cash flows in cash and liquidity management. You can search for cash flows using fields such as value date from, value date to, snapshot date A/B, snapshot time A/B, display currency, exchange rate type, and bank account currency. The **Data Analysis** tab displays data in such a way that you can compare forecasts across different snapshot time stamps and assess forecast records against actual cash flow data. The **Query Information** tab offers parameters and settings used to generate the results. Thus, by analyzing variations between forecast records and real financial outcomes, you can refine your predictive models and improve future cash flow forecasting.

Cash Flow – Detailed Analysis (F0740)

Application Type: Analytical, Web Dynpro

This app provides a comprehensive overview of daily cash inflows and outflows, allowing analysis over recent weeks or months for all subsidiaries and liquidity items, thereby providing a company's overall liquidity position. You can search for details using parameters such as conversion date, display currency, conversion type, liquidity item hierarchy, calendar day, company code, bank account currency, and the hierarchy variable for day. The app further enhances the cash flow position by providing cash flow forecasting, which aids short-term and long-term financial planning through historical trend analysis. In addition, you can identify extraordinary and abnormal cash flows, ensure accuracy and compliance, and consider liquidity planning. The app offers multidimensional analysis from various perspectives, including viewing the cash inflow, cash outflow, and cash net flow confirmed by banks; reporting in specified currencies and dates; and drilling down by calendar day and other dimensions. In addition, the app supports technical features such as data export, email functionality, and creating tiles with current selection criteria.

> **Tip**
>
> Consult SAP Note 2512842 for troubleshooting guidelines if the app does not display data.

Cash Flow Statement – Indirect Method (W0054)

Application Type: Web Dynpro

In accounting, the *indirect method* is an approach to preparing the cash flow statement in which you adjust the net income for noncash transactions and changes in balance sheet accounts to determine the cash flow from operating activities. This app generates a cash flow statement based on the indirect calculation method. The app allows for customization by adding columns or rows for detailed account analysis. The cash flow statement total is calculated by summing the cash flow from operating activities, the cash flow from investing activities, and the cash flow from financing activities, along with the operating cash flow derived from net profit, noncash adjustments, and balance sheet changes. The calculation is based on

semantic tags assigned to relevant general ledger accounts. Thus, the app provides real-time details of cash flow movements.

> **Tip**
>
> If the standard cash flow delivered does not fit your needs, you can create customized report layouts with revised formulas in the Custom Analytical Queries app (F1572) by copying existing CDS view C_CASHFLOWINDIRECTIFRS.

> **Transaction Codes**
>
> The corresponding transactions in SAP GUI are Transactions S_ALR_87012271 and S_ALR_87012272.

Cash Pool Transfer Reports (W0129)

Application Type: Web Dynpro

This app provides a daily report on concentration amounts within a cash pool, allowing you to track fund transfers between header accounts and subaccounts over a specified period in cash and liquidity management. You can search for cash pool transfers using search parameters such as start date, end date, and cash pool name. The **Data Analysis** tab lets you view confirmed bank transfers, as well as cash in transit, while utilizing dynamic dimensions for adjustments. In addition, you can filter, sort, and drill down into data, extend row and column dimensions using the **Navigation Panel** function, and navigate to other cash management apps through the **Jump To** option.

> **Tip**
>
> After executing the cash concentration in the Manage Cash Concentration app (F3265), this report displays cash pool transfers once payments are made and confirmed via bank statement postings.

Cash Position (S/4HANA) (F1737)

Application Type: Analytical

This app allows you to check forecasted cash positions for the current date by country, company code, and currency. Cash position data is calculated based on memo records and various data sources from the One Exposure from Operations hub. Depending on your assigned role, you can display cash positions by several analytical dimensions, switch between charts and tables, and export search results to a spreadsheet. In addition, the app comes with analytical dimensions, including bank country/region, bank, bank group, company, and various bank account currencies.

Transaction Code

The corresponding transaction in SAP GUI is Transaction FF7A.

Central Finance – Navigation between OP Source System and Central Finance (F5253)

Application Type: Transactional

This app helps with navigation between a source system and the Central Finance system, supporting navigation in both directions. To differentiate terms in Central Finance navigation scenarios, the app distinguishes between the navigation source system and navigation target system. In scenarios in which navigation from one Central Finance system to multiple source systems is involved, there's only one supported navigation option for the unique combination of the Central Finance system and source system at a time. For the target system, different options can be configured for varying navigation options. Once the first navigation target system is set up with RFC destinations and alias mappings, the selected navigation option applies universally to all tar-

gets. For instance, if the **UI5 External** option is unavailable due to a navigation target system lacking SAP Fiori, a **Web GUI** scenario can be configured instead.

Tip

Consult SAP Note 3318736 for detailed guidance and currently supported navigation scenarios.

Change History (F4273)

Application Type: Transactional

This app allows revenue accountants to track and review changes made to revenue contracts and performance obligations. The app tracks changes in specific fields of revenue contracts and performance obligations, grouping them by document number and sorting them by default in descending order. Document numbers are system-generated to specify contract changes. With this app, you will see logs organized by system-generated document numbers and comprehensive details of the changes. In addition, you can search for change logs for revenue contracts or specific fields, such as contract status. The app displays results such as the change date, transaction type, triggering change type, change context, and the person who made the change. The app also provides search options that include attributes like the revenue contract, field name, manual transaction, changed on, and changed by, with fuzzy search available for fields such as change document and performance obligation name.

Change Request for Bank Account Approval (WDA_FCLM_BAM_CHGREQ)

Application Type: Web Dynpro

This Web Dynpro–based app helps the approval process for modifying bank account details within SAP's bank account management (BAM) framework. The app allows you to submit requests for creating, updating, or closing bank accounts while ensuring compliance with financial governance policies. The app features an automated approval workflow that routes requests to designated cash managers or approvers. In addition, the app maintains audit records for regulatory compliance, integrates with cash and liquidity management for financial control, and provides real-time tracking and notifications of approval statuses.

Check Cash Flow Items (F0735)

Application Type: Transactional

This app allows cash managers and cash management specialists to analyze cash flow item details after reviewing liquidity forecast and cash position reports for a selected value date. You also can search with other parameters such as company code, transaction currency, bank account, planning level, certainty level, and bank account currency. With this app, you can track and trace cash flow items from integrated source applications, viewing the line-item details of original documents like journal entries and bank assignments. The app provides features such as listing aggregated cash flows, filtering items by various dimensions, monitoring the payment status in bank communication management (BCM), and navigating to line-item details from forecasts. In addition, you can check the original document information, adjust account assignments, and export search results to a spreadsheet. The app also supports email functionality and creating tiles based on the current selection criteria. As a prerequisite to display data, Customizing activity **Define Basic Settings** must be configured. Moreover, you must be assigned the FQM_FLOW authorization object to display or edit cash flow liquidity item activities.

Tip

The app allows you to adjust actual cash flows that occur after the defined block date, which is set in the Set Block Date for Actual Cash Flows app (S_ER9_11002196). For an actual cash flow to be adjusted, SAP recommends that its value date be later than the defined transaction date, and its posting date must be later than the defined posting date.

Transaction Code

The equivalent transaction in SAP GUI is Transaction FF7AN.

Check Liquidity Items on G/L Accounts (F6459)

Application Type: Transactional

This app displays default liquidity items assigned to general ledger accounts in the Define Default Liquidity Items for G/L Accounts app (F6459). The app lists general ledger accounts based on the chart of accounts or company code, displaying cash-relevant details such as general ledger account ID, name, type, account group, balance sheet account indicator, and open item management status. In addition, you can view additional fields including the account type, account group, chart of accounts, reconciliation ID, and more. The **Detail** page retrieves original entries linked to selected general ledger accounts, ensuring visibility into the predefined **Liquidity Item** settings.

Clean Up Productive Data (F5728)

Application Type: Transactional

This app allows you to delete revenue contracts and associated revenue accounting items (RAIs) linked to header IDs created incorrectly in revenue accounting and reporting (RAR). You can clean up RAIs and relevant revenue contracts originating from sender components such as sales and distribution, billing and revenue innovation management, and external sources. If revenue contracts have no postings, the program deletes the data directly. For contracts with postings, the program reverses general ledger postings to maintain reconciliation between the general ledger and revenue accounting subledger, while deleting all other data. To avoid data inconsistency, you must ensure no new RAIs are created during or after the cleanup process. Following the cleanup, SAP recommends an operational load be executed to retrieve all data from the sender component.

Tip

For classic SAP Revenue Accounting and Reporting, you must consider that this app is applicable only to company codes set to **Productive** for all accounting principles.

Transaction Code

The corresponding transaction code is Transaction FARR_PROD_CLEANUP in SAP GUI.

Clear G/L Accounts – Manual Clearing (F1579)

Application Type: Transactional

This app allows you to manually clear general ledger account open items that did not clear automatically. The app lets you find open items by using various search criteria, such as company code or general ledger account; clear multiple open items manually; enter characteristics for profitability-related postings; create notes and attachments while posting; simulate the resulting journal entry; export data to a spreadsheet; and share open item data via email.

Tip

The app doesn't support accounts defined as automatic postings in the general ledger master data (table SKB1, field XINTB). For example, if the **Automatic Posting** flag is enabled for goods receipt/invoice receipt (GR/IR) accounts, then the app excludes such accounts. In such cases, SAP recommends manually clearing GR/IR accounts with the Clear GR/IR Clearing Account app (MR11) or the Reconcile GR/IR Accounts app (F3302).

Transaction Code

The corresponding transaction in SAP GUI is Transaction F-03.

Clear Incoming Payments (F0773)

Application Type: Transactional

This app allows you to clear receivable payments manually, such as open incoming payments for customer invoices, especially when automatic clearing fails due to missing customer information. You can search for incoming payments using fields such as company code, posting date, customer, and assignment. The app offers features that include viewing open incoming payments in the My Open Worklist Version 2 app (F1611A), obtaining a list of open items for clearing payments, adding or changing discounts, creating residual items, posting incoming payments to general ledger or customer/supplier accounts, and using customer search criteria for matching open items. In addition, the app supports creating dispute cases, using promise-to-pay information, searching open items with fuzzy logic, entering profitability characteristics, viewing withholding tax, creating notes and attachments, simulating journal entries, and exporting the open items list to a spreadsheet. You can edit simulated clearing documents, save clearing proposals, and clear open items for customer accounts or down payments.

1

Tip

The app doesn't include open follow-on documents (identified with follow-on document type BSEG-REBZT). In addition, the app limits the number of items to 10,000 to avoid performance bottlenecks.

Transaction Code

The corresponding transaction in SAP GUI is Transaction F-32.

Clear Outgoing Payments (F1367)

Application Type: Transactional

This app allows you to manually clear payable payments, such as open outgoing payments for supplier invoices, when automatic clearing is not possible. You can search for incoming payments using fields such as the company code, posting date, vendor, and assignment. The app offers features such as viewing open outgoing payments in the My Open Worklist Version 2 app (F1611A), obtaining a list of open items for clearing, adding or changing discounts, creating residual items, posting outgoing payments to general ledger accounts or on account, searching for open items using fuzzy logic, entering profitability characteristics, viewing withholding tax, creating notes and attachments, arranging open items by invoice reference, simulating and editing journal entries, saving clearing proposals, and clearing open payments with matching items. In addition, you can clear open items for supplier accounts or down payments and export the open items table to a spreadsheet.

Transaction Code

The corresponding transaction in SAP GUI is Transaction F-44.

Co-Liability Worklist (FICA_COLIABILITY_WL)

Application Type: Web Dynpro

This app helps you manage and track co-liability transactions in contract accounting. The app provides all pending co-liability worklists for multiple business partners for you to review and take necessary action. The app helps you define your own classification criteria to assign work items to specific worklists. The app integrates with other SAP components such as general ledger, customer relationship management, and public section contract accounts. In addition, you can predefine distribution rule percentages to distribute receivables amounts to business partners. Thus, the app acts as a centralized location to track and manage various types of coilability for both shared and third-party liability.

Collection Progress (F1738)

Application Type: Analytical

The analytical app displays the **Collection Progress** KPI, allowing you to monitor overall payment collection progress and the performance of different collection specialists and groups. The app offers features that include viewing the percentage of accumulated promise-to-pay amounts since the last generation of the collection worklist; filtering collection progress by segment, by group, by collection specialist, or by priority; and comparing current versus target rates. In addition, this KPI helps accounts receivable managers manage their teams, with the goal of achieving close to 100% collection progress by the end of each day. The app also supports SAP Jam integration for comments and emails, uses the CDS view C_APCSHDISCUTILIZATION, and has caching enabled for better performance.

Combine Revenue Contracts (F4069)

Application Type: Transactional

This app allows revenue accountants to merge two revenue contracts into a single target contract. As a result, the app simplifies revenue accounting management. To combine revenue contracts using this app, you can search and select two contracts to combine, and then specify the target revenue contract, reason code, and effective date in the popup screen. Once the app is executed, the source contract is removed, while the target contract remains. As a prerequisite, contracts that are being combined must share attributes like accounting principle, company code, customer, transaction currency, local currency calculation method, level for posting liabilities and assets to posting table, type of financial valuation object, and receivable adjustment account. They must also meet specific criteria, such as being created using optimized contract management and not being soft-deleted or terminated. Also, price reallocation occurs based on whether the change type is prospective or retrospective, with retrospective changes including catchup calculations for recognized revenue. In addition, postings on the source contract are reversed and reposted under the target contract ID based on granularity, reconciliation key status, and account determination rules.

Tip

Besides defining reason codes in the Customizing activity, you must implement the FARR_ BADI_CHANGE_TYPE_DETN BAdI to determine whether each reason code represents a retrospective or prospective change.

Commodity Hedge Management Cockpit – Analytical View (F5654)

Application Type: Transactional

This app provides commodity traders with an overview of the current hedging situation and prospective trends for selected exposures. The app provides key insights such as exposures, assigned hedges, and target quotas for hedge management and hedge accounting. In addition, you can analyze quotas, hedged quantities, and trades, drilling down by years, hedge books, periods, or trades, with data displayed as a table or bar chart. In addition, the app highlights trades linked to plan exposures, and compares hedging situations across different hedge books. The app supports weighted averages for designated cash flow hedge prices and hedged quotas are calculated based on exposures. The app also features fast trade capture for entering commodity swaps and forwards directly.

Commodity Hedge Management Cockpit – Overview (F5656)

Application Type: Transactional

This app provides commodity traders with a comprehensive overview of the current hedging situation for selected exposures and prospective trends, enabling quick identification of areas requiring action. Thus, the app allows you to identify areas requiring action quickly. The app offers important insights such as current exposures, assigned hedges (e.g., swaps, forwards), and target quotas in commodity hedge management and commodity hedge accounting. The app further displays hedged quantities, details of trades linked to plan exposures, and a comparison of hedging situations across hedge books. In addition, the app integrates exposures, hedges, and hedging relationships into a single view while displaying three currencies: **Target Crcy** (from exposure position), **Risk Currency** (exposure amount currency), and **C-Crcy** (hedging entity's or the company code currency). Moreover, it supports all roles in the foreign exchange (FX) risk management process by providing fast, reliable information and configurable data. The app comes with drilldown capabilities that support analysis across different aggregation levels (years, hedge books, periods, and trades) using **Table View** and **Chart View**.

- **Commodity Hedge Management Cockpit – Overview**
  This app shows an overall hedging situation monitoring. It provides a portfolio view of exposures and hedges and supports drill-down capabilities.

- **Review Hedge Constellation Details**
  This app performs detailed hedge structure analysis (relationships). It focuses on specific hedge constellations and provides a detailed linkage view and attribute analysis.

- **Commodity Hedge Management Trader Cockpit**
  This app supports trade execution and position management. It focuses on trader's order book and market activities, as well as order management, trade execution, and market data.

Commodity Hedge Management Trader Cockpit (F5658)

Application Type: Transactional

This app provides traders with a quick overview of trade orders defined in the Manage Hedge Specifications app (F5661), allowing immediate action to conclude trades for open exposure quantities requiring hedging. Thus, the app acts as a central location to manage trading and hedging activities related to commodities. The app's functionalities include displaying hedge specification orders and target quotas, identifying open orders and areas requiring attention, navigating referenced hedge specification orders and trades, and directly creating commodity swaps or forwards from within the app.

Let's compare the Commodity Hedge Management Cockpit – Overview (F5656), Review Hedge Constellation Details (F5657), and Commodity Hedge Management Trader Cockpit (F5658) apps:

Commodity Order Fill Enrichment (F5173)

Application Type: Transactional

Financial information exchange (FIX) is a standardized protocol used for the electronic communication of trade-related information. This app is designed to add subaccount IDs to commodity derivative order FIX messages, which are required for creating commodity futures transactions. The app's primary function is to fill out missing data from the initial order details received from external systems such as brokers via FIX messages. If brokers receive order requests without subaccount ID information, this app allows you to retroactively supply the IDs, either automatically or manually. By enabling automatic subaccount determination via the **AutoDet SA** flag in the Customizing activity **Define Derivative Order Fill Packet Types**, you can process single or multiple fills simultaneously and assign valid subaccount IDs. Once this step is completed, commodity futures transactions in SAP Commodity Risk Management can be automatically created based on the system setup.

Commodity Price Risk Hedge Accounting Application Logs (F6251)

Application Type: Transactional

This app allows you to display and filter all application logs related to commodity price risk hedge accounting. The app supports logs under the **CMM_HEDGE** category, including subcategories such as counterdeal requests, dedesignation requests, foreign exchange (FX) plan exposure, commodity hedge constellations, commodity hedge specifications, late designations, migration requests, offsetting FX hedge requests, planning date file uploads, planning data, plan exposures, reclassification requests, and target quotas. With this app, you can display the logs based on search criteria such as dates, specific hedging instruments and type of log. Thus, the app offers a central point to display all application logs related to commodity price risk hedge accounting.

Compare Source/Target Accounting Principle – For Classic Contract Management (FARR_TRANSITION_COMP_REPORT)

Application Type: Web Dynpro

This app lets you analyze differences in revenue recognition between original (source) and adjusted (target) accounting principles under different accounting standards in revenue accounting and reporting (RAR). The app's features include comparing different accounting principles for revenue recognition, expense allocation, and overall reporting, highlighting key differences between the source and target accounting principles. The app integrates with classic contract management for comparison reasons. In addition, the app provides an audit trail of changes in accounting standards to determine the impact of revenue recognition,

expenses and revenue reporting. The app supports and complies with IFRS 15 and ASC 606 accounting standards.

Comparison of G/L Accounts – Revenue Accounting and G/L (FARR_RECON_ACCOUNT_RA_GL_AP)

Application Type: Web Dynpro

This app helps reconcile general ledger accounts and revenue accounting entries in revenue accounting and reporting (RAR). The app's features include transaction-level details to analyze and to detect discrepancies, supports error identification and resolution, and maintains accurate revenue reporting. The app supports and complies with IFRS 15 and ASC 606 accounting standards. This Web Dynpro app supports UI customizing for adding customer fields to contract and performance obligations.

Consolidated Balance Sheet (W0059)

Application Type: Web Dynpro

This app provides a comprehensive view of the balance sheet for a specified consolidation group, fiscal year, and period in SAP S/4HANA Finance for group reporting. The app displays amounts in group currency across all posting levels and compares balances across years. In addition, you can analyze balance sheet data, compare year-over-year trends, and drill down into subitems for detailed analysis.

Tip

This app has been replaced by the Group Data Analysis app (W0135) as a successor app. Therefore, this app is recommended to be used in earlier versions, especially prior to 1909.

Consolidated P&L by Nature (W0070)

Application Type: Web Dynpro

This app provides a structured view of profit and loss (P&L) accounts, categorizing them by the nature of expenses rather than functional areas in SAP S/4HANA Finance for group reporting. The app provides period-specific analysis across fiscal years, consolidation groups, and reporting periods in group currency, supporting multilevel reporting across all posting levels.

Note

This app has been replaced with Group Data Analysis (W0135) as the successor app. Therefore, this app is recommended to be used in earlier versions, especially prior to 1909.

Consolidation Data Release Cockpit (F2591)

Application Type: Transactional

As part of group reporting activities, the app allows local consolidation accountants from countries to view and verify financial data of specified consolidation models and entities before releasing the data. The app presents data in **REPORTED DATA**, **VALIDATION**, and **RELESE HISTORY** tabs. Based on the status of the last released data, the system controls whether currency translation, validation, and release are possible for new posting data, and presents released and to-be-released amounts in both local currency and/or group currency. The app provides you with a data status overview with lock status, period status, adjustment status, and new posting detection, detailed reporting of financial data by fiscal period, running currency translation and validation, and a release process with history tracking. In addition, the app allows you to reverse the last released data and replace certain obsolete transaction codes.

Consolidation Data Release Monitor (F2620)

Application Type: Transactional

As part of real-time group reporting, this app allows group accountants to monitor the status of released data for specified consolidation models and fiscal periods. You can search for consolidation data using parameters such as model, version and group. The app provides you with functionality to perform mass actions, such as period initialization, currency translation, data validation, release, defer, and approval for all or selected entities in sequential order. The system controls each task by verifying the results of preceding tasks to ensure flawless data processing. The app provides you with information such as a data status overview for each entity, showing lock status, period status, adjustment status, and new postings. The app supports the two types of data release requests: periodic release request (PRR) and adjustment release request (ARR), each with specific requirements and actions. This ensures the data is complete and accurate before proceeding with consolidation.

Contract Balance (W0155)

Application Type: Web Dynpro

This Web Dynpro–based app presents the opening and closing balances of receivables, contract assets, contract liabilities, deferral revenue, and unbilled receivables from customer contracts in revenue accounting and reporting (RAR). You can view balance amounts across multiple dimensions and currencies, including display currency, document currency, company code currency, and local currencies. The app's **Display Mode** allows you to display contract balances for by period, by quarter, and by year. In addition, the app supports multidimensional analysis, allowing you to customize rows and columns using the **Navigation** function, apply filters, and adjust exchange rate types. This app is delivered using Web Dynpro data grid; it offers visualization in **Chart View**, **Table View**, or a combination of both, and includes **Jump To** functionality for seamless navigation to other SAP Fiori applications.

> **Tip**
>
> This app provides a static snapshot of contract balances, including receivables, contract assets, contract liabilities, deferral revenue, and unbilled receivables, whereas the Contract Balance Movements app (W0154) offers the ability to track changes in contract balances over a reporting period.

Contract Balance Movements (W0154)

Application Type: Web Dynpro

This Web Dynpro app provides you with an explanation of significant changes in contract asset and liability balances during the reporting period in revenue accounting and reporting (RAR). The app displays opening and closing balances, revenue recognition from contract liabilities, changes in the time frame for receivables and performance obligations, asset impairments, and cumulative catch-up adjustments. In addition, you can analyze contract balance movements across multiple currencies, including display currency, transaction currency, company code currency, and local currencies. The app supports configurable filters such as company code, accounting principle, fiscal year, fiscal period, display currency, and exchange rate type. By default, the app presents quarterly changes in the company code currency using a structured tree format. Also, you can adjust dimensions using rows and columns of the **Navigation Panel** function, switch between **Chart View** and **Table View**, and leverage additional functions such as **Jump To** for navigation to other relevant apps.

> **Tip**
>
> As a prerequisite to use this app, you must process revenue contracts using the Calculate Contract Liabilities and Contract Assets app (F3875) for the to-be-disclosed periods.

Contract Balance Reclassification (W0176)

Application Type: Web Dynpro

This Web Dynpro–based app allows you to classify balance sheet items related to revenue contracts into short-term and long-term categories and is executed monthly, quarterly, and yearly. The app presents reclassification balance amounts for the reporting period, including total, short-term, and long-term amounts in document currency using measures such as total amount in document currency, short-term amount in document currency, and long-term amount in document currency. The app determines the short-term amount in document currency using the first increment with the shortest duration, while all remaining increments are classified under the long-term amount in document currency calculation. Also, reclassification follows first in, first out (FIFO) consumption, reflecting future events' impact on current balance sheet items, such as future fulfillments affecting contract liability reclassification and planned or assumed invoices influencing contract asset reclassification. In addition, you can add additional dimensions using the **Navigation Panel** function for multidimensional analysis. The app supports and complies with both the IFRS 15 and ASC 606 accounting standards.

> **Tip**
>
> As a prerequisite to using the app, you must ensure the aging increment is appropriately defined in the **Define Aging Increments for Advanced Valuation** activity and assigned to an accounting principle in the **Assign Aging to**

1

Accounting Principles activity. These Customizing activities are carried out under IMG menu path **Financial Accounting • General Ledger Accounting • Periodic Processing • Advanced Valuation in General Leder Accounting • General Settings**. The **Aging Category** should be set to **REC Reclassification** to enable accurate classification of balance sheet items.

Contract Change History (FARR_CONTRACT_CHG_HISTORY)

Application Type: Web Dynpro

This app allows you to track and review modifications to revenue contracts over time in revenue accounting and reporting (RAR). The app logs updates made to contract details, including performance obligations, transaction prices, and financial postings, while maintaining version control for tracking and audit purposes. As a result, the app provides a structured audit trail for compliance and reporting to comply with IFRS 15 and ASC 606 accounting standards. Thus, the app provides you with version control, which helps to track changes and thereby prevent discrepancies in revenue recognition.

Contract Comprehensive View (FARR_CONTRACT_BALANCE_SAMPLE)

Application Type: Web Dynpro

This app provides an in-depth analysis of revenue accounting contracts, displaying performance obligations, transaction prices, and revenue recognition status in revenue accounting and reporting (RAR). In addition, the app offers visibility into financial postings, including accruals and adjustments, while ensuring compliance with IFRS 15 and ASC 606 standards. Moreover, the app is integrated with Business Rule Framework Plus (BRF+) to support rule validation and maintenance for contract monitor-

ing. Users are recommended to validate contract balances against revenue postings regularly to align IFRS 15 and ASC 606 accounting standards.

Create Business Partner Contact (FKK_WDY_BP_CONTACT)

Application Type: Web Dynpro

This Web Dynpro–based app allows you to create and manage contact persons such as sales reps and account managers related to business partners in contract accounting. With this app, you can define contact persons, edit existing contacts, delete contacts, search specific contacts, assign specific roles, and establish relationships with business partners. You can use the **Relationships** function to assign to link to a business partner. In addition, the app supports role-based access control

Create Correspondence (FKK_WDY_CORR_MANUAL)

Application Type: Web Dynpro

This Web Dynpro–based app allows you to generate correspondence for invoices, payment reminders, and customer notifications in contract accounting. The app supports template-based formatting for standardized messaging, integrates approval workflows, and maintains audit records for tracking correspondence history. The app's features include creating, modifying, and monitoring correspondence requests, and attaching relevant documents while leveraging multichannel delivery options such as email, print, and electronic formats. The app synced with financial records by validating correspondence against predefined business rules. Also, the app links correspondence details to corresponding invoices, payments, and customer accounts. You can leverage SAP-delivered correspondence types or create your own to meet business needs.

Create Correspondence (F0744)

Application Type: Transactional

This app allows accounts receivable accountants or finance users to preview, email, print, fax, and download correspondence related to customers and suppliers. The app allows you to create different types of correspondence, automated processes, customization options, preview and print functionality, tracking and monitoring, and integration with other SAP modules. The app allows you to create documents such as payment advice, dunning letters, account statements, and other correspondence to communicate with business partners.

The app can be launched either from the SAP Fiori launchpad or other SAP Fiori apps such as Process Receivables Version 2 (F0106A), Display Customer Balances Version 2 (F0703A), Display Supplier Balances Version 2 (F0701A), Manage Customer Line Items (F0711), and Manage Supplier Line Items (F0712). Depending on your role assignment, you can create, preview, email, fax, print, or download correspondence as a PDF file. As a result, the app streamlines communication related to financial transactions thereby improving efficiency and maintaining professional correspondence records.

Transaction Code

The equivalent transaction in SAP GUI is Transaction F.64.

Create Correspondence (Version 2) (F0744A)

Application Type: Transactional

With this app, you can preview, email, print, and download correspondence related to customers and suppliers. It can be launched from the SAP Fiori launchpad or other Fiori apps, such as Manage Customer Line Items and Manage Supplier

Line Items. Key features include creating correspondence, previewing it, emailing, printing, or downloading it as a PDF file. Advanced parameters allow you to adjust the appearance and add more details to your correspondence documents.

Transaction Code

The equivalent transaction in SAP GUI is Transaction FB12.

Create Credit/Debit Memos (Mass Processing) (F3068)

Application Type: Transactional

As part of accounts receivable and accounts payable processes, this app allows you to efficiently create credit memos or debit memos for large volumes of invoicing documents, allowing adjustments to invoiced amounts without altering the original invoice. You can search, filter, and sort invoicing documents based on various criteria while excluding those already associated with a credit or debit memo. For example, you can search for credit and demos using fields such as invoicing document, business partner, contract account, invoicing type, document date, and invoiced on. If reversing the invoicing document is not feasible, the original document along with the credit or debit memo defines the effective invoiced amount. The app supports batch creation for up to 500 documents, provides error status visibility, and allows entry of fixed or percentage-based amounts for memos.

Before saving, you can calculate the total amount, access the Manage Credit/Debit Memos app (F2389), and modify created memos before approval. Navigation options, such as hiding filters or customizing columns, are dependent on roles and authorizations, enhancing flexibility and usability. One major advantage is that the app allows you to perform bulk processing of transactions in a single run.

Tip

As a prerequisite, business functions **SAP FI-CA** (Contract Accounting for Receivable and Payable) and **SAP CI** (Convergent Invoicing) must be activated.

Create Payments (F3648)

Application Type: Transactional

This transactional app allows you to create and manage payments using SAP-delivered templates or by creating new ones. You can save draft payments for later editing, process payment orders, or cancel and discard them. The app's features include adding attachments, intermediary agents via the **Maintain Bank Chain** action, and tax-related data using the **Edit Tax Fields** action. To begin with, you can select a clearing area and payment order type or use a template. In addition, you can specify general order details such as priority and due date, originator account information, and recipient payment item details like account data, references, and remittance data. For check processing, you can enter check-specific data such as check numbers and delivery methods. After entering the required data, the system validates it and displays errors or warnings instantly. Optionally, you also can simulate payment processing to verify enrichment, validation, and routing without altering the order. Once validated, you can save the order, save it as a template, or process it directly. Moreover, simulation runs through all processing steps and displays errors or warnings encountered.

Tip

To support displaying regulatory reporting and structured remittance information, the Create Payments app (F3648) helps capturing these details during payment creation manually, while the Manage Payments app displays the information originate either from other SAP or non-SAP systems or from manual entries.

Create Refunds for Digital Payments (F7326)

Application Type: Transactional

This app helps create refunds for payment card transactions processed through SAP S/4HANA Cloud for customer payments or the **SAP Digital Payments: Payment Card Authorizations for Customer Line Items with Standard Card** job in the Schedule Accounts Receivable Jobs app (F2366). You can track the refund process, including its relationship with customer invoices, payment documents, and refund records. The app's functionalities include searching for refundable payments, generating partial or full refunds—only when the previous refund is completed, and utilizing the **Payment Card Settlement** job template in the Schedule Accounts Receivable Jobs app (F2366) for finalization. In addition, the app provides detailed process flow, allowing you to navigate from the refund record to related invoices and payments. You can track refund status with **Pending Settlement**, **Settled**, **Reversed**, or **Failed**. Also, you can access reversal details using the Manage Journal Entries app (F0717). In addition, refund creation logs are available within the app for troubleshooting activities.

Tip

As a prerequisite, the app supports refunds only to be executed by the payment service provider if the SAP digital payments add-on is activated. To use this functionality, a separate license for the add-on is required. Consult SAP Note 3431302 for additional details on refunds for digital payments.

Create Single Payment (F0743)

Application Type: Transactional

This app allows you to make direct payments to suppliers without invoices and pay open supplier line items. For direct payments, you specify supplier and bank details, and the amount, then create the payment, which posts as a down payment request and initiates the payment run. When paying open supplier line items, you select the items through the Manage Supplier Line Items app (F0712) and create the payment, which automatically fills in the payment information. Open items clear upon successful completion of the payment run, and you receive a confirmation email. As a prerequisite, the Customizing activity via Transaction FBZP is required to define which house bank accounts are eligible for payments, including ranking orders, available amounts, and value dates. The app then uses this setup to automatically select the most suitable bank during payment creation. The email functionality in your system is configured to ensure you receive emails. In addition, the app allows you to execute payments without invoices, trigger payment processes for open items, make payments in foreign currencies, attach files, create payment files, receive detailed email logs, and print payment summaries.

Tip

This app can only process one invoice and make one payment at a time. For paying multiple invoices, SAP recommends the Manage Automatic Payments app (F0770) for automatic payments or the Post Outgoing Payment app (F1612). In addition, the generation of payment advice is not supported. Rather, SAP recommends using Create Single Payment (F0743) or Manage Automatic Payments (F0770). Moreover, tax jurisdiction, which is used to determine tax rates in particular nations or areas, such as the United States, is not supported.

Transaction Code

The equivalent transaction in SAP GUI is Transaction F-47.

Credit Limit Utilization (S/4HANA) (F1751)

Application Type: Analytical

With this app, credit controllers can get a detailed analysis of a business partner's credit limit utilization. The analytical app displays the **Credit Limit Utilization** KPI, indicating the utilization of a business partner's credit limit as a percentage and an absolute amount. You can identify business partners exceeding the defined threshold value and view credit limit utilization and credit exposure by business partner, credit segment, country, and region. In addition, you can drill down to analyzing the top 10 business partners with the highest credit limit utilization and credit exposure. In addition, you can perform tasks such as viewing the number of business partners above the threshold on the KPI tile, calculating credit limit utilization by comparing granted and used credit limits, and analyzing top business partners by credit limit utilization and exposure. The app also supports customizable threshold values, display currencies, and exchange rate types. The app is enabled with caching for improved performance.

Tip

The behavior of credit segment 0000 differs from other credit segments in terms of calculation and display across different transactions and apps. For credit segment 0000, the credit exposure is calculated in Transaction BP dynamically. On the other hand, this calculation is not performed in the SAP Smart Business modeler app. As a result, segment 0000 is not displayed in the SAP Smart Business modeler app by design.

Credit Line Analysis (W0051)

Application Type: Analytical, Web Dynpro

This app provides an overview of total credit line trends, available amounts, and utilization across money market and trade finance transactions for informed funding decisions in the treasury and risk management (TRM) functionality. The app offers KPIs such as **Total Credit Line**, **Available Amount**, and **Utilized Amount**. These KPIs can be displayed for a specific period such as start date or end date, company code, reporting currency, reporting frequency, exchange rate type, and counterparty. In addition, you can view data in **Chart View** or **Table View**, analyze utilization rates, adjust filters dynamically, drill down into transaction-level details, and navigate to related apps such as Display Facility (TM_63) for further analysis.

Currency Exchange Rates (F3616)

Application Type: Transactional

This app allows you to create, change, and monitor currency exchange rates. The app provides the capability to create and delete exchange rates, set validity periods, view all active exchange rates, copy and modify existing rates, and choose between **Standard** and **Full Overview** display variants. The app also offers functions to check for duplicates and inconsistencies with **Check Currency Pairs**, convert amounts using the **Currency Converter** functionality, and view exchange rate trends for currency pairs over the last 7 days, 30 days, or 12 months.

Tip

If the app displays a warning message for certain quotation entries, the issue may stem from inconsistencies in the quotation data itself rather than missing entries in Customizing Transaction ONOT (maintenance view V_TCURN). As a result, the app doesn't allow you to copy or edit entries. To resolve such errors,

you must update the relevant Customizing tables to ensure alignment with quotation types.

Transaction Code

The corresponding transaction in SAP GUI is Transaction OB08.

Currency Translation – Difference Analysis (CCONS_FPM_OVP_IS_COS_BY_CU (CCONS_FPM_OVP_IS_COS_BY_CU))

Application Type: Web Dynpro

This app allows you to analyze currency translation differences arising from periodic currency conversions across consolidation units in financial consolidation. You can search for currency translation differences using parameters such as fiscal year, posting period, and consolidation unit. The app is useful to reconcile profit and loss impact due to exchange rate fluctuations. Thus, the app helps short term financial analysis to determine currency translation effects. The app's features include displaying currency translation difference details at consolidation unit level, analyzing foreign currency fluctuations and their impact on consolidated units, and navigating to individual translation reserve components for further analysis. The app leverages currency translation methods to determine the currency exchange translation differences.

Currency Translation – Reserve Analysis (CCONS_FPM_OVP_CUR_RESERVE (CCONS_FPM_OVP_CUR_RESERVE))

Application Type: Web Dynpro

This app allows you to analyze currency translation reserves across consolidation units in financial consolidation. You can search for currency translation reserves using parameters such as fiscal year, posting period, and consolidation unit. This app tracks the cumulative effects of foreign currency fluctuations over time. Thus, the app is useful for long term reporting of balance sheet items that accumulate over a period. The app's features include displaying currency translation reserve details, analyzing foreign currency fluctuations and their impact on consolidated units, and navigating to individual translation reserve components for further analysis. The app leverages currency translation methods to determine the currency exchange translation differences.

Data Analysis (CCONS_FPM_OVP_TOTALS))

Application Type: Web Dynpro

This Web Dynpro–based app in financial consolidation allows you to analyze totals of consolidation data using various dimensions such as consolidation unit. company code, segment, and profit center. In addition, you can drag and drop additional dimensions into rows and columns from the **Navigation Panel**. The app also supports drill down capabilities along with export capabilities for further analysis.

Data Analysis – Reporting Logic (W0056)

Application Type: Web Dynpro

This app allows you to analyze consolidated data by providing access to granular information within the consolidation database from the perspective of consolidation groups in group reporting. Based on the specified fiscal year, fiscal period, and consolidation group, you can display amounts for all relevant financial statement (FS) items across local, group, and transaction currencies, along with details on posting levels, consolidation units, partner

units, and accounting master data. In addition, the app allows drill-through to the Display Group Journal Entries – With Reporting Logic app (F3831) for line item–level insights. The technical name of the CDS data view is `CCSTOTAL-SRR01Q`, and by prefixing `2C`, users can derive the analytical query name for the report.

> **Note**
>
> The Group Data Analysis app (W0135) replaces the Data Analysis – Reporting Logic app from SAP S/4HANA 2023 release onwards.

Days Beyond Terms (S/4HANA) (F1739)

Application Type: Analytical

The analytical app displays the **Days Beyond Terms (DBT)** KPI, providing insight into customers' payment history and collection efficiency. A high DBT figure indicates delayed payment collection. The app highlights when predefined thresholds are exceeded. In addition, you can view DBT figures in charts or tables by account group, accounting clerk, calendar month/year, company code, country key, customer, customer classification, display currency, exchange rate type, reconciliation account, or region. For new businesses, the **Days Sales Outstanding** KPI may be more helpful. The app provides you with additional features such as viewing DBT figures for the last 12 months, setting thresholds, and customizing display currency and exchange rate types. A DBT figure with a minus sign indicates that payments were made earlier than the specified payment terms.

> **Transaction Code**
>
> The corresponding transaction in SAP GUI is Transaction S_ALR_87012167.

Days Payable Outstanding Analysis (S/4HANA) (F1740)

Application Type: Analytical

This app allows you to drill down to the top 10 suppliers with the highest or lowest days payable outstanding, displaying results in charts or tables based on company code, supplier, country of the supplier, and timeline. The app delivers features such as analyzing days payable outstanding for the last 12 months, viewing the top 10 suppliers by the highest/lowest KPI values, examining days payable outstanding by company codes, and filtering by supplier. In addition, the app supports SAP Jam integration for comments and emails. The app uses CDS view C_APDAYSPAYOUTST, and has caching enabled for better performance.

Days Payable Outstanding – Detailed Analysis (F2688)

Application Type: Analytical

With this app, accounts payable users can measure how long, on average, it takes a company to pay its suppliers after receiving invoices. You can view KPIs by company code, supplier, and country. The information is presented in a graphical form with options to view the days sales outstanding (DSO) by company code or by company code by time. It's a critical metric in assessing a company's liquidity, payment behavior, and the efficiency of its accounts payable process. With this analytical app, you can perform a detailed analysis of days payable outstanding (DPO) by using predefined steps to view DPO by company code, supplier, and supplier country. The app allows you to monitor trends over time and refine their analysis with various filters. The app provides you with key features that include saving frequently used analysis paths and filtering by company code, supplier, accounting clerk, country, and period range.

Days Payable Outstanding – Indirect Method (F2895)

Application Type: Analytical

With this analytical app, you can view a company's payable outstanding (DPO), or the average number of days it takes a company to pay their suppliers. The app allows you to see a company's DPO by time, company code, and for all suppliers or the top 10 suppliers by highest DPO. The app calculates DPO on an aggregate level over the last 12 months using rolling monthly averages for accounts payable balance and purchases. In addition, you can specify parameters for calculating these averages, with flexibility in the number of months used, affecting monthly variance.

Days Payable Outstanding – Indirect Method – Detailed Analysis (F2896)

Application Type: Analytical

This analytical app allows you to perform a detailed analysis of days payables outstanding (DPO). Predefined analysis steps allow you to view the DPO by time, by company code, supplier, and supplier country. In addition, you can focus on your analysis using filters such as accounting clerk, company code, country, date range, region, and supplier. The app provides key features that include analyzing purchases and payables, saving frequently used analysis paths, and applying various filters to drill down into the data.

Let's compare the Days Payable Outstanding – Indirect Method (F2895) and Days Payable Outstanding – Indirect Method – Detailed Analysis (F2896) apps:

- **Days Payable Outstanding – Indirect Method**
  This app provides a high-level overview of a company's DPO by displaying the metrics by date, company code, and supplier. It caters to higher management, especially for providing quick and aggregated insights into accounts

payable performance. The app is used by management for strategic decision-making related to cash flow and vendor management.

- **Days Payable Outstanding – Indirect Method – Detailed Analysis**
  This app provides detailed and granular analysis of a company's DPO with drilldown capabilities to provide deeper insights. It caters to analysts or managers in accounting or procurement teams, allowing them to view and investigate late payments or understand payment terms. The app is used for troubleshooting issues related to payments or to understand data patterns in order to improve accounts payable processes.

Days Sales Outstanding – Detailed Analysis (F2687)

Application Type: Analytical

With this analytical app, you can perform a detailed analysis of a company's days sales outstanding (DSO) using predefined steps to view DSO by company code, due period, customer, and country. The app presents DSO in graphical format. For example, you can view the DSO by company code. The app allows you to monitor revenue and overdue receivables over time and refine their analysis with various filters. The app provides key features such as saving frequently used analysis paths and filtering by company code, country, region, customer, accounting clerk, and period range.

Days Sales Outstanding (S/4HANA) (F1741)

Application Type: Analytical

The analytical app displays the **Days Sales Outstanding (DSO)** KPI, showing the average number of days it takes for a company to collect receivables. A high DSO figure indicates delays in collecting payments and potentially too lenient credit terms. In addition, the app highlights when thresholds are exceeded. Also, you

can view DSO figures in charts or tables by account group, accounting clerk, calendar month/year, company code, country key, customer, customer classification, display currency, exchange rate type, general ledger account, reconciliation account, region, or special general ledger. For established businesses, the **Days Beyond Terms** KPI may be more relevant. The app provides key features such as viewing total DSO figures, analyzing top 10 customers by DSO, filtering DSO by company, and viewing DSO by period. Moreover, the app provides threshold warnings, supports SAP Jam for comments and emails, allows customization of display currency and exchange rate type, and includes caching for better performance.

Transaction Code

The corresponding transaction in SAP GUI is Transaction S_ALR_87012167.

Debt and Investment Analysis (F3450)

Application Type: Analytical

This app provides the treasury team with a comprehensive visual overview of outstanding debts and investments for their company and subsidiaries, with nominal amounts displayed in the default currency. Analysis can be customized using filtering options in chart view or table view, such as book values in position currency or net present values in valuation currency. The **Adapt Filters** option include parameters like key date, display currency, exchange rate type, debt/investment type, interest category, issuer, and rating agency. You can switch between chart types, adjust dimensions (e.g., product type or security account), and access detailed transaction lists. Moreover, the app offers navigation options to allow review of original transactions and collective processing for financing decisions. This app empowers corporate treasury with precise monitoring of debts and investments in a central location.

Debt and Investment Maturity Profile (F3130)

Application Type: Analytical

With this app, treasury risk managers can manage the maturity schedules of debt and investment portfolios. The app provides a visual overview of outstanding debts and investments for a company, illustrating nominal amounts and their time until maturity. The maturity profile summarizes the value distribution across specific maturities and supports corporate treasury in closely monitoring debt and investment profiles. In addition, the app features customizable filtering options via the filter bar, allowing grouping by maturity date, company code, counterparty, product type, and nominal currency. Also, the app offers both **Chart View** and **Table View**, enabling drilldown analysis by period, chart type selection, and detailed transaction navigation for granular insights and decision making. The app uses CDS view `C_MaturityProfile`.

Deficit Cash Pool (S/4HANA) (F0517A)

Application Type: Analytical

This app helps you quickly identify cash pools with deficits and the total deficit amount, allowing for prompt responses and smart fund allocation among cash pools. The app presents deficit cash pool information by cash pool and by bank account and company. The app supports displaying deficit amounts by cash pool and bank account, switching between charts and tables, exporting search results to a spreadsheet, and sending emails with a URL to replicate current selection criteria. In addition, you also can create tiles with default settings and benefit from caching for improved performance, with manual and automatic data refresh options. As a prerequisite, you must have defined and assigned bank account balance profiles to cash pools. To manage these profiles effectively, SAP recom-

mends assigning them to the cash pools using the Manage Cash Pools (Version 2) app (F3266A). Also, the app calculates the certainty levels of actual cash flows and intraday memo records based on the bank account balance profile assigned to each cash pool.

Transaction Code

The equivalent transaction in SAP GUI is Transaction FTE_BSM.

Define Bank Account Settings – Bank Statements (F5488)

Application Type: Transactional

This app allows you to configure settings for integrating end-of-day electronic bank statements with cash management at the bank-account level. As a result, the settings determine how bank statement data is consumed by cash management activities. With the app, you can retrieve default company code settings, adjust settings for individual bank accounts, and define posting categories for end-of-day and intraday statements. As a prerequisite, you must define company code settings in the Customizing activity **Define Settings for Bank Statements**. The app then retrieves all bank accounts belonging to the company code by selecting the **Retrieve Default Settings** button. For defining the posting category, which determines whether bank statements get processed, the app allows either with default values configured or by specifying settings for individual bank accounts. In addition, the app offers options that include enabling or blocking posting, transferring data such as balances, line items, or both, and excluding updates for cash management. Balance update methods, such as manual entry or spreadsheet import, can be activated. In addition, the app supports bank statement forwarding, allowing statements to be transferred to other systems for processing, ideal for scenarios involving subsidiary companies.

Tip

When no appropriate settings are defined either in this app or in the Customizing activity **Define Settings for Bank Statements**, the system applies default logic for bank accounts: **Posting Category (End of Day)** is set to **Posting to Be Processed, Transfer Mode (End of Day)** to **Bank Statement Items Only, Posting Category (Intraday)** to **No Posting Processing,Bank Statement Forwarding** to **No Forwarding, Enable Balance from Manual Entry** is disabled, **Enable Balance from Spreadsheet Import** is disabled, and **Enable Balance from EPIC** is disabled.

For posting intraday bank statements after importing, the app supports format camt.052 only.

Define Bank Account Settings – Instant Balances (F7805)

Application Type: Transactional

This app allows you to define instant balance settings for selected bank accounts. You can switch instant balances for bank accounts using the **Switch On** or **Switch Off** buttons and view the **Account Switch Status**, which can be **On, Off,** or **On (Not Monitored)**. In addition, you can search bank account balances using search parameters such as account number, account switch status, house bank enablement, company code, house bank, bank country/region, and bank number. As a prerequisite to enable bank instance balances monitoring, you must set up a connection to SAP Multi-Bank Connectivity and activate instant balances for the house bank via the Manage Banks – Cash Management app (F1574A).

Note

The app only supports selected member banks of SAP Multi-Bank Connectivity; check scope item Bank Integration with SAP Multi-Bank Connectivity (16R) for further details.

Define Bank Transfer Templates (F3759)

Application Type: Transactional

This app allows cash management users to create, edit, or delete templates for recurring bank-to-bank transfers, streamlining the process and reducing errors for frequent transfers. The app uses predefined templates to streamline the process, enabling you to create multiple transfers in a batch efficiently. With the app's predelivered templates, you can be used in the Make Bank Transfers – Create with Templates app (F3760) to batch-create multiple transfers efficiently. In addition, the app provides options to search for existing templates using filters or fuzzy search functionality. The app uses CDS view C_PaymentMethodVH.

Transaction Code

Transaction OT81 is used for manual bank-to-bank transfers, allowing you to post transfer entries directly between bank accounts. On the other hand, this app is tailored for recurring bank-to-bank transfers.

Define Consolidation Units (F4685)

Application Type: Transactional

As part of SAP S/4HANA Finance for group reporting, this app allows you to manage consolidation unit master data individually by displaying, changing, creating, or deleting them. The app's features include filtering view settings like local currency and upload method, creating new consolidation units or copying existing ones with options to copy data for specific fiscal years, and deleting units across all versions after a where-used check confirms they are not in use. To create a consolidation unit, you can select the **Create** option, enter the name of the unit, or copy an existing unit to create a new one. To delete, select at least one consolidation unit to activate the **Delete** button. To modify the hierarchy of consolidation units, SAP recommends utilizing the Manage Global Hierarchies app. In addition, to display the time- and version-dependent attribute values of consolidation units, you must select one consolidation version along with a fiscal year and period. These filter values are initially derived from the global parameter settings but can be modified as needed.

> **Tip**
>
> The local currency of a consolidation unit can only be changed at the first period of the year. During balance carryforward, the system translates old local currency values into the new currency, requiring a currency translation method to be assigned. However, changing the local currency may cause inconsistencies if data already exists for the consolidation unit using the previous local currency.

> **Transaction Code**
>
> The corresponding transaction in SAP GUI is Transaction CX1M.

Define Financial Statement Items (F3297)

Application Type: Transactional

This app allows administrators of group reporting to display, update, and manage financial statement (FS) items and their properties based on specified criteria. You can filter and display FS items along with details like the consolidation chart of accounts, FS item type, breakdown category, elimination attribute, currency translation, and FS item role attribute. In addition, sorting and filtering options are available, and you can drill down into each FS item to review assigned breakdown category fields, required fields for consolidation tasks, allowed values, and multilingual descriptions. Also, you can modify properties by editing and saving changes or create new FS items by defining their properties or copying existing ones, including

time- and version-dependent data. The app provides additional functions, which include uploading, downloading, and importing FS items and checking their usage through the **Check Where Used** feature.

SAP introduced the new FS target attribute, offsetting target, with SAP S/4HANA 2023, which helps with the reclassification of negative balances to the appropriate offsetting item through the reclassification rule. Also, the app is flexible and can provide display-only access or create/edit access to certain users.

> **Tip**
>
> FS item selection attributes and FS item target attributes value assignments are dependent on time and version. When creating, displaying, or modifying FS items, changes apply only to the selected version (and related versions) and take effect from the specified fiscal year and period onward. In addition, statistical FS items, automatically generated by consolidation of investments, are crucial for recording historical values required for future processes like divestiture. These items are identified by the dollar ($) prefix and SAP recommends not to modify them, as any changes will be overwritten by consolidation of investments.

Define Master Data for Consolidation Fields (F3007)

Application Type: Transactional

As part of SAP S/4HANA Finance for group reporting, this app allows group reporting administrators to define consolidation-specific master data values alongside the standard master data in SAP S/4HANA Cloud products, enabling you to enhance the consolidation process and generate more insightful analysis reports. With the **Enable Master Data** and **Enable Hierarchy** options, you can define consolidation master data fields for supported master data types. The app offers key features include creating master data for consolidation-specific fields, modifying or deleting existing master data, and accessing the Manage Global Hierarchies app (F2918) to define hierarchies for master data types that support hierarchies.

> **Tip**
>
> The links delivered with this app are context-sensitive, directing you to a filtered page tailored to the selected master data type in consolidation. For instance, clicking the link for **Functional Area** leads to a page listing hierarchies specific to the consolidation functional area, where only the consolidation functional area hierarchy type can be created. These links are available only for master data types with the **Enable Hierarchy** option selected in the **Define Consolidation Master Data Fields** configuration activity.

Define Matching Methods (F3862)

Application Type: Transactional

As part of SAP S/4HANA Finance for group reporting, this app simplifies reconciliation by facilitating robust matching method configuration and management for intercompany transactions. As part of corporate close, the app allows you to create, display, modify, or delete matching methods for reconciliation processes. You can assign a unique ID and description, choose storage types (master data or configuration data), and configure data sources for matching context. For intracompany matching, two data sources may be required, and filters can be applied to restrict the data scope, supporting up to 100 filters for performance efficiency. With this app, matching rules can be created with various match types, including exact match, auto-assign, group as matched, group as assigned, and auto-assign as exception, each tailored to specific reconciliation needs. In addition, rules

may include matching expressions for comparing data values, and you can manage data slices for segmentation and grouping, with options for aggregation and item matching. The app allows rule activation/deactivation, sequence adjustment, and copying existing rules. Also, you can drill down to details or delete unused methods, though deletion is restricted if matching documents have been generated.

> **Tip**
>
> If you observe data inconsistencies or missing data in ICMR, consult SAP Notes 3114279 and 3113859 for remediation options.

Define Posting Rules – Intercompany Matching and Reconciliation (F4773)

Application Type: Transactional

This app allows general ledger users managing intercompany transactions to define rules for automatic postings triggered by intercompany matching and reconciliation (ICMR) processes, such as variance adjustments for accounting documents and elimination postings for group reporting. ICMR allows you to establish rules for posting differences identified during the matching process to specific general ledger accounts automatically. The app's features include viewing, filtering, and downloading existing posting rules, creating rules using predelivered scenarios, editing rules by modifying header fields, control options, and item fields, and deleting existing posting rules. To create a posting rule, you can either copy an existing rule or create one from scratch. You can choose **Create** and select a predelivered automatic posting scenario, such as **SA001 (Adjustment Postings for Accounting)** or **SC001 (Elimination Postings for Group Reporting)**, or a custom scenario. Based on each scenario, the app determines the available fields and options for the rule.

Tip

The app changes are carried out in a Customizing client to transport changes to other clients. As a result, in the production client, you have read-only access to the configured rules. To make changes with the app, you must require a business role that includes the ICA – Intercompany Configuration (SAP_FIN_BC_ICA_CONFIG) business catalog.

Define Reconciliation Cases (F3863)

Application Type: Transactional

This app helps with the creation, management, and deletion of reconciliation cases to streamline data analysis and reporting in SAP S/4HANA Finance for group reporting. Reconciliation cases ensure precise, tailored data management, enhancing reporting accuracy in group reporting. When creating reconciliation cases using this app, you can specify unique 5-digit alphanumeric IDs, descriptions, and storage types (master data or configuration data) to define cases. In addition, matching methods can be assigned exclusively to reconciliation cases, ensuring alignment between storage types and filters. The app also provides optional settings, which include hierarchies for matrix reconciliation and display options such as year-to-date or period-based views. Also, display groups can be created with tolerances, difference calculations, converted measures, and filters for opposing data views, enabling dynamic analysis with functions like date-based filtering. The app supports pairing display groups intercompany netting views. Moreover, you can copy cases, adjust filters, or inherit settings from leading groups for consistency. While intersections between data sides should be avoided, a leading display group calculates balances in reports. Reconciliation cases also can be deleted, except when associated with matching documents generated.

Define Selections (F3725)

Application Type: Transactional

As part of SAP S/4HANA Finance for group reporting, this app allows you to create data selections by applying filters to master data fields such as financial statement (FS) items, document types, and posting levels, restricting the data range as needed. The app's functionality includes creating, editing, or copying selections by defining filters in the **Selection Expression** section, using criteria like value ranges, hierarchy nodes, and attributes. You can use multiple filters for the same field using **OR** logic if they share the same operator, or **AND** logic if both **Include** and **Exclude** operators are used. In addition, **Create Selection** can include custom fields, and hierarchies with time dependencies can be used by specifying valid periods. Moreover, **Create Selection** can be activated or deactivated, with mass activation option for bulk actions. In addition, the app supports viewing filtered values and checking where selections are used across various configurations. Selections can be saved as **Draft**, **Active** versions, or **Inactive** based on usage.

Define Validation Methods (F2655)

Application Type: Transactional

This app allows you to create or update validation methods that may contain multiple groups of validation rules. You can enter general information for a new validation method, add groups under the default header row, and assign existing validation rules to these groups. You can search for validation methods with **Status** field. In addition, if needed, you can drag and drop rules to change their sequence or group assignment. Also, editing or copying existing methods is possible, except for SAP Best Practices content. Once a method is created, you can save it as a draft or an active version and later deactivate it if necessary. Methods can be in draft, active, or inactive status. You also can use the **Mass Activation** feature to schedule background jobs for bulk activation.

> **Tip**
>
> This app is used to create or update validation methods that may contain multiple groups of validation rules. On the other hand, the Define Validation Methods (RTC) app (F2598) is part of the real-time consolidation solution, which integrates SAP S/4HANA and SAP Business Planning and Consolidation (SAP BPC).

Define Validation Methods (RTC) (F2598)

Application Type: Transactional

As part of real-time consolidation (RTC), to verify reported financial data, a set of defined rules must be assigned to a validation method, which is then assigned to specific entities for data checking. You can search for valuation models using parameters such as model, method ID, and method name. With this app, for a specified consolidation model, you can create a validation method, assign multiple validation rules to it, or adjust the rule assignment for an existing method. In addition, the app allows you to test rules against sample data before applying them to actual financial data, ensuring correctness.

Define Validation Rules (F2627)

Application Type: Transactional

To ensure the integrity of the consolidation process, with this app, group reporting administrators can define validation rules specifying the logical conditions that relevant financial data. You can search for validation rules with the **Status** field. These defined validation rules can be utilized to verify reported data, standardized data, and consolidated data throughout the consolidation process. It allows you to create and

manage validation rules, improving compliance with reporting standards and the quality of financial reporting. The app comes with features such as defining customizable validation rules, using the intuitive rule expression language, testing rules in simulation mode, integrating with other apps, managing existing rules, assigning rules to specific entities, and implementing a comprehensive validation process.

Define Validation Rules (RTC) (F2575)

Application Type: Analytical

With this app, you can define a series of rules for validating reported and standardized data throughout the consolidation process. Rules and result messages are created using an intuitive rule expression language. Also, you can run these rules in simulation mode for specific criteria to verify their definitions. The app allows you to perform tasks such as creating validation rules with defined logical conditions, setting control levels (hard or soft), saving rules as raw or active versions, enabling or disabling active rules, and associating existing reports with validation rules for drill-through functionality. In addition, the app supports transporting and activating rules in the backend system for specified consolidation models.

> **Tip**
>
> For further guidance on the SAP HANA rules framework, go to *https://help.sap.com/hrf10*.

Delete Credit Management Data (F4802)

Application Type: Transactional

This app allows credit controllers to schedule jobs for various types of data in credit management, utilizing the app's delivered job templates to delete data relevant to credit management processes. This is essential for data housekeeping and maintaining system performance considering data volume. Users are recommended to use available job templates to delete credit exposures for business partners. Using this app, you can select **Create** option and choose the desired template in the **Job Template** field. In addition, you have the option to create an application job by selecting business partners, credit segments, and credit exposure categories. You then can proceed with the **Delete Line Items** option to delete credit exposures or deselect it for a test run. In addition, you can choose **Only Accounts with Liability Error** to limit deletion to accounts with liability errors during credit commitment updates. Finally, the job can be scheduled.

Transaction Code

The corresponding transaction in SAP GUI is Transaction UKM_COMMITMENTS.

Depreciation Lists (F1616)

Application Type: Transactional

This app supplements the profit and loss statement by enabling you to analyze the interest and depreciation of fixed assets. You can search for depreciation information using parameters such as fiscal year, to-period, depreciation variant, company code, ledger, depreciation area, currency type, asset, sub-number, asset class, and depreciation key. The app is primarily used for depreciation reporting purposes. With this app, you can report depreciation values as a total or by type, such as ordinary depreciation, special depreciation, unplanned depreciation, write-ups, and transferred reserves (deferred gain). Also, the status of depreciation and asset attributes, like the depreciation key and useful life, help you to analyze the reported figures.

Tip

The app uses the key date, which determines the date for which time-dependent data is selected. The default value is the current date. If no value is entered, the system selects today's date automatically. To select data for previous fiscal years, months, or years, you must enter the relevant date in the **Key Date** field.

Transaction Code

The corresponding transaction in SAP GUI is Transaction AR03.

Disaggregation of Recognizable Revenue (W0156)

Application Type: Web Dynpro

Disaggregation of revenue refers to breaking down total revenue into meaningful categories for further evaluation. This Web Dynpro–based app provides you with disaggregate revenue or cost across various categories such as business partners, profit centers, fulfillment types, and sales organizations in revenue accounting and reporting (RAR). The app allows you to display recognizable revenue for a specific accounting period using multiple dimensions, including all standard fields in the revenue accounting subledger, such as company code, customer, business partner number, performance obligation type, accounting principle, display currency, fiscal year, and fiscal period. You can perform multidimensional analysis by adding or removing rows and columns in the **Navigation** function, visualizing data in **Chart View**, **Table View**, or both, and utilizing **Jump To** for quick access to other apps.

Disaggregation of Revenue (W0157)

Application Type: Web Dynpro

Disaggregation of revenue refers to breaking down total revenue into meaningful categories for further evaluation. This app disaggregates revenue and cost across categories such as business partners, profit centers, fulfillment types, and sales organizations. You can display posted revenue for a specific accounting period using multiple dimensions in both document currency and company code currency, while posted costs can be analyzed in company code currency. The report, delivered via a Web Dynpro data grid, supports configurable selections for company code. You can perform multidimensional analysis by adding or removing rows and columns in the **Navigation Panel** function, visualizing data in **Chart View**, **Table View**, or both, and utilizing **Jump To** for quick access to other apps.

Tip

This app provides a broad view of posted revenue across multiple dimensions and currencies, whereas the Disaggregation of Recognizable Revenue app (W0156) offers the opportunity to focus on revenue that is eligible for recognition for a given time frame.

Display Accounting View of Purchase Order (F4700)

Application Type: Transactional

In this app, you can display a specific purchase order that is replicated from a source ERP system to the accounting view of logistics (AVL) in the Central Finance system. Here you can find important purchase order information, such as the purchasing document item text (EKPO-TXZ01), order quantity (EKPO-MENGE), and material group (EKPO-MATKL).

Transaction Code

In SAP GUI, details of replicated purchase orders can be accessed using Transaction FINS_CFIN_DIS_PO. Let's compare this app and SAP GUI Transaction FINS_CFIN_DIS_PO:

- **Display Accounting View of Purchase Order**
  Supports flexible placement of fields and navigation to other apps. However, navigation to the source system is not supported.

- **Transaction FINS_CFIN_DIS_PO**
  Doesn't support user adoption, and can't navigate to other apps. However, drill back to the source system is supported.

Display Accrual Object Items (F4899)

Application Type: Transactional

This app allows general ledger users to display a list of accrual object items with search criteria such as company code, accrual item type, key date, and lifecycle status. The general ledger provides tools to calculate, verify, and post accruals. As part of the accrual management process, accrual object items must be created to facilitate managing accruals to support revenue and expense postings in the correct accounting periods. You can view detailed information for each item, such as the validity period, start of life, end of life, and total accrual amount. The app's functionality includes customizable filters, a comprehensive table view of accrual object item details, and navigation to related apps for seamless access to additional functionalities. In addition, the app supports extensibility to add custom fields via the Customs Fields and Logic app (F1481).

Transaction Code

The corresponding transaction in SAP GUI is Transaction POACTREE03.

Display Additional Credit Information (F7367)

Application Type: Transactional

This app allows you to display additional information about business partners maintained within the credit profile or credit segment. You can search for relevant details using various filters, including business partner, credit segment, validity from, valid to, and risk class. The app provides an overview of all maintained information within credit management, covering both general data and credit segments. The app's search results are organized into various tab pages by category, including credit insurances, collateral, negative credit events, check exceptions, and additional credit limits.

In addition, you can navigate directly to the credit profile or credit segment in the Manage Credit Accounts app (F4596) for further analysis. Thus, with this app, you can regularly check for negative credit events or exceptions to identify potential risks early proactively.

Transaction Code

The corresponding transaction in SAP GUI is Transaction UKM_ADDINFOS_DISPLAY.

Display Bank Account Logs (F6777)

Application Type: Transactional

As part of cash and liquidity management, this app provides a structured overview of application logs for bank account management, enabling you to monitor errors effectively. The app supports viewing log details for various apps, including those related to activities such as bank control key migration, inactive account adaptation, bank account imports and exports, service billing file imports, payment approvals, and powers of attorney management. In addition, you can filter logs by severity, category, and subcategory, and search message texts for specific details for further troubleshooting.

Transaction Code

The corresponding transaction in SAP GUI is Transaction SLG0.

Display Bank Payment Approval Jobs (F6617)

Application Type: Analytical

This app allows accounts payable users to monitor the status of bank payment approval jobs running in the background, initiated from the Approve Bank Payments (Version 2) app (FO673A) when the **Approve in Background** function is enabled. Without this function activated, the app won't display any background approval jobs, thereby limiting visibility into automated payment processes. The app's features include searching for initiated background jobs and displaying their current status for tracking and management.

Display Billable Items (F2621)

Application Type: Transactional

With this app, contract accountants can search for, display, and process billable items for convergent invoicing, which form the basis for the billing process. You can search billable items using search parameters such as status, billable item class, billable item type, subprocess, source transaction type, billing document, contract account, business partner, contract, subapplication, and invoicing document. The app displays billable items in a table format. The app allows you to perform actions on a billable item apply to all billable items within the same data package or source transaction. In addition, the app provides features such as searching for billable items using various filter fields, transferring raw billable items to billable status, reversing billable items, excluding and restoring items from the billing process, executing billing, invoicing billing documents, creating clarification cases, displaying related record types and consumption items, and making changes to defined changeable fields.

Transaction Code

The corresponding transaction in SAP GUI is Transaction FKKBIXBIT_MON.

Display Billing Documents (F2250)

Application Type: Transactional

With this app, you can search for, display, and process billing documents for convergent invoicing, which are created during a billing run or uploaded from an external source. You can search for existing billing documents by searching with fields such as billing document, invoicing document, business partner, contract account, start of document period, and billed on. The app presents data in a grid format form with flexibility of adding additional fields as required. The app provides features such as searching and sorting billing documents, displaying lists of related consumption and billable items, performing plausibility checks, reversing or invoicing billing documents, creating clarification cases, setting invoicing locks, and managing attachments. In addition, you can collaborate with the other teams by sharing a specific billing document or list of documents by email or chat with Microsoft Teams. As a result, you achieve comprehensive management and processing of billing documents with this user-friendly app. Navigation options can be limited with appropriate roles and authorization object assignments.

Transaction Code

The corresponding transaction in SAP GUI is Transaction FKKINVBILL_DISP.

Display Business Transactions (F4294)

Application Type: Transactional

This app allows accounts payable and accounts receivable users to display business transactions for contract partners, providing a chronological overview of postings, master data changes, and correspondences in contract accounting. With this app, you can apply filters to enable searching for specific transactions and clicking on a transaction opens an object page with detailed information. The app serves as an alternative to the Display G/L Account Balance (New) app (F0707A) for viewing contract partner transactions.

Display Cash Pool Hierarchies (F6123)

Application Type: Transactional

This app allows cash managers to view cash pool hierarchies defined for cash pools with the service provider **Bank – Time Dependent** and **In-House Bank – Time Dependent** in cash management. Before using the app, as a prerequisite, you must ensure that cash pools with account assignments are defined in the Manage Cash Pools (Version 2) app (F3266A), including the assignment of one header account and one or more subaccounts, as the displayed hierarchies are based on these account assignments. The app's features include viewing cash pool hierarchies for a specific date, accessing detailed cash pool information through the Manage Cash Pools (Version 2) app, and exporting hierarchies to a spreadsheet.

Tip

The app excludes cash pools classified as **Company – Time Dependent**.

Display Collections Management Log (F6351)

Application Type: Transactional

This app allows you to display collections management logs, providing a comprehensive overview to quickly identify issues such as errors during worklist generation as part of the collections management process. The app's features include viewing log details, filtering logs by severity, searching for specific message texts, and displaying detailed information about messages. Thus, the app provides better visibility into the collections process, thereby improving a company's cash flow.

Display Correspondences (F3693)

Application Type: Transactional

As part of contract accounts payable and receivable, this app allows you to view the correspondence that you have created or printed, displaying data from the correspondence history. You can filter correspondence by type, receiver, contract account, and additional criteria like the date of issue. In addition, you can select single or multiple entries to either mark them for deletion or reprint them directly. For certain correspondence types, you can choose between test, repeat, or real printing options. The app uses CDS view C_CACorrespondenceHistory.

Display Consumption Items (F3410)

Application Type: Transactional

This app allows you to search, display, and manage consumption items for convergent invoicing, which serve as the foundation for the rating process. The app's features include searching with filters such as company code, status, and contract account, transferring raw consumption items to a **Rateable** status, reversing or excluding items from the rating process, and executing ratings. In addition, you can navigate to view item details, related errors, and billable items, or access the Display Billable Items app (F2621). Moreover, changes can be made to consumption items if specific fields are defined as editable for their status. In addition, the app allows navigation to primary billable items, sharing items via email or Microsoft Teams, and provides functionality based on user roles and authorizations.

Transaction Code

The corresponding transaction in SAP GUI is Transaction FKKBIXCIT_MON.

Display Credit Account Data (F4825)

Application Type: Transactional

This app allows credit controllers to display general master data of business partners alongside credit management master data. The app allows you to restrict searches by business partner, credit segment, or credit analyst, view business partners by credit segments, and navigate to other credit management apps using the business partner number.

Tip

As a prerequisite, the business partner role UKM000 is required for each credit segment being analyzed.

Transaction Codes

The corresponding classic credit management (CCM) transactions in SAP GUI are Transactions UKM_BP_DISPLAY and S_ALR_87012215.

Display Credit Exposure (F4826)

Application Type: Transactional

This app allows credit controllers to display the credit exposure of business partners, showing total exposure by credit segment. You can navigate entries to view detailed credit exposures for each business partner and drill down to the line-item level. In addition, the app provides an overview of the total credit exposure for business partners, categorized by credit exposure types and sales documents. This app is essential for credit controllers to monitor and manage credit limits and exposures effectively. The app's features include searching and filtering by business partners or credit segments, as well as restricting the search to a specific key date to obtain a snapshot of credit exposure for that date.

Transaction Code

The equivalent SAP GUI transaction is Transaction UKM_COMMITMENTS.

Display Credit Management Log (F2162)

Application Type: Transactional

This app provides credit controllers with detailed insights into credit decisions made within the system. With this app, you can efficiently display logs with severity and manage credit logs. It allows you to quickly check if a credit check failed and view detailed information about each failed credit check. The app supports error logs in multiple subcategories:

- Commitment update
- Credit check
- Credit management
- General information
- Integration of external credit agencies
- Master data related errors

The app provides features that include viewing detailed credit management logs, filtering logs by severity, searching for specific message texts, and displaying message details in a clearly structured overview. This ensures you have all the information you need at your fingertips.

Transaction Codes

The equivalent SAP GUI transaction in SAP S/4HANA is Transaction UKM_LOGS_DISPLAY. In SAP ERP, the corresponding transactions are Transactions S_ALR_87012215 and S_ALR_87012218. Customer credit can be displayed with Transaction FD32 for classic credit management.

Display Customer Balances (Version 2) (F0703A)

Application Type: Analytical

With this app, accounts receivable specialists and finance managers can obtain an overview of outstanding customer balances, helping manage receivables and track payments effectively. This app streamlines accounts receivable processes, providing a clear view of customer payment statuses and enhancing customer relationship management. You can search for customer balances using parameters such as company code, customer, city, clerk abbreviation, dunning procedure, and payment terms. The app allows you to view balances across different company codes and currencies, track open items and overdue amounts, filter balances by various criteria, analyzing aging reports, drill down into detailed transactions, and improve cash flow management by ensuring timely collections.

Display Customer List (F2640)

Application Type: Transactional

This app allows you to view all customer master data in one place. For instance, you can use the **Check Empty Fields** function to ensure that all customers have data entered for dunning procedure, dunning clerk, and dunning level, and the app will highlight any customers with missing information. You can search for a customer list using parameters such as company code, customer, city, clerk abbreviation, dunning procedure, and payment terms. In addition, you can quickly determine who in the company to contact regarding a particular customer, and SAP Collaboration Manager and the digital assistant provides an efficient communication channel for this purpose. The app can display all customer master data on one table with a wide range of filtering options, check data presence in multiple fields across all customers in one step, and quick access to contact information. The app doesn't support additional field attributes. However, the underlying CDS views ARUniqueCustomer and C_ARCustomer can be evaluated for extending custom fields.

Transaction Code

The corresponding transaction in SAP GUI is Transaction VCUST.

Display Dispute Management Log (F6395)

Application Type: Transactional

This app provides an overview of dispute management logs, enabling you to quickly identify errors, such as issues during the mass creation of dispute cases. The app's features include viewing detailed log entries, filtering logs by severity, searching for specific message texts, and displaying message details for further analysis. Thus, the app offers a structured overview to check errors that might have occurred during the mass creation of dispute cases, making troubleshooting more efficient.

Display Document Flow (F3665)

Application Type: Analytical

This app allows you to view original documents and their follow-on documents in a graphical format. In addition, the app lists all relevant business transactions and journal entries for a performance obligation. You can search for document flow using the document type and starting document. For example, for a revenue contract, the app includes key information such as the order date, sender component, order quantity, contractual price and so on. Also, the app is capable of searching revenue contracts, grouping performance obligations by default, checking related documents, and applying filters for a specific date range based on the document date.

Display Documents Related To Commodity Derivative Order Requests (F3513)

Application Type: Transactional

This app allows traders of derivative commodities to display, edit, and manage commodity derivative orders, providing an overview of all orders and detailed views of selected ones. You can create, copy, edit, or invalidate orders, resolve errors from data distribution to follow-on systems, transmit orders with a **Pending Transmission** status, and fetch the latest market or customer-specific prices. In addition, the app supports extensibility to add customer characteristics for enhanced functionality and reporting.

Display Dunning History (F2328)

Application Type: Transactional

The app provides a detailed view of the dunning activities performed on customer accounts to monitor, track, and analyze dunning processes over time. For example, the app provides collections specialists with an overview of dunned customers and their individual dunning history along with details such as dunning level, date, amount, and document references. You can search for dunning history using parameters such as account type, company code, customer, supplier, dunning level, and dunning run date. The app is designed to assist you in contacting customers to request payment of overdue receivables. With this app, you can perform tasks such as viewing an overview of all dunned customers, drilling down to see underlying transactions, filtering and sorting, accessing details of dunning notices at the line-item level, and displaying dunning letters sent to customers. This ensures collection specialists have all the necessary information to manage and follow up on overdue payments effectively.

Transaction Code

The equivalent SAP GUI transaction for this app is Transaction F150.

Display Dunning Run Exception (F7575)

Application Type: Transactional

This app allows you to review exceptions in the dunning process in contract accounting. The app provides a structured overview of dunning exceptions, enabling you to analyze the causes of failed attempts and take corrective action. You can search for dunning exceptions using search parameters such as dunning run date, dunning run ID, business partner, contract account, document number, net due date, date of issue, and exception reason. The app's functionalities include filtering exception lists by

business partner, dunning level, and execution date, reviewing impacted customer accounts affected by missing data or incorrect configurations, navigating to related transactions for resolution, and monitoring dunning proposal details. Thus, the app helps you to identify overdue accounts that were not successfully dunned.

Display Error Logs – Billable Items (F5494)

Application Type: Transactional

This app allows you to display error logs for billable items, aiding in the investigation of process issues as part of convergent invoicing in contract accounts receivable and accounts payable. Convergent invoicing allows companies to consolidate and manage a high volume of billable items. Error logging can be activated for the creation of billable items, the transfer of items to status billable, or both in configuration. The app provides filter criteria using parameters such as process ID, source transaction ID, business partner, contract account, and contract. The app's features include viewing error logs for these processes, deleting log entries, navigating them to related application logs, and sharing lists via email or Microsoft Teams.

Transaction Code

The corresponding transaction in SAP GUI is Transaction FKKBIXBIT_ERR_MON.

Display Error Logs – Consumption Items (F5493)

Application Type: Transactional

Both the Display Error Logs – Consumption Items (F5493) and Display Error Logs – Billable Items (F5494) apps allow you to view error logs for consumption and billable items, helping you investigate and resolve process issues in convergent invoicing. Consumption items are usage data or events that are processed in convergent invoicing. These items are typically unrated or rated based on predefined rules and are linked to billable items for invoicing purposes in contract accounting. The app offers search capabilities with fields such as process ID, source transaction ID, business partner, contract account, and contract. Thus, the app allows you to monitor and manage errors during the processing of consumption items across stages such as creation, rating, and status transitions (e.g., from **Raw** to **Rateable**). With this app, you can view, analyze, and resolve errors related to consumption items quickly, ensuring seamless invoicing operations.

Transaction Code

The corresponding transaction in SAP GUI is Transaction FKKBIXBIT_ERR_MON.

Display G/L Account Balances (New) (F0707A)

Application Type: Analytical

This allows you to check and compare balances, and credit and debit amounts of a ledger in a company code for each period of a fiscal year. The app also allows you to restrict data to a single general ledger account or other criteria (e.g., profit center). By using the **Compare** function, you can compare balances across multiple fiscal years. In addition, the app allows you to select data by ledger, company code, fiscal year, and other criteria, displays data in various currencies, and provides options to filter general ledger account balances by year-end closing postings.

Transaction Code

The corresponding transaction in SAP GUI is Transaction F.08.

Display Group Journal Entries (F2573)

Application Type: Analytical

As part of SAP S/4HANA Finance for group reporting, this app displays line items of posting documents with all posting levels (00-30), covering ad hoc data updates, adjustments, standardized data, eliminations, and consolidation entries. You can search for group journals using fields such as fiscal year, posting period, version, consolidation chart of accounts, consolidation group, and consolidation unit. The app presents data in a grid format with flexibility of adding additional dimensions. For posting level empty value (blank) and document type OF (real-time update from accounting), line items are directly read from table ACDOCA. The app provides features such as the ability to list journal entries with relevant filters, navigate between group journal entries (table ACODCU) and merged journal entries (tables ACDOCA and ACODCU) with a drill-through function, and drill down to the line items of group journal entries or accounting documents. The drilldown function is integrated with the Manage Journal Entries app (F0717) and requires the SAP_SFIN_BC_GL_JE_PROC business catalog for accessing accounting document line items. In addition, you can navigate to other analytical apps.

Display Group Journal Entries – With Reporting Logic (F3831)

Application Type: Analytical

In SAP S/4HANA Finance for group reporting, this app allows you to display line items of posting documents across all posting levels (00–30), including ad hoc updates, reported data adjustments, standardized data, eliminations, and consolidation entries, all aligned with new reporting logic. The app comes with two tables, Group Journal Entries (table ACDOCU) and Merged Journal Entries (ACDOCA). The **Cons. JE Item Origin** in the second tab identifies the source the origin of transaction. With this app, group accounting you can filter journal entries by consolidation units/groups, fiscal year/period, or hierarchies, such as consolidation unit, profit center, or segment hierarchies using the hierarchies defined with the Manage Global Hierarchies app (F2918). Virtual dimensions (e.g., fields with the _ELIM suffix) support elimination values at specific posting levels. In addition, you can drill down to group journal entry details in the consolidation journal database (table ACDOCU) or original accounting documents (table ACDOCA) based on document attributes, provided you meet certain prerequisites.

From group analytical reports, you can drill through to the app with prepopulated filters to view relevant journal entries and line-item details. This app offers integrated analysis, enhances reporting accuracy, and allows for detailed examination of posting documents in group reporting.

Transaction Code

The corresponding transaction in SAP GUI is Transaction CX56.

Display Head Office Receivables (F5401)

Application Type: Transactional

This app allows accounts receivable users to view customer line items of the head office debtor account associated with the currently processed business partner in the collection segment, utilizing the head office/branch relationship in accounts receivable accounting. In addition, the relationship is accessed via the **Display Head Office Receivables** button in the Process Receivables Version 2 (F0106A) app. The app's features include viewing invoices from the head office and related branches, accessing associated dispute cases, promises to pay, and resubmissions, examining company relationships, and reviewing key figures from credit management.

Display Interest Calculation Logs (F7615)

Application Type: Transactional

This app allows you to display logs from interest calculation reversals after using the **Reverse Calculated Interests** function in the Manage Interest Runs app (F4485). You can create, modify, and monitor interest calculation runs for customer accounts, including reversing calculated interests for adjustments with the Manage Interest Runs app. The app's functionalities include viewing reversal logs under the **FIAR_ INT** subcategory and navigating into detailed log records to analyze interest calculation adjustments.

Tip

If discrepancies are found in the logs, users are recommended to cross-check the Manage Interest Runs app to verify whether reversals and interest calculations are carried out correctly.

Transaction Code

The corresponding transaction in SAP GUI is Transaction FINTSHOW.

Display Interest Calculations (F4761)

Application Type: Transactional

This app allows contract accountants to analyze the interest amount for a specific interest document. The app's features include filtering to locate interest documents, viewing item details and their calculations by selecting a line in the overview, and managing documents by displaying, changing, or reversing them. For instance, you can search for interest calculations by various search criteria such as document, business partner, interest run ID. In addition, you can display or modify business partner data, access contract account details, and navigate to manage business partner items or view account balances. In addition, the app supports an authorization check for company code.

Display Invoicing Documents (F2048)

Application Type: Transactional

With this app, you can search for, display, and process invoicing documents for convergent invoicing, which contain the information needed for an invoice to be printed or sent to customers. You can search for invoicing documents using fields such as invoicing document, business partner, contract account, document date, official document number, and invoiced on. The app presents data in a grid format with flexibility of adding additional dimensions. The app brings features to you such as searching for invoicing documents using filter fields, displaying a print preview, printing documents, setting or removing printing locks, adding and viewing attachments, requesting corrections by creating credit/debit memos, displaying graphical overviews of previous invoice amounts for related contract accounts, and listing related consumption and billable items. In addition, you can reverse invoices, release preliminary invoices, and record customer complaints by creating dispute cases.

Transaction Codes

The app's features accomplished with corresponding SAP GUI transactions are Transactions VF05N (List of Billing Documents), FBL5N (Display Customer Line Items), VF03 (Display Billing Documents), and FB03 (Display Financial Accounting Document).

Display Invoicing Requests (F2622)

Application Type: Transactional

This is an essential app for managing and processing billable items within convergent invoicing. The app allows you to search for, display,

and handle billable items, which serve as the foundation for the billing process. In addition, the app offers functionalities such as searching and displaying billable items with detailed associated data, transferring raw items to billable status, reversing errors, excluding and restoring items from billing, executing billing, invoicing documents, creating clarification cases, and displaying related record types, navigating to billing documents, and consumption items.

Transaction Code

The transaction in SAP GUI is Transaction FKK-INV_MON.

Display Item Change Log (F2681)

Application Type: Transactional

This app allows you to view detailed changes made to all journal entries relevant to user role, displaying modifications to all log-enabled fields in the relevant documents. You can search for item change log using fields such as document type, changed on, company code, journal entry, customer, user, and field name. This app is useful for tracking item-level changes in documents for auditing, compliance, and analysis purposes. You can easily see what was changed, by whom, and when, providing demonstrable oversight of all changes within your area, which is useful for auditing. Key features include the automatic filtering of customer/supplier items based on your role assignment; the ability to see changes to general documents, parked documents, previously parked documents, recurring entry documents, and sample documents; and the use of SAP Collaboration Manager for quick communication with colleagues to clarify the reasons for document changes.

Transaction Code

The corresponding transaction in SAP GUI is Transaction S_ALR_87012293.

Display Journal Entries in T-Account View (F3664)

Application Type: Transactional

This app allows general ledger accountants to inspect how one or multiple journal entries impact ledger accounts where they are posted with T-account format, thereby facilitating a visual representation of journal entries. The visualization format aids you in understanding the flow of debits and credits across accounts, enhancing the analysis of financial transactions. The app is capable of displaying ledger accounts for selected journal entries, checking how line items affect debit and credit sides of ledger accounts, finding matching debit and credit entries, and displaying journal entries in either T-account view format or switching to table view.

In addition, you can navigate to the Journal Entry Analyzer app (FO956) from this app.

Tip

To display billing documents with general ledger document flow correctly, you must ensure the ICF node FIN_ACC_IMP_D is activated via Transaction SICF, as described in SAP Note 3493363.

Display Line Item Entry (F2218)

Application Type: Analytical

This app allows you to check general ledger account line items in the entry view using app's filters to display open and cleared items for open item-managed accounts or all line items of a general ledger account. You can select line items based on a key date or after the clearing date. In addition, you also can choose line items using company code, item type, posting date, general ledger account and status. With this app, you can display line items with or without the leading ledger, creating app and layout variants, displaying amounts in different currencies, and

filtering amounts in currencies without decimals. In addition, you can change assignments and item texts for line items, export results to a spreadsheet, and manage entries for various item types like normal, noted, and items without a leading ledger. Moreover, results can be exported to an Excel sheet for further analysis.

Tip

The app does not display custom fields created through extensions. To display custom fields, SAP recommends using the Display Line Items in General Ledger app (F2217). In addition, financial accounting documents with status **Posting in General Ledger Only** (BKPF-BSTAT = 'U') are not displayed with this app, as it is based on table BSEG, and items with status U are stored in Universal Journal table ACDOCA.

Transaction Codes

The corresponding transactions in SAP GUI are Transactions FALLO3 or FALLO3H.

Display Line Items – Cost Accounting (F4023)

Application Type: Analytical

This app allows you to view and manage general ledger account line items using app delivered advanced filtering, sorting, and grouping capabilities. You can display line items based on ledger or company code, or select items based on various selection criteria such as key date, posting date range, and fiscal periods. In addition, you can group, sort, and filter the results, and export them to Microsoft Excel for further analysis. Also, the app lets you display amounts in different currencies, making it easier to compare totals across multiple company codes with journal entries in different currencies for reporting for multicompany and multicurrency environments. In addition, the app allows you to create your own layout variants.

This app replaces the Controlling Document (S/4HANA) fact sheet app (F1720) as of SAP S/4HANA 1909.

> **Tip**
>
> The app does not display balance carryforward items (as they are technical system items rather than standard line items).

> **Transaction Code**
>
> The corresponding transaction in SAP GUI is Transaction FALL03.

Display Line Items in General Ledger (F2217)

Application Type: Analytical

The app allows you to check general ledger account line items in the general ledger view (Universal Journal entry view). The app comes with filters to display open and cleared items for open item-managed accounts, or all line items of a general ledger account. The app allows you to select line items based on various criteria for the posting date or time frame, including today, the last or next x days, quarters, months, and fiscal periods. In addition, you also can choose line items using ledger, company code, general ledger account, and status. If selecting all items, a posting date range can be used. In the results list, you can group, sort, and filter general ledger account line items using characteristics like general ledger account, segment, profit center, and so on. In addition, the app allows you to display additional characteristics and column order can be rearranged. You can view display amounts in a display currency, which is useful for reporting on several company codes with journal entries posted in different currencies, allowing an easy comparison of totals in one currency.

> **Tip**
>
> Like the Display Line Items – Cost Accounting app (F4023), this app also does not display balance carryforward items as they are technical system items rather than standard line items. If you plan to use the exported document as an input source for external consumption, note that the field labels may change. Consult SAP Note 2630594, which discusses known limitations for document export in SAPUI5 apps. SAP recommends using the G/L Account Line Items – Read (A2X) (technical name API_GLACCOUNTLINEITEM_SRV) API to extract large amounts of data from general ledger account line items for external consumption.

> **Transaction Codes**
>
> The corresponding transactions in SAP GUI are Transactions FAGLL03 or FAGLL03H.

Display Liquidity Hierarchy – China, Liquidity Item Hierarchy (WDA_FCLM_LQH (APP_CFG_FCLM_LQH))

Application Type: Web Dynpro

This app allows businesses to display liquidity items in cash and liquidity planning. You can define and modify liquidity item hierarchies to align with business needs and to align with liquidity management processes. In addition, the app provides structured liquidity reporting for forecasting accuracy and offers custom views and filtering options for enhanced reporting. You also can display the liquidity item hierarchy using the Cash Flow Analyzer app (F2332).

> **Tip**
>
> If you do not see liquidity items in the hierarchy nodes, you should check whether the liquidity items are configured, assigned with correct general ledger accounts and then maintained in the Manage Global Hierarchy app (F2918) for the liquidity item hierarchy.

Display Matching Items (F3869)

Application Type: Analytical

This app allows group reporting users to display all matching items and their details based on filter criteria such as reconciliation case, display group, and side of display group. As a result, reviewing and analyzing the results of the intercompany reconciliation matching process simplified with this app. The app offers mandatory filters in the header, including reconciliation case ID, and side of display, have preset comparison operators, while optional filters like fiscal year period, company, partner unit, processing status, and matching document offer additional flexibility. The app retrieves items from both data sources and the matching database table, showing processing statuses from **00 New** to **30 Matched**. In addition, you can view key details such as organizational units (e.g. company), document numbers, account numbers, currency amounts, processing statuses, assignment number, matching rules, and reason codes. Also, fields derived from the data source can be customized in the settings for hiding, sorting, or grouping, enhancing tracking and analysis of matching items.

Display Organizational Change Application Logs (F4720)

Application Type: Transactional, Reuse Component

This app allows you to view log details for organizational changes that have been activated, simulated, processed, or completed using the Manage Organizational Changes app (F4567) or related programs. You can review log details for all organizational changes by entering **FINS** in the **Category** field, **FINOC** in the **Subcategory** field, leaving the **External Reference** field empty, and selecting **Go** option. To view details for a specific organizational change, you can enter its identifier in the **External Reference** field.

Tip

When navigating from the Manage Organizational Changes app, the app populates the category and subcategory automatically. You can add the external reference to the table by clicking the settings icon and selecting **External Reference**.

Display Payment Card Data (F2935)

Application Type: Transactional

With this app, you can manage and monitor card payment transactions. This app is part of accounts receivable and provides a comprehensive view of card payments, including details of the card used, payment authorization, and settlement. The app allows you to display a list of card payments and related information, such as card details, payment authorization, and settlement. In addition, the app offers features such as finding individual payments and key information quickly using search filters like authorization date, card number, and company code. Also, the app clearly displays information about entry, authorization, settlement, and the settlement run beside payment and payment card details.

Transaction

The corresponding transaction in SAP GUI is Transaction FB03.

Display Payment Forms (F7105)

Application Type: Transactional

This app provides an overview of payment forms generated for receivables paid via payment links in contract accounting. The allows you to track the status of these payment forms and the payment status of grouped receivables. You can display payment forms data by searching with filter criteria such as business partner, contract account, and status. In addition, the app features a **List View** of all payment forms related to payments by link and detailed displays of individual payment forms, ensuring efficient payment tracking and management.

Display Payment Lists (F3917)

Application Type: Transactional

This app provides accounts payable and receivable accountants with an overview of payments created by a payment run, along with any payment exceptions. The app enhances payment functionality, and it ensures efficient tracking and resolution of payment-related processes. You can select a payment run (test or update) to view details, including total payments, payment orders, repayment requests, and bank amounts (net of cash discounts, withholding tax, and tolerances). Payments are grouped by format, company code, house bank ID, account ID, credit/debit memo, and payment method. In addition, the app displays exceptions such as balance zero, internal clearing, and payment cards. You can double-click to view detailed information for payment groups, individual payments, repayment requests, or payment orders. The app also highlights totals posted to general ledger accounts, cash flows in bank clearing accounts, and exceptions with reasons and logs.

> **Tip**
>
> This app requires contract accounting to be active in the system. To change the app's layout, consult SAP Note 3294344.

Display Periodic Accrual Amounts (F4900)

Application Type: Transactional

This app allows general ledger users to display a list of periodic accrual amounts with search criteria such as company code, review type, and approval status. This app is commonly used during month-end closing activities as part of accrual management. It helps you review and manage periodic accrual amounts, ensuring accurate financial reporting for the specified fiscal period. The app provides a detailed table view of periodic accrual amounts for the specified fiscal period and allows navigation to related apps for additional functionalities. The app's functionality includes customizable filters, such as fiscal year period, company code, accrual object, accrual subobject, start of life, end of life, and lifecycle status, ensuring a comprehensive accrual and review process.

> **Tip**
>
> As a prerequisite, the app displays data only if you've activated appropriate accrual objects and run the accrual proposal job (Transaction ACEPOSTINGRUN) for the period. In addition, the app does not provide the posted amount to compare to the proposed amount currently.

Display Posted Value Adjustments (F4824)

Application Type: Transactional

This app allows contract accountants to display and evaluate posted receivables adjustments, including doubtful entries and individual or flat-rate value adjustments for selected filter criteria. The app shows adjusted receivables based on the entered key date and valuation area, with additional selection criteria such as company code, business partner, contract account, and document number. For detailed information, you can expand the desired document from the list and click the arrow at the end of the row to access the detailed page of the value adjustment.

> **Tip**
>
> Value adjustments are posted in contract accounting and transferred as summary records to the universal journal. Then the Analyze Credit Loss Allowances app (W0134) allows you to view and verify total credit losses, with details available at the business partner level. To access posted amounts in contract accounting, SAP recommends using this app.

> **Transaction Code**
>
> The corresponding transaction in SAP GUI is Transaction FPRA.

Display Process Flow – Accounts Payable (F2691)

Application Type: Transactional

The app is a powerful tool for monitoring and managing the accounts payable process. It provides a comprehensive, visual representation of the end-to-end accounts payable process within SAP S/4HANA. You can search for process flows using the document type and purchase order. This app allows you to visualize the relationships between accounts payable-related documents such as purchase orders, goods movements, incoming invoices, journal entries, and clearing entries. You can display the process flow from a specific document and access details for individual documents with ease. In addition, with graphical displays, color-coding, and detailed drilldown capabilities, the app ensures transparency by offering insights into each step of the process, from invoice entry to final payment.

Display Process Flow – Accounts Receivable (F2692)

Application Type: Transactional

This app allows you to display the relationships between accounts receivable-related documents, including sales orders, goods delivery, billing documents, invoices, journal entries, and clearing entries. You can search for process flows using the document type and sales document. As a result, the app allows you to visualize the overall flow of activities related to the collection, payment, and reconciliation of customer accounts. The app provides key features that include the ability to display the process flow from a specific document and access detailed information for individual documents with ease. With this app, you can monitor and manage customer account activities and resolve issues promptly.

> **Transaction Code**
>
> The corresponding transaction in SAP GUI is Transaction S_ALR_87013433.

Display Purchase Order Accruals (F3928)

Application Type: Transactional

This app provides a comprehensive overview of purchase order accruals, including details such as lifecycle status, proposed and revised accrual amounts, review history, and posting history. As a result, all key information, from accrual generation to ledger postings, is accessible by general ledger accountants. The app identifies accrual subobjects with a combination of attributes such as purchase order number, line-item number, and account assignment. With this app, you can check accrual subobjects transferred from purchase orders, view accrual postings and their history by period, attach supplementary files, and analyze source systems for purchase orders replicated to the Central Finance system.

Transaction Code

The equivalent SAP GUI transaction code is Transaction POACTREE03. The app does not support all the features available in the SAP GUI transaction. Refer to SAP Note 3425792 for additional guidance on features that are not available with the app.

Display Returns (F3126)

Application Type: Transactional

Contract accounting is a subledger that is appropriate for processing large document volumes, which is important for certain industries. This app offers accounts receivable and accounts payable users detailed insights into returns data. The app offers the ability to view, manage, and track the entire returns process within contract accounting. For example, you can search for a list of returns using fields such as business partner, contract account, returns lot, return reason, return status, and posting date. Once a returns lot is successfully entered, closed, and posted, the system records the returns data in a history table (DFKKRH), which can be accessed through this app. The system stores returned process header and line-item details data in tables DFK-KRH and DFKKRP, respectively. In addition, the app allows you to filter returns based on business partner, contract account, and additional criteria, such as returns lot, return reason, and return status.

Tip

This app is exclusively available for industry solutions that use the contract accounting component, including SAP for Utilities, SAP for Public Sector, and SAP for Insurance. The app solely accesses postings and data from contract accounting, excluding data from other subledgers like accounts receivable and accounts payable.

Transaction Code

The corresponding transaction in SAP GUI is Transaction FPM4.

Display Saved Open Item Lists (F3638)

Application Type: Transactional

Contract accounting is a subledger accounting module tailed to handle high volumes of receivables and payables in SAP. This app allows reconciliation specialists to display open items for a specified key date and download the lists for reporting or explaining general ledger balances using detailed subledger data as part of period-end close. The app's features include viewing saved open item lists, displaying items in a list format, downloading lists directly as Microsoft Excel files or as CSV files generated by the Create Open Item List for Key Date app (FPO1), and deleting individual open item lists as necessary.

Tip

To prevent the use of specific lists, SAP recommends selecting the desired list and choosing the **Delete** option. Conversely, to remove lists that are no longer needed, SAP suggests utilizing the archiving object FI_MKKOPLS and the data destruction object FI_MKKOPREP_DESTRUCTION to archive open item lists that you no longer need.

Display Service Entry Sheet Accruals (F6107)

Application Type: Transactional

This app provides a comprehensive overview of service entry sheet accruals, including lifecycle status, proposed accrual amounts, and other vital information. Accrual subobjects are identified by the combination of purchase order number, line-item number, and account assignment. You can find service sheet accrual entries using search filters such as company code, accrual item type, and lifecycle status. In addition, the app allows you to check accrual subobjects transferred from purchase orders, view proposed accrual amounts based on unapproved service entry sheets, and review posted accruals for each fiscal period, ensuring access to key information from accrual generation to ledger posting. Thus, the app offers general ledger accountants a central place to monitor and review the accruals process related to services rendered and recorded in service entry sheets.

Display Shifted and Suspended Revenue History (FARR_SHIFT_HISTORY_AUDIT)

Application Type: Web Dynpro

This app provides detailed visibility into the history of revenues that have been shifted or suspended within contracts in revenue accounting and reporting (RAR). This happens due to various reasons such as deferred revenue to a future period, compliance issues, contract changes, or any other similar reasons. You can filter revenue history by various criteria such as company code, contract, performance obligation, business partner, or date. In addition, you can view the detailed audit trail of all shifted and suspended revenue entries, including time stamps, user, and reasons for adjustments. Thus, you can assess reasons for suspension or shifts to understand the overall impact of revenue recognition. The app supports and complies with IFRS 15 and ASC 606 accounting standards.

Display Status of Payment Documents (F7084)

Application Type: Analytical

This app provides an overview of payment documents and their statuses, allowing you to monitor individual payments and their corresponding batches. You can view payment document statuses and access detailed payment information such as bank details, amounts, accounts,

and business partners. In addition, you can navigate to related details such as payment batches, payment runs, or specific payment documents for further details. Thus, the app helps trace the lifecycle of payment transactions.

Display Supplier Balances (Version 2) (F0701A)

Application Type: Analytical

With this updated app, accounts payable specialists and finance managers can get an overview of outstanding supplier balances across different company codes and currencies. As a result, you can improve cash flow management by tracking outstanding payables, enhancing financial transparency, and supporting better supplier relationship management through timely payments and accurate financial reporting. The app helps you to analyze open amounts, overdue payments, and total liabilities; filter balances by supplier, company code, or fiscal period; and drill down into detailed supplier transactions for reconciliation. In addition, you can monitor outstanding supplier balances and overdue amounts to ensure timely payments and avoid late fees. Also, the app lets you drill down into specific supplier transactions for detailed insights and export supplier balance reports to Excel for further analysis or to perform reconciliation.

This app is the successor app for the deprecated Display Supplier Balances app (F0701).

Display Supplier List (F1861)

Application Type: Transactional

This app allows you to display and download a list of suppliers, using search filters to create custom lists for stakeholders and auditors. You can search for supplier lists using parameters such as supplier, company code, city, country, bank key and posting block. The app provides features such as viewing contact details for suppliers, creating custom lists of obsolete, blocked, or payment method-based suppliers, and accessing bank details and payment methods for suppliers. Typical use cases include searching for supplier details across company codes or purchasing organizations; verifying payment terms, bank details, or tax information; and analyzing supplier information as part of a performance strategy.

Transaction Codes

The app serves a single place to get information that can be found in multiple SAP GUI transactions: Transactions MK03 (Display Vendor (Purchasing)), XK03 (Display Vendor Centrally), BP (Business Partner), and S_ALR_87012086 (Vendor Data List).

Display Task List (FAGL_FCC_WD)

Application Type: Web Dynpro

This app allows you to view and manage financial close task lists in the SAP Financial Closing cockpit for SAP S/4HANA. The app provides an overview of assigned tasks, dependencies, and statuses within the closing cycle while ensuring that task lists meet predefined consistency requirements before generation. Access to specific task lists requires appropriate display permissions, and task lists are created from predefined templates that must be released before execution.

Display Treasury Alerts (F2025)

Application Type: Transactional

With this app, treasury and risk management (TRM) teams can monitor and respond to critical treasury-related alerts in real time. You can search for treasury alerts using fields such as issue message for, company code, city, transaction, transaction type, product type, counterparty, up to and including due date, house bank, house bank account, and classification. The app

presents alerts in a tabular column with important information such as transaction, message text, product type name, and transaction type name. Financial instrument processes encompass the release and settlement of financial transactions, payment and posting of cash flows, sending correspondence for financial transactions, and fixing of interest rates. To support and ensure timely execution of these process steps, you can use this app, which provides timely notifications when any activities are not executed as expected.

Transaction Code

The equivalent SAP GUI transaction is Transaction FTR_ALERT (Monitor Treasury Alerts).

Display Treasury Position Flows (F1754)

Application Type: Transactional

With this app, you can track and manage position flows for financial transactions and positions within the treasury subledger. You can search treasury posting flows using posted group, company code, valuation area, treasury ledger date, business transaction status, and transaction. Posting flows are presented in a column format with the flexibility of adding additional dimensions to the layout. This app allows treasury users to track position flows for financial transactions and positions by providing a list of flows for selected positions in the treasury subledger and detailed information on these flows. The app provides detailed information on posted, scheduled, and reversed flows with navigation options to original business transactions and to the Display Treasury Posting Journal app (F1755). In addition, you can navigate to master data such as business partner, futures account, securities account, and security class, displaying the count of position flows in the header, sorting, grouping, and subtotals. Moreover, the app offers the ability to trigger emails and export results to a Microsoft Excel spreadsheet.

Like Display Treasury Position Flows app (F1754), this app also excludes subpositions, meaning it does not select flows from hedge accounting for positions or value adjustment flows related to P-GAAP. Let's compare the two apps, Display Treasury Posting Journal (F1755) and Display Treasury Position Flows (F1754):

- **Display Treasury Posting Journal**
  This app focuses on posted flows and related accounts, thereby providing you with a comprehensive view of all postings. It displays actual accounting entries for treasury transactions.

- **Display Treasury Position Flows**
  This app tracks position flows, including posted, scheduled, and reversed flows. It exhibits a chronological view of how position value changes evolved over time.

Transaction Code

The corresponding transaction in SAP GUI is Transaction TPM13.

Display Treasury Position Values (F1867)

Application Type: Transactional

With this app, you can display a list of positions in the treasury subledger and their position component values on any selected key date. You also can search for treasury position values using posted group, company code, valuation area, and business area. The app includes position component values such as book value, purchase value, write-up/write-down in valuation, and position currency. In addition, the app provides features such as showing detailed information on selected treasury positions with navigation options (e.g., position flows, position indicator, and position management procedure), enabling navigation to master data (e.g., business partner, futures account, securities account, and security class). Also, you can perform tasks such as sorting, grouping, and subtotals, providing the count of positions in the

content area header, and enabling the triggering of emails and exporting of results to Microsoft Excel.

Tip

You must understand the differences between the Display Treasury Position Flows (F1754), Display Treasury Posting Journal (F1755), and Display Treasury Position Values (F1867) apps. While all three apps provide detailed information and navigation options, the Display Treasury Position Flows app (F1754) focuses on tracking position cash flows of treasury transactions, and the Display Treasury Posting Journal app (F1755) focuses on accounting postings of treasury transactions. On the other hand, the Display Treasury Position Values app (F1867) focuses on displaying current valuation of treasury positions or position component values on a selected key date.

Transaction Code

The corresponding transaction in SAP GUI is Transaction TPM12.

Display Treasury Position Values for Expected Losses (F3123)

Application Type: Transactional

With this app, you can display the expected losses in a company's treasury position at a selected key date. You also can search for the expected losses using product group, company code, valuation area, and business area fields. This is essential for treasury managers to manage and evaluate potential financial risks, ensuring that their organization has a clear understanding of its exposure to market fluctuations and other financial events that could affect the company's liquidity and assets. With this app, treasury users can display a list of treasury's subledger positions along with their component values, such as expected losses and position stage assignments, for any selected key date.

Also, the app offers detailed information on positions with navigation to related areas like position flows, indicators, and management procedures. In addition, you can access master data; sort, group, and calculate subtotals; view position counts; trigger emails; export results to Microsoft Excel; and analyze details like expected credit losses, foreign exchange (FX) effects, and position stages, streamlining treasury management.

> **Tip**
>
> The app does not support extensibility to add your own fields.

> **Transaction Code**
>
> The corresponding transaction in SAP GUI is Transaction TPM12.

Display Treasury Posting Journal (F1755)

Application Type: Transactional

The app allows you to view the posted flows of selected financial transactions within the treasury subledger. You can search for treasury journals using search fields such as product group, company code, valuation area, posting date, transaction, and posting status. Transactional line items are presented in a column format with important information such as general ledger account, amount in posting currency, amount (local currency), and account assignment reference with a flexibility of adding additional dimensions. This app allows you to display detailed accounting information of financial instrument postings by providing a list of posting line items and additional details of posted and reversed flows of financial transactions and positions. In addition, you can display detailed information on posted and reversed flows with navigation options to the original business transaction and the Manage Journal Entries app. Also, you can navigate to master

data like business partner and security class. The app also supports sorting, grouping, and subtotals, as well as triggering emails and exporting results to a Microsoft Excel spreadsheet.

> **Tip**
>
> This app excludes subpositions, meaning it does not select flows from hedge accounting for positions or value adjustment flows related to parallel GAAP (field label P-GAAP) valuation areas in treasury subledger accounting.

> **Transaction Code**
>
> The equivalent transaction in SAP GUI is Transaction TPM20.

Display User (FAGL_FCC_WD_USER)

Application Type: Web Dynpro

This app allows administrators to manage user assignments within financial close processes in SAP Financial Closing cockpit for SAP S/4HANA. The app provides an overview of assigned users, their roles, and access levels while ensuring proper authorization for task execution. The app integrates task lists to streamline monitoring and execution efficiently.

Display Userlist (FAGL_FCC_WD_USERLIST)

Application Type: Web Dynpro

This app provides a comprehensive overview of all users involved in financial close processes, displaying their roles, assigned tasks, and access levels in in SAP Financial Closing cockpit for SAP S/4HANA. The app ensures proper authorization for executing financial close activities and integrates users with task lists for streamlined monitoring and execution.

99

Display Write-Offs (F3125)

Application Type: Transactional

Contract accounts receivable and payable acts as a subledger and processes high document volumes typical in industries with numerous end customers while performing standard accounts receivable functions. With this app, accounts receivable and accounts payable users can get write-off data, which provides them with insights into a business partner's payment behavior. You can search for write-off information using search fields such as write-off document, document, company code, contract account, write-off amount, write-off-reason, date, statistical key, origin, reversed, and dunning procedure document. This is managed through automatic processes. The system generates write-off documents during both individual write-offs and write-off runs, recording each write-off and its reversal in the write-off history. Users with appropriate authorizations can update this history automatically during mass write-off operations. In addition, after write-off activities, you can display write-data using the Display Write-Offs app. The app offers filtering options based on criteria such as business partner, contract account, write-off document number, date, reason, or origin.

Tip

The app is exclusively available for industry solutions that incorporate contract accounting, including SAP for Utilities, SAP for Public Sector, and SAP for Insurance. The app solely accesses postings and data from contract accounting, excluding data from other subledgers like accounts receivable and accounts payable.

Transaction Code

The corresponding transaction in SAP GUI is Transaction FP04H.

Document Flow (Performance Obligation) (F4423)

Application Type: Transactional

This app lists all relevant business transactions and journal entries for a performance obligation, providing key details for each document, such as order date, sender component, order quantity, and contractual price. This app is part of revenue accounting and reporting (RAR). You can search for revenue contracts or performance obligations, grouped by revenue contract by default, and filter documents by date range. The app organizes documents into categories: order, fulfillment, invoice, cost correction, journal entries from revenue accounting, and journal entries from an operational document. Only relevant categories are displayed for a performance obligation, and fulfillment dates are shown for time-based or manually fulfilled obligations. In addition, you can navigate to related operational documents for detailed information by clicking the **Link** button on each document tab. The app's search features include filtering by attributes and fuzzy search for fields like company code and performance obligation name.

Document Management (FKK_WDY_DMS)

Application Type: Web Dynpro

This Web Dynpro–based app allows you to manage business documents in a secure storage environment in contract accounting. The app features include storing business documents with appropriate access control, tracking documents with version management, integrating with transactional documents such as invoices, payments, and customer accounts, approving changes via workflow function, and supporting multiple formats such as PDFs, word, and spreadsheets. As a prerequisite to use this app, you must maintain business partner master data, set up relevant roles for providing appropriate access control, define an approval hierarchy to support workflow function, and maintain storage-related configuration.

Doubtful Accounts Valuation (F3246)

Application Type: Analytical

This app allows accounts receivable managers to manage and assess allowances for doubtful accounts, ensuring compliance with GAAP or IFRS requirements for expected credit losses. The app offers a clear view of overdue receivables and their associated allowances by customer or by credit risks, with drilldown options to examine individual customer accounts at the journal entry level. The displayed allowance values are based on flat rate individual value adjustments (program SAPF107) created during valuation runs using Transaction F107 (FI Valuation Run). In addition, if you're using Transaction F103 (Transfer Postings for Doubtful Receivables; program SAPF103) or Transaction FAGL_104 (Provision for Doubtful Receivables; program FAGL_DR_PROVISION), SAP recommends utilizing the Allowance for Doubtful Accounts app (F2686). Also, the app provides features such as a chart view of overdue receivables and allowances by customer, country, credit risk class, or dunning block, along with detailed account-level insights.

Drill-Through Reports (F2638)

Application Type: Analytical

To enhance real-time consolidation, this app allows you to display a list of journal entries for a specified consolidation model and cut-off time. You can display document lists with filter fields such as model, version, entity, group account, fiscal period, and cut-off time. In addition, you can drill through from the account line item of reported data in the Consolidation Data Release Cockpit app (F2591) to the original financial journal entries. The app supports displaying amounts in various currencies at the entity and group account levels to facilitate currency translation.

Tip

If currency translation occurs in the consolidation system instead of the financial accounting system, the app supports only local currency amounts, which can be drilled through to the journal entry level within the reported financial data view. However, the group currency amounts cannot be traced back; it's a one-way process.

Dunning Level Distribution (S/4HANA) (F1742)

Application Type: Analytical

With this analytical app, you can get a comprehensive view of the **Dunning Level Distribution** KPI, displaying open dunning amounts per dunning level and customer in both chart and table formats for customers and vendors. The app presents data in graphical formats by customer, accounting clerk, company code, country key, region, and customer. As a result, you can monitor overdue receivables and take appropriate follow-up actions based on the assigned dunning levels. In addition, you can analyze data by the 10 customers with the highest open dunning amounts and drill down further by various criteria such as account group, company code, country key, and more. Moreover, the app offers features like viewing aggregated open dunning amounts, customizing the range of dunning levels, and analyzing the highest open dunning amounts by customer. It also supports follow-up actions through SAP Jam and email integration. The app is enabled with catching for enhanced performance, with data refresh options.

Transaction Code

The corresponding transaction in SAP GUI is Transaction S_ALR_87012167.

Application Type: Transactional

This app allows you to customize user-specific options for editing journal entries. For instance, if you prefer to enter journal entries only in the local currency, the user can select the **Doc in Local Crcy** option. As a result, the system will hide the fields that require entering foreign currency journal entries. If the **Amnt in Doc Currency** checkbox is selected, regardless of the chosen currency, the system displays the amount in the transaction currency (document currency) in apps such as Display Line Items in General Ledger (F2217) and Manage Journal Entries (F0717). The editing options for journal entries include setting documents in local currency, using alternative accounts, avoiding company code proposals, using foreign exchange rates from the first line item, not copying tax codes, allowing negative postings, displaying net amounts, setting documents by company code, showing amounts in document currency, avoiding special general ledger transactions, displaying periods, adding user invoice references, posting special periods, copying payment bases, setting default currencies, defining document types and dates, entering net amounts, and using payment references as search criteria.

Transaction Code

The corresponding transaction in SAP GUI is Transaction FB00.

Expense Report (FITE_EXPENSES)

Application Type: Web Dynpro

With this Web Dynpro–based app, you can manage employee expense claims and streamline submission, review, and approval processes in travel management. The app allows employees to enter travel costs, meals, accommodation, and other miscellaneous expenditures while supporting electronic receipt attachments for verification. The app further supports an integrated approval workflow that supports policy compliance, and automated calculations apply tax regulations, currency conversions, and reimbursement policies. In addition, the app integrates with the general ledger for expense posting and with the document management system (DMS) for storing receipts. The app's additional functionalities include pre- and post-trip accounting, corporate card settlements, and reimbursement tracking, alongside automated policy compliance checks and managerial audit reviews.

As a prerequisite to use the app, you must maintain employee master data such as business partner. In addition, for workflow approval, approval hierarchies are required. The app stores data in tables PTRV_HEAD, PTRV_EXPENSE, PTRV_PERIO, and FTI_EXPENSES.

Tip

The Expense Report app (FITE_EXPENSES) is useful for handling comprehensive expense reports that involve review and compliance checks, whereas the Express Expense Sheet app (FITE_EXPRESS_EXPENSES) allows for the kind of quick entry typically used for routine expenses.

Transaction Code

The equivalent transaction in SAP GUI is Transaction TRIP.

**Express Expense Sheet
(FITE_EXPRESS_EXPENSES)**

Application Type: Web Dynpro

As the name implies, this app provides an accelerated way to enter expenses for employees. You can quickly submit business-related expenses along with digital receipts via a simplified interface with predelivered templates. The app performs automated checks to ensure expenses entered comply with a company's policies. The app's features include employees creating expense sheets, selecting predefined categories to minimize input, and submitting reports for approval, either automatically or manually based on company policies. In addition, the app integrates with workflows to review expenses before approval.

To use this app, as a prerequisite, you must maintain employee master data in the business partner, configure financial posting integration in financial accounting, and define approval workflows. In addition, the app integrates with the general ledger for expense posting and with the document management system (DMS) for storing receipts. The app stores expense data in tables PTRV_EXPRESS, PTRV_HEAD, and PTRV_PERIO.

Tip

You can restrict the expense types using Customizing activity **Restrict Travel Expense Types** by navigating to IMG menu path **Travel Management • Travel Expenses • Dialog and Travel Expenses Control Dialog • Control**.

Transaction Code

The equivalent SAP GUI transaction is Transaction PR02.

FI-CA: FPM Application for Co-Liabilities (FICA_COLIABILITY_OVP)

Application Type: Web Dynpro

In contract accounting, this app helps you to manage shared financial liabilities within contract accounts. The app allows you to track and process co-liabilities across multiple business partners. The app's features include supporting detailed reporting, facilitating reconciliation, and allowing you to assess financial exposures across company codes. The app is integrated with the general ledger for posting liability-related transactions from the contract accounting subledger.

In addition, you can leverage authorizations to restrict sensitive liability transactions to comply with company policies.

FI-CA: FPM Application for Co-Liability Records (FICA_COLIABILITY_REC_OVP)

Application Type: Web Dynpro

In contract accounting, this app helps maintain detailed records of co-liabilities within contract accounts. The app features include maintaining detailed records of shared financial obligations between multiple business partners, supporting comprehensive reporting, facilitating reconciliation, and analyzing financial exposures across different entities. In addition, the app integrates with the general ledger to post liability-related transactions from the contract accounting subledger. The app also supports authorization controls to maintain data security.

Tip

While both apps FI-CA: FPM Application for Co-Liabilities (FICA_COLIABILITY_OVP) and FI-CA: FPM Application for Co-Liability Records (FICA_COLIABILITY_REC_OVP) support co-liabilities, the first app is more focused on liability management and processing, whereas the latter emphasizes record-keeping and reconciliation.

Financial Data Consistency Results (F7457)

Application Type: Analytical

This app allows you to visualize the results of Financial Data Consistency Analyzer (FDCA) reconciliation runs that have identified at least one inconsistency. The FDCA (Transaction FINS_FDCA_RUN) performs data consistency checks in general ledger accounting to identify and resolve common inconsistencies. The tool ensures financial accuracy by detecting discrepancies in master data, document postings, clearing balances, and ledger reconciliations. With this app, you can filter and focus on specific runs, with impacted business processes clearly displayed for each instance. Some inconsistencies can be corrected directly by navigating to the relevant transactional app. The interface includes a list of completed reconciliation runs filtered by consistency analysis run ID, company code, and execution date, with sorting and grouping options based on issue categories such as correctable, corrected, total, and other. For individual runs, you can access detailed views, including pie charts showing issue proportions, lists of impacted business processes, and multiple overviews such as issue group overview, issue criticality overview, issue correction status overview, and issue details.

In addition, you can review individual inconsistencies with Universal Journal entry attributes and navigate to specific journal entries through integrated apps such as Manage Journal Entries (New Version) (F0717A).

Tip

For additional information, refer to SAP Note 3256224, which offers answers to frequently asked questions.

Transaction Code

The corresponding transaction in SAP GUI is Transaction FINS_FDCA_RUN.

Financial Status (Book Value) (F2136)

Application Type: Web Dynpro

This analytical app displays the financial status of a company or group of companies on a specific key date and allows you to drill down to individual financial positions. You also can search by reporting currency, exchange rate type, and company code. These positions can originate from treasury position management,

bank account balances from cash management, or account balances from the general ledger. Each position represents either an asset or liability, structured by financial position groups. The app provides two tiles: Financial Status – Book Value (F2136), displaying the status based on book values, and Financial Status – Nominal Amount (W0122), displaying the status based on nominal amounts. You can calculate and display amounts in transaction or reporting currency, analyze financial positions by adding attributes, sort, group, and subtotal results, navigate to master data, and export results to Excel.

Financial Status (Nominal Amount) (W0122)

Application Type: Web Dynpro

This analytical app provides a comprehensive view of a company's financial status on a specific key date in treasury and risk management (TRM). The app further allows you to drill down into individual financial positions from sources such as treasury position management, One Exposure from Operations for bank account balances from cash management, or general ledger balances. In TRM, financial positions are categorized as assets or liabilities, structured into predefined groups for clarity. With this app, you can view nominal amounts for these sources using search parameters such as key date, reporting currency, exchange rate type, and company code.

Tip

Financial positions and position groups from facilities for credit lines in TRM are not supported in this app. These positions, representing credit lines, do not have a specified asset-liability indicator. To report them, SAP recommends utilizing the CDS view C_FinancialStatusQuery (Financial Status Query) for enhanced reporting needs.

FIN User Default Parameter Plugin (F1765)

Application Type: Transactional

The app is designed to allow you to set and manage default values for various parameters within the SAP S/4HANA system. This plugin is integrated into the SAP Fiori launchpad shell and provides a dialog where business users can enter their default values for parameters such as company code, currency, and fiscal year. These default values are automatically pre-filling commonly used values in various finance-related applications during runtime, reducing manual entry and improving efficiency. As a result, user consistency is maintained across different applications.

Foreign Bank Account Report (F1575)

Application Type: Transactional

With this app, you can identify foreign bank accounts owned by your company and employees with powers of attorney over these accounts. You can search for foreign bank account information using home country, company code, time range, and account range. The information from this app can be used for filing reports, such as the Report of Foreign Bank and Financial Accounts (FBAR) required for the US, or for analytical purposes. The functions of the app include checking foreign bank accounts for one or more company codes, displaying the maximum account value for each foreign bank account within a specified time range, applying minimum aggregate value conditions to determine reporting requirements, checking employees with powers of attorney, and exporting matching foreign bank accounts to a spreadsheet. The app also supports SAP Jam integration for sharing information and receiving feedback, sending emails with selection criteria URLs, and saving tiles with default settings.

Tip

The FBAR report is primarily a requirement for the US. It functions by specifying the home country as the US, subsequently displaying all bank accounts where the company's country is the US, and the bank's country is not the US.

Foreign Exchange Overview (F2331)

Application Type: Transactional, Analytical

With this app, treasury and financial risk management teams can display an overview of foreign exchange (FX)-related financial risks, including FX instruments, financial status, cash position, and liquidity forecast at a selected key date. You can search for foreign exchange information using the key date, display currency, exchange rate type, company code, and transaction currency. With this app, you can analyze several KPIs displayed as separate cards, such as financial status, credit line overview, cash position, liquidity forecast, FX forwards, FX options, nondeliverable forwards, and foreign exchange rate in real time across multiple currencies. This results in greater visibility of a company's exposure to currency fluctuations. In addition, you can specify default values for company code, exchange rate type, key date, and display currency, and filter the cards by various categories. Also, you can navigate from the cards to target apps for more detailed information and customize the overview page by managing and rearranging the cards.

Future Payables (S/4HANA) (F1743)

Application Type: Analytical

With this app, accounts payable users can analyze and view the top 10 amounts payable and check the number of open items for relevant suppliers. The app provides a view of future amounts payable in chart or table format according to various criteria including company code, supplier, country and region of the supplier, account group, and payment blocking reason. The app allows you to view accounts payable by predefined due periods, which can be customized, and analyze the top 10 suppliers with the highest amounts payable. In addition, you can filter accounts payable by payment blocking reasons and specify a key date for data analysis. The app also supports technical features such as posting comments via SAP Jam, sending emails, and enabling caching for better performance. To access the app, you must be assigned the SAP_BR_ANALYTICS_SPECIALIST business template role. The app is built based on CDS view C_APFUTUREACCOUNTSPAYABLE.

> **Tip**
>
> Default user values for display currency and exchange rate type set in the SAP Fiori launchpad are applied to the following standard analytical apps of the **Smart Business** KPI type in the Accounts Receivable Accounting: Overdue Receivables (F1747), Future Receivables (F1744), Total Receivables (F1748), Dunning Level Distribution (F1742), Days Sales Outstanding (F1741), and Days Beyond Terms (F1739) apps. SAP-defined default values for these parameters will be overridden by your custom default values set in the **Settings** option of the user action menu in the SAP Fiori launchpad.

> **Transaction Code**
>
> The corresponding transaction in SAP GUI is Transaction F.46.

Future Receivables (S/4HANA) (F1744)

Application Type: Analytical

This analytical app displays the **Future Receivables** KPI and allows you to view results in a chart or table based on due period, company code, or the top 10 customers with the highest amounts receivable. In addition, you also can drill down to see the number of open items for relevant customers. Also, you can analyze future amounts receivable by various filter criteria such as account group, accounting clerk, company code, country key, customer, customer classification, display currency, exchange rate type, general ledger account, net due interval, region, or special general ledger. In addition, the app supports posting comments via SAP Jam, sending emails, and enabling caching for better performance.

General Ledger Overview (F2445)

Application Type: Analytical

This app provides a comprehensive, dashboard-style overview of the general ledger. You can monitor general ledger accounting indicators and access the relevant general ledger accounting apps using search parameters such as display currency, key date, statement version, ledger, and company code. Upon searching, the app presents an overview of general ledger activities. To facilitate these features, the app features cards that can be conveniently rearranged according to user requirements by dragging and dropping cards to different locations. In addition, each card can be sized up or down. The app includes cards for recognized revenue, recognized cost of sales, journal entries to be verified, general ledger account balance, tax reconciliation balance, general ledger item changes, days sales outstanding, days payable outstanding indirect, my inbox, and quick links.

GR/IR Process Insights (F5796)

Application Type: Analytical

The app provides insights into the goods receipt/invoice receipt (GR/IR) account reconciliation process carried out via the **Process Items** function in the Reconcile GR/IR Accounts app (F3302). You can search for GR/IR information by company code and supplier. The app offers detailed insights to visualize reconciliation activities, identifies process improvement opportunities such as bottlenecks or delays, and allows retrospective reviews of purchasing document items alongside their history. The app provides insights overview data by reading table FINS_GRIRPROCHIS. This table stores historical data and GR/IR process insights. In addition, process owners can analyze process efficiency and define improvement activities. The app's features include detailed reconciliation analysis, filtering by organizational units, graphical and table views of process transitions, analysis by processing dimensions, and metrics like changes and average processing time. The app uses CDS view C_GRIRClrgProcessStatusChange.

Tip

As a prerequisite, you must run the Reconcile GR/IR Accounts app (F3302) prior to using this app.

Consult SAP Note 3326201, which offers documentation links for standard system behavior for GR/IR-related activities and troubleshooting.

Transaction Code

The corresponding transaction in SAP GUI is Transaction MR11.

Gross Margin – Presumed/Actual (F3417)

Application Type: Analytical

Predictive accounting leverages the most current data from SAP S/4HANA functionalities beyond finance, such as sales, integrated products like SAP Concur, or external systems, to forecast future results. Thus, it provides you with clearer insights into expected accounting outcomes at the end of a period or quarter for better decision making. For example, the app presents data in graph format for presumed margin by customer, presumed margin by product, presumed margin by profit center, and presumed margin by fiscal year. Thus, with this app, you are empowered with evaluation and analysis of predicted revenues, cost of sales, margins, and sales deductions derived from incoming sales orders. By leveraging predictive postings in the predictive ledger, based on sales orders, goods issues, and billing documents, it allows comparisons between presumed data, planned data, or data from prior periods. The app's key features include filtering revenue, cost of sales, and margins by time periods, organizational attributes, customer details, products, and accounting parameters. In addition, you can drill down into dimensions and measures for detailed journal entry information and navigate directly to profit centers, customers, or products from the list view.

Here's an example of predictive ledgers stacked on underlying ledgers and the type of data:

- Ledger OE (incoming sales), prediction data
- Ledger OC (management accounting), actual data
- Ledger OL (leading ledger), actual data

By selecting OE ledger, you can view both actual data from both OC and OL, along with predicted data in OE.

In addition, the app uses semantic tags to support financial KPIs that are calculated based on nodes in a financial statement version. To facilitate this, the following semantic tags are assigned to the FSV nodes:

- **RECO_COS**: Recognized cost of sale (COS)
- **RECO_REV**: Recognized revenue
- **REC_MARGIN**: Recognized margin
- **SALES_DED**: Sales deduction
- **BILL_REV**: Billed revenue

> **Tip**
>
> If data is not displaying as expected on the app, consult SAP Notes 2793172 and 3276276 for further guidance on troubleshooting tips.

Group Data Analysis (W0135)

Application Type: Web Dynpro

This app allows you to view consolidated financial data through the **Group View** in SAP S/4HANA Finance for group reporting. The app reflects consolidation groups and units, and the **Hierarchy View** organizes data by profit centers, segments, and consolidation units. You can search consolidated data using various search parameters such as version, ledger, consolidated chart of accounts, consolidation unit hierarchy, profit center hierarchy, and segment hierarchy. The app displays data in two tabs: **Data Analysis** and **Query Information**. You have flexibility to add additional dimensions to columns under the **Navigation Panel**. In addition, by selecting different financial statement (FS) item hierarchies, you can generate consolidated financial statements, including balance sheets and profit and loss statements, ensuring accurate financial reporting and analysis.

This app is the successor app for the following deprecated apps, which were based on the old reporting logic and are no longer available from SAP S/4HANA 2023:

- Data Analysis – Reporting Logic (W0056)
- Consolidated Balance Sheet (W0059)
- Consolidated Balance Sheet – Year Comparison (W0062)
- Consolidated Balance Sheet – By Movements (W0060)
- Consolidated Balance Sheet – By Subgroups (W0061)
- Consolidated P&L by Function – Year Comparison (W0063)
- Consolidated P&L by Function – By Subgroups (W0068)
- Consolidated P&L by Nature (W0070)
- Consolidated P&L by Nature – By Subgroups (W0072)
- Consolidated P&L by Nature – By Functional Areas (W0071)
- Consolidated P&L by Nature – Year Comparison (W0073)
- Interunit Reconciliation – Group View (W0099)
- Statement of Comprehensive Income (W0100)
- Statement of Changes in Equity (W0101)
- Interunit Reconciliation – Unit View (W0103)
- Cash Flow Statement (W0104)

> **Tip**
>
> The app is implemented with new reporting logic from SAP S/4HANA 1909. Prior releases must use the Consolidated Balance Sheet (W0059), Consolidated P&L Statements by Nature of Expense (W0074), and Data Analysis – Reporting Logic (W0056) apps.

Group Data Analysis with Reporting Rules (W0136)

Application Type: Web Dynpro

In SAP S/4HANA Finance for group reporting, this app allows you to view consolidated data through two perspectives:

- **Group View**: Based on consolidation groups and units
- **Hierarchy View**: Using hierarchies of profit centers, segments, and consolidation units.

You can display consolidated data using various fields such as version, ledger, consolidation chart of accounts, fiscal year period, period mode, consolidation group, consolidation unit hierarchy, profit center hierarchy, and more. By selecting reporting item hierarchies, you can generate rule-based consolidation reports, including **Statements of Changes in Equity, Statement of Cash Flows, Statement of Comprehensive Income, Consolidated P&L Statements by Function of Expense, and Interunit Reconciliations**.

Like the Group Data Analysis app (WO135), this app exclusively supports predefined rule-based reports. Let's compare the Group Data Analysis (WO135) and Group Data Analysis with Reporting Rules (WO136) apps:

- **Group Data Analysis**
  This app provides a general view of consolidated financial data, allowing you to analyze FS items across different dimensions. It offers flexible analysis using various filters.

- **Group Data Analysis with Reporting Rules**
  This app exclusively supports predefined rule-based reporting. You must select certain fields such as reporting item hierarchy, reporting rule version, and more.

Group Financial Statements Review Booklet (F6133)

Application Type: Analytical, Reuse Component

This app allows you to display, analyze, and validate group reporting data through an aggregated view of financial statements in SAP S/4HANA Finance for group reporting. It combines predefined business pages under key areas such as the consolidated balance sheet, profit and loss (P&L) statement, cash flow statement, and statement of changes in equity, forming a review booklet for financial reviews. The app's features include displaying and analyzing group reporting data at various levels of detail, exporting business pages to spreadsheets via the **Export to Microsoft Excel** option in the report menu, and navigating to other areas of group reporting.

> **Tip**
>
> This app leverages the new reporting logic and is compatible with SAP S/4HANA 1909 and later releases. However, for releases earlier than 1909 without the new reporting logic activated, SAP recommends using the Consolidated Balance Sheets (WO059), Consolidated P&L Statements by Nature of Expense (WO074), and Data Analysis – Reporting Logic (WO056) apps.

House Bank Account (S/4HANA) (F1759)

Application Type: Fact Sheet

The app provides you with a granular view of individual house bank accounts. It provides detailed information about the bank accounts associated with house banks, including account IDs, account types, and balances. The app includes the bank account number and IBAN or SWIFT/BIC. You can create, edit, and close bank accounts, as well as manage account hierarchies and overdraft limits. The app also supports the assignment of signatories and approval workflows, ensuring compliance with internal controls and regulations. In addition, you can generate reports and export data to Microsoft Excel for further analysis.

Let's compare the House Bank Account (F1759) and House Bank (F1758) apps:

- **House Bank**
  This app provides high-level overall management and an overview of house bank data. You can manage at the banking institution or bank entity levels.

- **House Bank Account**
  With this app, you can manage the details and operations of individual bank accounts. The app manages specific bank accounts under a house bank.

House Bank (S/4HANA) (F1758)

Application Type: Fact Sheet

A house bank is important bank-related master data that represents a company's bank account used for processing or receiving payments. SAP provides this app as an intuitive and centralized way to manage house banks and bank accounts as part of bank account management (BAM). The app provides a comprehensive overview of house bank data, and allows you to display an overview of house bank data, including house bank description, country/region, and city. The app provides key facts relevant in the business context, such as company code, bank key, and bank number. The app supports sorting, grouping, and subtotals, and allows you to trigger emails and export results to a Microsoft Excel spreadsheet for further analysis and reporting. In addition, the app serves as a starting point for navigating additional relevant information, like related business partners, master data, or documents. The app provides navigation to other apps with additional functions for editing or analyzing related business data.

Tip

This app allows house bank setup as master data rather than as a Customizing setting, thereby significantly improving the user experience for managing house banks.

Transaction Code

The corresponding transaction in SAP GUI is Transaction FI12.

Implement Powers of Attorney for Banking Transactions (F6374)

Application Type: Transactional

The app allows you to manage the implementation process of powers of attorney across various channels. The app presents details in the **Header**, **Power of Attorney**, and **Channel** tabs. In cash management, with this app, you can create and process implementations, assign channels, and manage correspondence objects that document communication with banks. For instance, each power of attorney implementation requires assigning channels and creating correspondence objects, which document communication with banks, such as outgoing notifications of authorizations and incoming confirmations. Implementations can be deleted if no channels are assigned, and no correspondence objects exist. On the other hand, the app allows revoking an implementation only under specific conditions, such as all outgoing correspondence objects being replaced, obsolete, or confirmed. If revoked unintentionally, the revocation can be withdrawn if the power of attorney is still active and valid. Overview tables display all implementations or correspondence objects, enabling you to process entries, change statuses, or create new implementations or correspondence objects. The implementation status can include **Active**, **Power of Attorney Revoked**, **Replaced**, or **Revoked**, depending on its lifecycle stage and correspondence status.

Import and Export Bank Accounts (WDA_FCLM_UPLOAD_DOWNLOAD (APP_CFG_FCLM_BAM_UPLOAD_DOWNLOAD))

Application Type: Web Dynpro

In bank account management (BAM), this Web Dynpro–based app helps you to migrate bank account data across systems. The app allows you to transfer bank account master data from both SAP and non-SAP systems into SAP S/4HANA Finance for cash management. In addition, the app helps with data migration, allowing you to import and export bank account details, including general information, payment approvers, overdraft limits, and house bank account connectivity. The app provides predelivered templates to upload an XML spreadsheet for master data, while bulk processing allows mass data transfers to streamline account setup and updates.

> **Tip**
>
> Prior to using this app, you must activate the WDA_FCLM_UPLOAD_DOWNLOAD service via Transaction SICF.

Import Bank Account Balances (F5172)

Application Type: Transactional

This app allows you to manage bank account balances updated by imported end-of-day bank statements and offers features for manual entry or importing balances from spreadsheets. The app's primary function is to efficiently update end-of-day balances based on electronic or manual bank statements. Specifically, the app allows you to update bank account balances by importing data from external sources, typically spreadsheets. In addition, the app supports manual entry for specific accounts, initial balance setup during implementation, and handling non-automated statements for accounts without electronic processing. The app supports the following bank account balance types:

- Ledger balance
- Value date balance
- Available balance

In addition, the app complements electronic statements by addressing gaps or errors and provides a snapshot of account balances for cash position monitoring and liquidity forecasting. Once the app successfully imports bank data, the system stores data in table FQMET_BALANCE.

> **Tip**
>
> If there is already a balance for the bank account (in table FCLM_ACC_BS_CONF) for the same validity period, the app does not support importing additional balances.

Import Bank Services Billing Files (F3002)

Application Type: Transactional

The purpose of this app is to facilitate the import of bank fee data into the cash manager's system for analysis. The app allows cash managers to import bank services billing files, adhering to the ISO 20022 Bank Services Billing (BSB) standard, to load bank fee data into their system for analysis. Upon importing bank services billing files, the app displays imported data in a tabular column with key fields such as statement group ID, sender, recipient, statement ID, account number, from date, and to date. The app offers features such as importing BSB files to collect bank fee data from company's banks and deleting source files to erase imported data when needed. In addition, the app supports deriving information based on multiple configurations for different currencies.

> **Tip**
>
> For additional information on the BSB standard, refer to the ISO 20022 website: *https://www.iso20022.org/*.

Import Consolidation Master Data (F3924)

Application Type: Transactional

This app allows group reporting users to import consolidation-related master data into the consolidation system. You can download predelivered templates provided with the app or an existing file, update or modify the master data, and upload the completed file back into the app for validation and database import. The app supports three master data types, including consolidation group structure (assigning consolidation units with group-specific settings), consolidation unit master data (managed by fiscal year and reporting logic), and financial statement (FS) items with attributes alternatively maintained in the Define Financial Statements Items app (F3297).

In addition, the app supports the new group reporting logic for SAP S/4HANA 1909 or later releases; activation for earlier releases requires an SAP case opened with the FIN-CS-COR component with SAP Support. Other master data types like account numbers or cost centers must be imported using the Import Master Data for Consolidation Fields app (F3071). Also, you can customize table columns but must save changes as a new view to retain them across sessions.

Import Credit Account Hierarchies (F6020)

Application Type: Transactional

Credit account hierarchies are essential for credit management processes as they define relationships between credit accounts at both higher levels and lower levels, thereby helping businesses evaluate credit exposure and limits on a group level or at different account levels. This app allows credit controllers to manage credit account hierarchies by maintaining higher-level and lower-level credit accounts for business partners. You can add or delete relationships (relationship category **UKM001**)

between credit accounts, upload changes via predelivered XML templates, evaluate these changes before importing, and discard saved drafts within the app.

Tip

Prior to importing a credit account hierarchy using the app, you must ensure the hierarchy structure has been defined using the Manage Credit Accounts app (F4596). In addition, the app supports up to 10 levels of hierarchy.

Import/Export Validation Settings (F3663)

Application Type: Transactional

As part of SAP S/4HANA Finance for group reporting, this app allows administrators to export selections, validation rules, and methods to a spreadsheet or import them from another system, facilitating the transfer of validated settings from a quality system to a production system. During export, you can specify the items to include, along with their referenced objects, ensuring only active items are considered. Imported items are assigned a draft status and categorized as **New for Import** or **Duplicated**, requiring validation or resolution of conflicts before activation. In addition, you can activate items individually or through mass activation using the app's delivered template **Mass Activation for Selections, Rules and Methods**. Moreover, the app allows simulation and validation of imported settings, ensuring a seamless integration process.

Import Financial Statement KPIs (F6599)

Application Type: Transactional

With this app, you can import financial statement (FS) data for business partners, enabling creditworthiness estimation in credit management. Uploaded KPIs are accessible in the Manage Credit Accounts app (F4596) under **Financial Statement Data** and can serve as parameters for scoring calculations in the Manage Formulas in Credit Management app (F6575). The app's key features include uploading and overwriting KPIs, supporting Microsoft Excel (XLSX) file imports, downloading templates with existing data, evaluating changes before import, and discarding saved drafts. In addition, the app allows you to evaluate uploaded changes before finalizing the import, helping to prevent errors in financial statement data. Thus, the app helps ensure creditworthiness calculations remain accurate and aligned with credit management processes.

Import Foreign Exchange Rates (F2092)

Application Type: Transactional

With this app, finance and treasury teams can efficiently import and update foreign exchange (FX) rates from external sources, financial institutions, or market data providers into the SAP system. Upon importing exchange rate data, the app displays uploaded data in **All**, **Error**, **Warning**, and **Correct** sections for ease of troubleshooting errors. As a result, the app ensures accurate and up-to-date exchange rates for currency conversions, financial reporting, and global transactions. The app is especially beneficial for multinational companies, which typically operate in multiple operating currencies. The app helps you comply with regulatory requirements, financial reporting standards (IFRS, GAAP), and treasury policies. With this app, you can import foreign exchange rates from a file into SAP. A downloadable template is delivered with the app for editing with the most up-to-date foreign exchange rates data, and you can upload files containing a maximum of 1,000 data records. Also, the app provides features such as downloading and editing the template, uploading files, checking data correctness (e.g., errors and warnings), and importing correct and warning data into the system. The app also supports sending emails with a URL for recipients to

check the app with the same selection criteria and saving custom tiles with the current selection criteria as default settings.

Import FS Item Mappings (F3335)

Application Type: Transactional

This app allows general ledger accountants to manage or view mappings between financial statement (FS) items and general ledger accounts using its import and export features for group reporting. To create or modify mapping revisions, you can download the template in the app and input details like consolidation chart of accounts, mapping ID (to define or create mappings), mapping revision (to differentiate or create new revisions), general ledger chart of accounts, general ledger account numbers, and FS item IDs. Once the file is prepared, you can upload the completed file, check for errors, and import the data after resolving issues. In addition, if needed, you can navigate to the Map FS Items with G/L Accounts app (F3333) for additional edits by selecting the arrow icon on any mapping revision row.

> **Tip**
>
> The export feature in this app mirrors that of the Map FS Items with G/L Accounts app, with an added option to include unmapped general ledger accounts in the exported list. To utilize this feature, simply check the **With Unmapped G/L Accounts** box when downloading the spreadsheet template.

Import Group Journal Entries (F3073)

Application Type: Transactional

As part of SAP S/4HANA Finance for group reporting, this app allows you to post multiple manual journal entries simultaneously to adjust financial data in alignment with group requirements. By selecting the appropriate posting type such as unit-dependent adjustments, two-sided elimination, or group-dependent adjustments, you can download a predelivered relevant template file, populate it with required data, and upload it back to the app. After verifying the data, entries are posted to the database, updating tasks in the Data Monitor and Consolidation Monitor based on the selected document type. The app features include choosing specialized templates, managing detailed header and line-item data, performing validations, and accessing logs for each posting, ensuring efficient and accurate consolidation processes.

> **Tip**
>
> SAP recommends imported journal entries are initially added as drafts in the Post Group Journal Entries app (F2971), identified with the **Draft Entry Source: 1 (Imported from File)**. These drafts are only removed after successful posting of the entries or upon uploading a new file.

Import Market Data (F2610)

Application Type: Transactional

This app allows you to import market data, such as foreign exchange (FX) rates, security prices, interest rates, basis spreads, credit spreads, FX swap rates, FX volatilities, and security factors, from a file into SAP from various sources. The app provides various templates that can be downloaded to maintain data consistency and accuracy. Updated files with up to 1,000 data records can be uploaded with this app. Also, you can download category-specific templates, enter both master data and transactional data, and ensure correct category numbers are assigned. Once template files are prepared and uploaded, the app validates uploaded data, displays errors and warnings, and allows importing valid data into the system. In addition, you can send emails with a URL to the app and save tiles with custom selection criteria.

> **Tip**
>
> The app supports decimal separated by dots. If commas are used to separate decimal points, the app may display the data as invalid.

Import Master Data for Consolidation Fields (F3071)

Application Type: Transactional

As part of SAP S/4HANA Finance for group reporting, this app streamlines the import of multiple consolidation-related master data entries into the consolidation system in one step. You can download a predelivered template file listing available master data types, populate it with the necessary information, and upload the completed file. The app then validates the data and imports it into the database, ensuring efficient and accurate consolidation. The app delivers multiple predelivered master data type templates such as account number, billing type, chart of accounts, controlling area, cost center, country key, functional area profit center, and segment for segment reporting.

Import Memo Records 2.0 (F6124)

Application Type: Transactional

Memo records 2.0 are the next-generation planning items built using structured memo record types and categories in cash management. Memo records represent expected incoming and outgoing payments without actual financial accounting postings. This app replaces the classic Memo Records app (FF63). It allows you to schedule jobs to import memo records from XLSX files, exclusively supporting memo records 2.0 in cash management. You can download predelivered templates to prepare memo record data, ensuring all mandatory fields (marked with an asterisk *) are entered and specifying a valid memo record type. Each import job can only include memo records of the same type. The app's features include creating jobs for memo record imports and monitoring their statuses, such as **In Process, Uploaded Successfully, Upload Failed**, or **Obsolete**. For failed jobs, you can review log details to identify issues and mark them as obsolete using the **Mark as Obsolete** button if no further action is required.

Tip

To use this app, you must complete the following Customizing activities:

- Define Memo Record Types
- Define Number Range for Memo Records
- Define Number Range for Memo Record Imports
- Manage Field Status Groups for Memo Record (optional activity)

To enable notifications, the entry **FCLM_MR_ MMRD_IMPORT_NOTIF_PROV** in the Manage **Notification Providers** Customizing activity is required, with the **Is Active** indicator set.

Consult SAP Note 3496240 for additional troubleshooting tips.

Import Objects for Organizational Change (F5487)

Application Type: Transactional

This app offers you the chance to import objects, such as work breakdown structure (WBS) elements and orders, for reorganization; these objects then are accessible only via the Manage Organizational Changes app (F4567). Using delivered file templates, you can input data for objects to reorganize, ensuring all cells are formatted as text to avoid errors during import. In addition, columns must align with the template structure, as any additional columns render the file unusable. Once the template is prepared, you can save the file, browse for it, and import it into the app. The app displays errors or warnings from the import process directly, allowing

you to perform detailed review by selecting message types. After successful import, you can view the objects in the Manage Organizational Changes app, where you can manually adjust profit center assignments or overwrite data by importing another file.

Tip

This app is available through the Manage Organizational Changes app (F4567) only.

Import Supplier Invoices (F3041)

Application Type: Transactional

This app allows you to efficiently handle the import and processing of supplier invoices, thereby integrating supplier invoice data into the system. With this app, you can streamline the process of uploading multiple supplier invoices into SAP efficiently. This ensures accurate data processing and simplifies supplier invoice management while providing tools for error validation and corrections when needed. The app allows you to import multiple supplier invoices into the system manually. This is accomplished with delivered template file in Microsoft Excel (XLSX) or Google Workspace format options. Once the template is downloaded, you can enter the invoice information such as header data and general ledger account items and upload the completed file back into the app. After importing supplier invoices, the app presents data in the **Worklist** view with invoice status. In addition, you can then check for errors, set the status of invoices to complete, and post them. In addition, you can navigate to create supplier invoice within the Manage Supplier Invoices app (F0859) to review details, correct issues, or post the invoices. Also, the app includes an option to calculate input tax automatically. To support this, you must ensure the accounts used to post the invoice set with **Input Tax Account** (symbol <) on the **Tax Category** field in the general ledger account master record.

> **Tip**
>
> This app does not support uploading purchase order-based invoices with reference to purchase documents. In other words, the app is limited in supporting non-purchase order invoices only.

Incoming Sales Orders – Predictive Accounting (F2964)

Application Type: Analytical

The app offers the ability to integrate predictive accounting capabilities with incoming sales orders. As a result, the app presents businesses with the ability to forecast and predict the financial impact of sales orders on revenue and costs in realtime, enabling more accurate financial planning and better cash flow management. For example, the app presents data in a graphical format for revenue by customer group, revenue by product group, and revenue by profit center. This is helpful for companies to get comprehensive insights into how incoming sales orders affect the company's financial situation with complex order-to-cash (OTC) processes. This analytical app allows you to assess predicted revenues, cost of sales, margins, and sales deductions from incoming sales orders using predictive postings generated in the predictive ledger. The app uses predictive analytics and accounting rules to forecast financial outcomes from sales order fulfillment and provides tools to analyze predicted cost structures, deductions, and orders. Additional features include filtering data by time periods and various organizational, customer, product, and accounting attributes, drilling down into detailed dimensions and measures, and navigating directly to journal entries, sales orders, customers, and products from the list view.

In addition, this app relies on financial KPIs derived from nodes in a financial statement version. These KPIs are computed using semantic tags assigned to the nodes, with each application requiring a specific subset. A semantic tag is a short text that represents a key figure in the app. The following semantic tags are essential for this app:

- RECO_COS: Recognized COS
- RECO_REV: Recognized revenue
- REC_MARGIN: Recognized margin
- SALES_DED: Sales deductions
- BILL_REV: Billed revenue

> **Tip**
>
> Consult SAP Note 3254104, which outlines the performance release restrictions.

Interaction History (FKK_WDY_INT_HIST)

Application Type: Web Dynpro

This Web Dynpro–based app allows you to track and manage interaction history in contract accounting. Interaction history refers to logged details of business partner interaction within contract accounting such as inquires, approvals, and correspondence exchanges. You can search with parameters such as business partner and business partner contact. As a prerequisite, you must maintain appropriate authorizations to view history and maintain an approval hierarchy to support the workflow function.

Interest Rate Overview (F3098)

Application Type: Analytical

This app provides an overview of financial risks related to debt and investments, including maturity profiles, total debts and investments by key date, and breakdowns by interest category and reference interest rate. Treasury users can analyze KPIs displayed as separate cards, such as **Current Interest Rate**, **Historical Interest Rates**, **Yield Curves**, and **Debt/Investment Maturity Profiles**. Default values for parameters like key date, display currency, and exchange rate type can be set for streamlined use. In addition, the app allows filtering by various categories, navigating to target apps for detailed insights, and customizing the overview page by managing and rearranging cards. It leverages the following CDS views to ensure robust data analysis and enhances financial risk management with tailored flexibility:

- C_MaturityProfile
- C_DebtInvestmentNominal
- I_DebtInvmtFixedVariable
- C_DebtInvmtRefInterest
- C_RefInterestRateHistory
- C_RefInterestRateRank

Invoice Processing Analysis (S/4HANA) (F1745)

Application Type: Analytical

With this app, you can get comprehensive insights into invoice processing activities. This app allows you to view the total amount of posted invoices and the total number of posted line items. The app allows you to analyze this data by month, supplier, user, and processing status, including **Free for Payment**, **Cleared**, **Blocked**, and **Parked**. In addition, the app supports posting comments via SAP Jam, sending emails, and allows caching for better performance. It uses CDS view C_APINVOICEPROCGANALYSIS for data processing.

Tip

The net due date calculation includes the grace period for outstanding invoices. This calculation determines when a payment is officially due and considers various parameters such as baseline date and payment terms, including a grace period configured in the system.

Transaction Code

The corresponding transaction in SAP GUI is Transaction FBL1N.

Invoicing Overview for Contract Accounts (F2473)

Application Type: Transactional

With this app, you can display an overview of documents in convergent invoicing related to specific contract accounts. You can search for invoicing data using fields such as business partner, contract, and invoicing category. The app presents results in multiple sections such as invoicing documents, billing documents, billing plans, billing items, and consumption items. The app provides you with features that include searching for and sorting contract accounts using filter fields, as well as displaying and navigating detailed information about invoicing documents, billing documents, billing plans, billable items, consumption items, credit and debit memos, clarification cases, charges and discounts, and provider contracts. In addition, you can filter the list, create filter variants, save lists or specific memos as tiles, and share details by email or Microsoft Teams. The available navigation options depend on your roles and authorizations assigned.

Journal Entries by Contract (FARR_RECON_FI_USER)

Application Type: Web Dynpro

This app provides a view of journal entries linked to revenue contracts in revenue accounting and reporting (RAR). The app lets you perform contract-level tracking of journal entries, helps with detailed financial analysis, and detects discrepancies for corrective action. You can search for journal entries linked to revenue contracts using parameters such as company code, fiscal year, posting period, revenue accounting contract, and customer. The app complies with IFRS 15 and ASC 606 accounting standards.

> **Tip**
>
> In addition, to use this app, you must activate the Internet Communication Framework (ICF) security service for FARR_RECON_FI_USER via Transaction SICF.

Journal Entry Analyzer (F0956)

Application Type: Web Dynpro

This app allows you to analyze journal entries posted in a selected period, providing an overview of aggregated amounts on general ledger accounts. Entries can be grouped by fields such as company code, fiscal year, posting date, business area, functional area, or profit center. The app lets you view journal entries by company code, fiscal year/period, ledger, and general ledger account number, sorted by company code, general ledger account number, and journal entry number. You can see additional information such as master data attributes or texts, and data can be grouped by external hierarchies. Also, you can specify a key date for time-dependent master data, drill down into entries for more details, and compare data in a pivot-style format. The app is based on CDS views I_GLAccountLineItemCube and C_JournalEntryItemBrowser to display results.

Liquidity Forecast (F0512A)

Application Type: Analytical

Businesses manage cash flow by looking at insights into expected incoming and outgoing payments. This app allows cash managers and cash management specialists to forecast liquidity trends for the next 90 days, with options to filter and drill down by various dimensions. With this app, you can display aggregated forecast cash flows and closing balances by company code and by liquidity item, filtering data by calendar day

and planning group, navigating to detailed cash flow items, switching between charts and tables, and monitoring past daily actual cash flows. In addition, the calculation logic is straightforward and can be easily followed. The numbers displayed on the app tile are identical to those in the top-right corner of the app. The forecasted amount is displayed in the bank account currency and is derived from transaction data provided by memo records and the One Exposure from Operations hub. Also, the app features caching, which improves performance, with data refreshable via an icon and automatically updated based on cache duration settings.

> **Tip**
>
> The app pulls data from relevant sources such as accounts payable, accounts receivable, bank statements, and so on. Therefore, you must ensure data sources are configured correctly.

> **Transaction Code**
>
> The equivalent transaction in SAP GUI is Transaction FF7A.

Liquidity Forecast Details (F0741)

Application Type: Analytical, Web Dynpro

This app provides a comprehensive overview of projected cash inflows and outflows, broken down by liquidity items, enabling cash managers to precisely see the source and destination of future cash by various dimensions. The app provides capabilities such as displaying opening balance, net cash flows, and closing balance; setting transaction currency and date for the overview; showing net cash flows in transaction and display currency; and adjusting display currency, liquidity item hierarchy, and date. Uses can search for forecast details using parameters such as start date, end date, and company code. The app displays results in three tabs: **Data Analysis**, **Graphical Display**, and **Query Information**.

In addition, the app's drilldown capabilities provide granular insight into the transactions comprising each liquidity item, highlighting potential risks and opportunities. Also, you can make comparisons with previous forecasts, and planned figures facilitate trend identification and forecast accuracy assessment. The app also allows you to customize the dimension layout by switching rows and columns and to switch between charts and tables for display mode.

> **Tip**
>
> As a prerequisite for F4 help to work, the app requires liquidity item hierarchies to be maintained via the Manage Liquidity Item Hierarchies app (F4966). In addition, you must be assigned the SAP_BR_CASH_MANAGER and SAP_BR_CASH_SPECIALIST roles. Consult SAP Note 3555551 for further details.

> **Transaction Code**
>
> The equivalent transaction in SAP GUI is Transaction FF7B.

Liquidity Forecast Details – Overview (W0125)

Application Type: Analytical, Web Dynpro

This app provides an overview and detailed breakdown of forecasted liquidity amounts across various dimensions in cash and liquidity management. The app allows you to display opening balances, net cash flows, and closing balances while setting transaction currency and date parameters for the overview. Uses can search for forecast details using parameters such as start date, end date, calendar day (hierarchy), and company code. The app displays results in three tabs: **Data Analysis**, **Graphical Display** and **Query Information**. In the **Detailed View**, you can configure display currency, liquidity item hierarchy, and date settings, as well as adjust the dimension layout by rearranging rows and columns. The **Graphical Display**

and **Query Information** tabs provide visualization and filter tracking. In addition, the app supports switching between **Chart View** and **Table View** for enhanced data visualization.

> **Tip**
>
> For troubleshooting various issues with liquidity forecast details, consult knowledge-based SAP Note 3438687.

> **Transaction Code**
>
> The corresponding transaction in SAP GUI is Transaction FF7B.

Maintain Decision Table (FARR_BRF_UI_LAUNCHER)

Application Type: Web Dynpro

This app allows you to configure and manage decision tables for revenue recognition in revenue accounting and reporting (RAR). The app supports defining, modifying, and validating business rules while ensuring compliance through consistency checks. In addition, you can import and export decision tables to spreadsheets for offline analysis, customize condition columns and column order for tailored rule configurations, and leverage Business Rule Framework plus (BRF+) for structured rule maintenance and execution. BRF+ provides a structured framework for rule maintenance and storage, utilizing an integrated BRF+ decision table expression. Using BRF+, you can check rules for consistency and errors, import and export decision tables into spreadsheets, and modify table settings by adding or removing condition columns or adjusting column order.

Maintain Dunning Notices (F3678)

Application Type: Transactional

Dunning is the procedure for reminding customers about overdue payments. With this app, you can manage the dunning notices process by generating several correspondences for different categories of dunning notices. Once dunning notices are generated, the app allows you to display details of the dunning notice on a page. Dunning notices are displayed in the **Details**, **Dunning Items**, **Dunning Activities**, **Dunning Reductions**, and **Correspondence** sections. The app allows manual intervention in dunning processes by recording all dunning data for each dunned item for outstanding debts. The app's functionality includes displaying dunning notices and their items, setting the next dunning level for specific items, and reversing dunning notices.

Moreover, the app offers to navigate to other relevant apps such as Display Dunning History (F3679) and Schedule Dunning Runs (F3676).

Transaction Code

The corresponding transaction in SAP GUI is Transaction FPM3.

Maintain Employee List (FITV_EMPLOYEE_LIST)

Application Type: Web Dynpro

This Web Dynpro–based app allows travel administrators to maintain employee list, which is later used in travel management. The app allows you to update employee personnel details, assign travel roles, filter employees for expense handling, define approval hierarchies, and enforce role-based authorization for travel approvals. With this app, administrators can create, modify, or remove employees, or submit expenses to other people in the hierarchy. As a prerequisite, employees must be created as business partners with an employee role. In addition, administrators can search for specific employees by applying various filters, such as cost center and personnel number.

Maintain Hierarchy (WDA_FCLM_BAM_HIER_MAINTAIN (APP_CFG_FCLM_BAM_REQOVERVIEW))

Application Type: Web Dynpro

This app allows businesses to define and manage hierarchical relationships between bank accounts within the bank account BAM) framework. You can create and modify hierarchical structures based on business needs, apply custom views and filter for financial management, and integrate with cash and liquidity management. The app also supports bulk account management and maintains audit compliance by recording changes.

Tip

To use this app, you must activate the WDA_FCLM_BAM_HIER_MAINTAIN service via Transaction SICF.

Maintain Master Data ID (F4522)

Application Type: Transactional

This app allows you to maintain master data IDs for dependent items, facilitating the inference of relevant master data required to create dependent items. Each master data ID must specify at least a business partner and/or contract account. The app's features include mapping business partners or contract accounts to master data IDs to ensure accurate inference for billable item creation and adding additional data as needed for dependent items. In addition, the app allows sharing master data IDs or lists via email or Microsoft Teams Chat.

Maintain Number of Units for UoP Depreciation (F4907)

Application Type: Transactional

This app supports units-of-production depreciation by allowing asset accountants to create usage objects, assign fixed assets to them, and manage their depreciation based on production output. Usage objects represent entities like an oil field, where all associated assets are depreciated proportionally to the expected total production and actual output per period. For example, the expected total production amount, referred to as the total number of units, represents the forecasted output over a defined period. The actual amount of oil extracted during a specific period, known as the actual number of units per period, reflects the realized production within that time frame.

The system calculates the remaining number of units by subtracting the cumulative number of units from previous time intervals up to a specific date from the total number of units valid on that date. This is calculated using predefined depreciation keys: **UOPT** (depreciation over the total number of units) and **UOPR** (depreciation over the remaining number of units). The app provides features for creating and managing usage objects, maintaining unit data for custom time intervals, assigning assets via manual entry or API, and viewing change documents. Lifecycle statuses such as **Created**, **Active**, and **Deactivated** govern the status and assignment of usage objects and their assets.

Tip

During the transfer of legacy data, if the system recalculates accumulated past depreciation using units-of-production depreciation, SAP recommends you input the number of units for the period starting from the acquisition date of the oldest asset up to the point of legacy data transfer.

Maintain Payment Approver (F1372)

Application Type: Web Dynpro

With this app, you can maintain a payment approver in multiple bank accounts. The app lets you replace a payment approver, add a payment approver, and revoke the authorization of a payment approver by changing the validity

period across multiple bank accounts. The app also supports using SAP Business Workflow to control these operations. This app is essential for implementing internal controls and ensuring that payments are properly authorized before being processed. In addition, it allows the setup of approval hierarchies and rules, automating the routing of payment requests to the appropriate approvers.

Maintain Payment Blocks (F3650)

Application Type: Transactional

This app allows payable specialists to display, create, edit, and delete payment blocks for currencies, banks, and countries/regions, effectively preventing payments from being processed. You can apply payment blocks currency blocks, country blocks, and bank blocks. For bank blocks, for example, you can apply payment blocks using critical fields such as bank country, bank key, BIC, valid from, and valid to. During enrichment and validation, payments are checked against these blocks. If a payment block is detected, the payment is halted and sent to the Repair Payments app (F3651), where it can either be rejected (blocked) or processed further.

Make Bank Transfers (F0691)

Application Type: Transactional

This app allows you to make transfers between bank accounts in your company by creating a payment request for each transfer. You can make bank transfers by providing important information such as amount, transfer date, payment method, note, from, and to. Depending on the company's process, you can either create and clear the payment request in one action or create, release, and clear it in separate steps. The app's features include initiating new transfers; entering bank account IDs, transfer dates, payment methods, amounts, and reference texts; releasing and clearing payment requests; creating transfers with templates; checking transfer details and application logs; and using fuzzy search for transfers. In addition, payment requests can be released and cleared automatically or by other users. Also, when you select a cleared bank transfer, the app allows you to navigate to display associated clearing document.

Tip

SAP supports transfers using currencies other than the bank account currency. In addition, the app supports only the exchange rate type M (average rate) for converting payments in foreign currency to the bank account currency. SAP recommends using the exchange rate type M and maintaining the corresponding exchange rates.

Transaction Code

The corresponding transaction in SAP GUI is Transaction FRFT_B.

Make Bank Transfers – Create with Templates (F3760)

Application Type: Transactional

This app allows you to create single or multiple bank transfers in a batch using predefined templates. Thus, the app streamlines recurring payment processes. Prior to use, templates must be set up in the Define Bank Transfer Templates app (F3759), and required configurations for bank transfers should be completed as outlined in the Make Bank Transfers app (F0691) prerequisites. The app's features include fuzzy search for locating templates and the ability to generate transfers by selecting a template, entering the amount and necessary details, and saving the entries.

Transaction Code

The corresponding transaction in SAP GUI is Transaction FRFT_B.

Manage Accounting Notifications (F7289)

Application Type: Transactional

This app lets you track and handle accounting-related alerts within accounting processes. The app allows you to monitor accounting notification statuses, such as payroll posting notifications received from SAP SuccessFactors Payroll, project edition. In addition, the posting logic in accounting varies based on the business transaction type. If a notification **Status** is **Failed** or **Not Posted**, you can select the affected items and trigger postings directly within the app. The app's functionalities include searching for accounting notifications, verifying journal entry postings for payroll items, reviewing payroll posting details in the worklist and on the details page, monitoring failed postings, and initiating postings for unresolved notification items.

Tip

Users are recommended to evaluate failed or unposted notifications to prevent delays in postings. For recurring postings, you can consider setting up automated monitoring for important alerts. Thus, you can intervene to avoid delays in postings.

Manage Asset Assignments to Purchase Orders (Central Finance) (F7892)

Application Type: Transactional

This app provides a worklist displaying replicated purchase order line items related to asset acquisition, along with their assignment status and acquisition posting status in Central Finance. Asset accountants use this app for period-end closing to review purchase order items and handle those with **Incomplete** or **Error** statuses. You can filter by default statuses (**Assignment Incomplete**, **Assignment Unconfirmed**, and **Error**) and navigate to the **Object Page** for detailed information on purchase order items, assigned assets, posted journal entries, and SAP Application Interface Framework error messages. During replication, configuration inconsistencies may occur due to company code, material group, or account assignment category changes that impact asset acquisition. You can create assets via the **Create Asset** function, assign or change assets for purchase order items, and confirm asset assignments after configuration changes. The app also allows you to mark purchase items as not relevant, preventing additional asset postings unless an acquisition has already occurred.

You can track changes using the **Display Change Document** function and navigate to apps such as Accounting View of Purchase Order (F4700), Manage Fixed Asset (F1684), Manage Journal Entry (F0717), and SAP Application Interface Framework messages.

> **Tip**
>
> As a prerequisite to use this app, central asset accounting must be activated for relevant company codes by following Central Finance IMG menu path **Central Finance • Central Finance: Target System Settings • Central Asset Accounting**.

Manage Assignment Proposals for Dispute Cases (F5999)

Application Type: Transactional

In SAP, proposals for dispute cases streamline the dispute resolution process by identifying potential matches between open items and dispute cases. This app allows you to display, edit, and assign assignment proposals of dispute cases to open items, based on proposals previously created using the **Create Proposal for Assignment of Open Items to Dispute Cases** job template. The app retrieves data from the latest assignment proposal, enabling you to revise

assignments, check for similar open items or matching dispute cases, and assign revised dispute cases to the open items.

Manage Automatic Payments (F0770)

Application Type: Transactional

This app allows you to facilitate the scheduling of payment proposals and direct payments, providing an overview of their status. It identifies overdue invoices and verifies complete payment information. You can create, copy, display, edit, and delete payment parameters; schedule proposals or payments directly; automatically generate payment media viewable in the Manage Payment Media app (F1868); and create payment advice. You also can search for existing payment runs using fields like run date, ID, created by, and company code. The app presents multiple **Standard** tabs such as **Parameters Created**, **Proposal Processed**, and **Payment Processed** for easier navigation. Integration with SAP Business Integrity Screening helps identify invoice irregularities. In addition, direct debit pre-notifications can be created, deleted, displayed, and reset. In addition, search functionality is available for proposals and payments. Also, you can view payment proposals and payment lists, check logs for scheduling validity, view payment items in a predefined currency, and access bank account details. Moreover, exceptional payment runs can be reset and rescheduled. SAP Jam integration allows for commenting, and email functionality is included.

> **Transaction Codes**
>
> The equivalent SAP GUI transaction is Transaction F110. The payment runs can be displayed with Transaction FBZ8. Consult SAP Note 2863917 when generating payment advice as the behavior varies between the Manage Automatic Payments app and Transaction F110.

Manage Balance Validation Rules and Groups (F6386)

Application Type: Transactional

As part of financial close and group reporting, this app allows you to display, modify, and create validation rules and rule groups for balance validation. The app's features include creating basic and compound rules, simulating validations, assigning validation rules to groups, and activating or deactivating rules and groups. Thus, the app helps you to organize rules for verifying the accuracy and consistency.

Tip

To gain write access, you need a business role containing the General Ledger – Validation Configuration (SAP_SFIN_BC_GL_VAL_CONF) business catalog. On the other hand, users with the General Ledger – Validation Results (SAP_SFIN_BC_GL_VAL_RESULT) business catalog has read access and can navigate from the View Balance Validation Results (F6387) app to view rule or group details.

Manage Bank Account Balances (F5175)

Application Type: Transactional

This app offers you the chance to manage bank account balances by viewing updates from imported end-of-day bank statements, manually entering balances, or importing them via spreadsheets. The app displays balances by reading table FQMET_BALANCE. If the balances are stored by bank statement dates already, then entries updated electronically cannot be modified manually or through spreadsheet imports. In addition, the app allows you to review balance history for bank accounts. The following bank account balance types are supported by this app:

- **Ledger balance**
  Represents the closing balance provided by banks in their statements. Supports the Manage Bank Account Balances app.

- **Value date balance**
  This represents the interest-bearing closing balance for a specific day, calculated based on the ledger balance. The value date balance is derived from the day's ledger balance by excluding bank statement items with value dates later than the bank statement date, including items from previous bank statements. Supports several key apps: Manage Bank Account Balances, Define Cash Position Profiles, Short-Term Cash Positioning, and Foreign Bank Account Report.

- **Available balance**
  This reflects the funds readily accessible to the account holder for immediate use. Supports several key apps: Manage Bank Account Balances, Define Cash Position Profiles, and Short-Term Cash Positioning.

Tip

For additional troubleshooting or to learn about the required Customizing activities to support this app, consult SAP Note 3334466.

Manage Bank Account Hierarchies (F4973)

Application Type: Transactional

This app lets cash management users create and edit bank account group hierarchies and their validity time frames. The app allows quick creation of new hierarchies based on existing ones to accommodate structural changes or evolving reporting needs. In addition, you can expand hierarchies by adding levels or importing nodes from other hierarchies. The app's functionalities include editing hierarchies in **Draft** or **Active** status, leveraging tools like fast entry, copy-paste, and drag-and-drop for efficient adjustments, and exporting hierarchies to spreadsheets for further use. Once bank hierarchies are created, the defined bank hierarchy can be maintained with the Manage Global Hierarchies app (F2918).

Manage Bank Accounts (F1366A)

Application Type: Transactional

Both house banks and house bank accounts are master data required to transact treasury transactions. The house banks of a company are where your company holds bank accounts. With this app, you can get an overview of bank accounts and maintain bank account master data according to your business needs. The app offers functionality such as searching for bank accounts using keywords and filters, creating new accounts, checking and editing bank account details, and copying, closing, and deleting inactive accounts and drafts all at once. The app presents standard views such as **Bank Hierarchy**, **Bank Account Group Views**, **Account List**, and **House Bank Account List** for easier navigation. House bank ID (ID category) and general ledger account are maintained under the **House Bank Account Connectivity** tab, which defines the connectivity path. In addition, you can create documents or URL attachments and search for attachments.

The app supports SAP Business Workflow or dual control for managing account processes, sending emails with URLs pointing to selected information, posting comments in SAP Jam groups, and creating tiles with default selection criteria.

Tip

Bank accounts with active workflow processes are not allowed for deletion. SAP does not recommend uploading personal data. The document management system (DMS) serves as the foundation for the bank account attachments. As a result, the Information Retrieval Framework (IFR) does not support retrieving personal data for documents based on the DMS.

Manage Bank Accounts – Bank Hierarchy View (F1366)

Application Type: Web Dynpro

The app allows you to create, change, display, and block general ledger accounts for business needs. The app provides a centralized view of general ledger account master data, including account number, description, account type, and other relevant attributes. As a result, this app streamlines the management of general ledger accounts, ensuring data consistency and accuracy in financial reporting. With this app, you can check bank accounts in the **Bank Hierarchy** view and self-defined views, as well as view bank account groups created before SAP S/4HANA 1909 for legacy data compatibility.

Manage Bank Chains (F4004)

Application Type: Transactional

This app allows accounts payable accountants to create and edit bank chains, which facilitate payments through up to three intermediary banks before reaching the beneficiary bank. The app allows the determination of intermediary banks based on predefined factors such as house bank, supplier or customer bank details, currency, and payment method supplement. In addition, you can filter existing bank chains using multiple filter criteria such as SWIFT code, IBAN, and business partner. During payment runs for open items or treasury and cash management requests, the system determines bank chains based on system settings and master data in the system. In addition, any changes to intermediary bank entries during payment proposal editing trigger redetermination of the bank chain, which is then displayed on screen. Typically, bank chains are used only for payment methods requiring them and are included when creating payment media. However, payments like checks do not utilize bank chains.

Tip

SAP recommends this app is specifically used for maintaining vendor-specific bank chains. In such cases, you must define the recipient's bank country, bank key, bank account, and specify the payment currency according to your needs. Previously, payments to foreign business partners required specifying the house bank and business partner's bank, leaving intermediary bank determination to the house bank. With this app, you can now define the complete bank chain. This results in faster payment processing and significant cost savings by reducing bank charges.

Transaction Code

The corresponding transaction in SAP GUI is Transaction FIBPU.

Manage Bank Fee Conditions (F3185)

Application Type: Transactional

As part of cash and liquidity management, cash managers streamline the management of bank fee conditions for specific bank services on a day-to-day basis to prevent errors or improper charges. With this app, you can create, edit, and delete conditions that define how bank fees are charged for specific bank services. The system uses these specific conditions to validate imported bank fee data, identifying errors or improper charges. Once conditions are defined, they can be assigned to bank services and validated in the Monitor Bank Fees app (F3001). The app's additional features include sending emails with URLs to selected information and creating tiles that save default selection criteria for easier access.

Tip

The app supports defining unit prices with standard decimal places. If you require unit prices of bank fee conditions with more decimal places, consult SAP Note 3291142.

Manage Bank Statement Reprocessing Rules (F3555)

Application Type: Transactional

This app allows you to create and manage processing rules for performing general ledger and accounts payable/accounts receivable account postings. The app's features include defining conditions and actions for processing rules, sharing manually applicable rules with colleagues, deprecating outdated rules, and using rules to automatically process payment advice. In addition, you also can download and upload rules across systems, automate rules for background processing of bank statement items, and manage templates for processing rules.

Tip

To display all conditions of a reprocessing rule in this app, you must have authorization object F_RP_BUKRS for respective company codes.

Manage Bank Statements (F1564)

Application Type: Transactional

This app allows you to manage both manual and electronic bank statements, providing an overview of all bank statements for house bank accounts and detailed information for each statement. You can search for bank statements using fields such as editing status, house bank, house bank account, bank statement date, company code, and statement status. The app presents bank statements in a grid format with the flexibility of adding additional dimensions. In addition, the app is enabled with features such as creating, editing, and posting manual bank statements; obtaining an overview of all bank statements for house bank accounts; posting bank statements with multiple items as a background process; searching for bank statements; uploading bank statement items from a spreadsheet file; deleting bank statements; displaying intraday bank statements; and reversing bank statements.

Tip

The app does not support deleting accounting statements unless the posting phase has not occurred yet. This is determined by status in table FEBKO-ASTAT. Allowed status are blank value (received), A (aborted), or 2 (saved).

Transaction Code

The corresponding transaction in SAP GUI is Transaction FF.6.

Manage Banks – Cash Management (F1574A)

Application Type: Transactional

With this app, you can display existing banks and add crucial data like house banks, business partners, and bank service mappings essential for cash management and payments. The app displays basic bank information (such as names, SWIFT codes, and addresses) for managing both standard and internal banks and supplementing banks with corporate data for cash management

and payment activities. In addition, the app displays important attributes, which include house banks, business partners (e.g., risk, netting, and contact persons), and bank service mappings for fee validation.

Authorization objects F_BNKA_MAO and F_CLM_BNK must be added to your role to work with standard banks. On the other hand, authorization object type F_CLM_IBNK is required to work with internal banks. For house banks, authorization object F_BNKA_BUK is required.

Tip

SAP recommends using the Manage Banks – Master Data app (F6437) to maintain basic bank information, and the Manage Bank Accounts app (F1366A) to view or edit bank accounts.

Manage Banks – Master Data (F6437)

Application Type: Transactional

This app allows you to display and manage master data for banks used by your company, business partners, customers, and suppliers. The app allows you to view key bank details, including account description, bank country/region, account status, currency, SWIFT/BIC code, bank number, and address information. In addition, you can create and edit both standard and internal banks, as well as international address versions. Additional cash management features are also available in the Manage Banks – Cash Management app (F1574A), which you can navigate to from this app. The app supports bank replication to other systems using SAP Master Data Integration and allows bank data import from external files through the Import Bank Directories app (BIC2S).

This is the successor of the deprecated Manage Banks app (F1574).

Manage Billing Plans (F2824)

Application Type: Transactional

The app is designed to provide an intuitive way to create, edit and monitor billing plans for customers. With this app, you can create, change, or display billing plans to schedule invoice amounts for regular and non-recurring payments. In addition, you can search, sort, and filter billing plans, create new ones, change their status, and delete them if permitted. For example, you can search for billing plans using parameters such as editing status, billing plan, business partner, contract account, status, valid from, next request, and valid to. The app presents billing plans in a tabular format with the flexibility of adding additional fields. Also, you can navigate specific billing plans to review details, add or change items, create adjustments or follow-on items, check items for correctness, and view pricing conditions. The app also supports filtering lists, hiding the filter bar, adjusting columns, adding attachments, and sharing billing plans via email or Microsoft Teams.

Transaction Code

The corresponding transaction in SAP GUI is Transaction FKKBIX_BILLPLAN.

Manage Business Partner Items (F2562)

Application Type: Transactional

This app allows you to manage business partner items and track, monitor, and manage open and cleared financial transactions related to customers and suppliers. On the initial app screen, you can restrict items according to various selection criteria and sort the list for easier follow-up actions. For example, you can search for business partner items using parameters such as business partner, contract account, document number, posting date, clearing status, amount, clearing reason, net due date, and statistical key figure. The presents result in a tabular column with **Change**, **Clear Write off**, **Reset Clearing**, and **Reverse** functions. In addition, you can use these features to change, write off, reverse, or clear selected items, with clearing only possible if the receivables or credit balance to zero. Also, you can reset clearing for selected items.

To access this app, you must have the relevant user roles and authorizations assigned to your user ID. For example, posting documents requires a reconciliation key, which is automatically proposed if assigned in the backend system via Transaction S_KK4_74002323.

Manage Cash Concentration (F3265)

Application Type: Transactional

This app allows cash managers to manage efficient cash concentration for cash pools, helping their company manage cash balances centrally. As a result, businesses can manage and track their cash concentration processes efficiently. This results in enhanced management of cash flows between various bank accounts and ensures that funds are moved efficiently to a central account. For cash pools with the service provider **Enterprise**, the app supports transfers between header accounts and subaccounts based on system proposals, along with simulation details for child cash pools for service provider **Bank**. For cash pools operated by banks, the app generates memo records to simulate cash movements. In addition, the app allows filtering by criteria such as concentration date and displays cash concentration details on a list page with application logs accessible via concentration IDs. As a result, the app enhances cash management by consolidating processes effectively.

Tip

Alternatively, the Schedule Jobs for Cash Concentration app (F3688) allows you to set up background jobs for cash concentration.

Consult SAP Note 3508926 on frequently asked questions related to this app. In addition, if you observe reconciliation differences between the Cash Flow Analyzer and Manage Cash Concentration apps, consult SAP Note 2978537 for troubleshooting tips.

Manage Cash Pools (F3266)

Application Type: Transactional

This app allows you to create, delete, and display cash pools, with the master data later utilized in the cash pooling feature to centralize total cash

balances and enhance liquidity management. The app's key functionalities include the following:

- Creating cash pools by entering details like description, pool type, and service provider
- Assigning header accounts and subaccounts through the Manage Bank Accounts app (F1366A)
- Viewing a list of cash pools to check individual details
- Filtering pools by criteria such as pool usage
- Deleting unused cash pools, which remain viewable in the list even after deletion

Manage Cash Pools (Version 2) (F3266A)

Application Type: Transactional

This updated app allows you to define, display, modify, and delete cash pools, comprising a header account and one or more subaccounts assigned for specific validity periods. You can create multicurrency cash pools by assigning subaccounts with different currencies and build multilevel cash pool hierarchies by adding a header account already used as a subaccount in another pool.

In addition, the app integrates with the Display Cash Pool Hierarchies app (F6123) to showcase cash pools in hierarchical order. On the other hand, the app integrates with the Short-Term Cash Positioning app (F5380), which allows analysis of cash positions using cash pool data. Also, the Manage Cash Pools app also supports the creation of classic cash pools for regular bank accounts and in-house cash pools for in-house bank accounts tied to advanced payment management.

This revised version brings enhanced features and better performance over the previous app, Manage Cash Pools (F3266).

Tip

Consult SAP Note 3440631 for known issues and possible resolutions.

Manage Chart of Accounts (F0763A)

Application Type: Transactional

The app allows you to display general ledger accounts based on their assignment in a financial statement version (FSV) or by the chart of accounts ID. In the FSV view, you can either display accounts in hierarchy or list view. The app presents chart of accounts data with **All Accounts**, **Balance Sheet**, **Profit & Loss**, **Unassigned Accounts**, and **Notes** sections. In addition, you can create an account by copying from an existing account, extend it to a new company code, and choose the FSV based on the assignment. Moreover, the app allows you to display all accounts, balance sheet accounts, profit and loss statement accounts, unassigned accounts, and accounts with zero balance (**Notes** tab). With this app, you also can search accounts by account number or account name, edit account attributes, and copy existing accounts or FSV assignments to create new ones.

Transaction Code

The equivalent transaction in SAP GUI is Transaction F.10.

Manage Checkbooks (F1577)

Application Type: Transactional

With this app, accounts payable accountants and account managers can monitor and manage checkbook usage within their organization efficiently. The app offers a comprehensive overview of the status of various checkbooks, including alerts for checkbooks nearing depletion, and allows immediate action on this information. You search for checkbooks using search criteria such as paying company code, house bank, bank account ID, checkbook ID, and checkbook name.

The app presents information in multiple sections such as **New**, **Normal**, **Depleted**, and **Exhausted**. The app offers key features such as creating, monitoring, editing, deleting, and splitting checkbooks; defining successor checkbooks for continuous operation; and reordering checkbooks. In addition, the app supports email integration and offers an export to spreadsheet option for streamlined reporting and analysis.

Transaction Codes

The corresponding transactions in SAP GUI are Transactions FCH1, FCH2, and FCH3.

Manage Checkbooks for China (F7015)

Application Type: Transactional

This app allows organizations to monitor and manage checkbook usage, providing a clear overview of checkbook statuses, including alerts for those nearing depletion. The app is available for accounts payable accountants and accounts payable managers and includes features such as creating, editing, deleting, reordering, and splitting checkbooks. In addition, you can define successor checkbooks to activate upon depletion, ensuring continuity in payment processes. You can filter checkbooks using search criteria such as payment company code, house bank, bank account ID, checkbook ID, and checkbook name. In addition, the app provides features such as email integration and an export-to-spreadsheet option for streamlined data management and reporting. A China-specific extension is available for localization.

Transaction Code

The corresponding transaction in SAP GUI is Transaction FCH3.

Manage Collection Contact Persons (F4708)

Application Type: Transactional

This app allows accounts receivable users to manage contact persons for business partners, which are later utilized in the collection process through the Process Receivables (Version 2) app (F0106A). To create a new contact person, open the business partner in the Process Receivables (Version 2) app and select **Create** in the **Collection Contacts** section. The app's features include searching and filtering contact persons by criteria such as business partner or collection segment, assigning default contact persons to business partners with multiple contacts, and navigating individual contact persons to manage their data.

Manage Collection Strategies (F2946)

Application Type: Transactional

Collection strategies in a company involve defining strategies and processes for collecting overdue payments from customers. This helps businesses manage cash flow and reduce bad debts across enterprise wide. This app helps you manage and customize collection strategies to meet your business needs. It allows you to prioritize items in a collection worklist, specify the currency for displayed items, and define aging intervals for sorting open receivables. You can search for collection strategies using parameters such as editing status, strategy currency, and created by user. In addition, you can copy and edit the base collection strategy, modify valuations, prerequisites, and conditions for collection rules, and adjust aging intervals. Also, the app offers additional features that include excluding items in legal processes, factoring in cash discounts, setting tolerance days for dunning notices, and managing the assignment of strategies to collection groups. It also lets you simulate a collection strategy for a business partner while keeping your primary strategy active.

Transaction Code

Although this task can be performed using SAP GUI via Transaction UDM_STRATEGY, the app offers a modern user-friendly interface to customize managing collection strategies by an end user.

Manage Collections Emails (F6120)

Application Type: Transactional

This app helps with email communication as part of the collections management process, allowing you to improve customer interactions by including various attachments such as billing documents and other supporting documentation. To facilitate collections emails, the app provides you with an intuitive layout with sections such as **Header**, **Message Text**, **Correspondence**, **Linked Objects**, and **Attachments**. The app is useful especially for situations where direct contact is unavailable, or time constraints prevent discussing all items. In addition, you can resend payment requests for all or selected open items, adapt email recipients, add CC and BCC recipients, use language-specific templates, manually edit message text, attach financial correspondences, include linked journal entry items in a single table, upload additional documents or links, and compress attachments into a ZIP file.

Tip

To send emails, you must complete correspondence types configuration. Consult SAP Note 3448054 for further guidance.

Manage Collections Master Data (F5400)

Application Type: Transactional

With this app, accounts receivable users can manage customer account attributes that are essential for controlling the collection of outstanding receivables. The app's features include displaying collection master data for business partners, modifying collection profiles, managing collection groups, assigning dedicated collection specialists, navigating collection segments, and handling temporary assignments of collection groups or substitutes for collection specialists. You can search for collection master data using parameters such as collection group, collection strategy, and segment. The app presents maintenance of master data in the **Assignment to Collection Segments** and **Change Documents** tabs. In addition, the app is capable of fetching both standard collection master data business partner role **UDM000** and custom roles.

Tip

Consult SAP Note 3508571 to understand when a collection specialist is assigned to a business partner at the master data level.

Transaction Code

The corresponding transaction in SAP GUI is Transaction UDM_BP.

Manage Clarification Cases (F2349)

Application Type: Transactional

With this app, you can search for, display, process, and complete clarification cases for convergent invoicing. The app provides features that include searching and sorting clarification cases using filter fields; reviewing details like master data, clarification reasons, status/lock information, and notes; and changing the status or

assignment of a clarification case. You can search for clarification cases with fields such as clarification case, status, clarification case category, clarification reason, business partner, contract account, created on, and last processor. Upon searching, the app presents data in **Open**, **Resubmission**, and **Completed** sections. In addition, you can upload and display attachments, navigate related documents, resubmit clarification cases, create tiles for direct access, and send emails with links to case details. The app also supports creating filter variants, hiding the filter bar, customizing column visibility, and sharing cases via email or Microsoft Teams, with navigation options depending on roles and authorizations.

Manage Commodity Counter Deal Request (F5659)

Application Type: Transactional

This app allows the creation of counterdeal requests for hedge accounting trades that result in gross or net overhedges, as well as for freestanding trades. The app provides an overview of counterdeal requests generated manually or automatically through dedesignation requests. In addition, it allows navigation to referenced dedesignation requests and supports the completion of counterdeal requests by entering commodity swap transactions using the fast trade capture function.

Manage Commodity Dedesignation Request (F5660)

Application Type: Transactional

This app allows you to create and process dedesignation requests for financial commodity trades tied to selected exposures, helping to address over-hedge situations. The app primarily catered to treasury teams, risk managers, finance controllers, or compliance and auditors to comply with IFRS or US GAAP. The app's features include evaluating the system's proposals of gains and losses for other comprehensive income (OCI) handling (OCI freeze or immediate reclassification) and counterdeal creation. In addition, the app allows analysis of the over-hedge situation before and after assigning trades for dedesignation.

Manage Commodity Derivative Order Request Brackets (F3514)

Application Type: Transactional

This app allows you to display, edit, and manage commodity derivative orders, offering an overview of all orders and detailed views of selected ones. You can search for commodity derivative order request brackets using search fields such as editing status, order bracket type, created on, and last changed on. Upon navigating further, the app displays order bracket information in **Order Request Bracket Information** and **Assigned Order Requests** tabs. In addition, the app allows you to create, copy, edit, or invalidate orders, address errors from data distribution to follow-on systems, transmit orders with a **Pending Transmission** status, and fetch the latest market or customer-specific prices.

Like the Manage Commodity Derivative Order Requests app (F3352), this app also supports extensibility to add customer specific fields for enhanced functionality and reporting. Let's compare the Manage Commodity Derivative Order Requests (F3352) and Manage Commodity Derivative Order Requests Brackets (F3514) apps:

- **Manage Commodity Derivative Order Requests**
  Suitable for individual order management. You can process, monitor, and resolve a single order.
- **Manage Commodity Derivative Order Requests Brackets**
  Supports handling grouped order requests.

You can organize and process multiple related orders collectively.

Manage Commodity Derivative Order Request Fill Packets (F3629)

Application Type: Transactional

This app allows traders of futures derivatives to display, edit, and manage commodity derivative orders, offering an overview of all orders and detailed insights into selected ones. It allows you to create, copy, edit, or invalidate orders, address errors from data distribution to follow-on systems, transmit orders with a **Pending Transmission** status, and fetch the latest market or customer-specific prices. In addition, the app supports extensibility to add customer-specific characteristics for enhanced functionality and reporting.

Manage Commodity Derivative Order Request Fills (F3229)

Application Type: Transactional

This app allows traders of commodity futures to display all available commodity derivative order fills in their system, with filtering options to meet specific needs. As a result, the app streamlines the trading process, providing real-time access to critical information. The app offers an overview of all order fills and detailed views of selected ones. If configured, the app allows the approval or rejection of order fills received from external systems, even when not linked to a commodity derivative order request. You also can search for derivative order request fills using parameters such as created on, commodity, fill packet status, and assigned subaccount. The app presents data in **With Order Request** and **Without Order Request** tabs.

Manage Commodity Derivative Order Requests (F3352)

Application Type: Transactional

This app allows traders of commodities to display, edit, and manage commodity derivative orders. As a result, the app acts as a single point for accessing and managing all order requests, thereby improving visibility and control. The app provides key features including an overview of all orders and detailed views of selected ones. You can search for commodity derivative order requests using search fields such as editing status, order request type, order request reason, and order request status. In addition, you can create, copy, edit, or invalidate orders, resolve errors from data distribution to follow-on systems, transmit orders with a **Pending Transmission** status, and fetch the latest market or customer-specific prices. Also, the app supports extensibility to add customer-specific fields for enhanced functionality and reporting.

Manage Commodity Hedge Books (F5793)

Application Type: Transactional

This app allows commodity risk managers to create and process hedge books at the level of specific company codes and, optionally, the derivative contract structure (DCS) in commodity hedge accounting. The DCS is a framework that allows you to manage and process derivative contracts. The app's functionalities include creating hedge books with a defined time structure, assigning an authorization group, deprecating hedge books, and reactivating previously deprecated hedge books. The app uses CDS view R_CmmdtyHedgeBookCmpltDetailTP.

Tip

As a prerequisite, master data such as a hedge book is required to define the periods for pricing and delivery expressed in months based on the requirements such as company code and DCS combinations. In addition, authorization objects CMM_CHB and CMM_CHEX must be assigned to your roles for defining hedge books.

Manage Commodity Hedge Specification (F5661)

Application Type: Transactional

This app allows commodity risk managers who manage hedge accounting to create and process commodity hedge specifications, providing long-term instructions for traders to optimize financial trades for hedging commodity price exposures. Thus, the app allows users to create strategic hedge specifications with trade orders for defined future periods and supports traders in making effective decisions to hedge exposures. By defining order quotas for utilization, the app aims to maximize the use of orders and hedge quotas while facilitating monitoring of hedge specifications through the trader's order cockpit. With this app, you can define hedge types to align with various commodity hedging strategies and assign them to relevant trading contract types to support seamless hedging. In addition, you can configure **Exposure Management 2.0** settings to align with risk management policies. Moreover, workflows approval process can be enabled for hedge specifications for the governance process.

Manage Commodity Hedging Area (F5662)

Application Type: Transactional

This app allows commodity risk managers to configure commodity hedging areas with settings for commodity hedge accounting. Thus, the app allows you to define a hedging area, which is essential for structuring and monitoring hedge relationships in commodity risk management. In SAP, the hedging area is an entity that represents a section of organization's hedging policy. This app allows you to define and manage hedging areas, which are essential for structuring and monitoring hedge relationships in commodity risk management. The app's functionality includes activating hedge accounting, specifying whether commodity foreign exchange (FX) hedging is applied, defining hedging classifications for commodity financial and FX trades by product type, setting the default dedesignation method, and disabling dedesignation for specific product types or at the company level, if needed. With this app, you can achieve full compliance with accounting standards for hedge accounting.

Manage Commodity Plan Exposure (F5664)

Application Type: Transactional

This app allows the management of commodity plan exposures derived from planning data or the manual processing of exposure data, which includes detailed information such as derivative contract specification ID (DCS-ID), market identifier code (MIC), and price type. The app's features include viewing generated or manually created exposures; sorting and filtering by parameters such as company code, DCS, validity, and hedge book; reviewing and processing exposures from plan exposures; manually creating or modifying exposures; releasing expo-

sures for hedge management and hedge accounting; handling incomplete or erroneous exposures; and checking the version of released exposures.

Let's compare the Manage Commodity Planning Data (F5663) and Manage Commodity Plan Exposure (F5664) apps:

- **Manage Commodity Planning Data**
  This app focuses on uploading and managing raw planning data. It helps with unloading (for example future purchase/sales forecasts of commodities). Data types can be uploaded such as quantities, values, and time frames (for example, unstructured/raw planning data).

- **Manage Commodity Plan Exposure**
  This app focuses on viewing and managing derived plan exposures. It derives information from planning data and then supports analysis. The app uses structured exposure objects already tied to risk types and profiles.

Manage Commodity Planning Data (F5663)

Application Type: Transactional

In commodity hedge accounting, this app allows commodity risk managers to upload commodity planning data via Excel files, reflecting planned commodity demand based on long-term contracts with suppliers or purchasers, forming the basis for generating commodity price exposures. You can achieve this through the app's predefined templates. The app's features include uploading and downloading data using a predefined Microsoft Excel template, marking planning data as relevant for hedge accounting or hedge management, making corrections by reuploading adjusted files, and generating related commodity plan exposures upon releasing the planning data. In addition, the app maintains the history of data uploads for compliance and audit reasons.

Manage Commodity Subaccounts (F3213)

Application Type: Transactional

For traders of futures for industries that deal with commodity trading, such as oil and gas, metals, and agricultural products, the app helps with processing and management of commodity subaccounts, including updating their status. You can search for commodity subaccounts using commodity and profit center fields. The app allows you to view subaccounts filtered by status, access detailed information for individual subaccounts, and create, modify, or update their lifecycle status. In addition, the app supports enhancements through key-user extensibility tools, which also can be used to create customer-specific characteristics.

Manage Consolidation Group Structure – Group View (F3733)

Application Type: Transactional

As part of SAP S/4HANA Finance for group reporting, this app helps with the management of consolidation unit assignments within consolidation groups, enabling you to define critical consolidation-relevant settings such as the first consolidation period, divestiture period, and consolidation method. The app allows you to assign, delete, and edit consolidation units, with the system providing warnings for conflicting assignments or potential data inconsistencies. In addition, the app allows you to set up flexible filtering and view customization, including setting default and public views.

The app's additional features include the ability to specify consolidation method changes, including changes at the beginning of a period, and seamless navigation to related apps like Manage Consolidation Group Structure – Unit View (F3766) and Define Consolidation Units (F4685) for comprehensive group structure management.

Transaction Code

The corresponding transaction in SAP GUI is Transaction CX1X.

Manage Consolidation Group Structure – Unit View (F3766)

Application Type: Transactional

As part of SAP S/4HANA Finance for group reporting, this app allows you to manage the assignment of consolidation units to consolidation groups and configure related settings, such as the first consolidation period, divestiture period, and consolidation method. The app's functionalities include displaying and assigning consolidation units to groups, where the fiscal year and period are automatically sourced from global parameters, with options to customize the filter settings for the current session. You also can search for consolidation group structures using search criteria such as editing status, consolidation version, consolidation group, and fiscal year and period. The app presents the assignment of units to group in a tabular column format with drilldown capabilities. In addition, you can assign multiple consolidation groups to a unit, specify relevant consolidation periods, and select the consolidation method. Moreover, the app supports viewing, editing, and deleting assignments, with warnings for conflicts or potential inconsistencies. Consolidation methods can be changed, but only once per period, and manual adjustments may be required in certain cases.

Also, you can navigate between unit and group views and access the consolidation unit's master data for further modifications. Let's compare the Manage Consolidation Group Structure – Group View (F3733) and Manage Consolidation Group Structure – Unit View (F3766) apps:

- **Manage Consolidation Group Structure – Group View**
  This app focuses on the overall structure and settings of consolidation groups. You can

define settings such as the period of first consolidation, divestiture, and consolidation methods for the group.

- **Manage Consolidation Group Structure – Unit View**
This app focuses on individual consolidation units within groups. You can edit assignments, review consolidation methods, and manage assignment time frames for individual units.

Transaction Code

The corresponding transaction in SAP GUI is Transaction CX1X.

Manage Contract Accounts (F5474)

Application Type: Transactional

This app serves as a centralized single location for maintaining contract accounts in contract accounting. The app's features include displaying a list of contract accounts, filtering the list based on search criteria such as business partner and company code, viewing contract account details by selecting them, modifying contract accounts via the app for changes, and creating new contract accounts through the dedicated app. This app allows you to access relevant functionality easily for efficient account management.

From this app, you can navigate to the other SAP Fiori–based SAP GUI apps such as Create Contract Account (Transaction CAA1), Change Contract Account (Transaction CAA2), and Display Contract Account (Transaction CAA3) with intent-based navigation configuration.

Manage Credit Accounts (F4596)

Application Type: Transactional

This app allows credit controllers to display and update credit data for business partners, including editing their credit profiles and recalculating

scoring. In credit management, the credit profile provides information on scoring, credit segments, blocked sales documents, and credit limit requests, highlighting credit exposure and utilization per segment. In addition, with this app, you can navigate global credit data, edit credit details, recalculate scoring, maintain external ratings, calculate credit limits, and simulate credit checks at the segment level. In addition, the app supports maintaining credit profiles, insurances, collateral, negative credit events, and check exceptions; creating and deleting credit segments; assigning credit analysts; managing credit blocks; and viewing credit limit requests, overdue sales document details, payment behavior, and key figures like days sales outstanding (DSO), cash discounts, and last 12 months' sales. In addition, the app supports adding notes at both global and segment levels.

Transaction Code

The corresponding transaction in SAP GUI is Transaction UKM_BP.

Manage Credit/Debit Memos (F2389)

Application Type: Transactional

With this app, you can create, change, or display credit and debit memos for an invoicing document to adjust the amount already invoiced to a customer without altering the original invoice. In cases where reversing an invoice document is not possible, the combination of the original document and the credit or debit memo determines the effective invoiced amount. The app provides key features such as searching and sorting memos using filter fields, creating credit or debit memos, changing their status or assignment, distributing amounts across items, and checking items for correctness. For example, you can search for credit/debit memos using search criteria such as editing status, credit/debit memo, business partner, contract account, invoicing document, status, category, type, reason, and created on. The app presents data in tabular format with functions **For Checking**, **Release**, **Reject**, **Reverse**, **Discard**, **Delete**, **Create**, and **Simulate Release**. In addition, you also can add and view attachments, navigate to specific memos to review details, and share memos via email or Microsoft Teams. The navigation options depend on user roles and authorizations.

Tip

To display data, business functions relevant for contract accounting for receivables and payables and convergent invoicing must be activated.

Transaction Codes

The corresponding SAP GUI transactions are Transaction FKKBIX_BILLREQ_MON (Display Billing Request), Transaction FB75 (Enter Credit Memo), and Transaction FB65 (Enter Debit Memo).

Manage Credit Limit Requests (F5602)

Application Type: Transactional

This app allows credit analysts to display, manage, and modify credit limit requests for business partners in credit management. In the request header, current credit management data provides insights, such as scoring trends, credit limit trends, aging grids with due and overdue documents, and counts of blocked credit decisions and broken promises to pay. The app's features include viewing all credit limit requests for business partners, creating and editing requests, maintaining general data, case details, assignees, and agreements, approving, rejecting, or voiding requests, accessing approval hierarchies and histories of past requests, and adding notes and attachments. To support all these functions, the app comes with multiple tabs such as **Docu-**

ment Overview, **Assignees and Agreements**, **Snapshot Data and Changes**, **Approval Hierarchy**, and **Credit Limit Request Log**.

Tip

To use this app seamlessly, you must complete prerequisite tasks such as:

- Assigning the business partner role UKM000 and maintaining credit segments, including score, risk class, and credit limit is transferred to credit limit requests during creation

- Configuring approval hierarchies and conditions via the Manage Credit Management Rules app (Transaction UKM_BRF_ CONFIG), including formulas for credit limit calculations

- Using the Manage Teams and Responsibilities app (F2412) to define authorized users, team owners, and members

- Setting up situation types for approvals with the Manage Situation Types app (F2947)

Transaction Codes

Equivalent SAP GUI transactions are Transactions UKM_CASE and S_SE3_50000106.

Manage Credit Management Rules (UKM_BRF_CONFIG (UKM_BRF_CONFIG_AC))

Application Type: Web Dynpro

This Web Dynpro–based app allows you to define approval hierarchies and configure formulas for credit limit calculations and scoring in credit management. The app requires appropriate access control, so only authorized personnel can approve credit limit requests and documented credit decisions. In addition, you can establish multiple approval processes based on conditions such as open credit amounts or document types. Also, the app also supports formula-based credit scoring, risk adjustments, and credit limit thresholds. Thus, the app allows businesses to tailor credit management rules according to their needs. The app can display processing steps to understand the sequence of steps followed. For example, you can approval group using predetermined criteria such as credit segment, requested amount, and risk class.

Manage Creditworthiness (F4408)

Application Type: Transactional

This app allows accounts payable and accounts receivable teams to display and edit the creditworthiness of business partners, which consists of manual and automatic values in contract accounting. The automatic creditworthiness value is calculated by multiplying credit numbers from relevant processes (such as returns and dunning notices) with time-dependent credit weighting and combining the results. The app determines creditworthiness using the following formula: *Automatic creditworthiness value × Percentage credit factor + Manual creditworthiness*. A value of 0 represents excellent payment history, while higher values indicate poorer payment history.

In addition, you can filter and view creditworthiness records using filter criteria such as editing status, business partner, postal code, city and risk class, monthly totals, and change history using filters such as business partner and risk

class. Also, the app supports manual edits, which include adjusting creditworthiness values, percentage creditworthiness factors, and creating or reversing records. For instance, fixed creditworthiness values can be released for updates when needed. The app also records all changes made, ensuring accurate tracking and management of creditworthiness. The app uses CDS views `C_CACreditWorthinessItemTP` and `R_CACreditWorthinessTP`.

Manage Custom Hierarchy Type (F3553)

Application Type: Transactional

Using this app, you can choose additional dimensions as hierarchy types for various dimensions to enable their maintenance within the Manage Global Hierarchies app (F2918). For instance, you can select an industry and set it as a hierarchy type to be supported in the Manage Global Hierarchies app. This functionality allows you to create and manage custom hierarchies tailored to specific business requirements, enhancing data organization and reporting capabilities. If a user wants to create a custom trial balance report with a specific hierarchy type, SAP recommends using this app to create a new hierarchy type.

Manage Customer Contacts (F4707)

Application Type: Transactional

This app allows accounts receivable users to create and manage customer contacts for business partners, initiated by collections specialists during customer interactions. These contacts are utilized in the collection process via the Process Receivables Version 2 app (F0106A), which serves as the starting point for creating customer contacts. As a result, the app allows you to centrally manage and maintain customer contact information, enabling a single point to key contact details. The app's features include searching and filtering customer contacts by various criteria, viewing open or finished contacts, navigating to individual contacts, displaying related objects like promises to pay or dispute cases, viewing contact summaries and notes, accessing attachments, and sending email recaps to customers. In addition, you can navigate to **Send Email** (via the Manage Collections Emails app [F6120]) and manage collection worklists from the header screen. The app uses CDS view I_CustomerContactType.

> **Tip**
>
> As of SAP S/4HANA 2023, the app does not support mass download for customer contact data, including contact notes.

Manage Customer Down Payment Requests (F1689)

Application Type: Transactional

Down payment requests act as triggers for the payment program to debit customer accounts or for the dunning program to dun customers for down payments. They do not update account balances. However, this app allows accounts receivable users to display and manage automatically created down payment requests based on sales orders, and to find and adjust these requests if, for example, a customer wants to negotiate the payment due date. You can search for down payment requests using parameters such as customer, company code, and posted by. If no suitable down payment request exists, you can create one manually, even in the absence of a sales order. Once a down payment is made, the corresponding request is cleared automatically, although manual clearance may be needed in cases of discrepancies or split payments. The app provides features such as viewing and exporting existing requests, modifying or creating new requests, and adding notes and attachments.

> **Tip**
>
> In this app, the tax reporting country attribute is derived from the company code and cannot be specified manually as of SAP S/4HANA 2023. For reversal of down payment request with multiple suppliers, consult SAP Note 3068500.

> **Transaction Code**
>
> The corresponding transaction in SAP GUI is Transaction F-37.

Manage Customer Line Items (F0711)

Application Type: Transactional

This app allows you to easily find customer line items for ad hoc requests or recurring reports using a wide range of search criteria. You can view all line items of a specific customer account or unallocated payments for a specific company code at the end of a period. In addition, you also can search for customer line items by status, open on key date, item type, and net arrears. To improve efficiency, you can personalize the table layout, predefine recurring queries, and save settings as variants. In addition, you can set or remove payment or dunning blocks, export data to a file, and collaborate with colleagues. The app also serves as a navigation target from other apps for drilling down into customer line items. Key features include finding customer line items, sorting and grouping data, exporting data, displaying dispute cases, setting or removing blocks, changing line-item attributes, creating predefined queries, and personalizing table settings. The app is extensible for further customization.

> **Transaction Code**
>
> The equivalent classic credit management transaction in SAP GUI is Transaction FBL5.

Manage Data Validation Tasks (F2653)

Application Type: Transactional

With this app, you can validate reported data in local, transaction, or tolerance currency, as well as quantity-type data, for a specified fiscal period and consolidation units in group reporting. You can search for data validation tasks using fields such as version, consolidation chart of account, consolidation group, consolidation unit, and period/year. The app displays results in a tabular format with the status of each validation task and with the flexibility of adding additional fields. For each consolidation unit, you can check the validation results, which are broken down into datasets verified against each validation rule, indicating whether the data passed or failed and the root causes of any failures. The app comes with features that include integration with the Data Monitor, where you can perform reported data validation tasks for multiple consolidation units in either a test run or an update run, and view the validation results in a list format. In addition, you can launch a new validation run; specify consolidation units, consolidation version, and fiscal year/period; and see essential information such as consolidation unit, period/year, status, last run time stamp, and commenting status.

Manage Dispute Cases (FI-CA) (F4443)

Application Type: Transactional

The app allows accounts receivable users to create and display dispute cases to address customer complaints regarding incorrect invoices or credit memos, missing credit memos, or payments in contract accounting. With this app, you can filter and search dispute cases using parameters such as case ID, case number, status, dispute type, category, and priority. You can also create new cases by specifying a business partner and relevant document, and manage them at the document level, even assigning multiple cases to one document. In addition, the app allows adding or removing documents, attachments, and notes within cases, and modifying attributes like status, reason, category, or contact person details. Also, you can upload attachments, add notes, and send emails using templates predelivered with the app, which can be customized for language and additional information. The app displays the disputed amount and currency, which are mandatory and derived from the corresponding document, ensuring alignment with its data. While creating dispute cases, the app also populates the contact person, email, telephone number, and fax number automatically from the business partner master data.

> **Tip**
>
> To use this app, you must first convert all existing notes and attachments using Transaction FPDM_CONV (Convert Dispute Subnodes) as a prerequisite.

> **Transaction Code**
>
> The corresponding transaction in SAP GUI is Transaction FPAR_DISPUTE. If you plan to use both Transaction FPAR_DISPUTE and the SAP Fiori app, the system automatically converts notes and attachments for display in the SAP Fiori app upon opening a dispute there for the first time. Conversely, while you can create and modify attachments and notes in both approaches, changes made in the SAP Fiori app will not be visible in the SAP GUI transaction due to technical limitations.

Manage Dispute Cases (Version 2) (F0702A)

Application Type: Transactional

With this app, you can analyze and process existing dispute cases related to open accounts receivable for customers. The app organizes dispute cases in **Case Overview**, **Assignments & Agreements**, **Amounts**, **Contact Information**,

Related Objects, Change Protocol, and Notes tabs. The app simplifies the dispute procedure for all relevant departments when customers deduct amounts from payments without prior agreement. The app allows you to display dispute cases with search criteria, add open invoices to dispute case, view details of individual cases, change attributes such as contact person, add notes and attachments, send emails, preserve drafts to prevent data loss, create customer-disputed objects, manage invoice references, and reopen closed cases to add additional data.

Transaction Code

The corresponding transaction in SAP GUI is Transaction UDM_DISPUTE.

Manage Documented Credit Decisions (F5587)

Application Type: Transactional

This app offers credit analysts and controllers an overview of essential data for assessing credit decisions for sales orders, deliveries, and service contracts, enabling you to check, release, or reject relevant documents in credit management. You can navigate to blocked sales documents or the credit check log and download snapshots of data supporting negative credit decisions to fulfill legal disclosure obligations. The app's features include managing credit decisions for sales orders, deliveries, service contracts, and service orders; finding, releasing, rechecking, or rejecting documents with credit blocks; maintaining general data, assignees, agreements, and notes; and accessing related logs and documents. This is accomplished via the app's header buttons: Edit, Release, Recheck, Show Blocked Document, and Show Credit Check Log.

Tip

When integrating external systems to facilitate the exchange of data, for example in a Central Finance scenario, RFC connections are necessary for remote calls. If released documented credit decisions are not transferred to credit management, consult SAP Note 3101401 for additional troubleshooting guidelines.

Transaction Code

The corresponding transaction in SAP GUI is Transaction UKM_MY_DCDS.

Manage Documents (FI-CA) (F6109)

Application Type: Transactional

This app gives contract accountants a centralized point of access for managing contract accounting documents, thus enabling them to display, create, change, or reverse documents, as well as view document changes and related business locks. The app's features include managing reconciliation keys, writing off documents, and accessing related objects, such as a business partner's contract account and account balance. In addition, using filters such as reconciliation key, journal entry date, and posting date, you can display specific document details, including disputes, dunning notices, business partner items, and general ledger line items. Also, the app supports navigation to other related apps for performing tasks such as managing reconciliation keys, business partners, and contract accounts, provided the required authorizations are in place.

Transaction Code

The corresponding transaction in SAP GUI is Transaction FPE1.

Manage Financial Transactions (F6157)

Application Type: Analytical

This app provides a centralized platform for managing financial transactions across all financial instrument groups. You can filter transactions by attributes, save variants for quick access, and navigate directly to apps to create, display, or process financial transactions. Thus, with this app, you can set filter criteria, apply them, and view or process transactions. For instance, the mandatory chosen key date, set to the current date by default, influences values such as nominal amount, total amount, or interest rate. You also can apply filters with fields such as company code, transaction, counterparty, product type, transaction type, active status, trader, and portfolio. The app's features include grouping and ungrouping transactions in the table, adapting filters by product category, and saving views as tiles on the SAP Fiori launchpad. The app also utilizes situation handling to inform you about matters requiring quick attention.

Tip

Although the app supports transaction types involving **Product Category 760 (OTC Options)**, not all product categories are supported. Over-the-counter (OTC) options include foreign exchange (FX) options, swaps, and security. On the other hand, the app does not support product categories such as commodity options and custom-defined OTC.

Transaction Codes

The related transactions in SAP GUI are Transactions FTR_CREATE, FTR_EDIT, and FTR_00.

Manage Fixed Assets (F1684)

Application Type: Transactional

In general, fixed assets are created and capitalized automatically through integrated procurement and invoicing processes, based on configuration. You can search for assets using various search parameters such as asset number, company code, and asset class. This app provides a comprehensive view of asset data in one place, allowing you to quickly assess an asset's status and adjust its valuation using a graphical lifecycle representation. The app provides key figures such as acquisition and production cost, accumulated depreciation, and net book value help you to understand asset valuation in detail and identify reasons for unexpected changes, such as special depreciation due to accidents.

You can access this app via the Display Asset Master Worklist app (F1592) or related reporting apps like Asset History Sheet (F1615), Asset Balances (F1617), and Depreciation Lists (F1616). The app provides key features such as viewing asset characteristics, parallel depreciation areas, key figures by area, related journal entries and transactions, lifecycle charts, fiscal year comparisons, master data, attachments, notes, and editing capabilities.

Tip

SAP recommends using the Manage Fixed Assets app (F3425) or the Manage Legacy Assets app (F7280) if universal parallel accounting business function FINS_PARAL-LEL_ACCOUNTING_BF is active. As a result of this activation, the older app, Manage Fixed Assets (F1684), is obsolete. In addition, to use relevant business catalogs and business groups as templates when the universal parallel accounting business function is active, consult SAP Note 3191517 for further guidance.

Manage Fixed Assets (New) (F3425)

Application Type: Transactional

This new app replaces the Manage Fixed Assets app (F1684) after enabling the universal parallel accounting (FINS_PARALLEL_ACCOUNTING_BF) business function, which is available from SAP S/4HANA 2022 onwards. This app consolidates all asset data into one view, offering a graphical representation of the asset's lifecycle for a quick overview of its status and valuation adjustments. Key figures such as **Acquisition and Production Costs**, **Accumulated Depreciation**, and **Netbook Value** provide detailed insights into asset valuation and help identify unexpected changes, like special depreciation due to recent incidents or changes in law. Fixed assets are typically created and capitalized automatically through integrated procurement and invoicing processes, if configured.

You can access the app via the Asset Master Worklist app (F1592), navigating from fixed asset master data or through reporting apps like Asset History Sheet (F1615), Asset Balances (F1617), Depreciation Lists (F1616), or Asset Transactions (F1614).

> **Tip**
>
> Consult SAP Note 2847767 to understand the new features, simplifications, and enhancements related to this app. To learn more about universal parallel accounting, consult SAP Note 3265275.

Manage Flexible Hierarchies (F2759)

Application Type: Transactional

With this app, you can create and modify hierarchies for cost center, profit center, and company code reporting, customizing them to meet reporting needs and business requirements of an entity. While creating a flexible hierarchy, you also can preview the edition with existing master data and upload master data. For creating a new hierarchy, the app organizes the data into **Basic Information**, **Filter Criteria**, **Structure**, and **Master Data** sections. In addition, you also can extend these hierarchies with custom fields. The app offers the functionality of creating and editing hierarchies, generating hierarchies based on master data attributes, performing mass updates and downloads of master data and attributes, and mass uploading extended attributes of master data used in hierarchies.

> **Tip**
>
> The app only supports standard predefined types such as profit center, cost center, and company code. The app does not support hierarchies created in all SAP modules or custom hierarchies created outside of SAP standard configuration.

Manage Formulas in Credit Management (F6575)

Application Type: Transactional

This app allows you to maintain formulas for calculating scoring and credit limits in credit management. You can define and activate formulas via the **Simulate** function, then assign them to calculation rules in the **Create Rule for Scoring** and **Credit Limit Calculation** configuration activities. In addition, the app allows formula simulation on selected credit accounts to validate results before implementation. The app displays results in the **Scoring Simulation Result** screen. The app's features include creating formulas for business partner scoring and credit limits, copying and deleting formulas, using adaptable formula templates, exporting formulas to spreadsheets, and navigating directly to the Maintain Formula Parameters (UKM_FMLA_PARAMETER) and Manage Credit Management Rules (UKM_BRF_CONFIG) apps.

Tip

You can focus on maintaining settings for **Formula: Parameter Evaluation**, which determines the calculation of scoring, and credit limit calculations, which help to fine-tune formulas for more accurate credit assessments.

Manage G/L Account Master Data (Version 2) (F0731A)

Application Type: Transactional

The app lets you display, create, and edit general ledger account master data in a user-friendly screen. With this app, you can add new accounts from scratch or based on existing ones and mass change account descriptions and other attributes. To facilitate maintenance, the app provides **General**, **Company Code Data**, **Controlling Data**, and **Where Used** tabs. Based on the role assigned, you can view general ledger account master data, create and edit general ledger accounts, and mass change descriptions and attributes from the chart of accounts, company code, and controlling area views. In addition, the app allows you to mass copy general ledger accounts in the same chart of accounts, make multiple copies of a general ledger account, and view where the account is used in financial statements or the Automatic Account Determination app (F1273).

Tip

The app has limitation of selecting up to 5,000 general ledger accounts for each mass change.

Transaction Code

The equivalent transaction in SAP GUI is Transaction FS00.

Manage Global Hierarchies (F2918)

Application Type: Transactional

The main purpose of this app is to manage global hierarchies for master data, which is often reusable and centrally maintained. The app allows you to either import hierarchies already created in SAP GUI or create hierarchies using this app directly. In addition, the app allows you to view, create, edit, activate, and deactivate hierarchies, as well as quickly create new hierarchies by importing or copying from existing ones. Also, you can check change logs and maintain node texts in multiple languages. You can perform tasks such as viewing and filtering hierarchies by criteria like type, status, hierarchy ID and validity time frame; managing hierarchies' validity; activating or deactivating hierarchies; generating simulated versions; exporting, editing, and importing hierarchies via a spreadsheet; importing from compatible applications; and maintaining node texts in multiple languages.

Let's compare the Manage Global Hierarchies (F2918) and Manage Flexible Hierarchies (F2759) apps:

- **Manage Global Hierarchies**
  This app is primarily used for predefined hierarchy types like cost centers, profit centers, general ledger accounts, etc. It's structured to represent the organizational setup of a company across different entities, regions, business units, and so on. The primary focus is on legal and financial consolidation.

- **Manage Flexible Hierarchies**
  This app is used to create flexible or custom hierarchies based on user-defined attributes or standard attributes. It's mainly catered to different organizational reporting requirements. The primary focus is to meet internal reporting needs.

Transaction Codes

The app combines several SAP GUI transactions:

- Cost center groups via Transactions KSH1, KSH2, and KSH3
- Profit center groups via Transactions KCH1, KCH2, and KCH3
- General ledger account and cost element groups via Transactions KAH1, KAH2, and KAH3
- Financial statement versions (FSVs) via Transactions OB58 and FSE2

Manage Hedge Constellations Tasklist (F5666)

Application Type: Transactional

This app allows commodity risk managers to manage tasks within the change process, created automatically or manually due to specific reasons or inconsistencies. Thus, with this app, you can manage approvals, adjustments, and reclassifications related to hedge constellations in hedge accounting. The app's features include addressing changes in trade counterparties, replacing foreign exchange (FX) trades linked to modified commodity trades, handling changes or reversals of commodity trades, resolving mismatches between FX trades and FX hedge requests, and managing late designation of commodity trades.

Let's compare the Monitor Hedge Constellation (F5665) and Manage Hedge Constellations Tasklist (F5666) apps:

- **Monitor Hedge Constellation**
  This app focuses on monitoring hedge constellation statuses for a list of all hedge constellations. It supports performance and feedback. You can track changes and updates to hedge constellations to support compliance and reporting.
- **Manage Hedge Constellations Tasklist**
  This app focuses on managing hedge constellations tasks. It supports the execution function and governance via workflows. You can track the history of task activities performed.

Manage Incoming Payment Files (F1680)

Application Type: Transactional

This app allows you to import and manage electronic payment files such as bank statements, payment rejections, intraday bank statements, lockbox batches, and payment files for advanced payment management. After importing, these files can be further processed using the Manage Bank Statements (F1564), Manage Lockbox Batches (F1681), or Manage Payments (F3647) apps. You can search for incoming payment files using filter criteria such as file type, status, imported by, and imported on. In addition, you can perform tasks such as reprocessing files that failed initial import, marking obsolete files that should be disregarded, and seamless navigation to related apps for further processing. The app requires appropriate formats customized in the **Configure Formats for Import** and **Define Parameter Sets** activities for import parameters.

Transaction Code

The corresponding transaction in SAP GUI is Transaction FF.5.

Manage In-House Bank Account Balances (F6341)

Application Type: Transactional

This app provides you with an overview of account balances (both incoming and outgoing transactions) across bank areas and account groups, structured hierarchically in the app's **Accounts** section. With the **Show Details** option, you can aggregate current balances by groups or bank areas, with subtotals and totals displayed, and separate balances for multiple currencies without conversion. In addition, you can expand the **Account Group** node to view included accounts or navigate to the Manage In-House Bank Accounts app (F5942) for details.

The **Details** page includes sections for **Transactions**, **Bank Statements**, and **Account Maintenance**, thus offering features such as viewing payment transactions with color-coded indicators, displaying bank statements based on defined frequencies, and accessing account settings without navigating elsewhere. You also can link to the Manage Payments app (F3647) for further transaction details from the header area. The **Account Maintenance** section provides essential details about account settings, including configuration specifics such as requirements for account balance, bank statements, and bank notifications, along with the defined frequency for these updates. As a result, this section allows you to quickly review account-related information without navigating to the Manage In-House Bank Accounts app (F5942). Also, the account balance field displays the current account balance, which is dynamically calculated based on bank statements and posted transactions, excluding future-dated items from the immediate balance. This app is based on CDS view C_IHBAccountBalanceTP.

Manage In-House Bank Account Overdrafts (F6340)

Application Type: Transactional

As part of in-house banking, this app allows you to view in-house bank accounts or account groups with balances that have breached the lower limits (overdraft) defined in the Manage In-House Bank Account Limits app (F5940). An overdraft refers to a situation where the balance of a bank account falls below zero or a predefined lower limit. The app supports both single level and group level accounts, displaying details such as account number, account description, limit amount, and limit group for grouped accounts. In addition, the **Details** page is divided into **Transactions**, **Limit**, and **Accounts** sections. The **Transactions** tab includes overdraft payment items with details like posting date, value date, and amount in account currency. On the other hand, the **Limit** tab provides information on limits such as object level, limit group, limit amount, currency, and validity period. The **Accounts** section lists accounts with details such as account number, valid from/to dates, and account balance amount. This app is based on CDS view C_IHBOverdraftTP.

Also, you can navigate to payment processing apps such as Revise Payment Proposals (FO771), Create Payments (F3648), Manage Payments (F3647), and Repair Payments (F3651) via links in the **Transactions** section.

Manage In-House Bank Account Templates (F6039)

Application Type: Transactional

As part of payments and bank communication, this app allows you to create, edit, or delete templates for in-house bank account creation. You can define attributes such as basic data, account balancing, bank statements, balance notification, and interest scale notification to save effort when setting up new accounts. The app supports you in creating templates from scratch or based on existing accounts using the **Save As Template** option. Once templates are created, these templates are accessible within the Manage In-House Bank Accounts app via the **Manage Template** option. Previously created templates can be searched with filter criteria such as bank area, editing status, template short description, template long description, and created by. In addition, you can edit templates on the **Details** page or discard drafts, and templates can be deleted from either the **Overview** or **Details** page.

Manage In-House Bank Accounts (F5942)

Application Type: Transactional

This app allows payment specialists to create, edit, deactivate, or close in-house bank accounts as part of payments and bank communication. Thus, the app streamlines the creation, editing, deactivation, and closure of in-house bank accounts in SAP. While managing in-house bank accounts, you can manage details such as account balancing, notifications (balance and interest scale), and bank statements. With this app, you can filter in-house banks with various search criteria such as bank area, editing status, account number, account currency, and account status. In addition, the app's features enable you to create accounts with mandatory fields such as in-house bank area and account holder, use templates to save time, and generate IBANs automatically. In the app, you can edit accounts in **Active, In Review,** or **Closure Released** status, and workflows control approval processes with statuses like **Active, In Review,** and **Closed**. With predelivered templates, the app supports mass account creation through file uploads, template management, and XML data downloads for integration with bank account management (BAM). Only **Active** status accounts influence processes, while the app discards draft versions. The app leverages CDS view C_IHBAccountTP to manage in-house bank account data.

Manage In-House Bank Application Jobs (F6653)

Application Type: Transactional, Reuse Component

This app allows cash management users to create, edit, deactivate, and close in-house bank accounts. With this app, you can define key account details such as in-house bank area, account holder, account currency, and opening date, with the option to reuse prefilled templates to streamline data entry. In addition, the app supports various functionalities, including

account balancing, balance notifications, interest scale notifications, and bank statement management, ensuring compliance and financial accuracy. Also, you can manage different account statuses, such as **Active, In Review, Closure Released,** or **Closed**, and initiate changes like setting an account **Inactive, Closed, Closure Released,** or **Prepare for Closure**. In addition, the app allows mass account creation via file uploads, user-defined templates, and XML-based downloads for bank account management (BAM) integration. The app supports workflow approvals, which govern account creation, ensuring a structured and compliant process.

Manage In-House Bank Conditions (F5941)

Application Type: Transactional

In SAP, in-house bank condition items refer to the financial conditions and parameters used to calculate interests based on account balance situations. For instance, the conditions include information such as interest type (credit, debit, overdraft) interest calculation method, interest lockout period, and interest calculation type. This app allows payment specialists to manage financial conditions for calculating interests based on account balance situations such as credit interests, debit interests, or overdraft interests at single account, group, or bank area levels, with defined validity periods. Within a condition, you can configure multiple interest types (credit, debit, or overdraft) and additional criteria like account currency for better maintenance. The app supports creating, editing, and deleting financial conditions, with workflows for approvals and two data version statuses (**Active** and **In Review**). In addition, single-level conditions take priority over group and bank area levels, and conditions breaching limits are flagged as overdrafts in related apps like Manage In-House Bank Account Balances (F6341) and Manage In-House Bank Payment Items (F6339). The app allows only one condition per validity period. The app uses CDS view C_IHBConditionTP.

Manage In-House Bank End of Day (F6338)

Application Type: Transactional

As part of in-house banking, this app allows payment specialists scheduling of end-of-day jobs to review the process steps executed for in-house banking related operations. As a result, the app complements the Manage In-House Bank Application Jobs app (F6653). The app provides job details with multiple tabs such as **Scheduling Options**, **Run Details**, **Steps**, and **Parameters**. With these tabs, the app provides an overview of executed process steps through reports such as account balancing, account closure, bank statement, foreign currency exchange, general ledger transfer, notifications, and payment transaction date. In addition, you can search for specific reports, view results in the **In-House Bank End of Day Processes** section, and adjust columns like bank area, process step, status, and total messages. The app is based on CDS view `C_IHBEndOfDayProcessLogTP`.

Moreover, you can access detailed logs for individual process steps via the Manage In-House Bank Application Jobs app (F6653).

Manage In-House Bank Fees (F6335)

Application Type: Transactional

The app is designed for treasury teams to streamline the management of fees associated with in-house banking. You can filter fee transaction details with search criteria using fields such as editing status, bank area, bank key, bank country/region, object level, and account number. On the **Object** page, the app data is provided with two tabs: **Fee Basic Data** and **Fee Amount**. As a result, the app allows you to define, edit, and monitor fee structures for transactions processed through in-house bank accounts. For instance, fees can be defined at three levels: single account, group (where the same fee applies to all accounts in the group), and bank area. An account can only belong to one level, checked in

the listed order, and the validity periods for single or group levels cannot overlap. Fee types include account maintenance fee (monthly) and transaction fees for domestic and international payments, categorized into normal or priority processing. In addition, the **Scale Fee** indicator allows defining interval-based fees (e.g., 12 payments at $1 each result in $12) or incremental fees (e.g., 4 payments at $5, 5 at $4, and 3 at $1 total $43). On the other hand, activity-based transaction fees use specific counters adjusted via configuration. During account balancing, the calculated total fees are charged for the respective balancing period. Thus, the app helps treasury teams to ensure accurate fee calculations and provides transparency in fee management, supporting seamless treasury operations.

Manage In-House Bank Interest Compensation (F6652)

Application Type: Transactional

As part of the centralized treasury process, this app allows accurate interest calculations and financial transparency across subsidiaries by defining interest compensation groups, setting interest conditions based on balances by bank area, and configuring validity periods for compliance. You can manage reference interest rates such as ESTER or SOFR, apply markups or markdowns for financial modeling, and implement lockout periods to stabilize rate applications. Thus, the app allows centralized treasury operations, ensuring efficient management of internal banking relationships.

Manage In-House Bank Limits (F5940)

Application Type: Transactional

This app allows payment specialists to manage account balances by setting lower limits at either a single account or group level, with specified values and validity periods. First, the app allows you to search for in-house bank payment items with multiple search criteria such as item number, account number, bank area, statuses, payment type, posting date, and value date. Next, the app supports two data version statuses, **Active** and **In Review**, for each limit. However, the system only recognizes the **Active** version, while the **In Review** version serves as a draft status that does not impact the limit-setting process. Also, you can create positive or negative limits and edit limits by triggering a workflow for approval, deleting existing limits, or withdrawing draft versions. When an in-house bank account has more than one limit, single account limits take priority over group limits, which apply to the total balance of grouped accounts. The app allows only one limit per validity period, and active limits supersede draft versions.

Integration with the Manage In-House Bank Account Workflows app (F6131) allows seamless approval processes. Moreover, the app flags accounts for breaching limits as overdrafts in the Display In-House Bank Account Overdrafts app (F6340).

Manage In-House Bank Payment Items (F6339)

Application Type: Transactional

This app allows you to monitor payments relevant to in-house bank accounts. The app can search with various filters such as item number, account number, posting date, value date, status, payment type, and company code. Payments are displayed in the list page categorized by posting processing status, including tabs for **All**, **Posted**, **Parked**, and **Errors**. With this app, you can navigate from **Account Management** to **Payment Processing** for detailed payment information. In addition, the app's item details, accessible via a link on the far right of each row, lead to the corresponding payment page in the Manage Payments app (F3647), with sections such as **General**, **Account**, and **Reference**. The header area indicates the payment type (e.g., clearing, originator, recipient, statement entry, or turnover). Items in the **Errors** tab require manual repair, with error details and a **Repair** option available on the **Details** page. Thus, this app

allows you to manage and monitor payment transactions effectively. This app is based on CDS view `C_IHBPaymentMonitorTP`.

Manage Installment Plans (F5720)

Application Type: Transactional

This app allows accounts receivable accountants to display, create, and modify installment plans related to customer payments, review installment plan history, and analyze installment plans as key figures in contract accounting. The app's features include user-friendly filtering and sorting for installment plans, viewing detailed information, editing open plans, and creating new plans. For instance, you can use filter criteria using parameters such as installment plan, status, business partner, contract account, start date, and end date. In addition, you also can access installment plan history by navigating to the Analyze Installment Plans (Version 2) (F2363A) app to get an overview of created plans.

From this app, you can navigate to transactions such as Create Installment Plan (Transaction FPR1), Change Installment Plan (Transaction FPR2), Display Installment Plan (Transaction FPR3), and Display Installment Plan History (Transaction FPRH).

Manage Interest Runs (F4485)

Application Type: Transactional

This app allows accounts receivable users to display and manage calculated interests for each business partner and currency. You can search for previously created interest runs using parameters such as interest document, company code, customer, and supplier. To use the app, output control must be activated for SAP Fiori apps as a prerequisite. In addition, interest document reversals are performed by the generic workflow user (**SAP_WFRT**), requiring the assignment of workflow template **78500143** for authorization. The app's features include displaying interests from interest runs executed via the Schedule Interest Calculation Jobs app (F4176), viewing associated journal entries, managing interest letters for both automatically or manually sent, sending/resending letters for interest documents, and reversing up to five interest documents. The app supports reversal of interest documents along with their related history.

Tip

While reversing interest documents, the app does not support reversing a complete interest run with this function; instead, individual interest documents generated by the interest run must be selected and reversed one by one.

Manage Invoice Prioritization Rules (F5470)

Application Type: Transactional

This app allows accounts receivable users to define rules for prioritizing invoices based on collection strategies, ensuring that the invoices displayed in the Process Receivables (Version 2) app (F0106A) are arranged by priority. As a result, the app helps collection specialists to focus on the most critical invoices during customer calls, especially when dealing with numerous invoices. The app's features include creating invoice prioritizations for each collection strategy, defining prioritization valuations, setting priority definitions, and establishing rules for each prioritization. These are organized into **Invoice Prioritization**, **Priority Definition**, and **Invoice Prioritization Rules** tabs. The app also allows you to edit prioritization rules during invoice processing as necessary. This app is helpful for accounts receivable accountants, especially for scenarios where multiple invoices are due, and businesses need to determine the order in which they should be processed to support an effective cash flow.

Manage Journal Entries – New Version (F0717A)

Application Type: Transactional

This app allows you to manage and analyze journal entries efficiently. It allows you to find, filter, sort, and group entries based on business needs. In addition, you can navigate detailed views to access transaction data, line items, associated entries, and related documents. You can search journal entries with search fields such as company code, ledger group, journal entry, journal entry date, posting date, fiscal year, fiscal period, journal entry created by, journal entry type, and journal entry category. The app presents results in the **Journal Entries** and **Predictive Journal Entries** tabs. Key functions include creating, copying, verifying, editing, and reversing journal entries. You also can initiate correspondence, analyze change and reversal logs, and display entries in t-account views or document flows. Also, the app allows you to display line items, access detail screens, create new entries, copy and edit entries, display parked documents, and more.

Manage Journal Entry Templates (F3803)

Application Type: Transactional

The app allows you to create and customize templates to streamline and simplify the entry of numerous journal entries to avoid manual errors. You can search for existing templates using filter criteria such as template ID, posting application, and access level. In addition, the app allows you to select specific fields and values for different uses, share templates with others in an organization, and set access levels for template usage. Also, you can perform activities such as searching for existing templates, creating new ones as needed, deleting templates, determining mandatory or read-only fields, choosing to share templates, setting access levels, and checking business logic before saving.

Tip

If standard templates are not visible for download, you must run the program S4_XPRA_JE_TEMPLATE via Transaction SE38, as indicated in SAP Note 3299445. Authorization object F_PST_TMPL with display (03) access must be added to your role prior to running the program. Consult SAP Note 2794539 for guidance on choosing the role.

Manage Legacy Assets (F7280)

Application Type: Transactional

This app helps asset accountants to manage legacy fixed assets within new asset accounting in SAP S/4HANA. The app allows you to create, display worklists, track, and maintain asset master data and values for various valuations from previous systems. The app's features include defining transfer parameters, specifying offsetting accounts, and managing transaction types for asset data migration. Thus, the app allows you to maintain accurate asset data during system transitions.

Tip

The app can be used only if universal parallel accounting business function FINS_PARALLEL_ACCOUNTING_BF is activated.

Transaction Code

The corresponding transaction in SAP GUI is Transaction AS91.

Manage Liquidity Item Hierarchies (F4966)

Application Type: Transactional, Analytical

This app allows cash managers to create and edit liquidity item hierarchies along with their validity time frames. You can search for liquidity item hierarchies with hierarchy ID, status, and validity date. In addition, you can add new time frames under existing hierarchy IDs to prepare for planned changes or quickly generate new hierarchies based on existing ones to address structural changes or evolving reporting requirements. The app's functionalities include creating and managing hierarchies, working with multiple time frames under specific hierarchy IDs, editing and saving hierarchies in draft or active status, deactivating liquidity items, leveraging tools like fast entry, copy-paste, and drag-and-drop for efficient adjustments, and exporting hierarchies to spreadsheets.

Tip

Cash and liquidity apps such as Cash Flow Analyzer (F2332) or Liquidity Forecast Details (F0741) use liquidity item hierarchies maintained via this app. To create liquidity items hierarchies, you must have the SAP_BR_CASH_MANAGER and SAP_BR_CASH_SPECIALIST roles.

Manage Liquidity Item Transfer (F6385)

Application Type: Transactional

As part of cash and liquidity management, this app helps with the manual transfer of actual amounts between liquidity items. The app allows you to process liquidity item transfer documents throughout their lifecycle, including creation, modification, copying, and deletion. Transactions posted through the app are ultimately saved in One Exposure from Operations (table FQM_FLOW). The app allows you to create, edit, copy, or delete liquidity item transfers while ensuring that the total balance for each transfer remains zero. Transfers can be initiated directly within this app or through the Cash Flow Analyzer (F2332) and Check Cash Flow Items (F0735) apps. When a transfer is created, liquidity items generate actual One Exposure from Operations flows, documenting shifts between items without necessarily altering the originals. Thus, the app allows managers to track and validate liquidity shifts without altering the original liquidity.

Each saved document is required to have a zero balance to ensure accuracy. If any flow is deleted or aggregated, the corresponding transfer is marked **Invalidated**, preventing further updates. **Invalidated** transfers cannot be deleted. In addition, the app's **Blocked Date Range** feature is used to prevent changes during reconciliation or reporting periods. Thus, this ensures data integrity and control over liquidity item adjustments. Also, you can create transfers by specifying required fields, including company code and transfer currency, with at least two line items containing liquidity item, amount, and value date, ensuring a total balance of zero. Additional fields can be configured via the **Settings** function.

Manage Lockbox Batches (F1681)

Application Type: Transactional

Checks are a common form of payment in the United States. Banks provide lockbox services, which allow customers to transfer their payments straight to the bank (lockbox), to process these payments quickly. The bank then scans the checks and provides the payee with a data file containing the check information in batches. With this app, you can manage lockbox batches efficiently, and banks can offer services for processing check payments. Customers send checks directly to the bank's lockbox, which scans and sends the check information in batches to the payee as a data file. The app provides the ability to search for and display batches from all your lockboxes and display individual incoming checks (lockbox batch items). You can search for lockbox batches using search criteria such as company code, house bank, lockbox number, status code, deposit date, batch number and latest lockbox batch. The app presents lockbox batches in a grid format with the flexibility of adding additional dimensions. The app uses a format that contains descriptions of the records in a lockbox batch.

Tip

To process intercompany clearing of lockbox items, consult SAP Note 3467772 for further guidance.

Transaction Codes

This app covers the functionality of several SAP GUI transactions, including Transactions FOEBL1 (Lockbox Total Report), FCHN (Overview of Outgoing Checks), FLB1 (Post Process Lockbox Data), FLBP (Post Lock Box Files), and FLB2 (Import Lockbox File).

Manage Manual Accruals (F5423)

Application Type: Transactional

In accruals management, this app allows you to create, import, edit, copy, and manage accrual objects, including changing their status. The app offers features such as displaying, editing, creating, and copying accrual objects, importing them using spreadsheet templates, and managing their status by pausing, suspending, or resuming accrual subobjects. The app is also capable of displaying derived values for opening, periodic, and closing accounts. In addition, you also can prematurely finish accrual subobjects through termination or undo termination. The app provides access to parameters assigned to accrual objects and a detailed summary of changes made to them. The app's predelivered templates (via the **Download Template** button) provide features such as importing include options for accrual objects with total amounts, periodic amounts, or updates to existing periodic accrual objects. You can manage the status of an accrual object by pausing, suspending, or resuming its accrual subobjects, which are items within the corresponding purchase order.

Accrual subobject statuses allow flexibility in accrual postings, ensuring accurate management of purchase order items. For instance, **Pausing Accrual Subobject** preserves its accrual balance, allowing accrual postings to resume later. **Suspending Accrual Subobject** sets its balance to zero with a release posting, but it can still be resumed for future accrual postings. On the other hand, **Resume Accrual Subobject** allows the continuation of accrual postings for the associated purchase order item.

Tip

A purchase order item linked to a paused or suspended accrual subobject can only be reviewed and approved once the accrual subobject has been resumed.

Transaction Codes

The corresponding transaction codes in SAP GUI are Transactions ACACTREE01 and ACACTREE02.

Manage Matching Assignments (F3870)

Application Type: Analytical

As part of intercompany reconciliation, this app allows general ledger accountants to execute matching runs, view results, make manual assignments, and monitor follow-up activities for data from underlying sources. As a result, the app allows precise, efficient management of matching assignments and associated processes for group reporting. You can specify filter values, such as company, matching method, fiscal year period, and the leading unit, with options to save filters as view variants. Matching items progress through processing statuses, from **00 New** to **30 Matched**, with statuses like **01 Roll-In** for items requiring manual assignment and **29 Auto-Adjusted** for temporary resolutions. The app displays results in tables for unassigned items (upper area) and assignments (lower area), showing details like processing status, matching rules, and reason codes. In addition, the app offers workflow features to manage statuses and adjust logs to support tracking of approval processes and automatic variance adjustments. Also, you can customize views, unassign items

for reprocessing, and analyze grouped items for effective reconciliation.

Tip

To select a default view, follow SAP Note 3460274 to understand standard behavior and additional guidance.

Manage Matching Documents (F3868)

Application Type: Analytical

This app allows users involved in intercompany reconciliation to monitor and manage the automatically generated matching documents. As a result, the app enhances functionality, ensures accuracy and improves efficiency in reconciliation processes. The app allows the display and management of matching documents based on a selected matching method, aiding data tracking and auditing. In addition, matching documents are automatically generated during each matching run or data upload, provided new posting data is detected. The app displays key details like business transaction type (e.g., **ROLL-IN**, **UPLOAD**, or **API**), data cutoff time, creation information, and deletion status. In addition, you can delete matching documents, unassign related data, and reprocess it during the next matching run. Deleted documents remain visible with a deletion flag for auditing purposes. Deletions should occur in reverse chronological order to maintain data integrity. Also, you can drill down to matching items, inheriting context values to view original posting documents, results, and line-item details within the Display Matching Items app (F3869).

Manage Memo Records (F2986)

Application Type: Transactional

Memo records are used to make temporary or non-posting records that do not directly affect the general ledger, but they are important for tracking, reporting, or informational purposes. The app is used to manage information or reminders or to make temporary adjustments or notes to track non-posting entries with its own table FCLM_MMRD. This app allows you to centrally manage memo records for expected cash flows that are not yet generated in the system, and these records are displayed in cash management apps like Cash Flow Analyzer (F2332) and Check Cash Flow Items (FO735).

You can search for existing memo records using search criteria such as editing status, status, company code, planning level, bank account, value date, currency, and certainty level. In addition, you can manually create memo records, for example cash trade request (FX), by entering details such as company code and value date or using SAP Collaboration Manager for seamless integration. The app's features include copying, distributing, or offsetting memo records, converting statuses between **Active** and **Inactive**, and autoexpiring records upon reaching their expiration date. The **Distribute** option allows you to create multiple memo records simultaneously by distributing an existing memo record. With this option, you can specify the number of records to be created, define the recurrence pattern, and decide whether to consider a working calendar during the distribution process. This feature streamlines the creation of recurring or similar memo records, saving time and ensuring consistency in cash flow management. In addition, you can view, modify, and delete intraday memo records from imported bank statements.

All memo record data is stored centrally in the One Exposure from Operations hub, ensuring accessibility across cash management apps. The One Exposure from Operations hub serves as a centralized repository for operational data essential to managing cash and liquidity.

Transaction Code

The equivalent transaction in SAP GUI is Transaction FF63.

Manage Memo Records 2.0 (F2986A)

Application Type: Transactional

This app allows you to create memo records for actual and forecasted cash flows not yet generated in the system. These records are displayed in cash management apps like Short-Term Cash Positioning (F5380), Cash Flow Analyzer (F2332), and Check Cash Flow Items (FO735) to provide an accurate view of your organization's cash position. You can search for existing memo records using search criteria such as editing status, status, company code, planning level, bank account, value date, currency, and certainty level. The app provides additional features, which include manual creation of memo records, importing records via the Import Memo Records 2.0 app (F6124), managing various types of memo records for both forecasted and actual cash flows, implementing dynamic field status controls based on memo record types and categories, and deleting invalid memo records, which are then excluded from cash management reports. The key difference between Manage Memo Records (F2986) and Manage Memo Records 2.0 (F2986A) lies in the enhanced features and flexibility offered by the latter. Let's compare these apps:

- **Manage Memo Records**
  This app enables the creation and management of memo records for actual and forecasted cash flows. It supports standard fields and controls and simple cash flow management scenarios.

- **Manage Memo Records 2.0**
  This app enables the creation and management of memo records for actual and forecasted cash flows. In addition, the app offers the ability to import memo records from spreadsheets. It supports dynamic field status controls based on user-defined memo record types and predefined categories. The app is suited for complex or large-scale cash flow management scenarios.

155

Tip

Consult SAP Note 3496240 to find additional relevant SAP Notes and restrictions related to this app.

Transaction Code

The corresponding transaction in SAP GUI is Transaction FF63.

Manage Migration Request (F5668)

Application Type: Transactional

This app helps with the creation of migration requests for commodity financial trades requiring retroactive designation. It automatically identifies and displays related transactions, including counterdeals, offsetting intercompany trades, intercompany trades, and foreign exchange (FX) transactions. The app's features include determining financial commodity trades for retroactive designation based on specified plan exposures, enabling selection from a proposed list of trades, simulating migration request creation, making trades from released migration requests accessible in the Execute Migration (CMM_MIGRATION) report, and displaying migration results and statuses in the application log. This report executes late designation for each trade in selected migration requests, designating relevant hedging instruments at the migration date.

Tip

In commodity price risk hedge accounting, late designation (process of assigning hedging relationships to financial trades retroactively) is performed solely as part of the migration process. The migration of on-behalf-of trades is supported only if the product type of external trades differs from that of the corresponding offsetting and intercompany offsetting trades.

Manage Organizational Changes (F4567)

Application Type: Transactional

This app allows you to manage the complete process of organizational changes, such as dividing, combining, or replacing profit centers assigned to objects like work breakdown structure (WBS) elements, projects, products, orders, and networks. With this app, you can define the effective date, company codes, and profit centers for reorganization, specifying affected objects for new profit center assignments. As a result of organizational change, stocks and balances of reassigned objects are reposted to new entities or objects. The app's features include entering effective dates, importing objects via templates, selecting specific company codes or profit centers, performing simulation and processing runs, and viewing organizational change history and related logs. The app's activities include creating and defining changes, optional simulation to preview transfer postings, activation of time-dependent profit center derivations, processing to reassign objects and post transfers, and final confirmation to release objects for future reorganizations. The app uses CDS view I_OrganizationalChange to display header data, including name, status, and effective date.

Tip

To create or process multiple organizational changes simultaneously, as a prerequisite, you must define the company codes and/or profit centers to include in each change. If no profit centers are specified, a company code can only be part of one incomplete organizational change at a time. If neither company codes nor profit centers are specified, all company codes in the system are automatically selected for the organizational change.

Consult SAP Note 3470484 for profit center reorganization for SAP S/4HANA.

Manage Outgoing Checks (F1578)

Application Type: Transactional

With this app, accounts payable accountants and managers can monitor and manage outgoing checks within their organization efficiently. You can search for outgoing checks using fields such as payment date, void date, encashment date, paying company code, house bank, bank account ID, and checkbook ID. The app presents outgoing check data in a grid format with the flexibility of adding additional dimensions. In addition, the app offers a comprehensive overview of the status of various checks, indicating whether they are new, voided, or cashed, allowing for immediate action. The app provides key features such as creating, editing, voiding, cashing, and reprinting checks. There is also a specific extension for China, providing tailored functionality. Also, the app supports email integration and offers an export to spreadsheet option for streamlined reporting and analysis.

Transaction Codes

This app covers the functionality of several SAP GUI transactions, including Transactions FCHN (Overview of Outgoing Checks), FCH1 (Display Single Check), FCH5 (Manually Create Single Check), FCH6 (Encash Check), FCH7 (Reprint Check), FCH8 (Void and Reverse Payment Document), FCH9 (Void Single Issued Check), and FCHG (Reset Encashed Check). However, the app does not support the functionality of Transaction FCHR, which is used for manual clearing during check cashing operations. You can consider an SAP GUI for HTML application for Transaction FCHR. Consult SAP Note 3037953 for further guidance.

Manage Outgoing Checks for China (F7014)

Application Type: Transactional

This app allows organizations to monitor and manage outgoing checks, providing a clear overview of their status such as **New**, **Voided**, or **Cashed**. Thus, you can take immediate action on outgoing checks. The app is available for accounts payable accountants and accounts payable managers roles, supporting functions such as creating, editing, voiding, cashing, and reprinting checks. For instance, you can filter outgoing checks with filter criteria such as payment date, void date, encashment date, paying company code, house bank, and bank account ID. Also, a China-specific extension is provided for localized functionality. In addition, the app includes email integration and an export-to-spreadsheet option for enhanced usability and record management.

Transaction Code

The corresponding transaction in SAP GUI is Transaction FCH1.

Manage Payment Advices (F2550)

Application Type: Transactional

With this app, accounts receivable and accounts payable users can manually create, view, edit, and delete payment advice, thereby ensuring payments are matched with invoice. This app serves a single place for displaying, searching, and managing customer or vendor payment advice. In addition, if machine learning is enabled for a customer's account, you also can import payment advice from PDF, TIF, JPG, PNG, and XLS files. You can search for existing payment advice using search criteria such as editing status, company code, payment advice type, customer ID, created on, and created by. The app displays payment advice in a tabular format with the flexibility of adding additional fields as necessary.

If contract accounting is enabled for customer accounts, you can use this app to manage payment advice for the contract accounting component. The app provides features such as creating, viewing, editing, and deleting payment advice, as well as importing them from various file formats.

Transaction Code

The corresponding transaction in SAP GUI is Transaction FBE1.

Manage Payment Agreements (F4629)

Application Type: Transactional

This app allows the definition of payment agreements and payment rules with banks for payment specialists. You can create, copy, and delete payment agreements and rules, which include general data, bank account details, format specifications, batch options, and grouping parameters. The app helps with payment processing by reconciling open items against payments and outlines how transactions are forwarded to house banks. The app provides four forwarding options for payment orders:

- **Forward Entire Order (B)** sends complete payment orders for bank communication management (BCM) approval without posting to additional ledgers
- **Forward by Originator Transaction (X)** splits payment orders by ordering party but doesn't support format-based forwarding
- **Individual Processing (I)** allows payment on behalf handling between head offices and subsidiaries
- **Batched Processing (C)** works like individual processing but groups items into batches based on set criteria

Outbound converters and formats must be assigned for processing, and grouping times can be scheduled for organizing transactions during the day.

Tip

SAP recommends using the Manage Workflows for Payment Agreements app (F5142) to set up the release process for payment agreements. Any modifications to payment agreements become effective only after the release workflow is completed. In other words, changes to payment agreements only take effect after the release workflow is completed.

Manage Payment Approval Rules (F6959)

Application Type: Transactional

This app allows you to create, modify, display, and delete payment approval rules, defining how authorized representatives can approve payments in cash management. Payment approval rules establish patterns for approver groups, specifying the number of approvers (one or two) required for different payment amounts. Once the approval rules are defined, these rules can be assigned to powers of attorney within the Manage Powers of Attorney for Banking Transactions app (F5742), linking approver groups to authorized representatives. In addition, you can view and filter lists of payment approval rules, create new rules, process existing ones, and delete rules with a **Create** status. The app also provides visibility into status transitions of payment approval rules such as **In Approval**, **Active**, and **Revoked**.

Tip

Payment approval rules can be defined and assigned to the powers of attorney for documentation purposes in the system, but they are not validated or enforced during payment processing.

Manage Payment Batches (F4039)

Application Type: Transactional

This app allows payment specialists to manage and process batched payments efficiently, thereby helping with the management and processing of batched payments for suppliers. You can search for payment batches by clearing area, batch status, and batch number. The app's key features include searching for payment batches, viewing detailed information for a specific batch, and navigating to related payment orders for streamlined processing. On the initial screen, you can apply multiple filter criteria such as payment run date and batch status. One key advantage of this app is it allows the bulk handling of multiple vendor payments at once, thereby improving process efficiency.

Manage Payment Blocks V2 (F0593A)

Application Type: Transactional

With this updated app, you can set and remove payment blocks on invoices or supplier accounts while utilizing search, sorting, and grouping functions to display invoices and view their status. The app offers features such as searching and filtering for supplier accounts and open invoices, using fuzzy search across two search bars, applying customizable filters using parameters such as supplier, company code, status and payment block reason, and displaying detailed information about supplier accounts and invoices. In addition, the app allows setting and removing payment blocks with a popup dialog for selecting block reasons and adding notes, enhancing the efficiency of managing payment blocks and invoice statuses.

This app replaces the deprecated Manage Payment Blocks app (F0593).

Transaction Code

The corresponding transaction in SAP GUI is Transaction FB02.

Manage Payment Items (F7051)

Application Type: Transactional

This app allows you to manage payment items by checking transaction status and details as part of payments and bank communication processes. The app supports searching for payment items, viewing their history, and displaying specific details. You can leverage filter options to quickly locate specific payment items based on fields such as status and paying company code.

In addition, authorized users can navigate from a payment item's details to related important apps, including House Bank Account (F1630), Manage Memo Records 2.0 (F2986A), Monitor Payments (F2388), Approve Bank Payments (Version 2) (F0673A), and Display Check Register (FCHN). The app supports managing both incoming and outgoing payments to streamline the tracking of payment items.

Manage Payment Media (F1868)

Application Type: Transactional

This app allows the transfer of data required for electronic payment transactions to banks via a data medium. A payment medium (electronic file) is created with each successful payment run and automatically saved as a TXT or XML file upon download, which is used to instruct banks to process payments. The app provides user with features such as viewing existing payment media with processing status, downloading and deleting payment media, and displaying and analyzing payment summary information corresponding to the data media. For example, you can search payment media using parameters such as reference number, bank country, run date, identification, and paying company code. The app displays payment files in **New**, **Downloaded**, and **Processed** categories. In addition, the app supports multiple payment formats such as SWIFT, ACH, SEPA, and so on. The app seamlessly integrates with bank communication management (BCM) for direct transmission to banks.

In addition, the app integrates with the Manage Automatic Payments app (F0770) and supports technical features like posting comments in SAP Jam and sending emails directly from the app.

Transaction Codes

Equivalent transactions in SAP GUI are Transactions FDTA (Administrate Data Medium), FBPM (Monitor Payment Media), and F110 (Automatic Payment Transaction).

Manage Payment Plans (F4806)

Application Type: Transactional

This app simplifies the management of regular payment runs, such as supplier payments and expense payments, by automating payment plans. You can search for payment plans using fields such as editing status, plan name, category, paying company code, and status. The app presents payment plan data in a grid format with the flexibility of adding additional dimensions. In addition, you can create payment plans by defining the category, payment method, and applicable business partners (suppliers or customers), and specifying the frequency, posting date, and document entry date. Factory calendars or customized calendars can be used as needed. Once payment plans are created, they can be activated, deactivated, edited, or deleted, with active plans running automatically as scheduled. The app's features include creating payment plans for different company codes and payment methods, categorizing plans (e.g., domestic or international payments), specifying recurrence (daily, weekly, or monthly), selecting items for payment runs, and shifting posting dates for specific recurrence types.

Additional technical features include SAP Jam integration for comments and emails, and

configuration options for payment plan categories and customized calendars. Payment runs for active plans appear in the Schedule Accounts Payable Jobs app (F2257) with distinct identification and universally unique identifiers (UUIDs). Deactivated plans remove jobs from the list, while reactivated plans retain the same UUID.

Tip

The app allows you to adjust the next posting date by a fixed number of days based on the last payment run's posting date. However, this feature is supported only for daily, monthly, and weekly recurrence types and is not applicable to payment plans using a custom calendar functionality.

Transaction Code

The corresponding transaction in SAP GUI is Transaction F110S.

Manage Payments (F3647)

Application Type: Transactional

This transactional app allows you to manage existing payment orders and items. It allows you to search for payment orders, check payment statuses, view payment history and details, recall payments, access purchase order expert, and view attachments and notes linked to payments. In addition, you can perform tasks such as approving payment batches and displaying payloads for successfully processed outbound orders. The app support field names to align with ISO 20022 terminology, layouts are tailored to specific payment scenarios, and the payment flow visualization has been optimized.

This app's features include predefined tabs for payments based on their status, navigation to corresponding batches in Monitor Payments (F2388) and Approve Bank Payments Version 2 (F0673A) apps, access to the bank and bank account of the payment originator, and links to bank messages in the Manage Bank Messages (F4385) and Manage Payment Media (F1868) apps. You also can navigate to related apps, such as House Bank Account (F5942), Manage Memo Records (F2986), Approve Payments Version 2 (FO673A), Check Register (FCHN), and Manage Journal Entries New Version (FO717A), provided you have the appropriate authorization.

Manage Pending Journal Entries (F5482)

Application Type: Transactional

This app allows general ledger accountants to view journal entries with errors that prevent posting to Central Finance and analyze their financial impact. The app provides a central worklist for reviewing pending entries during period-end close, enabling general ledger accountants to assess and resolve issues promptly. Journal entries from interface **AC_ DOC** replication can be filtered by criteria such as posting date, fiscal year, or company code, with results grouped by company code for evaluating business impact. In addition, you can navigate specific entries to review error messages, document details, and line items, using this information to determine appropriate actions. Also, the app integrates with the Manage Temporary Postings app (F6062) for creating or managing temporary postings linked to pending entries and offers filtering options based on the **Has Temporary Posting** field status. For related SAP Application Interface Framework messages in the namespace /FINCF, you can display links by adding the transaction ID field to the list of displayed fields.

Tip

The app might not display SAP Application Interface Framework messages with the **In Process** status. Consult SAP Note 3441148 for further guidance on this topic.

Manage Pending Journal Entries – Message View (F7149)

Application Type: Transactional

This app helps Central Finance users to review and resolve pending journal entries in SAP Application Interface Framework caused by validation errors or system checks. Thus, the app helps you ensure accurate financial close in Central Finance. You can filter failed entries by error messages, company codes, or other criteria for targeted issue resolution, while errors are grouped by message number to simplify corrections. The app provides detailed views of affected accounts, transactions, and error descriptions.

Moreover, the app integrates with the Manage Temporary Posting app (F6062), which allows corrective actions before posting entries. Thus, the app ensures compliance by highlighting discrepancies and validation failures when source system documents are transferred to Central Finance.

Manage Posting Period Variants (F2581)

Application Type: Transactional

This app allows you to manage account ranges, enabling the opening and closing of posting periods according to business needs. You can search for posting period variants using search criteria posting period variant and account type. The app offers you the ability to edit, copy, delete, split, and merge variant items. In addition, you can search for and view posting period variants and items, view associated company code and ledger information, and change the authorization group of variant items.

Tip

The app does not allow editing, merging, splitting, or deletion if a posting period variant is blocked. However, the app allows you to make copies.

Transaction Code

The corresponding transaction in SAP GUI is Transaction OB52.

Manage Posting Periods (F2293)

Application Type: Transactional

This app allows you to open and close posting periods to ensure postings are accounted for in the correct periods. You can schedule the opening of posting periods to run once or multiple times. You can search for current posting periods using posting period variant, fiscal year variant, account type, and account to fields. The app displays search results in **Posting Period View** and **Company Code View** sections for easier navigation.

To use its features, this app requires roles assigned to you with authorization object F_FAGL_OPV. Authorized users can view open posting periods from both posting period variants (PPV) and company code views, view which periods are open, shift open intervals manually, reset adjustment and controlling periods, schedule background jobs, and manage scheduled jobs by viewing, correcting, or deleting them.

Tip

To schedule special periods, SAP recommends scheduling all jobs separately. Consult SAP Notes 2896867, 2895403, and 3382230 for additional guidance on this subject.

Transaction Code

The corresponding transaction in SAP GUI is Transaction OB52.

Manage Posting Periods – Cost Accounting (F4684)

Application Type: Transactional

This app allows you to open and close fiscal periods for cost accounting transactions, ensuring that postings occur in the appropriate fiscal periods. With this app, you can display the status of fiscal periods, open or close fiscal periods by company code or for specific business transaction types, apply filtered criteria using fiscal year, fiscal period, ledger, business transaction type and status, and schedule background jobs to automate the process of opening and closing posting periods. As a prerequisite to use this app, the user role must be assigned with authorization object F_BTTYPE. In addition, a fiscal year variant must be assigned in the **Company Code Global Data** Customizing activity via Transaction OBY6.

Tip

If the app does not show any data, follow SAP Notes 3134145 and 3409836 for the recommended solutions.

Transaction Code

The equivalent transaction in SAP GUI is Transaction OKP1. As of SAP S/4HANA 2023, unlike SAP GUI Transaction OKP1, this app does not support keeping special periods from the previous year open independently if periods from the current year need to be closed.

Manage Powers of Attorney for Banking Transactions (F5742)

Application Type: Transactional

This app allows cash management users to manage powers of attorney for banking transactions within their company codes by defining principals, authorized representatives, payment rules, bank accounts, and permitted activities. In general, the power of attorney for banking transactions refers to the formal, legal authorization granted by a company to one or more designated individuals, allowing them to perform specific banking transactions on behalf of the company. The app offers you access to a list view of all powers of attorney, which can be filtered using search criteria such as editing status, power of attorney ID, and status. Once the list is displayed, you can create, edit, display, or delete powers of attorney, depending on the current status, such as **Created**, **In Approval**, **Active**, **Revoked**, or **Obsolete**, in the list. In addition, status transitions are automated based on user actions, ensuring adherence to dual control principles. Moreover, the app provides active powers of attorney benefit from automated bank account updates and can be modified or revoked as necessary. In addition, authorization concepts restrict access and data entry to specific company codes, with change documents maintaining a detailed history of modifications.

Tip

Deleting a power of attorney ID does not remove the associated change documents, as these records are retained for reference. The change documents can only be permanently deleted as part of the destruction process for powers of attorney for banking transactions using information lifecycle management (ILM) object FCLM_POA_DESTRUCTION.

Manage Promises to Pay (F4710)

Application Type: Transactional

This app allows accounts receivable users to manage customer promises to pay, created during the collection process via the Process Receivables Version 2 app (F0106A). With this app, you can display, edit, and withdraw promises to pay while leveraging features like searching and filtering by criteria such as customer number or case ID, viewing open or closed

promises, navigating to specific records, linking invoices and related promises, editing attributes like the promised amount, creating or modifying installment plans, adding notes, and attaching files.

Tip

The app does not support actions to withdraw or void a broken promise to pay, as only promises to pay with an **Open** status can be withdrawn.

Manage Provider Contracts (F6350)

Application Type: Transactional

This app serves as a centralized platform for maintaining provider contracts in SAP S/4HANA for billing and revenue innovation management. A provider contract encompasses all legally binding agreements between a customer and a company involving the provision and billing of services for a defined period in billing and revenue innovation management. With this app, you can display a list of provider contracts, filter results based on criteria such as business partner, provider contract, start date time, end date time, contract account and company code, and access detailed contract views by selecting a contract from the list. The app also allows contract modifications by choosing a provider contract, and clicking the **Change** option redirects you to the Change Provider Contracts app (FP_VT2).

Manage Provider Contracts for Sales Billing (F6404)

Application Type: Transactional

This app provides centralized access to provider contracts, enabling you to efficiently manage and review contract details in contract accounting. A provider contract is a legally binding agreement between a company and a customer for the provision and billing of services over a specified period. With this app, you can display a filtered list of provider contracts based on criteria such as business partner, provider contract, product, contract start at, contract end at, and company code. You can select a contract from the results list, which navigates you to the dedicated app for viewing provider contract details in sales billing.

Tip

Although both the Manage Provider Contracts (F6350) and Manage Provider Contracts for Sales Billing (F6404) apps offer similar benefits, the former handles broader contract management, while the latter is tailored for billing-related contract access.

Manage Reclassification Request (F5667)

Application Type: Transactional

In commodity hedge accounting, this app allows commodity risk managers to create and manage commodity reclassification requests. Reclassification refers to the process of moving amounts recorded in other comprehensive income (OCI) to profit and loss (P&L) accounts when the hedged transaction impacts P&L accounts, especially when commodity trade exceeds the total exposure quantity and results in an over-hedge. Thus, this step is required in cash flow hedge accounting to transfer accumulated hedging gains or losses from OCI to the income statement at the appropriate time to comply with IFRS 9. This app helps with initiating such requests to reclassify accumulated OCI amounts related to commodity cash flow hedgers.

The app's features include displaying reclassification requests previously generated through the Create Reclassification Requests (RCMM_CRE-ATE_RECLASS_REQUEST) report, manually creating new reclassification requests, releasing requests with the **Created** status, and navigating directly to associated dedesignation requests.

Manage Reconciliation Close (F4774)

Application Type: Transactional

As part of SAP S/4HANA Finance for group reporting, this app allows you to manage reconciliation close processes, view overall reconciliation status and differences by reason code, initiate approval workflows for reconciliation closure, and generate reconciliation statements for specific unit pairs and periods. The app's features include opening fiscal periods for reconciliation, viewing reconciliation status and difference breakdowns by reason code, adding comments, closing or reopening periods, and downloading reconciliation reports. As a prerequisite, this app requires activating the reconciliation close process for relevant cases via Transaction ICAARC. Differences are calculated based on reference group currency amount and displayed by tolerance levels (green, yellow, red). Reconciliation statuses include **Initially Closed**, **Open**, **Closed**, and **Pending Approval**, with workflows requiring approval before closure in some cases. The elimination status tracks tasks like balance sheet item eliminations. The app's **Details** view page shows breakdowns by reason code, allowing navigation to related apps for line-item analysis. In addition, you can reopen reconciliations to allow new data processing and download reconciliation statements as PDFs.

Tip

To enable workflow approval process, you must carry out the **Activating the Reconciliation Close Process** Customizing activity.

Manage Reconciliation Key (F4531)

Application Type: Transactional

This app allows you to manage reconciliation keys, which group contract accounting documents for transfer to the general ledger in contract accounting. With this app, you can manually create, display, close, and delete reconciliation keys, which remain open for postings until they are closed. Reconciliation keys can be aggregated under an aggregation key, reducing totals records and minimizing general ledger postings. In addition, the aggregated keys must be flagged and can only be transferred after aggregation. Also, the app provides an overview of reconciliation key usage, including transfer status, related transfer documents, and totals records. Before transferring totals records to the general ledger, a reconciliation key must be closed to prevent further postings or changes. In addition, some processes close reconciliation keys automatically, but manual closure is possible. For instance, the app supports closing reconciliation keys that are reserved for certain processes, such as the payment lot, when they result in technical problems such as terminations. The app supports the deletion of reconciliation keys so long as they don't contain any postings. The app's additional features include transferring totals records, reversing transfers (if already transferred), and displaying and checking totals records, which the system validates if present in the key.

> **Tip**
>
> The app allows you to navigate to important SAP GUI apps based on SAP Fiori with intent-based navigation configuration, including Create Reconciliation Key (FPF1), Change Reconciliation Key (FPF2), Display Reconciliation Key (FPF3), Display Documents for Reconciliation Key (FPT5), Display Posting Totals (FPT7), Aggregate Reconciliation Key (FPFAGG), Transfer Total Records to G/L (FPG1), and Reverse (FPG8). Thus, these apps facilitate the management and processing of reconciliation keys efficiently in contract accounting.

> **Transaction Code**
>
> The equivalent SAP GUI transaction is Transaction FPF2.

Manage Recurring Journal Entries (F1598)

Application Type: Transactional

A recurring journal entry represents repeated transactions, for example, deferring costs or revenues over twelve months. This app allows you to create, display, edit, and delete recurring journal entries, trigger due postings, and review the posting status and planned postings. The app triggers postings of journal entries using a template and recurrence rule. You can copy an existing entry as a template for new ones or create and modify recurrence rules. You also can search for existing recurring journal entries using parameters such as company code, recurring journal entry, and reference. In addition, the app allows you to personalize the data layout, predefine queries, save settings as variants, export data to spreadsheets, and share information via email.

> **Tip**
>
> The recurring journal entries created with SAP GUI Transaction FBD1 are not shown with this app and, as a result, they do not appear in the worklist. On the other hand, recurring journal entries created by this app cannot be processed with classic Transactions FBD2 and FBD5 due to the missing recurrence rules. SAP provides the report FDC_MIGRATE_RECUR_ACDOCS2FIORI to migrate recurring entries from SAP GUI to the SAP Fiori app. Consult SAP Note 2604731 for further guidance.

> **Transaction Code**
>
> The corresponding transaction in SAP GUI is Transaction F.14.

Manage Repayment Requests (F5754)

Application Type: Transactional

This app allows you to search, modify, or reverse repayment requests in contract accounting. You can initiate repayment requests within a payment lot when received payments cannot be assigned for that company. Using the app's filtering capabilities, you can search for repayment requests quickly using fields such as repayment request, status, net due date, IBAN, and account holder, and sort the results by various criteria, with options to customize filter variants, manage list columns, and hide the filter panel. In addition, you can select and view repayment request details and, if permitted by the request's status, reverse the request or adjust its due date and usage text. By default, the app displays only unexecuted repayment requests, but filters let you view reversed or executed requests as well.

Manage Resubmissions (F4709)

Application Type: Transactional

This app helps accounts receivable users to manage resubmissions for business partners created for open receivables, which are used in the collection process via the Process Receivables Version 2 app (F0106A). The app's key features include searching and filtering resubmissions by criteria such as business partner or resubmission reason, viewing **Open**, **Closed**, or all resubmissions, completing resubmissions, accessing individual resubmissions, displaying related journal entries (e.g., invoices), updating resubmission dates or reasons, adding or editing text notes, and attaching files.

Manage Revenue Accounting Items (F4270)

Application Type: Transactional

This app allows revenue accountants to search for and display postponed and processed revenue accounting items, providing item-level details such as related condition items. The app allows you to edit and process postponed items, as well as view detailed error logs. The app's features include advanced filtering, saving filter variants, customizing and grouping result columns, exporting data to spreadsheets while preserving display settings, and navigating to item details for further analysis. In addition, the app offers standard variants along with maintaining separate variants for additional filter capabilities. You can view logs for reprocessed revenue accounting items (RAIs) via program FARR_REPROCESS_POSTPONED_RAI. You can navigate operational documents, for example, the sales and distribution source document of a sales contract and a customer return document. The app also supports displaying custom fields for both postponed and processed order items.

Manage Revenue Contracts (F3883)

Application Type: Transactional

With this app, revenue accountants can manage revenue contracts and performance obligations (POBs) by performing tasks such as searching for contracts and displaying POBs in **List View** or **Hierarchy View**. You can search for revenue contracts by various parameters such as editing status, company code, accounting principle, revenue contract, customer, created by, and contract status. The app displays results, and you can review attributes like contractual price, standalone selling price, allocated amount, invoice amount, and effective quantity. In addition, the app allows editing attributes, creating POBs manually, adding attachments and notes, and navigating to related apps like Revenue Schedule (F3882) and Change History (F4273). The app's search functionality supports criteria like contract attributes, operational documents, and POBs, including fuzzy searches for fields like company code and accounting principle. The app displays results in plain list or detailed management views, with hierarchical structures for compound group POBs. Also, you can explore POB details, condition types, and accounts, ensuring effective contract and obligation management to support IFRS 15 and ASC 606 revenue accounting standards.

Manage Security Deposits (F4415)

Application Type: Transactional

This app allows accounts receivable users to display, create, clear, release, and reverse cash security deposits in contract accounting. You can search for specific security deposits such as business partner and contract account. In addition, you can create new security deposits by entering details like contract account, reason, amount, currency, and start date. On the other hand, clearing or releasing deposits removes payment restrictions, and partial releases allow offsets or payouts if a payment method is defined. Releases can be reset to reapply restrictions, and deposits can be reversed or adjusted by clearing unpaid portions or adding new requests. Security deposit requests also can trigger customer correspondence via customizable channels and print options.

Manage SEPA Mandates (F5158)

Application Type: Transactional

Single Euro Payments Area (SEPA) direct debit mandates are critical for businesses that collect payments directly from their customers' bank accounts within the Eurozone and other participating countries. This app allows you to create, display, or modify SEPA mandates for customers or business partners by navigating through dedicated apps such as Create SEPA Mandates (FSEPA_M1_PC), Display SEPA Mandates

(FSEPA_M3_PC), and Change SEPA Mandates (FSEPA_M2_PC). Each mandate is uniquely identified using the creditor ID and a 12-character mandate reference.

With this app, you can switch between applications within the selection criteria, tailored to SEPA's use across various system modules, and search for existing mandates by entering relevant data. Moreover, mandates can be created either from scratch or by copying existing ones, while individual mandates can be viewed and edited using the **Display/Change SEPA Mandates** screen. This app supports seamless integration and flexible management of SEPA mandates.

Tip

To prevent performance issues, SAP recommends limiting the display to no more than 250 items in the **Maximum Number of Results** parameter.

Manage Service Level Agreements (F6183)

Application Type: Transactional

This app allows payment specialists to create and manage service-level agreements (SLAs) for various types, including clearing area, customer segment, customer group, and customer ID (customer account). You can define rules based on company structure, with clearing area SLAs required if no structure is defined. You can view the details and settings of existing SLAs using search criteria such as clearing area, status, segment, and customer ID. The app's features include creating, editing, deleting, copying, and deactivating SLAs, as well as ordering activated SLAs for advanced payment management. In addition, the app allows specification of cut-off times for payment validation and adjustment, ensuring payments are processed on their due date. Also, customers can activate bank statement distribution, work with central incoming payments, and assign a business partner as the distribution owner and owner of the in-house bank account for posting payments.

Manage Substitution and Validation – For Asset Master Data (F6968)

Application Type: Transactional

This app allows you to define custom rules for asset master records in asset accounting. The app presents information in three sections: **General Information**, **Posting Information**, and **Additional Details** for intuitive maintenance. Validation rules enable the system to check user-entered values; thus, the app ensures data accuracy. On the other hand, substitution rules define conditions under which specific fields in asset master records are automatically filled or overwritten. These rules apply to manual asset master record creation and modification via the Manage Fixed Assets app (F3425) and to manual legacy data transfers using the Manage Legacy Assets app (F7280). As the rules apply to relevant apps, the app maintains consistent asset data.

Tip

The app is supported only if universal parallel accounting is activated via business function FINS_PARALLEL_ACCOUNTING_BF.

Manage Substitution and Validation Rules (F4406)

Application Type: Transactional

This app allows you to display, change, and create substitution and validation rules for selected business contexts and events. Using the app, you can create validation rules, create substitution rules, change rules, and display rules. The app supports rules that validate, derive, or replace values during data entry for relevant fields. In addition, the app supports business

contexts such as financial planning, journal entry items, market segments, goods receipt journal entry items, and service documents, with rules applied based on predefined contexts, which can't be modified. However, the app does support custom fields for user modification. During runtime, substitution rules automatically derive or modify values without messages, while validation rules verify entries and raise warnings or errors as needed. At runtime, these rules are triggered during postings or data entry using apps like Post Group Journal Entries (F2971).

Tip

In general, rules can be created or maintained by functional or advanced users with authorization object F_RE_SC_RU assigned to their roles.

Transaction Codes

The equivalent SAP GUI transactions are Transactions GGB0 and GGB1 for validation and substitution rules, respectively.

Manage Supplier Down Payment Requests (F1688)

Application Type: Transactional

A down payment request is a noted item and not a real financial accounting document. With this app, you can create down payment requests manually. Although the system typically generates them automatically based on supplier purchase orders, there are situations where you require manual intervention. For example, if a supplier requests a down payment outside the purchase order (e.g., via email or fax), then this app allows you to create it manually. The system automatically clears down payment requests once the payment run is complete, but manual clearance might be required in cases of payment discrepancies, such as partial payments. The app provides features such as viewing and exporting existing down payment requests, modifying or creating new requests, and adding notes and attachments to them. You can use the supplier, company code, and posted by fields to review existing down payment requests.

Tip

The app doesn't perform duplicate check for vendor down payment requests. The duplicate check is only done for sales-related items.

Transaction Code

The corresponding transaction in SAP GUI is Transaction F-47.

Manage Supplier Line Items (F0712)

Application Type: Transactional

The app empowers accounts payable or finance managers to handle ad hoc requests or recurring reports by locating supplier line items through comprehensive search capabilities, allowing you to find specific vendor line items using various search criteria. In addition, the app lets you view all supplier account line items or overdue supplier invoices for a company code, status, item type, and on a specific open on key date. The app allows you to personalize table layouts, predefined recurring queries, and saving settings as variants. Besides displaying data, you can set payment blocks, create manual payments, and export data for colleague sharing. The app also serves as a navigation point for drilling down into supplier line items from other apps. The app comes with features such as finding supplier line items, sorting and grouping data, exporting data with volume limitations, setting/removing payment blocks, changing line-item attributes, creating manual payments, and sending emails.

Tip

Consult SAP Note 3431912 on how to extend the Manage Customer Line Items (F0711) and Manage Supplier Line Items (F0712) apps using the Custom Fields and Logic app (F1481). The note offers guidance on adding custom fields to the apps with step-by-step instructions.

Transaction Codes

The equivalent classic credit management transactions in SAP GUI are Transactions FBL1 and FBLN.

Manage Target Quotas (F5795)

Application Type: Transactional

This app allows traders to define target quotas based on the derivative contract structure (DCS), providing guidance for effective hedging decisions in hedge accounting. Thus, the app helps traders make informed decisions on hedge management quotas and hedge accounting quotas. In hedge accounting, target quotas set limits on exposures to be hedged and can be defined for future periods at the level of company code and DCS. In other words, target quotas determine the maximum level of exposure that should be hedged, acting as a limit to guide hedging decisions. With this app, each target quota is uniquely defined by a specific combination of parameters such as company code, DCS ID, hedge book, period category, and buy/sell information. The app delivered with multiple tabs such as **General**, **Target Quota Periods**, **Status History**, **Version History**, and **Administrative Data** to display target quota data. You can create target quotas manually or via mass upload and distinguish between quotas for hedge management and hedge accounting. In addition, the app tracks revisions with a version number for every transaction. Also, the app records positions for each defined period within a year. With predelivered templates, you can download and upload target quotas and their positions using the **Upload Positions** and **Download Positions** options. The app uses CDS view C_CmmdtyHdgTargetQuotaDetailTP.

Manage Teams and Responsibilities for Journal Entry Verification – In General Ledger (F3932)

Application Type: Transactional

This app allows you to view and manage the hierarchical structure of teams, subteams, and superteams in the team categories such as general ledger journal entry verification (**FGJEV**), settlement management (**STLMT**), procurement (**PROC**), central procurement (**CPROC**), and extended production operations (**EPO**). This provides a clear overview and control of team responsibilities and organization. The app allows you to enter a team name, team description, team status, and team type to assign employees to different functions. In addition, you can restrict function by account type, company, company code, and cost center to receive and approve journal entry verification workflows.

Tip

To display business partners in the value help on the **Team Members** tab of this app, check the value of business partner assignment of identification type in table USR21. The value of this field must be either 02 or 04 (USR21–IDADTYPE = 02 or 04).

Manage Teams and Responsibilities – Intercompany (F6783)

Application Type: Transactional

This app allows you to define teams and assign functions for intercompany workflow agents in SAP S/4HANA Finance for group reporting. As a prerequisite, you must ensure relevant system users (Transaction SU01) have been set up as business partners. The app supports team creation for two key intercompany matching and reconciliation (ICMR) processes: variance adjustment and reconciliation close. Variance adjustment occurs when a matching assignment with an adjustment class is processed, which triggers approval workflows for team members with relevant roles. Teams must be assigned a team ID, name, and description, with defined company responsibilities and assigned functions such as **ICAM_SV** (approving variance adjustments) and **ICAM_SP** (supervising approvals). On the other hand, reconciliation close is initiated when a user finalizes period reconciliation for a trading unit pair, requiring approval workflows. In addition, teams must be assigned to reconciliation cases, companies, and partner units, with roles such as **ICAM_RCA** (approving reconciliation close) and **ICAM_RCC** (supervising approvals). Once these are set up, teams are enabled to process workflow approvals efficiently.

Manage Temporary Postings (F6062)

Application Type: Transactional

In Central Finance, this app allows the creation of temporary postings from pending journal entries by copying relevant data and references, with automatic or manual reversal options. You can search for temporary posting with fields such as editing status, sender logical system, company code, posting date, journal entry type, sender journal entry, sender financial year, and has temporary posting. The app presents pending journal entries in a tabular format with the flexibility of adding additional fields. The app supports limited editable fields for compliance, but changes can be made to general ledger account, customer, vendor, and cost center, with extensibility available for additional fields. Temporary postings are excluded from the central payment process to ensure payments are based on original open item data.

You also can navigate to SAP Application Interface Framework messages related to the pending journal entry for the temporary posting.

Tip

In the Central Finance scenario, temporary postings are intended to address missing master data or mapping, enabling a temporary posting. However, they are not meant to resolve conceptual issues, such as missing profit center information in a document split scenario. Consult SAP Note 3400784 for further information.

Manage Withholding Tax Items (F5428)

Application Type: Transactional

This app offers the ability to manage withholding tax items efficiently, thereby enabling statutory reporting and compliance. This app is offered as part of SAP Document and Reporting Compliance. Users like general ledger accountants, tax consultants, or accounts payable managers can include or exclude accounting documents from a tax reporting period by adjusting the withholding tax reporting date. The app allows you to search documents by various filter criteria such as company code, withholding tax reporting date, withholding tax type, withholding tax code, and business partner. Although posted documents cannot be revised after the reporting period closes, the app provides options to address errors by reversing the reporting period, posting a credit memo, or generating a withholding tax advanced compliance report in a correction run. For unreported documents, tasks include excluding the document, posting a credit memo or delta correction, or reporting the document in the next period. For incorrectly reported documents, you can post a credit memo or delta correction in a new period, adjust the withholding tax reporting date to the prior period, or perform a correction run for SAP Document and Reporting Compliance reports to include delta postings.

Tip

You must complete the prerequisite configurations for the app to display data correctly. First, to enable a tax item for inclusion or exclusion, the tax reporting date must be activated for the relevant company codes in the Customizing activity **Activate Reporting Date of Withholding Tax**. Next, the SAP_BR_AP_MAN-AGER authorization is required to display or change the reporting date for each company code. Lastly, you must execute the Migration Report for Withholding Tax Reporting Date (RPFIWT_MGRT_WHT_DATE) report by selecting either the document date or posting date as selection criteria.

Manage Workflows for Group Journal Entries (F6164)

Application Type: Transactional

In SAP S/4HANA Finance for group reporting, this app allows the maintenance of workflows for group journal entries, ensuring validation of manually posted entries within a company before they are committed to the database. With this app, you can define the required workflows for approval by approvers. By streamlining the review process with the four-eye principle, workflows adopt productive collaboration, helping teams ensure submitted group journals are error-free prior to posting. This principle allows a full review and approval process by at least two individuals. This requires one person to create or submit the journal entry and another to validate and approve it before it is posted to the database. As a result of this principle, accuracy and accountability are improved to support compliance per the company's accounting policies. The app's functionalities include checking workflow details such as status, validity period, and preconditions; creating new workflows by copying existing ones; activating and deactivating workflows; defining the order for activated workflows; and deleting deactivated workflows.

Tip

This app supports workflows that are specifically applicable to manually created documents processed through the Post Group Journal Entries app (F2971) or the Import Group Journal Entries app (F3073).

Manage Workflows for Journal Entry Verification – In General Ledger (F2720)

Application Type: Transactional

The app automates the process of verifying journal entries before they are posted to the general ledger. Thus, this app allows you to define and activate workflows for verifying general journal entries (Verify General Journal Entries – For Requester app (F2547A)) or currency adjustments via the Verify Currency Adjustments New Version app (F4670A) app. All journal entries that require verification will go through the workflow when they meet the prerequisite conditions. The app offers functionalities such as creating, editing, activating, and deactivating workflows; setting preconditions to initiate workflows; defining step sequences like verification and posting; and specifying exception handling scenarios.

Manage Workflows for Purchase Order Accruals Review (F3625)

Application Type: Transactional

This app allows you to configure workflows for reviewing purchase order accruals in the My Inbox – Review Purchase Order Accruals app (F3517). With this app, you can determine approvers, such as cost center responsible personnel or specific users, and initiate approval processes when activated workflows meet the defined start conditions. The app's features include setting start conditions like company code and accounting type, organizing the order

of start condition checks, activating or deactivating workflows, and modifying or creating workflows with custom properties, preconditions, and step sequences. In addition, the app ensures only one workflow is triggered when multiple workflows have matching conditions, allowing precise control over accrual review processes.

Manage Workflows – Intercompany Variance Adjustment (F5766)

Application Type: Transactional, Reuse Component

This app allows you to define and manage workflows for intercompany variance adjustments in SAP S/4HANA Finance for group reporting. The app allows you to create, customize, and sequence workflows, ensuring that variance adjustments are processed systematically and align with accounting policies of an organization. This app supports the configuration of approval steps, roles, and conditions to streamline the variance adjustment process as part of the intercompany reconciliation process. You can initiate variance adjustment via the Manage Matching Assignments app (F3870) when a user processes a matching assignment with a reason code linked to an adjustment class, which is a key element for classifying adjusting journal entries used during the consolidation process. In addition, to ensure auto-adjustments require prior approval, you can attach a workflow scenario, such as the standard scenario **WS78500087**, to the reason code. As a result of using this scenario, you can define workflows in the Manage Workflows – Intercompany Variance Adjustment app (F5766). Thus, you can activate the predelivered workflow **Intercompany Variance Adjustment Posting**, which can be used directly or customized as per your business requirements.

In a multilevel approval workflow scenario, SAP recommends adding step type **Post Variance Adjustments** as the last step of the workflow in case you need to perform adjustment postings. In case of multiple workflow activation, the system begins the one with conditions met. Likewise, SAP recommends making **Check Variance Adjustment Posting** the first step always.

> **Tip**
>
> As a prerequisite to use the workflow, you must maintain role assignment via the Manage Teams and Responsibilities – Intercompany app (F6783).
>
> This app currently supports the organizational dimension company and does not yet support other dimensions, such as consolidation unit.

Manage Workflows – Reconciliation Close Requests (F4993)

Application Type: Transactional, Reuse Component

With this app, you can initiate the reconciliation close process to close period reconciliation for a trading unit pair. SAP-delivered workflow approval can be activated to facilitate this process. To require approval before reconciliation close, you must configure the relevant reconciliation case in the Activate Reconciliation Close Process app (Transaction ICAARC). You can utilize standard workflows WS78500026 (for accounting) and WS78500027 (for consolidation), which are predelivered for this purpose. In addition, you can either use predelivered workflows or create custom ones on your own. Workflows must be followed sequentially, as the system skips follow-up workflows once a preceding one is triggered. To create a workflow, you must populate a name, validity dates, and add steps such as **Approve Reconciliation Close Request** (to allow approvers to accept or reject a close request) and **Confirm Reconciliation Close Request** (to execute the close without user input). Moreover, step recipients can be defined by role or team, though roles are recommended. Multilevel approval workflows can include additional approval steps as necessary per your

requirements. The final step should always be **Confirm Reconciliation Close Request** to execute the process.

> **Tip**
>
> As a prerequisite to support workflow functionality, you must maintain the teams and role assignment using the Manage Teams and Responsibilities – Intercompany app (F6783).

Manual Fulfillment (F3881)

Application Type: Transactional

This app allows revenue accounting users to manually fulfill performance obligations (POBs) when fulfillment is not triggered automatically by incoming events in revenue accounting and reporting (RAR). With this app, revenue accountants can specify fulfillment details, such as the quantity or percentage completed, and the system calculates the fulfillment accordingly. In addition, you can search for revenue contracts with manual fulfillment POBs, grouped by distinct or compound POB groups. Manual fulfillment of a POB occurs when the accountant manually specifies the quantity or percentage fulfilled, as it is not triggered automatically. Moreover, the app supports event-based fulfillment using delta quantities and percentage-based fulfillment using cumulative completion percentages. The app also supports adjustments, like reversing fulfilled quantities or reducing percentages. You also can view grouped POBs and navigate to related apps like Revenue Schedule (F3882) or Change History (F4273).

Manual Spreading (F6122)

Application Type: Transactional

This app allows revenue accountants to manage recognizable revenue amounts for multiple performance obligations (POBs) within a single contract in revenue accounting and reporting (RAR).

You can either manually edit these amounts directly on the screen or modify them offline via a downloadable spreadsheet, which can then be uploaded back into the system. The app supports spreading amounts from both open and closed periods. The app's features include resetting manually distributed revenue amounts to system default values, with necessary checks performed before saving. If revenue-related changes occur later, for example adjustments to POB duration or allocated amounts, the system replaces the manually captured revenue amounts with default values, potentially causing spreading conflicts. To address these conflicts, you can manually adjust the revenue amounts or restore the default system values.

> **Tip**
>
> With this app, you can control the revenue recognition pattern for time-based POBs manually. On the other hand, manual spreading is typically prohibited for acquisition cost POBs, POBs within a compound group, POBs lacking a start date, soft-deleted POBs, nondistinct POBs, as well as impaired contracts and those created using classic contract management (CCM).

Map FS Items with G/L Accounts (F3333)

Application Type: Transactional

This app allows general ledger accountants managing group reporting to display, create, or modify financial statement (FS) item mapping revisions, including relationships between FS items and general ledger chart of accounts. You can filter and display mapping revisions with details such as general ledger chart of accounts, consolidation chart of accounts, mapping ID, revision, and status. Statuses include **Draft** (unsaved changes), **Active** (saved), **Active with Draft** (saved but with unsaved edits), and **Assigned** (linked to an FS item mapping version and cannot be deleted). The app also shows the number of mapped and unmapped general ledger accounts, allows sorting and filtering of revisions, and provides detailed views for each revision, including mapped general ledger accounts.

> **Tip**
>
> If you want to delete all the FS item mappings in a mapping revision, a workaround via the Import FS Item Mappings app (F3335) can be used as an alternative option rather than the **Select All** option. Consult SAP Note 3487876 for further details.

Market Data Overview (F3941)

Application Type: Analytical

This app provides treasury teams with an overview of financial risks related to foreign exchange (FX) and market data. You can analyze KPIs such as **FX Spot Rate**, **Historic FX Spot Rate**, **FX Swap Rate**, **Historic Reference Interest Rate**, **Current Security Price**, and **Historic Security Price**, displayed on separate cards. In addition, you can apply filters using default parameters like key date, currency pair, and security class. Moreover, you can navigate from cards to detailed apps or market data for more information, customize the overview page using the **Manage Cards** function, and rearrange card positions for convenience. The app uses the following CDS views:

- C_FXSwapRateKeyDate
- C_FXSpotRateDeviation
- C_SecurityPriceReporting
- C_MarketDataOverViewSel
- C_MktDataRefIntrstRateRptg

You can also navigate to apps such as Entering FX Spot Rates via Currency Exchange Rates (F3616), Entering FX Swap Rates (TMDFXFP), Enter Reference Interest Rates (JBIRMC), and Enter Security Prices (FW17), ensuring a streamlined financial risk analysis process.

Migrate Attachments for Dispute Cases (F5100)

Application Type: Transactional

As part of effective dispute management, this app helps accounts receivable users migrate attachments related to dispute cases, ensuring their availability in the new Manage Dispute Cases (Version 2) (F0702A) app. After migration, attachments are duplicated to a secondary storage location while remaining accessible in their original location for use in the deprecated app. The app's features include scheduling the **Migrate Attachments** job with parameters such as case ID, customer, company code, case type, and status. As a result of migrating dispute case attachments, you preserve data via historical information during system transitions, support business continuity by providing access to documentation for smoother operations, and enhance compliance by maintaining a complete audit trail.

Tip

The SAP GUI report DISPUTE_MIGRATE_ ATTACHMENTS can be executed once to display dispute cases in the Manage Dispute Case app (F0702A).

Migrate to Global Hierarchies (F4965)

Application Type: Transactional

With this app, you can migrate the financial statement versions (FSVs) created in Customizing activity **Define Financial Statement Versions** via SAP GUI Transaction OB58 to global hierarchies. Once migrated, the FSVs can only be edited in the Manage Global Hierarchies app (F2918). Hierarchies migrated successfully are set to **Migrated** status and are visible in the Manage Global Hierarchies app. On the other

hand, if you want to revert the migration of hierarchies, you can choose **Undo Migration**. This action will set the hierarchies status from **Migrated** to **Not Started**. If errors occur during the migration process, you can choose the **Last Error: Show in Log** option to view them. If the hierarchy has already been replicated by Transaction HRRP_REP, the app prevents it from being migrated again. You must remove the replicated hierarchy from HRRP tables using Transaction HRRP_REP and migrate the hierarchy using the app.

Tip

Although migration of FSV hierarchies can be reverted if you prefer maintaining the hierarchies via SAP GUI transactions, SAP recommends using the Manage Global Hierarchies app for any future changes to ensure consistency and alignment with leading practices, as SAP Fiori represents the future of business user interfaces.

Transaction Codes

The app supports migrating hierarchies created via SAP GUI transactions such as Transactions OB58, FSE2, or HRRP_REP.

Monitor Bank Account Balances (F5176)

Application Type: Transactional

This app allows cash and liquidity management users to effectively monitor bank account balances updates. The app supports timely updates of bank statements for statements imported automatically or manually. The app's functionalities include filtering bank accounts based on their status and attributes, exporting reports to spreadsheets, and tracking balance updates. Before using the app, you must ensure that the bank accounts to be monitored have defined attributes in the **Bank Relationship** tab of the Manage Bank Accounts app (F1366A), such as interval and interval unit to set the frequency of balance updates. In addition, you can set a factory calendar ID for considering specific calendar schedules optionally. The app uses the One Exposure from Operations table FQM_FLOW to display account balances.

> **Tip**
>
> To monitor bank account balances through this app, consult SAP Note 3334466 to understand Customizing activities for initiating the balance upload (FQM_INIT_BALANCES) and other troubleshooting tips.

Monitor Bank Fees (F3001)

Application Type: Analytical

Organizations manage and monitor bank fees associated with various banking transactions on a day-to-day basis. This app allows you to monitor imported bank fee data and track service charges, taxes, and compensations with detailed drilldown dimensions. The app presents bank fees in bar charts and graphs for easier visualization. For example, you can view total amount by bank key, total amount by bank country, total amount by company code, and total amount by account number. Also, you can analyze unit prices, service volumes, and fees by comparing data across companies, regions, and banks or monitoring charges over time. It validates imported fees to detect errors or improper charges. The app offers additional features such as sending emails with app links, sharing comments in SAP Jam groups, and saving custom tiles with preset selection criteria for quick access. Thus, the app offers a centralized tool for monitoring, analyzing bank fees, and tracking service charges.

> **Tip**
>
> As a prerequisite to using this app, SAP recommends that the exchange rates for the relevant currencies are properly defined. This includes all the currencies mentioned in the imported source files and the display currency you intend to use for monitoring bank fees. Otherwise, some data may not be displayed accurately.

Monitor Change Requests (WDA_FCLM_BAM_REQOVERVIEW (APP_CFG_FCLM_BAM_REQOVERVIEW))

Application Type: Web Dynpro

This app offers a centralized view of bank account-related change requests within bank account management (BAM), enabling you to track, review, and manage modifications, approvals, and closures while ensuring compliance with financial governance policies. The app provides comprehensive request tracking, aligns with SAP's approval workflows for validation, and supports filtering by request type, status, or submission date. In addition, the app maintains audit records for compliance and delivers real-time status updates to monitor pending and completed requests.

> **Tip**
>
> To use this BAM app, you must activate the WDA_FCLM_BAM_REQOVERVIEW service via Transaction SICF.

Monitor Disconnection Requests (F4943)

Application Type: Transactional

Contract accounting allows you to block or disconnect services consumed by customers who have open receivables that are subject to the dunning process. This app allows contract accounting users to monitor the status of disconnection and reconnection requests, which can be grouped by the **Valid, Invalid,** or **Obsolete** statuses. Disconnection requests become obsolete when the originating dunning run is reversed before being sent to an external system, while invalid requests are those already sent externally. You can sort and filter requests by parameters such as company code or contract account, and view details such as dunned business partner items or existing reconnection requests.

In addition, you can navigate to related apps for additional information or data modification, such as Business Partner Financial Overview (F2429), Manage Business Partner Item (F2562), Display Accounting Balance (New) (F0707A), Manage Contract Account (F5474), and Display Provider Contract (FP_VT3).

Monitor GR/IR Account Reconciliation (F3303)

Application Type: Analytical

This app provides an overview of the status of open financial accounting items on goods receipt/invoice receipt (GR/IR) accounts requiring clarification and analyzing discrepancies between goods receipts and invoice receipts. This app is important, especially since it is used during periodic activities in the general ledger. Within this app, under the KPI **Purchasing Document Items**, there is a card titled **GR Amount Equals IR Amount**. This card displays the total count of purchasing document items where the goods receipts amounts match the invoice receipts amounts, grouped by the relevant dimension.

You can navigate to this app from the Reconcile GR/IR Accounts app (F3302) for detailed investigation and clarification of mismatched amounts or quantities. Also, you can leverage the app by analyzing reconciliation processes, viewing workloads, identifying surplus by company code or plant, tracking open items in financial accounting, and utilizing machine learning to prioritize and highlighting items with changes older than 180 days. In addition, the app can integrate SAP Jam and has emailing capabilities for communication among stakeholders.

Tip

The app does not support intercompany recurring journal entry postings.

Transaction Code

The corresponding transaction in SAP GUI is Transaction MR11.

Monitor Hedge Constellation (F5665)

Application Type: Transactional

This app provides an overview of inconsistencies in hedge constellations and allows direct navigation to the impacted commodity financial transactions for detailed examination in commodity hedge accounting. In SAP, hedge constellations refer to the structured relationships between exposures, hedging instruments (e.g., swaps or forwards), and hedging strategies within a hedging area (in an organization's hedging policy). The app's features include identifying missing intercompany trades during trading on behalf, inconsistencies between company codes of hedge execution and offsetting intercompany trades, invalid valid-from dates in hedge constellations, missing exposure items for offsetting intercompany trades, missing company codes for offsetting intercompany trades, and invalid company codes for commodity exposure items.

Monitor Instant Balances (F7804)

Application Type: Transactional

This app allows cash managers to retrieve and update instant balances for selected bank accounts directly via SAP Multi-Bank Connectivity. You can filter accounts using account number, company code, or house bank; view the latest available balances along with time stamps; and check details such as status and last request time. The app lets you navigate to the Define Bank Account Settings – Instant Balances app (F7805), which allows management of settings for instant balance tracking. In addition, you can customize fields displayed in the bank accounts table for better visibility.

To monitor instant balances using this app, you must carry out the necessary setup, which includes connecting to SAP Multi-Bank Connectivity, enabling instant balances for the house bank in the Manage Banks – Cash Management app (F1574A), and activating tracking via the Define Bank Account Settings – Instant Balances app (F7805), which has a **Switch On** or **Switch Off** button to enable or disable balances for selected bank accounts. Thus, the Monitor Instant Balances app provides real-time visibility to view account balances instantly.

Tip

The app only supports selected member banks of SAP Multi-Bank Connectivity; check scope item 16R (Bank Integration with SAP Multi-Bank Connectivity) for further details.

Monitor Master Data Distribution – to SAP CC (F4194)

Application Type: Transactional

This app allows you to monitor the distribution status of master data objects from the SAP S/4HANA Cloud system to convergent charging in SAP S/4HANA billing and revenue innovation management. Master data distribution is required in an integrated scenario involving convergent charging, which manages service pricing and invoice recipient determination. This app allows monitoring of distribution status and allows manual restart of incomplete distributions based on system-written distribution records. In addition, the app checks the successful distribution of provider contracts and related master data; displays overviews of incomplete, failed, and successful distributions; and lets you filter results by provider contract or distribution date and time. Distribution statuses include **Distribution Successful, Distribution Incomplete** (repeating possible), **Distribution of Master Data Change Failed** (previously successful but failed), and **Distribution Failed** (requires sender system correction). Distribution records are shown as flow charts with nodes for master data objects, each with statuses like **Distribution Scheduled, Change Scheduled, Distribution Successful, Change Failed**, or **Distribution Failed**. You can navigate to nodes for detailed information.

Monitor Payments (F2388)

Application Type: Transactional

With this app, accounts receivable and accounts payable specialists and treasury users can display an overview of their payment batches and view the statuses of batches and individual payments at various processing stages. You can search for payments using fields such as paying company code, payment run date, payment run ID, and payment batch alert. The app presents results in a tabular format with multiple tabs such as **All Batches, New, In Approval, Approved, Sent to Bank, Completed**, and **Exceptions**. In addition, the app provides features that include performing remediations to fix common issues with payment batches; viewing the history of a payment batch from creation to completion; editing due dates and instruction keys; accessing details for banks, accounts, and payments; seeing contact information for all batch approvers; and navigating to the details of the payment medium file if it has been created.

Transaction Codes

The equivalent SAP GUI transaction is Transaction BNK_MONI. For people transitioning from SAP GUI, this app includes similar features as those in Transactions FBL1N (Vendor Line-Item Display), FBL5N (Customer Line-Item Display), FBZ0 (Payment Proposal), and FBZ5 (Display Payment List).

Monitor Plan Exposures (F6001)

Application Type: Transactional

This app provides a detailed overview of commodity plan exposures for selected combinations of company codes and derivative contract structures (DCSs). The app displays plan exposures based on various filter criteria such as company, DCS, hedge book, valid at, and delivery period. You can view exposure commodity quantities in **Table View** or **Chart View** formats for the current date or future dates; filter data by hedge books and delivery periods; choose specific levels of detail; toggle chart views or select chart types; and export the data to Microsoft Excel for further analysis. In **Chart View**, you can perform tasks such as including chart elements based on the chosen filters, displaying the legend with **Show Legend** option, and switching between chart types. On the other hand, with **Table View**, you can perform tasks such as drilling down to the node of the selected dimension to view granular details of a specific plan exposure, navigating to the underlying derivative contract, and checking the total of the plan exposure quantities. The app uses root CDS view C_CmmdtyHdgPlanExpsrQuery.

Tip

The app displays only released plan exposures (which have been formally approved and marked as ready for use in hedge accounting).

Monitor Predictive Accounting (F3828)

Application Type: Analytical

For predictive accounting reporting, correct and complete data is essential. Errors or incomplete source documents and configuration settings can hinder the generation of predictive journal entries, thereby affecting the quality of predictions for revenue, cost of sales, and margins. This app identifies and analyzes errors in source documents or configurations within the predictive accounting process. You can address these errors, improving data quality and enhancing predictions. The app presents predictive accounting data errors in graphs and bar charts with important information such as number of erroneous orders overall, number of erroneous orders by simulation step, and number of erroneous orders by message. The app's features include a tile showing prediction quality over the past seven days, with success rates calculated by the percentage of source documents processed. In addition, the app offers a **Standard** view listing source documents with errors, sortable by error type, process step, or error age, and provides filters for quick navigation. You can analyze top error types by system error category, process status, creation step, or error age. Moreover, you can exclude documents that can be reprocessed as needed, and manual checks can be triggered to validate adjustments for accurate processing.

Monitor Revenue Contracts (F5014)

Application Type: Transactional

This app allows you to track and manage revenue contracts effectively in revenue accounting and reporting (RAR). As a result, the app offers a centralized and user-friendly way to oversee revenue contracts, track their progress, and support accurate and timely revenue recognition. With this app, you can search for contracts associated with specific operational documents and view key details, such as the contractual price, pending days, manually created or fulfilled performance obligations (POBs), frozen POBs, and postponed items. In addition, the app offers features such as clicking on manual fulfillment counts to address POBs in the Monitor Revenue Contracts app (F5014) or resolving conflicted contracts through the Work List for Conflicted Contracts app (F4271). Also, the app supports navigation to other apps such as Revenue Schedule (F3882) and Allocated Amount Explanation (F4424).

The app's advanced search options include attributes like company codes, operational documents, and POB names, complemented by a fuzzy search across fields such as company code, business partner, and customer. The app's key data points include postponed items, indicating failed inbound processing, pending days for conflicted contracts, and contract end date, which reflects the maximum end date of time-based POBs or remains empty otherwise.

My Bank Account Worklist (F0736)

Application Type: Web Dynpro

This app allows cash management specialists to check and process bank account-related requests requiring their approval or processing in a centralized view. The app provides features such as approving or processing requests for opening, changing, or closing a bank account; maintaining a payment approver in multiple accounts; and reviewing bank accounts. In addition, the app provides details like account balance, available funds, and transaction history, eliminating the need to navigate multiple systems. Also, the app offers real-time monitoring of account balances and transactions, highlighting exceptions or discrepancies for prompt corrective actions. You also can forward requests to colleagues if needed and monitor all related change requests in your area of responsibility.

Tip

The app only supports workflow requests generated using the old workflow template WS74300043.

My Credit Card Transactions (F6506)

Application Type: Transactional

As part of travel management, this app allows you to view and manage credit card transactions efficiently. In addition, the app allows deletion of private expenses, assignment of transactions to existing expense reports, and creation of new expense reports based on selected transactions, streamlining financial tracking and expense management. The app requires the business traveler role to be assigned to the user.

Transaction Code

The corresponding transaction in SAP GUI is Transaction PR05.

My Dunning Proposals (F2435)

Application Type: Transactional

This app is designed to streamline and automate the dunning process within accounts receivable departments. With this app, you can create dunning proposals and review the dunning notices before sending them to customers. You also can search for dunning proposals using search fields such as company code, customer, delivery lock, and dunning procedure. The app presents search results in a tabular format with the flexibility of adding dimensions. If you decide not to send a dunning notice to a particular customer for business reasons, the app allows you to set a dunning block on that notice, excluding it from the batch of notices being sent. The app provides features such as creating new dunning proposals daily, previewing dunning notices, setting dunning blocks, and sending dunning notices. The app displays dunning proposal ID, company code, customer number and customer name, total overdue amount, dunning level, dunning date, and status, which is useful for decision making. In addition, the app allows you to send dunning notices to customers by email.

My Inbox – Approve Travel Expenses (F0410A)

Application Type: Transactional

The My Inbox apps let you make key decisions on mobile or desktop devices, processing both standard and custom workflow tasks defined in the backend system. This app features an **All Items** inbox for easy task management, the option to create scenario-specific tiles, and an **Outbox** for viewing completed and suspended tasks. The app provides functionalities that include processing tasks from various providers, customizing decision options, performing standard and mass actions, adding comments, managing attachments, creating substitution rules, and extensive task management capabilities. In addition, the app supports communication, detailed task history, filtering options, and customizable views, all aimed at efficient task handling within the SAP Fiori launchpad. Also, the app supports customization for UI extensions, expert user views, and alternate task organization.

> **Tip**
>
> The app does not support open task features on mobile devices.

> **Transaction Code**
>
> The corresponding transaction in SAP GUI is Transaction SBWP.

My Inbox – Approve Travel Requests (F0411A)

Application Type: Transactional

Like the My Inbox – Approve Travel Expenses app (F0410A), this app allows you to make important decisions and process standard or custom workflow tasks on mobile or desktop devices. The app provides an **All Items** inbox for managing tasks via a preconfigured tile in the SAP Fiori launchpad, the ability to define scenario-specific workflows, and an **Outbox** for completed and suspended tasks. Key features of the app include processing tasks from SAP and third-party providers, customizable decision options, standard and mass actions, comment and attachment management, substitution rules, communication via email or SAP Jam sites, extensive task browsing and filtering, and customizable views for efficient task handling. The app also supports detailed task history and the creation of alternate views for expert users.

> **Tip**
>
> The app does not support open task features on mobile devices.

> **Transaction Code**
>
> The equivalent transaction in SAP GUI is Transaction SBWP.

My Inbox for Convergent Invoicing (F7031)

Application Type: Transactional

This app provides a centralized interface for managing and approving convergent invoices in convergent invoicing under SAP S/4HANA billing and revenue innovation management. The app supports complex billing scenarios involving multiple services and charges. In addition, you can streamline invoice processing with workflow-based approvals, ensuring compliance with predefined business rules. The app enhances visibility and tracking, offering detailed invoice insights such as billable items, adjustments, and approval history. Moreover, the app integrates with contract accounting to provide accurate invoicing for subscription-based and usage-based services. The app provides customizable filters and notifications to help prioritize tasks and ensure timely approvals.

Transaction Code

The corresponding transaction in SAP GUI is Transaction FKKBIXBIT02_TRANS_MA.

My Inbox for Manage Bank Accounts (F2797)

Application Type: Transactional

The app simplifies and centralizes approval and review process for managing bank accounts, thereby reducing redundant manual steps. With this app, you can handle bank account-related work items, including creating, editing, closing, and reopening accounts. You can process workflow tasks such as opening or modifying a bank account, maintaining payment approvers, and handling account closures or reopening. In addition, the app allows you to approve, reject, claim, release, suspend, and forward tasks, as well as sort, group, and search for tasks, view task-specific details, add attachments and comments, and send emails with URLs for selected information. In addition, if your company uses SAP Jam, the app allows you to post comments about tasks in SAP Jam groups.

Transaction

The corresponding transaction in SAP GUI is Transaction SBWP.

My Inbox (Group Journal Entries) (F6165)

Application Type: Transactional

This app is used for the approval process for workflows related to group journal entries in SAP S/4HANA Finance for group reporting. It allows you to review and approve workflows directly from your inbox, streamlining the validation process for manually posted group journal entries before they are committed to the database. The Workflows for Group Journal Entries app (F6164) defines the required workflows for approval, whereas this app complements workflow approvers tasks using My Inbox functionality.

Tip

If journal entries are not displayed in the app, you must maintain relevant workflow task types in Transaction SWFVMD1 with the appropriate component name.

My Inbox – Review Purchase Order Accruals (F3517)

Application Type: Transactional

This app lets purchasing users review and adjust purchase order accruals, allowing accrual amounts to be reviewed and approved prior to posting in the system. Reviewers can validate the system-proposed periodic planned costs and make necessary adjustments as needed. Periodic accrual amounts are calculated as the difference between planned costs, determined periodically using an accrual method, and actual costs, based on receipt or invoice values for goods or services. The app's features include viewing total accrual amounts for purchase order items, revising planned costs, and confirming the review of accrual objects before posting.

My Open Worklists (Version 2) (F1611A)

Application Type: Transactional

With this app, you can display the results of all open worklists and their corresponding items created in various apps such as Analyze Overdue Items (Version 2) (FO860A), Analyze Payment Locks (Version 2) (F6134), Analyze Dunning Locks (Version 2) (F1655A), and more. This app allows you to track the working status, set worklists and their items to obsolete, and personalize table layouts. The app delivers features such as displaying open worklists and their items, showing results in different currencies, setting worklists to obsolete, calculating amounts based on status, personalizing tables by adding or removing columns, exporting data to spreadsheets, and forwarding worklist information via email. In addition, upon completion of all items in a worklist, you will receive a notification in the SAP Fiori launchpad indicating that the worklist has been closed.

My Sent Requests (Version 2) (F1371A)

Application Type: Transactional

With this app, you can track bank account-related requests. The app allows you to check the status, actions taken, next processor, creation date, and last change date of your requests; view change request details and approval history; cancel unprocessed requests; navigate bank account details; and filter or search for requests by ID or title. In addition, the app allows you to send URLs to selected information, post comments in SAP Jam, and save the current selection criteria as a tile.

My Travel and Expenses for Business Traveler (F6190)

Application Type: Transactional

This app allows employees or travel assistants to view and submit travel and nontravel expenses within travel management. The app's features include creating, editing, and submitting expense reports; simplifying report creation through the **Copy Trip** function; managing attachments at both the expense report and receipt levels; capturing receipts using a device camera; and assigning expenses via the receipt buffer with multiple selection. You can access various expense types, including per diems and mileage, assign attendees to expenses, itemize receipts with detailed fields, and deduct personal expenses. In addition, the app supports multidestination and multicurrency reports, cost splitting across multiple cost objects, and handling reports for other employees. The app offers additional capabilities, including saving and editing drafts, managing per diem advances, viewing credit card receipts prior to expense creation for personal expense deletion, and editing foreign currency rates.

My Travel Plans (F6768)

Application Type: Transactional

This transactional app allows real-time travel management on desktops and mobile devices. You can create, display, delete, and cancel travel plans, facilitating bookings for flights, hotels, cars, or trains through a third-party partner, GDS (Global Distribution Service). In addition, the app also allows browsing existing travel plans in PDF format and includes reservation details, ticketing, and pricing information. Also, the Travel Plans for Travel Assistant app (F6771) supports agents in booking trips on behalf of others.

Transaction Code

The corresponding transaction in SAP GUI is Transaction TP01.

My Travel Requests for Business Traveler (F6015)

Application Type: Transactional

As part of travel management, this app allows you to manage travel requests in real time using a desktop or mobile device, and it also allows travel assistants or travel agents to handle requests for individuals under their responsibility. The app offers the chance to create or copy travel requests with additional destinations, cost assignments, estimated costs, advances, travel services, and planning. Requests can be submitted with attachments to an organizational manager, browsed in detail, edited and resubmitted, or deleted. In addition, travel assistants or travel agents can use this app or use the Travel Requests for Travel Assistant app (F0409B) to create requests on behalf of others. For example, you can search for travel requests with filter criteria using parameters such as

employee name, status, start date, and end date. On the resulting screen, the app displays travel requests with important information such as trip number, reason, destination, country, status, start date, end date, and total cost.

Tip

SAP recommends a separate role created by an HCM administrator for access to this app and the My Travel and Expenses for Business Traveler app (F6190).

Transaction Code

The corresponding transaction in SAP GUI is Transaction PR05.

Navigation for RAI Monitor (FARR_RAI_MON_NAV)

Application Type: Web Dynpro

This app provides an intuitive interface for accessing and managing revenue accounting items (RAIs) in revenue accounting and reporting (RAR). The app allows you to search, filter, and navigate RAIs based on predefined criteria. For example, you can search for RAIs with parameters such as company code, customer, and contract. The app displays transaction details, processing status, and financial impact while supporting error identification and resolution. The app complies with IFRS 15 and ASC 606 accounting standards.

Tip

To use this app, you must activate the Internet Communication Framework (ICF) security service for FARR_RAI_MON_NAV via Transaction SICF.

New Versus Solved Disputes (F2523)

Application Type: Analytical

With this analytical app, accounts receivable users can display the **Created Versus Solved Disputes** KPI, which shows the number of dispute cases created and those closed or confirmed today. In addition, you can compare the number and amounts of created and solved disputes and drill down by status, processor, customer, or dispute case. The app provides functionality such as seeing the number of disputes created and solved today, comparing clarified amounts to new amounts, filtering by various categories, drilling down by coordinator, customer, status, or dispute case, and selecting different time periods for drilldowns. Caching is enabled for better performance, with manual and automatic data refresh options based on the cache duration in the tile configuration.

Let's compare the New Versus Solved Disputes (F2523) and Solved Disputes (F2521) apps:

- **New Versus Solved Disputes**
  This app includes both newly created and resolved disputes over a time frame. Its focus is on creation and resolution.

- **Solved Disputes**
  This app only shows resolved disputes. Its focus is on an overall view of disputes resolved.

Observe and Monitor Deal Requests (F5655)

Application Type: Transactional

This app allows traders of commodities to track and review the details of all requests created in commodity price risk hedge accounting. The app allows you to monitor various request categories in a comprehensive way, including migration requests, dedesignation requests, counterdeal requests, reclassification requests, and foreign exchange (FX) hedge requests. As a result, you can verify deal requests comply with internal policies and external regulations.

Tip

To display the complete data, SAP recommends assigning the business object CMM_OBSRV to the user.

Open Disputes (F1752)

Application Type: Analytical

This app is designed to provide a streamlined and intuitive interface for managing and tracking dispute cases in the accounts receivable area. It helps accounts receivable users to monitor and resolve outstanding disputes, ensuring better cash flow management and customer relationships across different customers and company codes. With this analytical app, you can display the **Open Disputes** KPI to view disputes that are not closed, confirmed, voided, deleted, or canceled. You can search for open disputes using company code or reason. The app allows you to view these open disputes in a chart or table, categorized by company code, customer, processor, and dispute reason. With this app, you can perform tasks such as viewing the total amount assigned to open disputes on the KPI tile and viewing the number of disputes by reason, processor, and the top 10 customers with the highest disputed amounts. In addition, you can customize the display currency and exchange rate type for conversion. The app is enabled with caching to improve performance, with options to refresh data manually or automatically based on cache duration settings.

Organizational Change Report – Financial Data (W0175)

Application Type: Web Dynpro

This app provides a comprehensive view of financial data impacted by organizational changes in the general ledger. The app presents organizational change data once an organizational change has been simulated, processed, or completed. With this app, you can view affected financial data, drill down to line-item details, and navigate to related apps. The app displays the impact of organizational change using measures such as amount before organizational change (balance of postings before the effective date), transferred amount by organizational change (identified by transaction category **OCTP**), and amount after organizational change (adjusted balance post-change). Business transaction category **OCTP** refers to organizational change transfer posting. By default, these measures are available in company code currency and global currency, with an option to add functional currency via the **Navigation Panel**. In addition, transfer postings before the effective date is classified under **OCTP**, whereas postings on or after the effective date are categorized as organizational change correction postings (**OCCP**) and are no longer reflected in the app. However, SAP recommends analyzing organizational change corrections posting (**OCCP**) as a result of structural changes like profit center reorganizations using the Profit Centers – Actual (FO959) or the Display Line Items in General Ledger (F2217) apps. Thus, the app allows you to analyze the impact of organizational change such as profit center reassignment.

> **Tip**
>
> The app presents measures in multiple currencies such as the company code and functional currency. However, when aggregated, only one currency is displayed, typically based on ascending alphabetical order. To avoid this issue, SAP recommends filtering for one or more company codes that share the same currency.

Organizational Change Report – Master Data (F4868)

Application Type: Transactional

This app allows you to manage organizational changes by splitting, combining, or replacing profit centers linked to objects such as work breakdown structure (WBS) elements, projects, products, orders, and networks. You can define the effective date, company codes, and profit centers impacted by the change, and specify the affected objects and their new assignments. The app's functionality includes setting a reorganization date, specifying objects for reassignment, importing data via templates, and selecting relevant company codes or profit centers. In addition, the app allows you to simulate, activate, process, and complete organizational changes while providing access to audit history, related reports, and error logs. The app supports the use of the CDS view I_OrganizationalChange to display header data changes for details such as name, status, and effective date.

> **Tip**
>
> To create or process multiple organizational changes simultaneously, you must specify the company codes and/or profit centers to include in each change. If no profit centers are specified, each company code can only be associated with one incomplete organizational change at a time. If neither company codes nor profit centers are specified, all company codes in the system are automatically selected for the organizational change.

Organizational Change Report – Master Data List (F7047)

Application Type: Transactional

This app provides a hierarchical overview of object master data impacted by organizational changes such as changes in cost centers and profit centers. You can search for organizational changes on master data using search fields such as organizational change, root object type, dependent object type, old profit center, new profit center, and status. The app presents results in structured tree format to navigate relevant details. The results displayed include important information such as object, old profit center, new profit center, and status. Reporting of changes is available once an organizational change has been simulated, processed, or completed. For instance, in case of profit center reassignment, the app status reflects the organizational change stage, with no reassignment occurring during simulation and a successful reassignment expected after processing. In addition, you can view affected master data by expanding or collapsing list levels for root and dependent object types. The app also provides access to additional actions for individual list levels by either right-clicking on desktop computers or long pressing on touch-enabled devices. Moreover, you can check reassignment statuses in the **Status** column.

In addition, the app supports navigating to the Manage Organizational Changes app (F4567) and other related apps for further processing. Thus, the app provides information such as who made changes, when they were made, and what specific changes occurred to support compliance and governance purposes.

Tip

Although both the Organizational Change Report – Master Data (F4868) and Organizational Change Report – Master Data List (F7047) apps track organizational changes, the former app provides a structured hierarchy, while the latter app presents a detailed list format for granular analysis.

Overdue Payables (S/4HANA) (F1746)

Application Type: Analytical

This app allows you to check the overdue payable amounts to a company's suppliers by supplier company code, supplier group, supplier, and reason for payment block. In addition, you can monitor the status of overdue payments for critical suppliers, identify potential risks, and notify responsible persons to take action. Also, the app displays both overdue and not-yet-overdue payable amounts, distinguishes between

critically and uncritically overdue amounts, and allows analysis by various views, including company code, supplier group, payment block reason, and supplier. It breaks down overdue amounts into four intervals before today. For example, if the interval is set to 30 days, the app segments overdue payables into four buckets: 0–30, 31–60, 61–90, and 91–120 days before today. The app also lets you specify a key date for data analysis, with additional technical features like posting comments via SAP Jam and sending emails. The app uses CDS view C_APOVRD and has caching enabled for better performance.

Tip

The app defaults to limiting data records to 200 entries in the drilldown view. This is standard behavior as the app is primarily designed to display KPI values. To understand how this app calculates critical and uncritical overdue amounts, consult SAP Note 2977812.

Transaction Code

The corresponding transaction in SAP GUI is Transaction F.46.

Overdue Receivables by Risk Class (F2539)

Application Type: Analytical

With this app, you can display overdue receivables by company code, country key, customer, and risk class, thereby helping them to analyze risk diversification and support segmentation, credit, and payment term decision making. In addition, the app has features such as viewing the top three risk classes by overdue amounts on the launchpad tile, selecting from existing variants (for example, company code and risk class, due period and risk class), and drilling down for more precise analysis by adding further dimensions and measures. This serves as a comprehensive tool for managing and evaluating overdue receivables by risk class. In addition, caching is enabled for better performance, with data updated automatically and options for manual refresh.

Overdue Receivables in Dispute (F2540)

Application Type: Analytical

With this app, you can display the amounts of overdue receivables in disputes by company code, reason, processor, country key, and customer. You can perform tasks such as viewing the total amount of overdue receivables in dispute, drilling down for a more precise analysis by various dimensions such as dispute reason, processor, and root cause. The app is enabled with caching for better performance. Also, the app provides options to manually refresh data and automatically update it based on the cache duration in the tile configuration. The app is primarily used for displaying overdue receivables, and not necessarily to provide a comprehensive view of all receivables. Thus, the app allows you to prioritize open disputes that delay the payment of overdue invoices.

Overdue Receivables (S/4HANA) (F1747)

Application Type: Analytical

This app allows you to view the **Overdue Receivables** KPI and analyze the 10 highest overdue receivables by customers, thereby enabling quick action to reduce the highest amounts owed to your business. You also can analyze overdue receivables by company code or accounting clerk, and view data in a chart or table based on company code, customer, country and region of the customer, and accounting clerk. The app provides features such as viewing the current overdue receivables status as a percentage, analyzing receivables by predefined due periods, and analyzing the top 10 overdue amounts by customers, company codes, and accounting clerks. In addition, the app supports SAP Jam for comments, and email notifications.

Tip

Like the Overdue Payables (S/4HANA) app (F1746), this app also defaults to limiting data records to 200 entries in the drilldown view. This is standard behavior as the app is primarily designed to display KPI values rather than providing granular details.

Transaction Codes

The corresponding transactions in SAP GUI are Transactions FOEBL1, FBL5N, and S_ALR_87012178.

P&L by Function of Expense – By Consolidation Units (CCONS_FPM_OVP_IS_COS_BY_CU (CCONS_FPM_OVP_IS_COS_BY_CU))

Application Type: Web Dynpro

This app provides a structured view of profit and loss (P&L) data by functional expense category using a cost of sales approach at the consolidation unit level in financial consolidation. The app allows group accountants to analyze cost distribution across entities by functional categories such as cost of sales and administrative expenses. In addition, the app supports intercompany reconciliation to align expenses across companies. The app also provides drilldown capabilities for further analysis of specific cost categories.

P&L by Function of Expense – Year Comparison (CCONS_FPM_OVP_IS_COS_YOY_BY_CU (CCONS_FPM_OVP_IS_COS_YOY_BY_CU))

Application Type: Web Dynpro

This app provides a comparative view of profit and loss (P&L) data year over year at the consolidation unit level in financial consolidation. The app's features include comparing consolidation data across fiscal years at the consolidation unit level, identifying cost fluctuations, and performing intercompany reconciliation. Thus, you can find trends, growth, and deviations in revenue and cost.

> **P&L by Nature of Expense – By Functional Areas (CCONS_FPM_OVP_IS_FA_BY_CU (CCONS_FPM_OVP_IS_FA_BY_CU))**
>
> Application Type: Web Dynpro

This app provides a structured view of profit and loss (P&L) data by functional area and by functional expense category using a cost of sales approach at the consolidation unit level in financial consolidation. The app allows group accountants to analyze cost distribution for cost categories such as salaries, depreciation, and material expenses using functional areas such as administration and production. In addition, the app supports intercompany reconciliation to align expenses across companies. The app also provides drilldown capabilities for further analysis of specific cost categories. This app uses the old reporting logic for handling financial consolidation in the previous group reporting solutions such as SAP Business Planning and Consolidation (SAP BPC).

> **P&L by Nature of Expense – Year Comparison (CCONS_FPM_OVP_IS_YOY_BY_CU (CCONS_FPM_OVP_IS_YOY_BY_CU))**
>
> Application Type: Web Dynpro

This app provides a structured view of profit and loss (P&L) data by nature of expense classification such as salaries, depreciation, and material consumption in financial consolidation. The app provides a year-over-year comparison to understand changes in each type of expense over time. In addition, the app supports intercompany reconciliation to align expenses across

companies. The app also provides drilldown capabilities for further analysis of specific cost categories. This app uses old reporting logic for handling financial consolidation in the previous group reporting solutions such as SAP Business Planning and Consolidation (SAP BPC).

> **Payment Factory Component for My Inbox (F3705)**
>
> Application Type: Transactional

Using this app, payment specialists can streamline payment processes by integrating payment-related workflows into the Payment Factory Component for My Inbox app (F3705). The app allows you to manage and approve master data changes or payment related activities efficiently, providing a centralized platform for handling payment workflows. It supports features such as task prioritization, decision making, and seamless integration with backend systems to ensure smooth cash payment operations. By using this app, you can enhance your payment management capabilities and improve overall operational efficiency in cash management.

> **Payment Statistics (S/4HANA) (F0693A)**
>
> Application Type: Analytical

With this app, you can display the KPI **Payment Statistics** to check the amount and number of critical payments made within 90 days through bank communication management (BCM) using various filtering criteria. The app allows you to view critical payments by status, company, house bank, processing days, payment method, and currency. In addition, critical payments include new payments, those in the approval process, sent to banks, and payments with exceptions. The app also supports posting comments on SAP Jam and sending emails. Also, you can navigate to other linked apps directly, provided those apps are already enabled in the system. If the linked apps are not enabled, SAP recommends implementing them along with this app. For example, one such linked app is Check Cash Flow Items (F0735).

Payments Analyzer (F4040)

Application Type: Analytical

This app allows accounts payable teams to analyze payments by displaying key figures in various charts and tables formats. You can search for payments using parameters such as display currency, exchange rate type, exchange rate date, clearing area, created on, processing status, released status, and in repair. The app presents payment items in both graphical and tabular formats to analyze payments. As a result, you can monitor the status of outgoing payments to vendors and suppliers, quickly identify and resolve payment failures caused by issues such as incorrect bank account details or insufficient funds, and analyze payment data to optimize workflows and eliminate bottlenecks. In addition, you can select key figures and time periods based on your requirements, filter and aggregate payment data, convert payment amounts to different display currencies on the spot, and save visualizations as tiles in the SAP Fiori launchpad for easy access.

Performance Obligation (FARR_POB_DETAIL_OVP)

Application Type: Web Dynpro

A performance obligation (POB) is a promise in a contract to transfer a distinct good or service to a customer or business partner in revenue accounting and reporting (RAR). This app provides a comprehensive view of POBs within contracts. The app allows you to track POB fulfillment details, manage revenue recognition timelines, and analyze transaction details. You can view fulfillment status, transaction prices, price allocation, and standalone selling price using search parameters such as POB, contract

status, business partner, customer, and company code. The app is integrated with other functionalities such as sales and distribution and financial accounting. In addition, the app complies with IFRS 15 and ASC 606 accounting standards.

Performance Obligation Structure (FARR_POB_MGMT_OVP)

Application Type: Web Dynpro

This app provides a structured view of performance obligations (POBs) within revenue contracts in revenue accounting and reporting (RAR). The app allows you to create and manage a hierarchical framework for POBs, organizing them into parent-child relationships that reflect the bundling of sales deals. The app organizes POBs hierarchically, tracks revenue allocation based on predefined rules and standalone selling prices, and monitors fulfillment using event-based or time-based criteria. In addition, the app is integrated with revenue accounting to comply with IFRS 15 and ASC 606 accounting standards.

Post Asset Acquisition – And Quantity Adjustment (F6488)

Application Type: Transactional

This app allows you to process asset accounting transactions, including acquisition, credit memos, post-capitalization, and quantity adjustments. You manage this process with the app's **General Information**, **Posting Information**, and **Additional Details** sections. Acquisitions involve posting purchase documents for newly acquired assets, while post-capitalization corrects acquisition costs, such as capitalizing mistakenly posted expenses. Credit memos adjust asset purchases and invoices, whether from the current or prior fiscal year. Quantity adjustments modify asset quantities without altering book values. Thus, this app streamlines asset postings for accurate financial management.

Tip

As a prerequisite, this app is available only if the universal parallel accounting business function (FINS_PARALLEL_ACCOUNTING_BF) has been activated.

Post Currency Adjustments (F1606)

Application Type: Transactional

The app allows you to adjust account balances in one or more currencies, which is particularly useful in volatile exchange rate environments and during period end. The app contains the **Header**, **Attachments**, **Notes**, and **Balances** tabs to post currency adjustments. In addition, the app allows you to enter amounts for specific currencies directly during posting, which then posts to the Universal Journal (table ACDOCA)—not to table BSEG, which is the basis for open item management. In addition, the app provides options for saving drafts before posting changes, adjusting balances in foreign currencies, and adding attachments and notes to explain postings. Moreover, the app allows adjustments to freely defined currencies.

Tip

The app does not support posting adjustments to general ledger view-only transactions—for example, foreign currency valuation (BSTAT = U).

Transaction Code

Transaction FBB1 in SAP GUI can provide a similar functionality, but the transaction is limited to three currencies and it doesn't support ledger-specific postings or freely definable currencies. Unlike Transaction FBB1 in SAP GUI, the app does not provide document currency value. The app merely translates the foreign currency into the local currency.

Post General Journal Entries (F0718)

Application Type: Transactional

This app allows you to manually enter or upload postings to general ledger accounts via a template file. This is useful for users who want to make adjusting entries during period-end closing or record transactions without source documents (e.g., expenses paid with petty cash). The app contains the **Header**, **Attachments**, **Notes**, and **Balances** tabs to post journal entries. In addition, the app allows you to autocalculate taxes when entering line items. Moreover, the app provides additional functions such as adding account assignments (for example, cost center, orders, or work breakdown structure (WBS) line items) and adding notes or attachment to substantiate postings.

Tip

To activate negative posting for automatically created items, you must select **Negative Posting** indicator in the Edit Options for Journal Entries – My Settings app (F2130). In addition, the Manage G/L Account Master Data (Version 2) app (F0731A) controls whether general ledger accounts appear in the Post General Journal Entries (F0718) app. Also, accounts will be excluded from search results if they are blocked for posting or if general ledger accounts are set with the **Posting Automatically Only** flag, assigned a **Recon. Account for Acct Type** field, or restricted to general journal entries.

Transaction Codes

The equivalent transactions in SAP GUI are Transactions F-02, FB50, and FB01.

Post Group Journal Entries (F2971)

Application Type: Transactional

As part of SAP S/4HANA Finance for group reporting, this app helps with consolidation tasks, such as intercompany eliminations and consolidation of investments, by supporting both automatic and manual postings for financial statement imbalances, deferred income tax, or balance adjustments. You can enter group journals in **General Data** and **Line Items** tabs. In addition, the app integrates with the Data Monitor and Consolidation Monitor, enabling tasks like manual postings, eliminations, and group share entries while checking and updating task statuses. You can create journal entries by specifying consolidation details, entering amounts in local or group currency, and leveraging system-assisted field updates and validation checks. The app offers additional features such as balancing line items, attaching documents, and copying existing entries streamline the process. In addition, you can post deferred taxes with appropriate attributes, assign financial statement (FS) items, reverse journal entries individually or in bulk, and save or delete draft entries before posting. This ensures accurate and efficient consolidation processes.

Transaction Code

The corresponding transaction in SAP GUI is Transaction CX_50.

Post Incoming Payments (F1345)

Application Type: Transactional

With this app, you can post and clear a single incoming payment in one step. If payments aren't received through electronic bank statements, you can enter the payment data manually and search for matching open items. Then the system ideally proposes matching items for one-step posting and clearing. If clearing isn't possible, you can post the payment on account

or to a general ledger account. The app provides features such as obtaining a list of open items, adding or changing discounts, creating residual items, defining business partner account postings, posting payments to general ledger accounts, using promise-to-pay information, searching open items with fuzzy logic, entering profitability characteristics, viewing withholding tax, creating notes and attachments, simulating journal entries, and exporting the open items list to a spreadsheet. In addition, you also can make additional account assignments for bank fees. The **Select More** option provides additional open items to be added in addition to previous open items. The app also offers supplier and customer search help.

Tip

This app can't be used to post an incoming payment, to clear customer invoices containing withholding tax, nor to post an incoming payment and clear customer invoices from multiple customer accounts when there is a payment difference that cannot be directly assigned to the invoices. For these situations, SAP recommends using the Post Incoming Payments – For Customers app (F1345) instead. In addition, the app doesn't allow you to edit line-item text in simulation mode. SAP recommends using the Clear Incoming Payments app (F0773) for this purpose.

Transaction Code

The corresponding transaction in SAP GUI is Transaction F-04.

Post Outgoing Payments (F1612)

Application Type: Transactional

With this app, you can post and clear a single outgoing payment in one step. While outgoing payments are typically processed automatically based on payment proposals, you can manually enter payment data for immediate processing.

The app allows you to clear outgoing payments with open items or post them on account or to a general ledger account. The app's features include obtaining a list of open items for clearing, adding or changing invoice discounts, creating residual items, posting to a general ledger account with account assignment, searching for open items using fuzzy logic, creating notes and attachments, entering profitability-related characteristics, arranging open items using invoice references, simulating journal entries, clearing payments with selected open items, exporting the open items list to a spreadsheet, and making additional account assignments such as for bank fees.

Transaction Code

The corresponding transaction in SAP GUI is Transaction F-07.

Post Tax Payables (F1597)

Application Type: Transactional

This app offers the chance to process tax amounts by posting balances from input and output tax accounts to a tax payable account, without needing a tax code, thereby allowing them to make manual adjustments to tax accounts directly. The app contains the **Header, Attachments, Notes**, and **Balances** tabs to post

tax payable adjustments. In addition, you can make manual adjustments and post in all relevant ledgers. Moreover, the app allows saving drafts before posting, posting tax account balances, and adding attachments and notes to explain postings. Moreover, the system supports postings to all relevant ledgers for the accounts you're responsible for.

Transaction Code

The corresponding transaction in SAP GUI is Transaction FB41.

Price Allocation (F5448)

Application Type: Transactional

This app allows the allocation of contractual prices across performance obligations (POBs) in a revenue accounting contract, typically based on standalone selling prices in revenue accounting and reporting (RAR). However, the system also supports alternative price allocation methods to address specific business scenarios. In addition, the app allows revenue accountants to adjust allocated amounts for multiple POBs within a single contract manually. Revenue is recognized over time or at a point in time, depending on the nature of each POB or contract requirements. This is critical especially for determining the amount of revenue to be recognized for each POB as the company fulfills contract obligations. As a result, the app helps users to comply with revenue recognition standards such as IFRS 15 and ASC 606.

Price Allocation (FARR_ALLOC_PRICE_OVP)

Application Type: Web Dynpro

This app helps with the allocation of transaction prices across performance obligations (POBs) within revenue contracts in revenue accounting and reporting (RAR). The app distributes revenue based on predefined rules and standalone selling prices, integrates with Business Rule Framework plus (BRF+) for maintaining allocation logic, and applies consistency checks to verify accuracy before the execution of revenue processes. The app complies with IFRS 15 and ASC 606 accounting standards. This differs from the other Price Allocation app (F5448) because that app focuses on specific contract adjustments, whereas the Price Allocation app (FARR_ALLOC_PRICE_OVP) focuses on broader management of price allocation.

Tip

To use this app, you must activate the Internet Communication Framework (ICF) security service for FARR_ALLOC_PRICE_OVP via Transaction SICF.

Process Collections Worklist (F0380)

Application Type: Transactional

This app allows collection specialists to manage and prioritize collections worklists, focusing on the most urgent customers with the highest or longest overdue amounts. With this app, you can display your collections worklist, access other worklists, assign worklist items to other users, and create customer correspondence. The app presents collection worklists in the **My Open Items**, **My Completed Items**, and **Unassigned Items** tabs for intuitive navigation. Thus, the app increases efficiency in collections management, reduces days sales outstanding (DSO), improves customer relationships through structured follow-ups, and enhances visibility into receivables performance.

Transaction Code

The corresponding transaction in SAP GUI is Transaction UDM_SPECIALIST.

Process Free Form Payments (F2564)

Application Type: Transactional

With this app, you can manage payment requests by displaying a comprehensive list and viewing detailed information for each individual request. You can search for free form payments using fields such as payment required, company code, payee name, payment currency amount, payment currency, created by, enhanced on, and status. The app presents results in tabular columns with the flexibility of adding dimensions and drilldown capabilities. Depending on your role, the app allows you to release or reverse payment requests. Releasing a request initiates the payment run and creates the payment batch, ensuring a smooth and streamlined payment process. Also, you can view detailed information about payment requests, releasing payment requests to initiate the payment run and create the payment batch, reversing payment requests if necessary, navigating clearing journal entry/accounting document smart link, and displaying and using payee aliases to simplify the identification and management of payees. These features ensure that you have full control over payment requests, enhancing the efficiency and effectiveness of your payment management process.

Transaction Code

The corresponding transaction in SAP GUI is Transaction FIBLFFP.

Process Hedge Requests – Balance Sheet FX Risk (F4764)

Application Type: Transactional

This app allows treasury users to display, edit, release, or delete balance sheet exposure hedge requests. The app's features include filtering hedge requests using search criteria such as hedge request ID, status, company code, or snapshot ID; editing, releasing, or deleting hedge requests with status **Created**; and undoing the release of requests with status **Released** if the associated trade request remains unfulfilled. In addition, you can view detailed information, including general data, account assignments, and action history, and navigate related trade requests or the underlying snapshot. Releasing a hedge request automatically generates a trade request, and you can adapt filters or analyze fulfillment statuses through fields like open amount and declined amount.

Tip

Consult SAP Note 3095561 for balance sheet foreign exchange (FX) risk hedge management for features and enhancement-related notes.

Process Overhedged Plan Exposures (F6003)

Application Type: Transactional

As part of commodity risk management, this app allows you to address over-hedged plan exposures by creating dedesignation requests, initiating counterdeal requests, and adjusting reclassification requests to resolve the over-hedge situation effectively. The app comes with filter criteria using common parameters such as company, derivative contract structure (DCS), hedge book, and order year. The app displays plan exposures in **List View** for over-hedged plan exposures. In addition, you can use **Create Dedesignation Request**, **Create Counter Deal Request**, and **Edit Reclassification Request** options for handling the planned exposure volume decreases or the hedge limit ratio decreases.

Tip

As a prerequisite, in addition to needing security access with authorization object CMM_ CHDD, SAP recommends creating two versions of plan categories using the Mange Commodity Plan Exposure app (F5664) to support plan exposure volume fluctuations.

Process Receivables (Version 2) (F0106A)

Application Type: Transactional

With this app, collection specialists can manage and collect open receivables efficiently. By providing a comprehensive list of all open receivables related to a specific customer, the app allows you to document subsequent actions taken. The app allows you to display customer information (such as open receivables, credit memos, and contact history) by business partner, collection segment, and risk class. You also can view the overall status of customer receivables, manage head office or branch relationships, send invoices and correspondence via email or print, display the aging of receivables, and create promises to pay, dispute cases, resubmissions, and notes at the customer level. The app can be accessed directly from the SAP Fiori launchpad or through the Process Collections Worklist app (F0380).

Process Snapshots – Balance Sheet FX Risk (F4763)

Application Type: Transactional

This app allows treasury users to manage snapshots generated using the Take Snapshots – Balance Sheet FX Risk app (FXM_SNAP) by checking, updating, releasing, or deleting them. With the Take Snapshots – Balance Sheet FX Risk app, you can take a snapshot of balance sheet exposures and the corresponding hedges on a key date. The Process Snapshot app's features, on the other hand, include displaying snapshots and items, manually creating snapshot items for those with the **Created** status, and releasing snapshots, with the option to undo accidental releases if no hedge requests are based on them. In addition, you can delete snapshots with the status **Created** and navigate related hedge requests for snapshots with the status **Released**. In addition, the app offers filters such as snapshot ID, key date, status, and others to locate snapshots, while the **Details View** provides comprehensive information on snapshot items. Moreover, manually created items must be released before releasing the snapshot, and accidental items or snapshots can be deleted as needed. Released snapshots allow access to related hedge requests displayed in the Process Hedge Requests – Balance Sheet FX Risk app (F4764).

Tip

To process or display a snapshot or snapshot items, for additional dimensions, you must select a single data source only. For instance, the app uses the planning level, which is relevant for data sources, especially for financial accounting and the One Exposure from Operations hub.

Processing Days of Open Disputes (F2522)

Application Type: Analytical

The analytical app displays the **Processing Days of Open Disputes** KPI, representing the average processing days of all open disputes that are currently open. The app includes disputes in **Open** and **New** statuses. In addition, you can compare the processing days of open disputes with the disputed amount and drill down by coordinator, customer, status, or dispute case. This helps monitor the workload of collection specialists and identify areas needing attention. The app's features include seeing the total disputed amount for cases due soon, comparing open disputed cases with their amounts, filtering by various categories, and drilling down by coordinator, customer, status, or dispute case. The app uses the CDS view C_OPENDISPPROCESSINGDAYS, and caching is enabled for better performance, with an option to refresh data manually or automatically based on the cache duration in the tile configuration.

Promises to Pay (F1753)

Application Type: Analytical

With this analytical app, you can display the **Promises to Pay** KPI, which displays open promises to pay as of the current date. You can view promised amounts in different periods, as well as overdue and broken promises by company code, customer, country and region of the customer, and collection specialist. The app provides features such as viewing the total amount of open promises to pay on the KPI tile, checking open promises within specific future periods, customizing up to three periods for viewing, and analyzing promises by customer and collection specialist. In addition, you can view broken promises and customize the display currency and exchange rate type for conversion.

1

Quick Create Free Form Payment (F3038)

Application Type: Transactional

The purpose of this app is to enable you to efficiently create free form payments with minimal effort. This app streamlines the payment creation process by allowing you to quickly specify key details like payment request type, amount, and recipient information. It is especially useful for handling one-time or non-standard payments without predefined templates, ensuring a smooth and user-friendly experience. SAP Collaboration Manager, the digital assistant, and conversational AI enable you to create free-form payments directly from your current SAP Fiori screen using the Create Free Form Payment app (F3038), without needing to navigate away. This seamless integration simplifies payment creation without leaving your current screen.

Quick Create Memo Records (F3029)

Application Type: Transactional

This app simplifies the process of creating memo records for expected cash flows for cash management teams. The app is designed for quick and efficient entry of essential details like company code, value date, bank account, and liquidity item to generate memo records. By focusing on key fields, the app minimizes input

time while maintaining accuracy, making it ideal for you to quickly register expected cash flow entries. Memo records created through this app are integrated with One Exposure from Operations and are reflected in cash management reports such as Cash Flow Analyzer (F2332) and Check Cash Flow Items (F0735), ensuring visibility and consistency in cash flow analysis. This app allows you to centrally manage memo records for expected cash flows not yet generated in the system. Memo records automatically expire after their expiration date, and active records can be manually converted into archived records. All memo records are stored in One Exposure from Operations, ensuring seamless integration with cash management tools.

Quick Create Payment Advice (F2790)

Application Type: Transactional

Payment advice generally refers to a document or note that is sent to a vendor or customer to inform them of a clearing open item, detailing the payment amount, the payment date, and other relevant transaction details. In SAP, payment advice can be created in multiple ways, including while processing bank statements or manually processing open items to indicate the current processing status of open items. With this app, accounts payable users can manually create, view, edit, and delete payment advice.

If machine learning is enabled for the customer account (using cash applications service and requires a separate license), you also can import payment advice from PDF, TIF, JPG, PNG, and XLS files. In addition, if contract accounting is enabled, you can manage payment advice for contract accounting.

Transaction Code

The corresponding transaction in SAP GUI is Transaction FBE1.

Quick View Payment Advice (F2789)

Application Type: Transactional

Payment advice generally refers to a document or note that is sent to a vendor or customer to inform them of a clearing open item, detailing the payment document number, payment amount, the payment date, and other relevant transaction details. With this app, accounts payable users can search and view payment advice. In addition, the app provides vendor/customer information and bank account details. In addition, if contract accounting is enabled, you also can search for and view payment advice for contract accounting.

Let's compare the Quick View Payment Advice (F2789) and Quick Create Payment Advice (F2790) apps:

- **Quick View Payment Advice**
  This app allows you to search for and view existing payment advice. It's primarily used when you want to track or verify payments.

- **Quick Create Payment Advice**
  This app allows you to create and post new payment advice. It's primarily used when making a payment.

Transaction Code

The corresponding transaction in SAP GUI is Transaction FBE3.

Rebuild Credit Management Data (F4234)

Application Type: Transactional

This app enables credit management users to schedule jobs for rebuilding data that's essential for credit management. This process may be required when changes are made to customer master data, credit limits, or risk categories. As a result, this app helps you ensure that all credit-related information is up to date and synchronized for business partners. The app's key features include the ability to reconstruct credit exposure for specific business partners and credit segments and to recreate payment behavior summaries. Once the job is scheduled, the system writes logs to spool during job execution. The app also supports performing mass rebuild of credit management data—for example, more than 2,000 business partners in a single run.

Reconcile Cash Flows – Intraday (F3418A)

Application Type: Transactional

This app allows you to manually reconcile intraday bank statement flows with forecasted cash flows, enabling timely identification of unfinished or unknown payments and eliminating duplicate forecast items for accurate cash positions. Reconciliation can be performed at the bank account level, selecting accounts with an **Unreconciled** status, or at the item level, comparing flows side by side. You can decide whether to reconcile unmatched flows directly or create residual flows to track unreconciled items, recorded as memo records with the residual forecast category. In addition, the app lets you monitor reconciliation statuses, undo reconciliations by resetting items to **Unreconciled**, and check results in the Short-Term Cash Positioning app (F5380).

Reconcile Cash Flows – Intraday Memo Records (F3418)

Application Type: Transactional

This app allows you to reconcile intraday memo records from intraday bank statements with forecasted cash flows, enabling comparison before end of day. It helps identify unfinished or unknown payments and eliminates duplicate forecast items for an accurate cash position. The app's features include viewing automatically reconciled records, triggering on-demand reconciliation using Transaction FCLM_CR_INTRAM (Reconcile Intraday Memo Records), and manually reconciling or undoing reconciliations to reset items to open status. In addition, the app monitors reconciliation statuses across bank accounts, with results reflected in the Cash Flow Analyzer app (F2332).

When a bank statement is imported, the system initiates automatic reconciliation. Upon importing, the system first checks table FEBCL to identify existing mappings between intraday memo records and corresponding financial accounting documents. Next, it checks the One Exposure from Operations table (FQM_FLOW) and matches remaining open intraday memo records with open forecasted cash flows based on criteria like company code, bank account, account currency, value date, amount, and tolerance group settings. The reconciliation results are then updated in the cash reconciliation table, where matched records are displayed as **Reconciled** in this app. Moreover, the **Adapt Filter** option allows detailed analysis of cash flows and account balances from various perspectives.

This app is improved with a new reconciliation model over the Reconcile Cash Flows – Intraday Memo Records app (F3418). Let's compare the Reconcile Cash Flows – Intraday (F3418A) and Reconcile Cash Flows – Intraday Memo Records (F3418) apps:

- **Reconcile Cash Flows – Intraday**
  - Reconciliation at the bank account level supported
  - Automatic reconciliation is not supported
  - Residual reconciliation supported
  - Tolerance settings not supported
- **Reconcile Cash Flows – Intraday Memo Records**
  - Reconciliation at the bank account level not supported
  - Automatic reconciliation is supported
  - Residual reconciliation is not supported
  - Tolerance settings supported

Reconcile Cash Flows – Memo Records (F4867)

Application Type: Transactional

This app allows manual reconciliation of memo records generated from end-of-day bank statements with forecasted cash flows, thereby allowing comparison between bank statements and forecasts to identify unfinished or unknown payments and remove duplicate forecast items for accurate cash positions. The app's features include viewing reconciled memo records and forecasted cash flows, performing automatic reconciliations using Transaction FCLM_CR_MMRD, creating and reconciling manual memo records, undoing reconciliations to reset items to open status, and monitoring the reconciliation status of bank accounts. Reconciliation statuses indicate whether memo records, forecasted cash flows, or both remain unreconciled, or if all items are fully reconciled. This is represented with four statuses: **Reconciled**, **Open End-End-Of-Day Memo Records**, **Open Cash Flows**, and **EoD Memo Records**.

This app is deprecated as of SAP S/4HANA 2023 and replaced with Reconcile Cash Flows – Intraday Memo Records (F3418).

Reconcile GR/IR Accounts (F3302)

Application Type: Analytical

Reconciling goods receipt/invoice receipt (GR/IR) accounts can be a tedious task due to differences between invoice amounts and corresponding purchase orders and goods receipts. The app enhances financial statement accuracy by helping you spot issues early on, thereby reducing open items. This app helps you reconcile these accounts by compiling a worklist of open items that need further analysis. It supports creating worklists based on various criteria, displaying items as charts or lists, and using smart facts to describe business situations. In addition, with this app, you can view details of purchasing documents, filter and save worklists, and reduce open items by assigning statuses and entering comments. When you observe no differences between the goods receipt and invoice on the **Intelligent Goods and Invoice Receipt Account Reconciliation** screen, you should consider running an automatic clearing job from the Schedule General Ledger Job app (F1927). The app uses CDS view C_GRIRAccountReconciliation.

Tip

The **Perform Write-Off** button in the app is only active if you select one or several purchasing document items with status values such as **Perform Write-off**, **Perform Write-off of Goods**, or **Perform Write-off of Delivery Costs**. By choosing one of these status values, you can specify whether to write off differences on the goods account, the delivery account, or both accounts.

To learn more about intelligent goods and invoice receipt account reconciliation using smart facts, consult SAP Note 2953157. To understand the app's logic behind purchasing document item selection, consult SAP Note 3490523.

Reconcile SD Billing Documents (F4866)

Application Type: Transactional

This app helps with the reconciliation of billing documents from sales billing and contract accounting documents, ensuring successful postings and consistency. The app identifies differences and provides an overview of the generated billing, print, and contract accounting documents. By default, data from the past week's billing date is selected, with filter options available for customization. For instance, you can apply filter criteria using fields such as billing document, billing type, billing date, business partner, and company code. Also, the app cate-

gorizes contract accounting documents into multiple tabs: **Not Transferred** (billing documents not transferred), **Inconsistent** (transferred documents with amount differences), and **Completed** (successfully transferred and consistent documents). You can analyze inconsistencies by viewing billing and contract accounting document details or master data using document numbers.

Reconciliation Status Overview and Balances (F3865)

Application Type: Analytical

This app offers SAP S/4HANA Finance for group reporting users a comprehensive overview of the intercompany reconciliation progress, allowing them to identify and address discrepancies promptly. The app allows you to monitor line item-level matching status and aggregation-level reconciliation status for a specified reconciliation case, fiscal year/period, and organizational units in group reporting. You can apply filters such as reconciliation case, fiscal year period, and organizational unit hierarchies, with leading display group and leading amount set as default filters. Matching status categories include **Initial**, **Not Assigned**, **All Assigned**, and **All Matched**, reflecting document-level progress, while reconciliation balance status uses a color-coded system (green, yellow, red) based on aggregated amount differences compared to predefined tolerances. In addition, **Reconciliation Close Status** tracks whether trading pairs are **Open** or **Closed**, **Initial**, **All Open**, **Partially Open**, or **All Closed**. In addition, the app has a **Hide/Show Initial Matching Status** option to filter out the table lines that have the **Initial** matching status. The app updates statuses based on the latest cutoff time, dynamically flags new postings, and supports detailed data analysis and reconciliation reporting through features like hierarchy views and item status filtering.

Reconciliation – Subledger and G/L (FARR_RECON_POSTING_GL_NEW)

Application Type: Web Dynpro

This app helps with financial reconciliation by comparing subledger entries with their corresponding general ledger transactions in revenue accounting and reporting (RAR). The app allows you to filter and sort, provides customizable reporting tools for further data analysis, integrates SAP financial accounting for revenue posting, and offers a user-friendly SAP Fiori–based interface for streamlined reconciliation and audit readiness. During reconciliation, if there are no differences found by the app, the system displays result status with a **No differences found** message. In addition, the app supports and complies with IFRS 15 and ASC 606 accounting standards.

> **Tip**
>
> To use this app, you must activate the Internet Communication Framework (ICF) security service for FARR_RECON_POSTING_GL via Transaction SICF.

Release Cash Flows (F3446)

Application Type: Transactional

This app allows cash managers to review and release cash flows integrated from various sources, including SAP and non-SAP systems (via web service CASHFLOW_IN). By validating cash flow information, cash managers can make informed decisions on cash position analysis. The app's features include checking the release status of cash flows (**Released** or **Unreleased**), displaying unreleased cash flows in a hierarchical list with filters like liquidity item and origin system, reviewing cash flow details such as amount and date, manually releasing cash flows, and navigating to the Manage Memo Records app (F2986) to create memo records.

For instance, after a memo record is created in a subsidiary, the cash flow appears in an unreleased status. Upon verifying and confirming the amount and date, you can release the cash flow, which then becomes visible in the Cash Flow Analyzer and Check Cash Flow Items apps.

> **Tip**
>
> As a prerequisite for manual release, SAP recommends completing Customizing activity **Define Criteria for Manual Release**.

Remaining Performance Obligation (W0159)

Application Type: Web Dynpro

This Web Dynpro–based app reveals the total transaction price allocated to unsatisfied or partially unsatisfied performance obligations (POBs) by the end of the reporting period in revenue accounting and reporting (RAR). The app presents nonrecognized revenue and cost across multiple currencies, including display currency and document currency, categorized into event-based and time-based amounts. You can analyze these figures using various dimensions from the revenue accounting subledger, such as company code, accounting principle, fiscal year, display currency, exchange rate type, and POB type. You can add or remove dimensions into rows and columns for multidimensional analysis using the **Navigation Panel** option, visualize data in **Chart View** or **Table View**, or both. In addition, you can leverage the **Jump To** feature for seamless access to other relevant apps.

> **Tip**
>
> The app displays total nonrecognized revenue amounts up to the reporting period, which is converted from the document currency to the display currency using exchange rate type M, based on the current system date.

Remaining Performance Obligation – Total Open Revenue (FARR_TOT_OPENREV)

Application Type: Web Dynpro

This app provides visibility into outstanding revenue commitments linked to performance obligations (POBs) in revenue accounting and reporting (RAR). Typically, this includes revenue that has been contracted but not recognized in the system yet. The app's functionalities include offering detailed view of all remaining POBs, providing contract-level insights, and filtering and sorting remaining POBs with parameters such as contract type, customer, contract, POB status, and date ranges. The app also provides total open revenue items graphically to visualize trends and patterns. The app supports and complies with IFRS 15 and ASC 606 accounting standards.

Remaining Performance Obligation – With Time Bands (W0158)

Application Type: Web Dynpro

This Web Dynpro–based app provides details of outstanding revenue expected to be recognized using time bands linked to fiscal periods, quarters, or years. You can search for remaining performance obligations (POBs) using search parameters such as company code, accounting principle, from fiscal year, from fiscal period, to fiscal year, to fiscal period, display currency, and exchange rate type. Upon selecting the selection criteria and pressing **Go**, the app presents to-be recognized revenue in display currency and to-be recognized revenue document currency for upcoming reporting periods, allowing analysis across multiple dimensions such as POB type. You can add or remove dimensions into rows and columns for multidimensional analysis using the **Navigation Panel** option and visualize data in **Chart View**, **Table View**, or both. In addi-

tion, you can leverage the **Jump To** feature for seamless access to other relevant apps. The app uses the CDS view P_RASumRevenuePerPeriod.

> **Tip**
> This app presents the timing of revenue recognition using a predefined quantitative time bank such as from fiscal year, from fiscal period, to fiscal year, and to fiscal period, whereas the Remaining Performance Obligation app (W0159) provides a snapshot of remaining POBs without time-based periods.

Repair Payments (F3651)

Application Type: Transactional

The transactional app allows you to address payments that cannot be processed due to incorrect information and have been transferred to exception control. You can correct errors or trigger reactions, such as returning payments to the sender. In addition, the app displays erroneous payment orders and items in their work basket, which can be filtered by specific filter criteria. **Exception Details** include information on correcting errors, with suggested corrections for some cases. Ignored errors lead to reprocessing, and persistent issues return to exception control. Moreover, you can reassign payment orders or items to another work basket, reserve them for exclusive processing, or remove reservations as needed. The app's functions include rejecting or reprocessing orders, simulating processing to check enrichment and validation steps, recalling orders, accessing **Open PO Expert** for advanced editing, and displaying file handler data via the **Show File Handler Data** option.

> **Tip**
> As a prerequisite for automatic reaction types, you must complete the **Define Response Types** Customizing activity for exception handling.

Replicate Runtime Hierarchy (F1478)

Application Type: Transactional

Maintaining hierarchies is always a challenging activity for large enterprises that have hundreds of hierarchies. Adding new hierarchies or changing attributes in other applications requires synching. This app allows you to manually replicate set-based and financial statement hierarchies to backend database tables after creating or editing them in other apps or SAP GUI. In addition, it enhances the performance of hierarchy-related analytic apps. SAP recommends periodic replication to save different hierarchy snapshots and view selection based on replication date and time to sync data correctly.

> **Tip**
>
> To replicate set-based hierarchies, SAP recommends completing Customizing activities **Set Report Relevancy for Hierarchies** and **Replicate Runtime Hierarchies** (Transaction HRRP_REP).

Repost Revenue Postings (F3877)

Application Type: Transactional

This app allows you to repost revenue postings that have been reversed previously. As a result, the app streamlines the revenue accounting process, maintaining compliance with accounting standards while providing traceability and efficiency in managing revenue postings. The app ensures accurate adjustments by allowing you to correct errors in previous postings and generate new run IDs for reposted entries. The app includes two tabs for managing these processes, and each reversal or reposting generates a new run ID. Reversal applies to all contracts posted in the prior job, while reposting uses the reversed run ID. To initiate, use the respective **Reverse** or **Repost** buttons. For example, if discrepancies in general ledger accounts are identified post-posting, the affected run ID must

first be reversed to allow corrections before reposting. This ensures accurate adjustments while maintaining traceable run IDs.

> **Tip**
>
> This app is part of the generic SAPUI5 job scheduling framework. The app catalog (SAP_FIN_FARR_REVNUE_REPOST) must be activated in the backend system.

Reprocess Account Determination (F4187)

Application Type: Transactional

This app allows you to reprocess account determination for single contracts or multiple contracts by applying a new assignment of general ledger accounts to revenue accounting contracts. Typically, this activity is carried out when changes happen in the general ledger, new general ledger accounts are created, or errors are observed during account determination. The app automatically selects the relevant logic for contract management and classic contract management (CCM) contracts. Next, the app ensures accounts are updated efficiently based on system validation rules. Accounts are redetermined similarly to newly created contracts or performance obligations (POBs), including accounts such as recognized revenue, receivable adjustment, contract asset, contract liability, and others. The early termination expense of revenue is updated only if an account number is already maintained for it. Contract status is not considered as the app applies redetermination to all contracts, regardless of whether they are in process or completed. This is to accommodate any subsequent status changes to contracts. For balance sheet accounts, updates occur only if postings have not been transferred to the general ledger; otherwise, they remain unchanged. For profit and loss accounts, reconciliation key statuses C or R are not updated, but new entries use the newly determined accounts. Early termination expense accounts are updated only if stored in the contract entity.

Reprocess Bank Statement Items (F1520)

Application Type: Transactional

This app allows you to reprocess bank statement
items that the system cannot automatically post
and clear. Bank statements can be entered automatically or manually, with rule-based processing for automatic assignment and clearing. If
automatic processing fails, manual reprocessing
is necessary. The app lists of bank transactions
require manual intervention. In addition, the
app allows you to reprocess items, mark them as
reprocessed, and provide reasons for reprocessing. The app brings features such as an overview
of reprocessed and nonreprocessed items, manual or rule-based reprocessing, assigning payments to accounts and clearing open items,
posting payments, and reversing multiple bank
statement items through mass reversal of journal entries. Attachments also can be added to
bank statement items. With these features,
accountants resolve exceptions to ensure accurate and timely bank reconciliation, especially
during period-end closing.

Reprocess Contracts (F5237)

Application Type: Transactional

As part of revenue accounting and reporting
(RAR), this app allows the selection of revenue
contracts and the reprocessing of any previously processed revenue accounting items
(RAIs) within them. The app allows the accounting team to redetermine attributes of existing
contracts and performance obligations (POBs)
based on updated Business Rule Framework
plus (BRF+) rules and configurations. If incorrect
configuration attributes or BRF+ decision table
rules were applied during the creation of contracts and POBs, reprocessing ensures alignment with revised determination rules. Using
this app, you can start reprocessing by entering
multiple selection criteria. As a result, the system identifies impacted contracts, reprocesses
their RAIs using the latest BRF+ decision tables,
and updates contract and POB attributes accordingly.

Reprocess Lockbox Items (F3398)

Application Type: Transactional

This app helps with the reprocessing of lockbox items that the system cannot automatically post and clear. When the system cannot automatically post and clear customer open items, manual handling of lockbox items is required by a cash management specialist. Although rule-based processing assigns and clears payments automatically, unsuccessful cases require manual intervention. You can search for lockbox items using fields such as account type and customer. In addition, the app allows you to manually reprocess lockbox items, mark them as reprocessed, and provide reasons for reprocessing. The app's key features include viewing processed and unprocessed lockbox items, reprocessing items manually or with rules, assigning payments to customer/supplier or general ledger accounts, clearing open items, posting payments, and performing mass reversals of multiple lockbox item postings. The app supports both open items for both internal and external customers.

Tip

To perform intercompany clearing of lockbox items using processing rules, consult SAP Note 3467772 further guidelines. As a prerequisite, lockbox-relevant configurations such as Transactions OB10 (Define Lockbox Accounts), OBAX (Define Lockbox Posting Data), and OBAY (Define Lockbox Control) are maintained to automate or perform manual handling of customer payments received through lockboxes. Additional configurations may be required using Transactions OBXI and OBXL, which are used for the configuration of a cash discount and assigning general ledger accounts to reason codes for the posting of payment differences.

Reprocessing Rate of Incoming Payments (F1619)

Application Type: Analytical

With this app, you can display and analyze how many incoming payments (bank statement items) have been manually reprocessed after automatic payment clearing was unsuccessful. Manual reprocessing is often required due to insufficient reference information, incorrect payment amounts, or incomplete automatic posting settings. This app provides detailed information to improve receivables management automation. You can run analyses by reason, by company code, by customer, or bank, and take steps to save processing costs, such as emailing customers or informing management about incorrect settings. The app considers only completed incoming payments from the last 12 months and calculates the reprocessing rate as a percentage using the following formula:

[Number of incoming payments that were manually reprocessed ÷ Total number of incoming payments (both manually reprocessed and automatically cleared)] x 100%

For example, if 10,000 incoming payments were received and completed over the last 12 months and 250 were manually reprocessed, the reprocessing rate is 2.5%.

Reset Cleared Items (F2223)

Application Type: Transactional

With this app, you can reset the clearing of line items and reverse the clearing entry if needed. This function is applicable to line items of customer accounts, supplier accounts, and general ledger accounts managed on an open item basis. You can reset cleared items using the app's **General Information**, **Accounts**, **Cleared Items**, and **Open Correspondence** tabs. The app provides features that include resetting clearing of line items when a payment has cleared the wrong items and reversing the clearing entry for incorrect payments. In addition, if there is an exchange rate difference in the clearing entry, reversing the entry is mandatory.

Tip

The app does not support mass reset of clearing items due to handling of follow-on items. In other words, you can reset only one reset per invoice. Consult SAP Note 2682947, which describes this limitation.

Transaction Code

The equivalent SAP GUI transaction is Transaction FBRA.

Revenue Accounting Contract (FARR_CONTRACT_MGMT_OVP)

Application Type: Web Dynpro

This app provides a centralized view of revenue contracts in revenue accounting and reporting (RAR). The app displays key attributes such as performance obligations (POBs), transaction prices, and revenue recognition status while offering insights into revenue postings, accruals, and adjustments. The app ensures compliance with IFRS 15 and ASC 606 accounting standards, tracks contract modifications with a structured audit trail, and features an SAP Fiori–based interface for intuitive monitoring of contract details and financial transactions. Thus, the app allows you to manage revenue contract details and relevant financial transactions.

Revenue Accounting Overview (F4067)

Application Type: Analytical

This app provides revenue accountants with analytical KPIs, such as recognized revenue, based on selected parameters like display currency,

company code, accounting principle, and fiscal year, with quick links to related apps. The app includes the following cards:

- **Recognized Revenue Overview** compares year-to-date revenue across fiscal years and navigating to future revenue analysis
- **Top 5 Business Partners/Customers** identifies key partners by contractual price and recognized revenue
- **Top 5 Best Sellers** highlights performance obligations (POBs) by revenue and contractual price
- **Revenue Trend** shows recognized revenue for the next 12 periods
- **Number of Revenue Contracts by Status** categories contracts by completion status
- **Recognized Revenue by Timing** displays revenue based on timing criteria, such as point-in-time goods transfer or service delivery over time

The app also supports navigation to related applications for deeper analysis and business insights.

Revenue Catch-Up (W0169)

Application Type: Web Dynpro

Revenue catch refers to the adjustment of recognized revenue due to contract changes such as contract terms or performance obligations. This Web Dynpro–based app provides visibility into recognized revenue and revenue catch-up in revenue accounting and reporting (RAR). Using the app, you can analyze revenue catch-up by period, quarter, or year, with the option to add additional dimensions from the **Navigation Panel** for further analysis, such as revenue catch-up per profit center. You can filter data by selecting parameters like company code, accounting principle, fiscal year, fiscal periods, display currency, exchange rate type, exchange rate date, and display mode. The default view presents aggregated recognized revenue and revenue catch-up in both document and display currencies, organized by quarter.

Tip

To understand how the system calculates the revenue catchup, consult SAP Note 2845644.

Revenue Catch-Up (FARR_REVENUE_CATCHUP)

Application Type: Web Dynpro

This app provides visibility into recognized revenue adjustments over specific periods in revenue accounting and reporting (RAR). A catch-up calculation triggered in RAR when a contract change occurs under specific conditions. For example, a catch up happens when a performance obligation (POB) is fully fulfilled or when the change type is defined as a retrospective change in the contract settings based on contractual terms. The app tracks catch-up adjustments by fiscal period, quarter, or fiscal year. You can customize reporting by company code, contract, customer and fiscal year. The app supports and complies with IFRS 15 and ASC 606 accounting standards. In addition, for cumulative catchup, you can use the reports FARR_CALC_TRANS_CATCHUP_CONTR and FARR_CALC_TRANS_CATCHUP for individual contracts and mass contracts, respectively.

Revenue Explanation (F4272)

Application Type: Transactional

This app allows revenue accountants to verify details like recognized revenue, fulfillment triggers and quantities, attributes like allocated amount and fulfilled quantity, revenue catch-up amounts, and events that trigger revenue recognition in revenue accounting and reporting (RAR). In addition, you can review price allocation results by searching for performance obligations (POBs) grouped by fiscal year periods and analyzing recognized revenue. With this app, you can analyze revenue recognition based on adjustment reasons such as fulfillment, catch-up, and contract modification.

Each adjustment reason offers detailed insights, such as fulfillment entries, change logs, or allocated amount explanations. The app's advanced search features include filters and fuzzy search using attributes like company code, customer, or accounting principle ensures precise revenue analysis for both distinct and compound group POBs.

Revenue Posting Job Monitor (FARR_JOB_MONITOR)

Application Type: Web Dynpro

This app allows revenue accountants to track and manage revenue posting jobs in revenue accounting and reporting (RAR). The app helps with monitoring revenue accounting tasks such as transfer revenue, calculate contract liabilities and contract assets, revenue posting run, revenue posting reversal. The app provides real-time visibility into job execution statuses, including success, failure, and pending tasks, while facilitating error handling and troubleshooting for failed postings before reprocessing. Also, the app is integrated with revenue accounting to comply with IFRS 15 and ASC 606 accounting standards. In addition, the app supports the automated scheduling of revenue accounting-related tasks as required.

Revenue Schedule (F3882)

Application Type: Transactional

This app allows you to view and analyze revenue schedules and fulfillments. It comprises two key sections: **Revenue Schedule Summary** and **Revenue Schedule Details**, providing a comprehensive overview and detailed insights into revenue-related data. The app provides revenue accountants with an overview of revenue recognition for a revenue contract, including details like total revenue, recognizable revenue, posted revenue, open revenue, revenue catch-up, total cost, recognizable cost, posted cost, and open cost. In addition, you can view aggregated amounts for recognizable revenue, open revenue, recognizable cost, and open cost are presented up to the current period. The app also displays revenue recognition, open revenue, and fulfillment progress status for each performance obligation (POB) by accounting period. To verify revenue recognition, you can use this app to search for contracts. Classic revenue contracts navigate to a classic revenue schedule, while optimized contracts open a new window. In the results list, you can review revenue recognition and fulfillment progress for distinct POBs and compound group POBs, where compound group POB revenue aggregates nondistinct POBs. Clicking a compound group POB for a specific fiscal year/period enables a detailed examination of nondistinct POB revenue recognition.

Revenue Schedule (FARR_POB_REV_RECOG_OVP)

Application Type: Web Dynpro

This app allows you to view and analyze revenue schedules and fulfillments in revenue accounting and reporting (RAR). A revenue schedule in RAR defines the timing and amount of revenue recognition for each performance obligation within a specified period. Thus, it serves as a structured plan that lays out when revenue is recorded in financial statements based on the fulfillment of contractual obligations. The app tracks revenue postings based on predefined rules, links schedules to specific obligations for accurate allocation, and supports reconciliation and adjustments to address discrepancies. Also, the app complies with IFRS 15 and ASC 606 accounting standards.

Revenue Variance (F3072)

Application Type: Analytical

With this analytical app, you can get insights into the differences between planned and actual

gross sales revenue, or between actual sales revenue within a specified time frame and the previous year. You can search for actual revenue and plan revenue information with search parameters such as object currency, fiscal year, fiscal period, planning category, company code, ledger, country key, product sold group, product sold, and sales organization, and customer. The app presents search results in **Benchmark Tables**, **Benchmark Charts**, and **Calculated Effects** tabs. In addition, the app highlights how volume, price, and mix effects contribute to sales revenue variance and considers both sales revenue for with and without quantity. This is helpful for sales managers in planning and forecasting. The app features benchmark tables and charts that display data by country, customer, distribution channel, or product sold for organization division or for entity. Also, you can compare actual versus previous year or plan data, specify the comparison time frame, and utilize various filters to focus on key data for management reporting. In addition, an effects chart shows how different factors impact revenue variance.

Reverse and Repost Revenue Postings (FARR_REV_REP_AP)

Application Type: Web Dynpro

This app allows you to correct revenue postings by reversing inaccurate entries and generating new transactions in revenue accounting and reporting (RAR). The app's features include reversing revenue postings that were posted incorrectly; reposting revenue with correct details; and performing detailed analysis on both reversal and reposting activities, including time stamps, user information, and reasons for adjustments. Thus, the app allows revenue accountants to manage the reversal and reporting of reposting process. We recommend reviewing revenue postings routinely to catch any discrepancies and follow up on corrections. The app supports and complies with IFRS 15 and ASC 606 accounting standards.

Reverse Document (F4494)

Application Type: Transactional

This app allows accounts receivable and accounts payable users to perform the reversal of both contract accounting documents and archived contract accounting documents. You can search documents that require reversals using search fields such as document, reference document, and origin. The app supports reversing open items, cleared items (with an automatic or manual clearing reset while retaining account distribution or creating a new open item), open repetition items (either all or specific dates), and cleared or partially cleared repetition items (with clearing reset options similar to those for cleared items). For documents included in clarification cases, the system automatically reverses the related clarification documents along with the selected document.

Reverse Revenue Postings (F3878)

Application Type: Transactional

This app allows you to reverse revenue postings from previous runs and repost them for accurate adjustments. Reversal applies to all contracts linked to a specific run ID, generating a new run ID upon completion. Reposting, also based on the run ID, assigns another new ID. The app comes with the **Revenue Posting Reversal** and **Repost UI** tabs, which offer important functions that can be triggered by selecting the respective **Reverse** or **Repost** buttons. These features streamline revenue accounting by ensuring compliance with standards, facilitating corrections, and maintaining traceability throughout the adjustment process.

Review Balance Sheet FX Risk (F1588)

Application Type: Transactional

With this app, treasury risk managers can understand and analyze their company's exposure to foreign exchange (FX) risk by company code or transaction currency. This app allows you to quickly see the amount of absolute balance sheet net open exposures, absolute balance sheet exposures, and absolute hedges at the company code level on a specific key date. In addition, you can branch into a company code to obtain detailed views of the **Origin** and structure of balance sheet exposures and hedges for various transaction currencies, including the resulting net exposures. For key figures, further drilldown to the single line-item level is possible, displaying specific line items for financial accounting balances, open line items for financial accounting open items, contributing cash flow items for One Exposure from Operations, and financial transactions for transaction management. The app supports exporting detailed data to a worksheet, providing calculated balance sheet FX risk hedges and exposures at the key figure level in both transaction and display currencies. However, totals at the key figure group level and net exposures at the company code level are not exported.

Review Bank Accounts (F1370A)

Application Type: Transactional

This app allows cash managers or reviewers to review assigned bank accounts and initiators to track the status of their review requests. Reviewers can search for tasks, navigate bank accounts, complete reviews, and add review notes, which are visible in the Manage Bank Accounts app (F1366A). In addition, initiators can search for their requests and monitor the review status. For example, you can search for bank accounts with search criteria such as only in review account, bank, country, contact person, and account holder. Both reviewers and initiators can track all review requests with this app. For example, reviewers can display the notes set by the review initiator using the **Show Note Initiator** button. Additional features include sending emails with URLs for selected information, sharing comments in SAP Jam, and creating tiles with default selection criteria.

Review Hedge Constellation Details (F5657)

Application Type: Transactional

This app allows you to review the details of existing hedging constellations and initiate manual changes for predefined actions. The hedging constellations reveal structured relationships such as similarities or share common characteristics between exposures, hedging instruments (e.g., futures, options, swaps), and hedging strategies. Hedge constellations can be added to the hedge constellation task list for future processing. The app's key features include an overview of hedge accounting-relevant data and trades for commodity price exposures; the ability to select hedge constellations for late designation; navigation to underlying plan exposures, commodity financial trades, foreign exchange (FX) trades, and FX hedge requests; and manual task creation to modify a trade's business partner or to replace a referenced commodity FX trade.

Let's compare the Commodity Hedge Management Cockpit – Overview (F5656) and Review Hedge Constellation Details (F5657) apps:

- **Commodity Hedge Management Cockpit – Overview**
  This app offers a summary of hedging situations at a high level with aggregated data. It emphasizes an overview and analysis of hedging data, including exposures, assigned hedges, target quotas, and hedged quantities. You can drill down at aggregated levels such as years, hedge books, and periods.

- **Review Hedge Constellation Details**
  This app provides a granular view of how commodity hedges are structured and connected to the underlying risks. It focuses on specific hedging relationships or hedge structures. You can understand individual hedges and underlying transactions.

Review Purchase Order Accruals for Cost Accountant (F3552)

Application Type: Transactional

This app allows cost accountants to review purchase order accruals and adjust revised costs in transaction currency before posting for a fiscal period. The app provides search criteria to review purchase order accruals using parameters such as company code, cost center, cost center group, supplier, review type, purchase order, and sales order. The app presents results in three tabs: **All**, **Unreviewed**, and **Review Finished**. In addition, reviewers can validate and adjust system-proposed periodic planned costs as needed. Depending on your role, tasks include reviewing purchase order accruals at the cost object level, viewing accrual amounts in group currency, confirming reviews, adjusting accrual amounts, and mass confirming revised amounts. The app's additional features include viewing source systems of purchase orders replicated to the Central Finance system and receiving email notifications for accrual reviews with direct navigation to the app.

The My Inbox – Review Purchase Order Accruals app (F3517) offers a workflow-based accrual review directly from your inbox, whereas the Review Purchase Order Accruals for Cost Accountant app (F3552) is tailored for cost accountants to review and adjust purchase order accruals at a detailed level. Moreover, the latter app allows you to perform more comprehensive accrual management tasks.

Tip

The app supports certain planned accrual item types, defined in configuration (Transaction ACEIMG), that can be activated for the review process. Consult SAP Note 3579074 for a possible workaround.

Review Service Entry Sheet Accruals (F6108)

Application Type: Transactional

Service entry sheet accruals help with the calculation and posting of accruals by transferring relevant data from service purchasing and recording within materials management to accruals management, where purchase order items are automatically converted into accrual subobjects. This app allows general ledger accountants who support overhead accounting to review periodic accruals of service entry sheets for cost objects under their responsibility. The app allows you to process service entry sheets in **General Information** and **Items** tabs. You can check proposed accrual amounts and mark their review before the system posts them. Periodic accrual amounts are calculated as the total amounts of unapproved service entry sheets. Depending on your assigned role, you can review purchase order accruals at the cost object level and view accrual amounts in both company code and global currencies.

Tip

As a prerequisite for the app to function seamlessly, you must complete Customizing activities **Adjust Transfer Priorities for Accruals** and **Define Accrual Item Types and Methods for Purchase Orders of Service**. Optional configurations include account determination, accrual item type definition, and Customizing checks, based on business needs.

Revise Payment Proposals (F0771)

Application Type: Transactional

This app allows you to monitor and revise payment proposals and the details of open items, ensuring that all payments are made correctly, on time, and in compliance with company policies. The app provides a centralized view of all payment media, including their status—such as

Created, **Printed**, and **Sent**—and allows for actions like printing, downloading, and canceling payment files. Also, you can search and filter payment media based on various criteria, ensuring efficient tracking and processing of outgoing payments. To revise payments, the app presents payment proposal information in **Payments** and **Exceptions** tabs. The app's key features include displaying open payment proposals, analyzing exceptions in the log, editing proposals to change payment methods and other details, blocking and unblocking items, reallocating items, mass blocking items, sorting payment amounts, and calculating total payment amounts by various criteria. The app also supports posting comments in SAP Jam and sending emails.

Transaction Code

The equivalent transaction in SAP GUI is Transaction F110.

Run Business Reconciliation in Background (F4841)

Application Type: Transactional

This app allows system administrators or revenue accountants to schedule recurring jobs to identify inconsistent data between sender components and revenue accounting and reporting (RAR) for optimized contract management (OCM). By default, the app reconciles data from sender components like sales and distribution and contract accounts receivable and payable. The app's selection criteria include company code, accounting principle, sender component, and source item logical system, with optional filters for revenue accounting contract and header ID. In addition, app's technical parameters include block size for mass selection, external ID, and problem class. The background job processes OCM data, and then app saves results to table FARR_D_BIZ_RECON, while batch execution issues are logged in Transaction SLG1.

Sales Accounting Overview (F3228)

Application Type: Analytical

This app provides easy access to essential information such as sales revenue, overdue receivables, aging analysis of invoices, cash flow contribution by sales, and other KPIs within the sales accounting area. You can search for sales accounting data using dimensions such as company code, display currency, planning category, fiscal year period, sales organization, and material group. The app presents data using various cards. In addition, it serves as a central source of information for sales accountants and includes a variety of filters and built-in navigation to related apps. Also, the app can be used to predictive accounting by enabling semantic tags in financial statement version (FSV). As a prerequisite, predictive accounting needs to be active to display data in the incoming sales orders and gross margin cards. Thus, the app offers insights into current sales trends.

> **Tip**
>
> Consult SAP Notes 2793172 and 2802696 for further guidance on enabling predictive accounting and to ensure data is displayed in this app.

Schedule Accounts Payable Jobs (F2257)

Application Type: Transactional

With this app, you can schedule periodic activities such as sending payment advice to payees, payment proposal requests to suppliers, and payment remittance request status updates to suppliers, or send them via SAP Business Network. The app's functionalities include adjusting withholding tax information, automatic scheduling of the payment program, creating positive payment files, and reversing payment runs. The app provides the following templates to schedule jobs:

- Adjustment of Withholding Tax Information to Relevant Types
- Automatic Scheduling of the Payment Program
- Creation of Positive Pay Files
- Recreate and Change Withholding Tax Data with Withholding Tax Rate of 0%
- Reverse Payment Run
- SAP Digital Payments: Advice Processing
- Create and Send Payment Advices to Payees
- Send Scheduled Payment Advices to Payees

> **Tip**
>
> To use some job templates, a separate license may be required.

Schedule Accounts Receivable Jobs (F2366)

Application Type: Transactional

This app allows accounts receivable specialists or accountants to automate key accounts receivable processes, enabling you to schedule, monitor, and manage periodic financial jobs related to customer invoicing, payment processing, dunning, and account reconciliation. This results in many benefits such as increasing efficiency, reducing manual effort, and improving cash flow management, thereby ensuring the timely execution of financial processes. With this app, you can schedule accounts receivable jobs using the provided templates and scheduling options. It offers access to various job templates, such as the **Automatic Reprocessing of Bank Statements** via SAP Cash Application, allowing you to efficiently manage and automate accounts receivable processes. The app is delivered with both standard templates and licensed add-on templates along with scheduling options. For example, **Payment Card Authorizations for Customer Line Items with Standard Card** allows payment card authorizations if the digital payments add-on is activated—a separate license is required.

Transaction Code

The corresponding transaction in SAP GUI is Transaction FCC1.

Schedule Accruals Jobs (F3778)

Application Type: Transactional

The app allows you to schedule accruals processes using predefined templates, such as **Purchase Order Accruals – Post Periodic Accruals** and **Service Entry Sheet Accruals – Propose Periodic Accruals**. The app supports important tasks like transferring purchase orders to the accrual engine, proposing periodic accruals, simulating accrual postings, and posting periodic accruals. To utilize the templates, you must register them in the Customizing view APJ_C_SCOPE by running Transaction SM30. You can choose APJ_C_SCOPE for table/view to maintain or add entries for adding relevant templates like **SAP_FIN_GL_ACE_POSTING_RUN**. After adding entries, you must save the changes to activate the templates.

Schedule Asset Accounting Jobs (F1914)

Application Type: Transactional

The app allows you to schedule and automate essential asset accounting jobs, such as depreciation runs, period-end closings, and other recurring tasks in asset accounting. For instance, this app allows you to schedule asset accounting jobs using provided templates and scheduling options. You can use job templates such as **Recalculate Depreciation** and **Depreciation Posting Run**, scheduling the job start time or starting it immediately, and defining job recurrence with a recurrence schedule, including options for handling nonworking days.

Transaction Codes

This app can perform functionality of SAP GUI transactions such as Transaction AFAB to execute the depreciation run, Transaction ASKB to transfer asset balances to the general ledger, Transaction AJAB to handles year-end closing for asset accounting, and Transaction AR31 recalculates depreciation.

In SAP GUI, depreciations are posted through a background job that uses a parameter to calculate the period to be depreciated, with table TVARVC updating the period. However, when scheduling the job using this app, there is no such option to select this parameter. To schedule jobs with this functionality, SAP recommends using classic Transaction SM36.

Schedule Balance Validation Jobs – G/L Accounting (F7470)

Application Type: Transactional

This app allows you to initiate immediate or recurring validation jobs for selected entities using predelivered job template **Balance Validation for G/L Accounting**. The app's key functionalities include creating single or recurring validation jobs, monitoring job statuses (**Finished**, **In Process**, or **Failed**), and reviewing job logs via **Information** or **Error** icons. Also, you can make entity selection with wide range of fields such as company codes, profit centers, and cost centers. Validation results can be checked in the View Balance Validation Results app (F6387), where each row represents an entity validated against a specified rule group. In addition, the app integrates with SAP Advanced Financial Closing, allowing balance validation jobs to be included as closing tasks using models **010035000** (balance validation in company code currency) or **010035001** (balance validation in group currency), with rule group parameter **P_GRP_ID** defined for execution.

Tip

To integrate and use SAP Advanced Financial Closing, a separate license is required.

Schedule Collections Management Jobs – Mass Changes (F3918)

Application Type: Transactional

This allows accounts receivable users to efficiently schedule jobs for managing collections processes. It therefore streamlines the management of collections tasks, ensuring efficiency and organization within the accounts receivable department. The app allows you to schedule essential jobs in collections management, including assigning collection specialists, assigning collection profiles to business partners, changing collection segment data for business partners, and deleting transactional data.

Tip

This app requires maintaining job templates in maintenance view APJ_C_SCOPE via Transaction SM30. Consult SAP Note 3418953 for further guidance.

Transaction Code

The corresponding transaction in SAP GUI is Transaction UDM_BP_PROF.

Schedule Credit Management Jobs (F3748)

Application Type: Transactional

The app allows credit management users to schedule jobs in credit management to update credit master data of business partners. You can select from various predelivered job templates, including basic tasks like assigning credit management roles, updating payment behavior

summaries, repeating credit checks, and verifying configurations, as well as advanced tasks like updating scores, credit limits, and scheduling credit checks for service transactions. In addition, the app supports credit integration by importing credit information from credit agencies. Consult SAP Notes 3580496 and 3586428 for displaying advanced key figures in this app.

Tip

This app is built upon the framework of the Application Jobs app (F1240).

Transaction Codes

Mass update of credit limit for business partners can be done with SAP GUI Transaction UKM_MASS_UPD3. On the other hand, Transaction UKM_MASS_UPD4 is used for mass updates to the check rule in credit management whereas Transaction UKM_MASS_UPD5 is used for mass updates to assign the UKM000 role to business partners especially during migration. However, this app offers a wider range of credit management functionalities, including scheduling and integration tasks.

Schedule Dispute Management Jobs (F3578)

Application Type: Transactional

This app allows accounts receivable users to schedule jobs for dispute management functionality in SAP S/4HANA, including the automatic creation and write-off of dispute cases. Dispute cases can be created for residual items from automatic incoming payments, check deposits, or postprocessing, and non-collectable amounts can be written off automatically, closing the cases. The app's features include defining scheduling options such as start date and recurrence, configuring output control (e.g., nondeductible input tax adjustments), setting posting parameters like posting date and alternative tax accounts, and customizing output lists for value-added tax (VAT) or input tax line items.

Tip

If you face issues with missing templates, consult SAP Note 3418953 to maintain missing entries in the maintenance table APJ_C_SCOPE.

Transaction Code

The corresponding transaction in SAP GUI is Transaction FDM_AUTO_CREATE.

Schedule General Ledger Jobs (F1927)

Application Type: Transactional

This app allows you to schedule general ledger jobs using the delivered templates. For example, you can schedule activities such as depreciation calculation or perform foreign currency valuation. The app supports scheduling periodic activities:

- Advanced foreign currency valuation
- Analyzing GR/IR clearing accounts
- Automatic clearing
- Balance carryforward
- Financial statement data according to eXtensible Business Reporting Language (XBRL) taxonomy
- Post various entries (e.g., general journal entries with auto-reverse, recurring journal entries), handle provisions for doubtful receivables, and regroup receivables/payables

In addition, the app supports defining scheduling options (start date, recurrence), output control settings, posting parameters, and output list configurations.

This app is recommended as a replacement for the Carry Forward Balances app (F1596), which was deprecated with SAP S/4HANA 2022.

Tip

Consult SAP Note 3363404 for SAP's recommendations on carrying out balance carryforward in an SAP Fiori environment.

Transaction Code

Alternatively, you can use the HTML SAP GUI Transaction FAGLGVTR to schedule as a background job.

Schedule Inference Jobs – Intelligent Intercompany Document Matching (F5773)

Application Type: Transactional

This app automates intercompany reconciliation by scheduling inference jobs that analyze and match documents based on predefined rules and intelligent algorithms. As a result, manual effort is reduced and intercompany reconciliation and matching processes are enhanced. You can define scheduling parameters, execution times, and recurrence patterns, while detailed logs facilitate troubleshooting of errors. The app helps you handle high volumes of intercompany transactions.

Schedule Interest Calculation Jobs (F4176)

Application Type: Transactional

This app allows accounts receivable users to schedule and automate interest calculation processes for various items. As a result, the app simplifies interest calculation by providing streamlined scheduling and monitoring tools, enhancing accuracy and efficiency. The app allows you to define selection criteria using fields such as customer account, company code, and interest indicator; specify the interest calculation to (interest calculation period); and manage job execution in the background. In addition, you can restrict interest calculation to special general ledger transactions or use dynamic selection options available in the job. You also can define the calculation period and specify forms for printing interest letters, including issue dates. To print interest letters, you can start the program either in an update run or test run. In a test run, the system provides an overview of interest due, excluding items without interest unless filters are removed, thereby you can validate information prior to running update run mode. You can access detailed interest calculation data by double-clicking a row.

Tip

The app excludes from interest calculation for blocked items. However, interest block is not considered by the account balance interest calculation. You must make an entry in the **Interest Block** field to block an item to avoid calculation of interest on the app's selection screen.

As a prerequisite for accessing job templates for this app, you must maintain APJ_C_SCOPE via Transaction SM30 by adding SAP_FIN_AP_ ITEM_INT_CALC and SAP_FIN_AR_ITEM_INT_ CALC entries.

Transaction Code

The corresponding transaction in SAP GUI is Transaction FINT.

Schedule Job for Insert House Bank and House Bank Account Data in Table BSEG (F7188)

Application Type: Transactional

This app allows you to schedule jobs that insert house bank and house bank account data into the accounting line-item table (BSEG). Thus, the app allows you to sync historical cash management data with cash management. To use historical cash management data in cash

management apps, you must run the program FCLM_UPDATE_HBKID_HKTID to insert house bank and house bank account data into table BSEG before using the apps. This program needs to be executed once for historical transaction data and the program displays the number of updated rows upon completion. In addition, the app has been enhanced with the **Additional Bank Account Derivation** option, which allows updating accounting documents with house bank and account data. However, this feature is unavailable if the **Disable Addit. Bank Acct Derivation Model** data model is selected in the cash and liquidity Customizing activity. Moreover, if house bank data is missing from bank statements, payments for open items, or payment requests, the app derives it based on the general ledger account. If the general ledger account is not a cash account, the app will be used for derivation.

Tip

For cash accounts, specific rules apply. For instance, petty cash accounts are excluded from updating. For bank reconciliation accounts, the app derives data directly from general ledger account. In addition, for bank sub accounts, the app uses sub-general ledger accounts for derivation. The system prioritizes a bank account management (BAM) connectivity path for derivation; if unavailable, it falls back on general ledger master data.

Schedule Job for Rebuild Flow Types in Accounting Documents (F6799)

Application Type: Transactional

This app automates the reconstruction of flow types within accounting documents (table BSEG) in cash management. You can schedule jobs to rebuild flow types when adjustments in accounting scope or liquidity analysis are required. The app features include automated flow type reconstruction for maintaining financial reporting consistency, flexible scheduling options for optimized job execution, integration with cash management for accurate liquidity analysis, and error handling with detailed logs for transparency and troubleshooting. The app primarily helps finance teams manage liquidity and cash flow tracking. As a result, you can use this app to ensure flow types are correctly rebuilt to maintain precise financial positions.

Schedule Job for Rebuild Tool for Liquidity Item (F6770)

Application Type: Transactional

For effective cash flow management and financial planning, this app allows scheduling rebuild jobs for liquidity items (table `fqm_flow`) data using predelivered templates and customizable scheduling options. You can define scheduling parameters, set configurations for line items and account assignments in accounting documents, and monitor job status. In addition, this app supports deriving liquidity items for previously posted line items, ensuring liquidity data is available even when automatic derivation is absent. The system retrieves relevant data from table `BSEG` and updates it for direct consumption by other apps. The app comes with two operational modes. The **Initial Load** option serves first-time processing of large datasets without derivation functions, using query sequences and adjustable package sizes for performance efficiency. On the other hand, the **Rebuild** option is used for refining liquidity item data after the initial run, supporting derivation functions and selective processing of line items based on criteria.

This app can be used in place of the deprecated Adjust Assigned Liquidity Items app (F3627).

Schedule Job for Value Date Correction (F6679)

Application Type: Transactional

With this app, you can automate value date adjustments for financial transactions. This app is typically used as part of period-end closing. As a result, the app helps to achieve accurate cash flow forecasting and payment processing, especially when a company has a high volume of transactions. Incorrect value dates can impact liquidity planning and financial reporting. The app's features include job scheduling options for automated adjustments based on predefined rules, integration with cash forecasting, error handling with execution logs, customizable scheduling, and compliance support to maintain regulatory alignment. You can use app's the three step approach to schedule a job: **Template Selection**, **Scheduling Options**, and **Parameters (Optional)**. In addition, the app's dynamic scheduling options help you filter specific selection criteria such as company code, fiscal year, posting date, general ledger account, house bank, and account ID. Thus, the app optimizes treasury operations by reducing manual interventions.

Schedule Jobs for Bank Account Balance Initialization (F7146)

Application Type: Transactional

With this app, you can schedule jobs to identify bank accounts whose balances may not appear in cash management, even after bank statements have been imported. Thus, the app ensures accurate short-term cash positions by replicating the latest end-of-day bank statement balances into cash management. The app allows you to schedule jobs that identify such accounts and replicate their latest end-of-day bank statement balances to cash management.

You can initialize bank account balances through scheduled jobs or directly via Transaction FCLM_BAL_INITIALIZE. In addition, jobs can be started immediately or scheduled, with options for recurrence. Once parameters are set and validated, scheduled jobs are listed in the application jobs for monitoring. You also can view the results of scheduled jobs for tracking purposes.

Schedule Jobs for Bank Statement Monitor (F6866)

Application Type: Transactional

This app allows job scheduling for the Bank Statement Monitor app (F6388) using the app's pre-delivered templates and customizable scheduling options. You can create single or recurring jobs for determining the status of end-of-day bank statements, specifying parameters such as start date, time zone, and recurrence pattern. The key date parameter allows flexibility in defining the reference date for status determination. In addition, you can filter bank statements by company code, house bank, and account, or leave fields empty for broader selection. Also, email notifications can be configured for failed or canceled jobs, with recipients specified with multiple languages support. Once jobs are scheduled, the system processes monitoring data based on the defined parameters. This helps you track and reconcile bank statements.

Schedule Jobs for Cash Concentration (F3688)

Application Type: Transactional

This app allows cash management users to create and schedule cash concentration jobs using delivered templates and provided options, with the ability to define jobs for each cash pool and check their status. In addition, you can specify the current date or a future date as the value date for scheduling jobs. As a prerequisite, prior to using the app, you must ensure that you have the necessary authorization objects (S_PROGNAM and S_BTCH_JOB) in place and the entry **SAP_FIN_ CLM_CASH_CONCN** is maintained in the Customizing activity under **Activation of Scope-Dependent Application Job Catalog Entries**. The app's features include creating and scheduling

recurring jobs, initiating them immediately or at a later time, and viewing the results in the **Application Jobs** list page.

Tip

Consult SAP Note 3508926 for troubleshooting tips for various symptoms when using cash concentration apps.

Schedule Jobs for Customer Line Items Mass Change (F7224)

Application Type: Transactional

As the name of the tile implies, this app allows you to edit more than 50 customer line items simultaneously. This is especially useful to improve efficiency in managing customer line items. However, this is not a standalone app as the mass change jobs can only be initiated from the Manage Customer Line Items app (FO711). With this app, you can track changes through the application log, monitor job statuses, and cancel long-running jobs to maintain process control and optimize performance.

Schedule Jobs for Organizational Changes (F4754)

Application Type: Transactional, Reuse Component

This app allows you to view log details for organizational changes that have been activated, simulated, processed, or completed using the Manage Organizational Changes app (F4567) or related programs. To view all organizational changes, you must enter **FINS** in the **Category** field, **FINOC** in the **Subcategory** field, leave the **External Reference** field empty, and select **Go**. For a specific organizational change, you are required to input its identifier in the **External Reference** field.

Tip

When navigating from the Manage Organizational Changes app (F4567), the app populates the category and subcategory automatically. You can add the external reference to the table by clicking the settings icon and selecting **External Reference**.

Schedule Jobs for Supplier Line Items Mass Change (F7225)

Application Type: Transactional

This app allows mass editing of more than 50 line items simultaneously. However, the app is not a standalone application. The mass-change job must be initiated from the Manage Supplier Line Items app (F0712). The app's functionalities include performing bulk edits, accessing the application log, monitoring change job statuses, and canceling long-running processes.

Tip

You can select more than 50 line items from the Manage Supplier Line Items app (F0712) to trigger a mass job to be scheduled as a background job. In addition, you can monitor the mass change logs after execution to verify the results by using the **Show Details** option or the **View Mass Change Logs** button.

Schedule Matching Run Jobs – By Company (F7669)

Application Type: Transactional

This app allows you to schedule intercompany document matching jobs using the predefined job template **Matching Run – By Company**. While creating jobs, you have options to choose scheduling options and to restrict data scope with parameters such as matching method, fiscal year, fiscal period, and company, and data

cutoff. You can launch matching run jobs, monitor their status (**Finished** or **Failed**) via the **Status** column on the job list page, and review matching results upon completion. When a job is marked **Finished**, you can access detailed matching results by selecting the corresponding icon in the **Results** column.

Schedule Matching Run Jobs – By Consolidation Unit (F7670)

Application Type: Transactional

This app allows you to schedule document matching jobs between consolidation units using the predefined job template **Matching Run – By Consolidation Unit**. While creating jobs, you can choose scheduling options and restrict the data scope with parameters such as matching method, fiscal year, fiscal period, and consolidation chart of accounts, consolidation version, consolidation unit, and data cutoff. You can launch matching run jobs, monitor their status (**Finished** or **Failed**) via the **Status** column on the job list page, and review matching results upon completion. When a job is marked **Finished**, you can access detailed matching results by selecting the corresponding icon in the **Results** column. A matching document is created for each unique combination of leading unit and mandatory filter fields within the data source. Then these documents can be accessed and reviewed in the Manage Matching Documents app (F3868).

Schedule Matching Run Jobs – General (F5011)

Application Type: Transactional

This app allows flexible scheduling of document matching jobs by composing SQL filter strings using the SAP-delivered template. The app's functionalities include creating jobs based on templates such as **Matching Run or Matching Run – For Consolidation, Schedule Matching Run**

Jobs – By Company, and **Schedule Matching Run Jobs – By Consolidation Unit**; scheduling options for immediate runs, future dates, or recurring patterns; and specifying parameters such as the matching method and filter string. In addition, the app's advanced options allow automatic triggering of follow-ups configured through reason codes, viewing matching run statistics, setting data cutoff points, and defining a safety interval to account for database update delays. Moreover, you can monitor job statuses, view matching run results, and access details like document counts, newly rolled-in line items, and assigned items, with insights into system processing times for reading, matching, and writing data.

This app replaces the Schedule Inference Jobs – Intelligent Intercompany Document Matching app (F5773), which is now deprecated.

Transaction Codes

The equivalent SAP GUI programs are as follows:

- ICA_MATCHING_RUN (Schedule Matching Run Jobs)
- ICA_MATCHING_RUN_AFCA (Schedule Matching Run Jobs – By Company)
- ICA_MATCHING_RUN_AFCC (Schedule Matching Run Jobs – By Consolidation Unit)

You can schedule these programs via Transaction SM36. Consult SAP Note 3389652 for further guidance on additional templates and required SAP Notes to use the app.

Search Payments (F2588)

Application Type: Transactional

With the Search Payments (F2588), Search Payments in Lots (53978), and Search Payments in Payment Runs (F3977) apps, you can search and analyze for incoming or outgoing payments in real time in contract accounting. With the Search Payments app, you can perform activities such as searching for payments from both

payment runs and lots using fields such as document date, payment account, and transaction currency; enabling navigation to specific payments, payment orders, or repayment requests; and accessing clarification processing.

Although all three apps offer the ability to filter, sort, and customize the display of payment information, there are some key differences:

- **Search Payments**
  This app includes both incoming/outgoing payments from payment runs and lots. It's used in general payment tracking in accounts receivable and accounts payable.

- **Search Payments in Lots**
  This app is limited to searching payments from lots with more specific filter criteria. It's used to manage manual batch payment, manual payment entry, or check deposits.

- **Search Payments in Payment Runs**
  This app focuses on payments from payment runs. It's used for large volumes of payment processing.

Transaction Codes

The equivalent transactions in SAP GUI are Transactions FP30 and FP30H.

Search Payments in Lots (F3978)

Application Type: Transactional

This app allows you to search exclusively for payments within payment lots using more specific filter criteria than the Search Payments app (F2588) in financial contract accounting. With this app, you can navigate individual payments to review details and access clarification processing for payments that require further resolution. You can search for payments in payment lot using dimensions such as origin, clarify, account holder, IBAN, credit card number, not to payee, document, value date, amount, and transaction currency. The app displays payments in a tabular column format with the flexibility to add dimensions. In addition, the app supports manual batch payment, manual payment entry, or check deposits. Moreover, the app provides filter criteria to select payments such as account number, IBAN, amount, and value date.

Transaction Codes

The equivalent transactions in SAP GUI are Transactions FP30 and FP30H, used to perform a user-defined search for payments within payment lots.

Search Payments in Payment Runs (F3977)

Application Type: Transactional

The app allows you to search for incoming or outgoing payments runs in financial contract accounting. The app offers functionalities such as selecting open items for payment, determining payment methods and bank details, posting payments, and creating payment media. In general, this app is used for searching for large volume of payment processing. In addition, the app provides filter criteria to select payments such as paying company code, business partner, country, and payment run date.

Transaction Code

The corresponding transaction in SAP GUI is Transaction FP31.

Search Revenue Contracts – For Classic Contract Management (FARR_CONTRACT_SEARCH_OVP)

Application Type: Web Dynpro

This app allows you to locate and analyze revenue contracts within the classic contract management (CCM) framework in revenue accounting and reporting (RAR). The app offers advanced

search capabilities to filter contracts by attributes such as customer, contract type, transaction price, and performance obligations (POBs), while providing a detailed contract overview, including revenue recognition status and financial postings. The app complies with IFRS 15 and ASC 606 accounting standards. In addition, the app provides audit trail access for tracking modifications and features an SAP Fiori–based interface for intuitive contract search and analysis.

Set Substitution/Validation Logging (F4945)

Application Type: Transactional

This app allows configurations to manage logging for substitution or validation rule execution by enabling or disabling it for specific users or time periods and setting the logging level. The app's functionalities include viewing or filtering enabled log settings by event ID, rule execution user, or logging level; creating and instantly enabling new log settings; deleting settings to disable logging; and navigating to substitution and validation rule execution logs. In addition, when configuring logs, you can specify the event, logging level (e.g., detailed, failures only or all instances), logging end time (defaulted to 24 hours but customizable), and restrict logging to certain users or apply it to all users.

Shift Contracts to Next Period (FARR_CONTRACT_SHIFT)

Application Type: Web Dynpro

This app allows you to transition revenue contracts to the next financial period while ensuring compliance with IFRS 15 and ASC 606 standards in revenue accounting and reporting (RAR). The app helps with period transitions, aligns financial postings with the updated period to prevent discrepancies, and incorporates error handling and validation checks to resolve potential issues before execution of revenue processes. In addition, reconciliation key statuses such as **Failed (F)**, **Transferred (P)**, **Replaced (R)**, and **Open (O)** could determine whether revenue contracts shifted. Thus, the app helps you to identify discrepancies early, preventing errors in revenue recognition.

Tip

SAP recommends not posting one contract in the current period and posting it the next period due to shifting of reconciliation key status. If you encounter a reconciliation key with status **Failed (F)**, **Transferred (P)**, or **Open (O)**, consult SAP Note 3033691 for further remediation.

Short-Term Cash Positioning (F5380)

Application Type: Transactional, Analytical

This app allows cash and liquidity management users to view cash positions by location, company, and currency, with data calculated based on profiles defined in the Define Cash Position Profiles app (FCLM_CP_PROFILE). The app's functionalities include displaying cash positions by various dimensions like country/region, company code, bank account, and currency, simulating cash concentration for cash pools from the Manage Cash Pools (Version 2) app (F3266A), and enabling custom scaling to adjust the display of amounts. The app's prerequisites include defining cash position profiles in the Short-Term Cash Positioning app, setting up bank account details in the Define Bank Account Settings – Bank Statements app (F5488), and performing bank account balance initialization as necessary. In addition, the app supports exporting search results to spreadsheets and utilizing hierarchy sources to aggregate balances in cash pool header accounts for simulation purposes. The **Enable Custom Scaling** function allows scaling amounts and specifying decimal places for precise value representation using the scaling

factor. The app displays data by reading from table FQMET_BALANCE.

Tip

Consult SAP Note 3371497 to understand the Customizing activities and troubleshooting tips for this app.

Simulate Revenue Posting Run (FARR_ACCR_RUN/FARR_ACCR_RUN_AP)

Application Type: Web Dynpro

Revenue accounting and reporting (RAR) is an add-on to SAP financials that functions as a subledger, providing detailed revenue recognition and compliance capabilities within financial processes. This app allows you to preview revenue postings before executing actual accrual runs in RAR. The app ensures compliance with predefined accounting principles and company code authorizations while identifying potential posting errors for resolution. In addition, this app is integrated with the accrual engine to support accurate revenue recognition and provides an SAP Fiori–based interface for streamlined monitoring and adjustments.

Solved Disputes (F2521)

Application Type: Analytical

This analytical app displays the **Solved Disputes** KPI, which refers to dispute cases that are closed or confirmed to customer invoices and payments. The app shows the amount generated by all solved disputes in the last 12 months. You also can display solved disputes in other categories such as by processor, by company code, or by customer. In addition, you can drill down by processor, coordinator, customer, and dispute case, and compare the actual solved disputes of a category in a specific period to the average resolve days. The app features include seeing the number of solved disputes in the last 12 months, comparing current period solved disputes to the average, filtering by various categories, and drilling down by coordinator, customer, or dispute case. Also, the app has caching enabled for better performance, allowing immediate data refresh. The data is also updated automatically based on the cache duration in the tile configuration.

Start Revenue Posting Run (F3876)

Application Type: Transactional

This app helps with revenue postings for selected contracts, ensuring compliance with accounting standards such as IFRS 15 and ASC 606 for revenue accounting. With this app, at the time of posting run, the system processes contracts sequentially and posted entries are recorded in the general ledger, while contracts with errors are skipped. In addition, the app can perform a simulation with filters like company code, accounting principle, fiscal year, and posting period. Simulation results can be displayed by general ledger account, performance obligations (POBs), or posting pairs. To post, you need to switch to posting mode, specify company codes (or ranges), fiscal year, posting period, and other criteria like contracts. In the selection screen, you have option to enable the **Posting Check** indicator to validate contracts individually. On the other hand, the system does not post skipped contracts. In addition, you can close the revenue accounting period using the **Close Period for Rev** option. Upon posting, general ledger lines are aggregated based on account assignments, or optionally, by debit/credit indicators.

> **Tip**
>
> You must carry out Customizing activity **Switch on Posting Optimization** to support optimization for aggregation and subledger use to manage detailed data postings.
>
> Consult SAP Note 3357659 for additional guidance on revenue accounting and reporting (RAR) for SAP S/4HANA 2023.

Statement of Changes in General Ledger (W0162)

Application Type: Web Dynpro

This app allows the analysis of various statements of changes, including those related to equity and provisions/accruals in the general ledger. You can filter data by company code, ledger, fiscal year, statement of changes type, profit center, and segment. The account level is determined by semantic tags assigned to report rows, while report columns display the opening balance based on carryforward transaction type, configured transaction types in a definable order, and the total account balance. In addition, the app supports drilldown capabilities that allow navigation to account line items via linked apps like the Display Line Items in General Ledger app (F2217). In addition, you can drill down into profit center or segment details by selecting **Drilldown** and adding specific fields for further analysis.

> **Tip**
>
> As a prerequisite to use the app, first, you should assign semantic tags to the financial statement version or general ledger accounts using the Manage Global Hierarchies app (F2918). You must complete the **Define Semantic Tags for Financial Statement Versions** activity in the general ledger Customizing under **Financial Accounting • General Ledger Accounting • Master Data • G/L Accounts • Financial Statement Versions • Semantic Tags for Financial Statement Versions**. In addition, you should complete the **Define Structure for Statement of Changes in General Ledger** activity to assign relevant general ledger accounts using semantic tags and transaction types for the statement of change (SOC) type under **Financial Accounting • General Ledger Accounting • Information System**.
>
> As of SAP S/4HANA 2023, the app does not support user extensibility fields.

Submit Bank Account Applications (F5861)

Application Type: Transactional

This app allows cash management specialists to submit and track bank account applications, and upload attachments using the predefined **SAT** document type in the document management system (DMS). You can receive automated email notifications for approvers, check application statuses, view approval steps, access information about the created bank account, and send mail to approvers. The app also allows you to withdraw applications awaiting approval or post-approval before account activation and delete unsubmitted applications. As a prerequisite, you need to have authorization object F_CLM_BAOR assigned to your roles.

> **Tip**
>
> SAP recommends that personal data should not be uploaded, as retrieval for DMS-based documents is not supported by the Information Retrieval Framework (IFR).

Substitution Rules for Planning Levels – Treasury Flows (F7689)

Application Type: Transactional

This app allows you to display, modify, and create substitution rules for assigning cash management planning levels to internal flows generated for cash management-relevant transactions in treasury and risk management (TRM). The app is accessible only if you utilize the simplified data transfer process from TRM to the One Exposure from Operations hub. In addition, the app is built on the framework used in the Manage Substitution/Validation Rules app (F4406).

The app supports key functionalities such as creating, modifying, displaying, and transporting substitution rules, along with extensibility features. You can review detailed execution logs

using the Substitution/Validation Log app (F4886), but logging must first be activated via the Set Substitution/Validation Logging app (F4945). Once activated, you can view execution logs in the Substitution/Validation Log app (F4886).

As a prerequisite, you must define all planning levels in the Customizing activity **Define Planning Levels**. You also must apply to planning level derivation only if **Derivation** is selected in the **Derivation Category for Planning Level** field within the **Basic Settings for Cash Management Integration** Customizing activity.

> **Tip**
>
> During runtime, substitution rules are executed in alphabetical order of the rule name, as all defined rules modify the same field. You must ensure that data used in substitution rules comply with legal and business requirements. In addition, applied rules and target field values are visible to users with treasury business roles, so sensitive data should not be included.

Substitution/Validation Log (F4886)

Application Type: Analytical

This app allows you to display detailed logs for substitution or validation rules at runtime for specific events. The app provides features such as viewing log lists and details; filtering logs by event ID, user, run executed on, and run executed at; combining multiple logs; and exporting logs to a spreadsheet. In addition, you can select a log row, which opens a detailed page showing the processing flow for triggered rules in a tree structure. Logs provide information on the processed event (e.g., table FINS_ACC_JEI_1 for journal entry item), the input document (e.g., table ACDOCU for group reporting journal entries), rule content depicting internal processing, and excerpts of the backend substitution or validation script. Moreover, the app's results indicate whether preconditions were met, fields validated or substituted, and outcomes of the rule application.

Supervise Collections Worklist (F2375)

Application Type: Transactional

With this app, you can display an overview of all open receivables for collection, supervise the daily progress of collections, manage your workloads, and redirect your efforts as needed. The app provides features such as displaying a prioritized list of urgent customers, viewing daily progress with various filtering options, and assigning or changing the assignment of collection specialists to open receivables. The app presents collection worklists in both graphical and tabular formats for intuitive analysis. This helps ensure efficient and effective management of the collection process.

Tip

Consult SAP Note 3508571 for details on collection specialist assignment for a customer/business partner. To see full historical notes, SAP recommends using the Process Receivables (Version 2) app (F0106A).

Transaction Code

The corresponding transaction in SAP GUI is Transaction UDM_SUPERVISOR.

Supplier Payment Analysis (Manual and Automatic Payments) (S/4HANA) (F1749)

Application Type: Analytical

With this app, you can get a comprehensive overview of your supplier payment activities. This app allows you to view payment data by dimensions such as company code, supplier, currency, and user. It displays aggregated

payment documents for a specific period defined in the KPI, using different colors for automatic (green) and manual (blue) payments. Also, you can filter data by company code, by supplier, by currency and other dimensions. The app supports SAP Jam for comments, email notifications, and uses CDS view C_APMANUALPAYMENTS for data processing. Caching is enabled for better performance, with options to refresh data immediately. The app displays data from automatic payments that are posted via Transaction F110, while manual payments are all other relevant transactions such as F-48 or F-58.

Transaction Code

The corresponding transaction in SAP GUI is Transaction FBL1N.

Supplier Payment Analysis (Open Payments) (S/4HANA) (F1750)

Application Type: Analytical

This app provides a comprehensive overview of open payments to the accounts payable manager, displaying payment data by company code, supplier, currency, and user. It shows aggregated data for all unreconciled payment documents, color-coded (green for down payments and blue for other open payments). It distinguishes payments using posting keys and transaction types, including residual payment documents and partial payments, ensuring accurate financial management. In addition, you can filter data by company code, supplier, or currency. The app supports SAP Jam for comments, email notifications, and uses CDS view C_APVENDOROPENITEMS. Caching is enabled for better performance, and data can be refreshed immediately.

Transaction Code

The corresponding transaction in SAP GUI is Transaction FBL1N.

Task Logs (F2795)

Application Type: Transactional

As part of group reporting, with this app, you can check the logs of each consolidation task during test and update runs by specifying the task category, consolidation unit or group, version, date range, or users. The app displays tasks in a list, and you can navigate to detail screens for categories such as manual posting, data collection, currency translation, reclassification, data validation, and the release of the Universal Journal. Logs display details including consolidation unit/group, overall status, document type, error messages, and journal entry line items. The app also offers additional features, including filtering by last log, grouping log items by certain criteria, viewing log header messages, displaying input fields, and exporting log items to a Microsoft Excel file for further analysis.

Tax Declaration Reconciliation (F2096)

Application Type: Analytical

This app allows tax managers, accountants, and financial reporting teams to reconcile tax data for tax declarations. With this app, you can check tax-relevant postings to be declared in a value-added tax (VAT) return. The app compares both SAP and submitted returns to tax authorities and provides any discrepancies found during reconciliation. The app is available for the general ledger accountant (SAP_BR_GL_ACCOUNTANT) and external auditor (SAP_BR_EXTERNAL_AUDITOR) roles. For users assigned with the external auditor role, a time frame restriction can be imposed to limit the range of visible data. The app provides features such as verifying that tax base amounts reported in the VAT return tax boxes correspond with VAT amounts, and generating lists for tax boxes, tax codes, tax rate, transaction key, tax base amounts, reported tax amount, calculated tax amounts, and differences.

Tax Reconciliation Account Balance (F2095)

Application Type: Analytical

This app displays general ledger balances on tax-relevant accounts. It's especially beneficial for external auditors who want to search for balances by tax attributes, such as tax code, for a selected period or a range of dates. You also can search for tax account balances using parameters such as company code, journal entry, fiscal year, posting date, and general ledger account. The app presents search results in a tabular format with the flexibility of adding fields as necessary. For instance, the app is particularly useful in Europe to show the value-added (VAT) tax amount. With this app, you can search and compare general ledger balances by various dimensions, such as company code, fiscal year, posting date, and tax general ledger account. Upon searching, the app reads data from table BSET. In addition, the app provides a searching capability with tax-relevant attributes, such as tax jurisdiction or tax code. In addition, you can limit your search to only items with differences via the difference mark filter criteria and validate calculated VAT tax amounts by general ledger account.

Tip

This app is not supported in company codes set up in certain countries, such as Brazil, Canada, China, India, and the United States.

Total Receivables (S/4HANA) (F1748)

Application Type: Analytical

With this analytical app, you can display the **Total Receivables** KPI, thereby allowing you to view the total receivables amount by various dimensions and filter the results according to different criteria. You can drill down to analyze total receivables by account group, company code, company code currency, country key, customer, customer classification, display currency, exchange rate type, general ledger account, reconciliation account, region, or special general ledger. The app offers features such as viewing the total receivables amount for today on the KPI tile; viewing by due period; top 10 customers, company code, or accounting clerk; and filtering results by company code, customer, customer country and region, accounting clerk, or due period.

Tip

Consult SAP Note 2688661 to get step-by-step instructions for selecting all customers, drilling down to the customer level, and exporting to Microsoft Excel for further analysis.

Transaction Code

The corresponding transaction in SAP GUI is Transaction S_ALR_87012167.

Track Supplier Invoices (F0217)

Application Type: Transactional

This transactional app allows you to efficiently monitor the status of supplier invoices, including detailed invoice information and reasons for unpaid invoices. The app helps with follow-ups by providing relevant contact information. With this app, you can display outstanding amounts, invoice numbers, payment status, general invoice information, and attachments. In addition, you can review payment history, sort invoices by due date or payment status, search master data, switch display modes, email invoice details, save invoice details as tiles, and post comments on SAP Jam. Also, the app can cater to a wide range of both financial non-finance users, including accounts payable managers, cash managers, and purchase managers.

Transfer Business Partner Items (F4398)

Application Type: Transactional

This app allows accounts payable and receivable accountants to transfer open items posted to a business partner or contract account in a specified currency in contract accounting. The app comes with separate tabs that display transferable, nontransferable, and all open items matching the applied filters, with explanations provided for nontransferable items, such as credit items or those locked for clearing or posting. In addition, you can apply filters such as mandatory business partner or optional document type. To transfer items, select them, provide the target business partner, contract account, or contract in the dialog, and confirm by clicking the **Transfer** option.

Tip

As a prerequisite to support transfers via this app, the Customizing settings under **Define Specifications Dependent on Transfer Reason** enable defining whether credit can be transferred.

Transfer Revenue (F3879)

Application Type: Transactional

The app allows the transfer of revenue and cost to the revenue accounting subledger while calculating exchange rate differences from foreign currency revaluations, including generating invoice records for simplified invoice processes. As a prerequisite, the app must be executed before the Revenue Posting Run app (F3876) and it performs a validation check during execution. If you select production mode, the app cannot run for future periods, and if the accounting principle requires foreign currency revaluation at the actual rate, it operates period by period.

For contracts handled through simplified invoice processes, it transfers invoice correction postings to the revenue accounting subledger. The app can be run manually for contract changes on an ad hoc basis or scheduled as a recurring background job based on business requirements and data volume.

Travel and Expenses for Travel Assistant (F0584B)

Application Type: Transactional

With this new app in travel management for SAP S/4HANA, you can view and submit travel and non-travel expenses. As a travel assistant, you can create, change, and submit expense reports, simplify report creation with the copy trip function, manage attachments, capture receipts using a device camera, access all expense types, assign attendees, itemize receipts, deduct personal expenses, manage multidestination and multicurrency reports, split costs, handle reports on behalf of others, save and edit draft reports, manage per diem advances, and edit foreign currency rates. For example, you can browse existing travel requests with various parameters such as employee name, status, start date personnel number, and cost center. As a prerequisite, SAP recommends maintaining employee business partner, info types, and workflow hierarchy.

Transaction Code

The equivalent transaction in SAP GUI is Transaction PR05.

Travel Assistant Work Center (FITV_POWL_ASSISTANT)

Application Type: Web Dynpro

This app provides travel assistants with a centralized place to manage travel-related tasks on behalf of employees in travel management. The app includes key information such as travel requests, expense claims, and travel approvals. In addition, you can delegate expense reporting and perform audit and policy compliance checks. As a prerequisite, an employee master record must be set up in the business partner employee role. In addition, based on the requirements, an approval hierarchy may be required to use the workflow approval feature. The app reads data from tables PTRV_HEAD and PTRV_PERIO, as well as dependent tables.

Travel Plan (FITP_PLANNING)

Application Type: Web Dynpro

This app helps with planning and managing travel activities such as creating travel plans or itineraries and estimating travel cost in travel management. Once a travel request is approved, the traveler can book flights, hotels, rental cars, and rail services in a travel plan by connecting to an external reservation system while ensuring compliance through automated policy validation and budget checks. The app enables checking company-specific rates and discount agreements.

Transaction Code

The equivalent transaction in SAP GUI is Transaction TP20. As a prerequisite, employees must be maintained as business partners.

Travel Plans for Travel Assistant (F6771)

Application Type: Transactional

With this app, employees or travel agents can manage travel plans in real time via desktop or mobile devices, offering functionalities to create, display, delete, and cancel plans in travel management. A travel plan helps with booking services such as flights, hotels, cars, and trains through a third-party Global Distribution Service (GDS). The app allows you to browse plans in PDF format, add reservation details with ticketing and pricing information, and modify existing plans when applicable. In addition, with this app, travel agents can book trips on behalf of others while maintaining the same set of core features.

Transaction Code

The corresponding transaction in SAP GUI is Transaction TP01.

Travel Request (FITE_REQUEST)

Application Type: Web Dynpro

A travel request contains all relevant trip details for approval, advance payment, or booking in travel management. The app supports the entire business travel process, from request submission and planning to booking, approval, and final settlement, thereby eliminating paper forms or email requests. Employees can submit a travel request including relevant details to their managers to review them for approval. You can browse existing travel requests with various parameters such as employee name, status, start date personnel number, and cost center. In addition, the travel requests are integrated with travel expenses for tracking purposes. The travel request includes key details such as trip dates, destinations, itinerary, cost centers, required advances, estimated costs, and necessary travel services (flights, hotels, car rentals, rail). If booking authorization is restricted for employees, a travel assistant (travel agency) or designated personnel can manage reservations based on the travel request. Because travel requests feed directly into travel expenses, they automatically transfer trip details, requested advances, and cost assignments. In addition, travel planning helps with the planning function and online booking, with optional integration into funds management.

Travel Requests for Travel Assistant (F0409B)

Application Type: Transactional

With this transactional app, you can manage travel requests in real time via desktop or mobile devices. Travel assistants also can manage requests for others. In addition, you can create new or copy existing travel requests, submit requests with attachments, and browse, edit, resubmit, and delete requests. Also, you can navigate to the Travel Requests for Travel Assistant app (F0409B), which allows travel agents or assistants to manage travel requests on behalf of others. You can browse existing travel requests with various parameters such as employee name, status, start date personnel number, and cost center. On the resulting screen, the app displays travel requests with important information such as trip number, reason, destination, country, status, start date, end date, and total cost. Also, you can perform tasks such as additional destinations, cost assignments, estimated costs, advances, travel services, and planning.

Transaction Code

The corresponding transaction in SAP GUI is Transaction PR05.

Traveler Work Center (FITV_POWL_TRIPS)

Application Type: Web Dynpro

This app provides employees with a centralized place for managing travel-related tasks in travel management. The app's features include managing employees' travel requests, expense reports, and trip approvals. With this app, employees can create, modify, and submit travel requests, track trip status, and manage expenses with visibility into historical, ongoing, and planned travel activities. You can browse existing travel requests with various parameters such as employee name, status, start date personnel number, and cost center. In addition, employees can delegate certain tasks. The app integrates with the general ledger for seamless integration. As a prerequisite, an employee must be set up in a business partner with an employee role. In addition, baseline configuration for the employee master along with relevant info types must be set up.

Treasury Position Analysis (W0049)

Application Type: Web Dynpro

This analytical app provides a high-level aggregated view of position values for all treasury positions on a user-selected key date, with the ability to drill down into detailed position-level data. With this app, you can perform treasury position analysis using other search parameters such as planned data included, valuation area, company code, product type, and portfolio. In addition, you can add fields to display in columns by selecting dimensions under **Available Fields** in the **Navigation Panel**, thus providing flexibility. The app features treasury position analysis, covering all positions, accounting views, over-the-counter (OTC) transactions, securities, and listed derivatives. In addition, you can generate ad hoc reports, refine results by adding attributes, and navigate to master data or originating financial transactions using the **Jump To** feature. The app's additional capabilities include sorting, grouping, subtotal calculation, and exporting results to Microsoft Excel for further analysis.

Besides this app displaying all treasury positions, the following four apps listed display the treasury position values for selected views or functions:

- Treasury Position Analysis – Accounting View (W0118)
- Treasury Position Analysis – OTC Transactions (W0121)

- Treasury Position Analysis – Securities (W0120)
- Treasury Position Analysis – Listed Derivatives (W0119)

> **Tip**
> For this app to work, you must ensure the FTR_FPM_FINANCIAL_STATUS service is active in Transaction SICF.

Treasury Position Analysis (Accounting View) (W0118)

Application Type: Web Dynpro

The app provides a structured view of treasury positions specifically related to listed derivates in treasury and risk management (TRM). You can view book values for all treasury positions from a listed derivatives focused. In addition, the app allows you to carry out multidimensional analysis across company codes, valuation areas, and product types. For example, you can search treasury position analysis with search parameters such as key date, planned data included, valuation area, company code, product type, and portfolio. The app's features include performing liquidity analysis for short-term obligations, predicting cash flows based on historical trends, and evaluating risk management for interest rate, exchange rate, and credit risk.

> **Transaction Code**
> The corresponding transaction in SAP GUI is Transaction TPM12H.

Treasury Position Analysis (Listed Derivatives) (W0119)

Application Type: Web Dynpro

The app offers a structured accounting perspective on treasury positions in treasury and risk management (TRM). You can view book values for all treasury positions from an accounting point of view. In addition, the app allows you to carry out multidimensional analysis across company codes, valuation areas, and product types. For example, you can search treasury position analysis with search parameters such as key date, planned data included, valuation area, company code, product type, and portfolio. The app's features include performing liquidity analysis for short-term obligations, predicting cash flows based on historical trends, and evaluating risk management for interest rate, exchange rate, and credit risk.

Transaction Code

The corresponding transaction in SAP GUI is Transaction TPM12H.

Treasury Position Analysis (OTC Transactions) (W0121)

Application Type: Web Dynpro

The app offers a structured accounting perspective on treasury positions, especially on over-the-counter (OTC) transactions in treasury and risk management (TRM). You can view treasury positions for non-exchange-traded derivatives such as swaps, forwards, and other similar products. In addition, the app allows you to carry out multidimensional analysis across company codes, valuation areas, and product types. For example, you can search treasury position analysis with search parameters such as key date, planned data included, valuation area, company code, product type, and portfolio. The app's features include performing liquidity analysis for short-term obligations, displaying position values from both an accounting and risk management standpoint, predicting cash flows based on historical trend, and evaluating risk management for OTC transactions including interest rate, exchange rate, and credit risk.

Transaction Code

The corresponding transaction in SAP GUI is Transaction TPM12H.

Treasury Position Analysis (Securities) (W0120)

Application Type: Web Dynpro

The app offers a structured accounting perspective on treasury positions for security in treasury and risk management (TRM). You can view treasury positions especially for security instruments such as bonds, equities, and other similar instruments. In addition, the app allows you to carry out multidimensional analysis across company codes, valuation areas, and product types. For example, you can search treasury position analysis for securities with search parameters such as key date, planned data included, valuation area, company code, product type, and portfolio. The app's features include performing liquidity analysis for short-term obligations, predicting cash flows based on historical trends, and evaluating risk management for interest rate, exchange rate, and credit risk.

Transaction Code

The corresponding transaction in SAP GUI is Transaction TPM12H.

Treasury Position History (F3966)

Application Type: Analytical

This app allows treasury accountants to analyze changes in treasury position values, such as book value and amortized acquisition value, over single or multiple periods defined by a key date and selected time frames like end of last year, year to date, or last four quarters. Position values can be reviewed at a high aggregation level or broken down into dimensions like company codes, currencies, valuation classes, and account assignment references for detailed reconciliation insights between the treasury subledger and general ledger. The app's filters also allow you to set calculation dates and select treasury positions based on criteria like company code, commitment business partner, product type, position currency, and valuation area, along with display currency and exchange rate type. The app presents results in **Chart View** or **Table View**, with options to drill down into detailed treasury positions or financial transactions contributing to the values. In addition, the app offers customizable views that include chart types such as a bar chart or column chart, enhancing usability and analysis depth.

Trial Balance (F0996)

Application Type: Web Dynpro

The app displays ledger balances in a debit and credit column format for a given period. The header section of the app contains basic selection fields, such as ledger, company code, and date range. In addition, the app provides additional filters in the following categories: general ledger account, account assignment, partner fields, public sector management fields, and so on. The results are displayed in multiple tabs: **Data Analysis**, **Graphical Display**, and **Query Information**. The **Data Analysis** tab shows the balances for the filter criteria selected. The **Graphical Display** tab shows a visual view of the trial balance.

Transaction Code

The corresponding transaction in SAP GUI is Transaction FAGLB03.

Trial Balance Comparison (W0097)

Application Type: Web Dynpro

The app displays trial balances for a ledger and company code by comparing two reporting years. This app includes selection criteria fields similar to the trial balance. In addition, the app provides additional filter categories: basic, general ledger account, account assignment, partner fields, and public sector management fields. Basic filter criteria include ledger, company code, and comparative posting date range. The **Data Analysis** section shows comparative balances. You can select a general ledger account hierarchy from the general ledger account filter criteria based on financial statement version (FSV), which is useful to display trial balances.

Tip

With this app, you can configure the trial balance display to reflect day-based reporting or period-based reporting, with the **Default Values** setting enabled to show the balance based on period-based reporting. However, you can adjust this preference in your profile settings based on your reporting needs.

Transaction Codes

The corresponding transactions in SAP GUI are Transactions FAGLB03 or F.01.

Upload Billable Items (F4664)

Application Type: Transactional

This app allows you to import billable items for convergent invoicing from an external source file in the contract accounting billing process. Billable items serve as the basis for creating documents in the billing process. The app offers a predelivered template in CSV format, which helps in the upload process. The app's features include selecting a local file from your computer for upload, adjusting the upload schema to match the file format (e.g., changing the separator), viewing a list of billable items, and specifying parameters such as immediate billing or saving items as raw data. In addition, you can simulate and execute uploads, modify field values in test systems, and save the original billable items as a local file. The app's **Simulate Upload** function allows you to process the file and identify potential errors or inconsistencies without saving the data to the system. After reviewing the simulation results, you can proceed with the actual **Upload** option to generate the billable items in the system.

Transaction Code

The corresponding transaction in SAP GUI is Transaction FKKBIXBIT_UPLOAD.

Upload Consumption Items (F5081)

Application Type: Transactional

This app allows you to import consumption items for convergent invoicing from external source files. The app is part of SAP S/4HANA billing and revenue innovation management. The app helps businesses to effectively upload consumption items for convergent invoicing. The app's functionalities include selecting a local file, adjusting the upload schema (e.g., changing the separator), reading and listing consumption items, and specifying parameters like saving all items as raw data. In addition, the app allows you to simulate or run uploads, edit individual field values in test systems, and save items with original values as a local file. The app's navigation options depend on your roles and authorizations. Also, you can create templates for consumption items.

Transaction Code

The corresponding transaction in SAP GUI is Transaction FKKBIXCIT_UPLOAD.

Upload Customer Open Items (F4051)

Application Type: Transactional

This app allows accounts receivable teams to upload multiple customer open items into the system in bulk. You can perform this task by downloading predelivered template file, entering the relevant data, and uploading the completed file back to the app. The template includes sections for **Header Data**, **Alternative Payer or One-Time Account**, and **G/L Account Items**. After uploading, the app generates a worklist of unchecked items, enabling you to check and post them via application jobs. During checks, the system flags errors on the line items with options to correct and recheck or navigate to the details page for relevant fixes. On the other hand, for items without errors, the system marks **Ready for Posting**. In addition, successfully posted items are removed from the worklist and can be viewed in the Manage Customer Line Items app (FO711) or the Manage Journal Entries app (FO717). The app also offers features like error tracking, job monitoring, and worklist updates for efficient processing.

Tip

This app excludes sales order–based invoices and limits uploads to 2,000 entries per file, with a posting cap of 999 entries per invoice—exceeding these requires file splitting. In addition, withholding tax is unsupported, so Transaction FB70 should be used for such cases. The app also lacks support for custom fields and key standard fields like trading partner, special general ledger indicator, and billing indicator, and does not allow multiple customers per document, cross-customer postings, or postings in currencies other than the document currency.

Upload Document for Bank Account (WDA_FCLM_BAM_UPLOAD_DOC (APP_CFG_FCLM_BAM_UPLOAD_DOC))

Application Type: Web Dynpro

This app allows you to attach and manage documents related to bank accounts in bank account management (BAM). The app allows secure document storage, integrates with approval workflows for account modifications, and provides search and retrieval functionality. In addition, you can access and maintain structured audit records to support regulatory adherence. The app also supports linking uploaded files to corresponding bank accounts for quick navigation.

Upload General Journal Entries (F2548)

Application Type: Transactional

Using this app, you can upload multiple general journal entries from a spreadsheet or CSV file. The app comes with the template for uploading journal entries. Once the template is populated and uploaded, you can directly post the general journal entries or route it for park, approve, and post. Parking is useful when you want to save the document to process it later. However, the system does not calculate taxes and does not perform any other checks as the document is not complete at that point.

If workflow is enabled for an approval route, the **Submit** action creates a parked document and then triggers a workflow event. You must have appropriate authorization access to perform this action. Once the general journal entries upload is submitted, you can see the current document posting status using the **Refresh** button. A successful load posts all relevant ledgers for the accounts you are responsible for. If errors are found during upload, you need to correct the errors, copy the batch ID from the initial upload to the upload file, and then repeat the upload steps.

The app is accessible from the SAP Fiori launchpad, but it also can be launched from the Verify General Journal Entries – For Requestor app (F2547A). In this case, you submit general journal entries for verification rather than posting them directly to the ledgers.

> **Tip**
>
> This app supports a maximum of 999 line items in a single journal entry upload action. However, the app supports posting more than 999 line items in the Universal Journal only. Consult SAP Note 2789867, which provides further guidelines on uploading over 999 line items and the limitations to be considered.

> **Transaction Codes**
>
> There is no equivalent SAP GUI transaction for this app. However, similar transactions in SAP GUI are Transactions FB50 and FB01.

Upload In-House Bank Accounts (F6038)

Application Type: Transactional

This app allows you to create multiple in-house bank accounts in a single upload using an Excel template. You can download the predelivered template, input relevant account information, and then upload it to the app. Errors in the file are displayed via the app's validation during upload, and successfully uploaded accounts are available in the Manage In-House Bank Accounts app (F5942). Alternatively, you can access this app directly through the **Upload Account** option in the Manage In-House Bank Accounts app.

Tip

Mass creation of bank accounts using this upload function is intended as a one-time operation, typically for migrating bank account data from a legacy system to in-house banking.

After completing the initial upload, SAP recommends performing activities related to additional bank account creation or modifications to existing accounts manually via the Manage In-House Bank Accounts app.

Validation Result Analysis (F3484)

Application Type: Analytical

As part of SAP S/4HANA Finance for group reporting, this app allows business analysts to perform analysis of local data for validation results—whether all or failed—by task type, period, or other criteria, using regular or visual filters, chart views, and table views. As an analytical list page (ALP) app, it offers features like a header area displaying micro charts for dimensions such as failed by version, failed by ledger, failed by fiscal year period, failed by consolidation group, and failed by task type. The content area provides hybrid, chart-only, or table-only visualizations. The chart view shows validation totals and failed counts, with drilldown functionality to explore details by dimensions like consolidation unit. The table view provides detailed rule-level validation results, enabling you to view specifics on a separate page or export data to a Microsoft Excel file for further analysis.

Verify Currency Adjustments New Version (F4670A)

Application Type: Transactional

With this app, you can create and submit currency adjustments for verification, view workflow status, edit and resubmit rejected adjustments, view workflow history, withdraw adjustments for editing before approval, and receive notifications when the status changes. With the Verify Journal Entries in General Ledger – For Requester app (F2547A), you can submit currency adjustment workflow requests for review before posting.

This app replaces the **Verify Currency Adjustments – For Requester** app (F4670) as of SAP S/4HANA 2023. The new app comes with an improved user experience and performance improvements. The deprecated app will be removed from the apps library with SAP S/4HANA 2025.

Verify General Journal Entries (F2547)

Application Type: Transactional

This app allows a workflow for reviewing and approving general journal entries before they're posted to the ledger. Requesters such as an accountant can create, display, edit, and submit journal entries for verification, track their status, and decide whether to edit and resubmit rejected entries. You also can copy, edit, and submit journal entries and receive notifications with the **DLM_FIORI_NOTIFICATION** role. As a processor, you can view, approve, reject, or suspend journal entries assigned to them in the inbox and outbox apps, with mass approval support. Notifications also can be received with the **DLM_FIORI_NOTIFICATION** role. The Verify General Journal Entries inbox and outbox apps require workflow configuration and defining responsibilities by key user. In addition, this app offers features such as submitting and editing journal entries, tracking status, receiving notifications, and viewing workflow logs, comments, and attachments.

Transaction Code

The corresponding transaction in SAP GUI is Transaction FBV4.

Verify General Journal Entries – For Requester (F2547A)

Application Type: Transactional

Using this app, you can create, display, edit, and submit journal entries for verification. Like the Upload General Journal Entries app (F2548), you can upload journal entries via Excel or CSV files and submit them for verification. With appropriate authorizations, you can track the status and change the history of journal entries, including submitted, rejected, deleted, waiting, and failed. In addition, you can edit and resubmit rejected journal entries, copy and edit existing entries, withdraw entries for editing before approval, and resubmit them. Also, you can receive notifications and email alerts for status changes and view workflow logs, comments, and attachments.

Tip

A limitation of uploading more than 999 line items applies to this app as advised in SAP Note 2789867.

Transaction Code

The corresponding transaction in SAP GUI is Transaction FBV4.

Verify Intercompany Variance Adjmt (F5764)

Application Type: Transactional, Reuse Component

The purpose of this app is to allow approvers to review and either approve or reject postings related to intercompany variance adjustments in SAP S/4HANA Finance for group reporting. Approvers can access tasks in the app's inbox, navigate to assignment details, and ensure adjustments align with accounting policies.

First, variance adjustment is initiated via the Manage Matching Assignments app (F3870) when a user processes a matching assignment with a reason code linked to an adjustment class, which is a key element for classifying adjusting journal entries used during the consolidation process. To ensure that auto-adjustments require prior approval, you can attach a workflow scenario, such as standard scenario WS78500087, to the reason code. As a result of using this scenario, you can define workflows in the Manage Workflows – Intercompany Variance Adjustment app (F5766). Thus, you can activate the predelivered **Intercompany Variance Adjustment Posting** workflow, which can be used directly or customized as per your business requirements. In a multilevel approval workflow scenario, SAP recommends adding step type **Post Variance Adjustments** as the last step of the workflow in case you need to perform an adjustment posting. If multiple workflows are activated, then the system begins the one for which the conditions are met.

Tip

This app currently supports the company organizational dimension; it does not yet support other dimensions, such as consolidation unit.

Verify Journal Entries in General Ledger – For Processor Inbox (F2728)

Application Type: Transactional

The app allows a processor who is responsible for reviewing and posting journal entries to manage and verify journal entries that are waiting for approval or require further processing before they're posted to the general ledger. With this app, a processor can view and verify journal entries. The app requires workflow configuration and appropriate authorization roles to control who can process the submitted journal entries. With necessary setup, the processor can verify, approve, reject, or suspend them. In addition, processors can track the status, forward entries for verification, claim tasks to become processors of the task, navigate via notifications, and receive system notifications for submission and posting to the ledgers in My Inbox. In addition, email notifications can be set up if necessary.

Verify Journal Entries in General Ledger – For Processor Outbox (F2729)

Application Type: Transactional

With this app, the processor can view already processed journal entries, such as waiting, suspended, and so on. The app requires workflow configuration and authorization roles to control who can process the submitted journal entries. In addition, processors can track the status, forward entries for verification, claim tasks to become processor of the task, navigate via notifications, and receive system notifications for submission and posting to the ledgers in My Inbox. In addition, email notifications can be set up if necessary.

View Balance Validation Results (F6387)

Application Type: Transactional

This app allows you to review validation runs executed individually or via scheduled jobs, determining whether data passed or failed predefined validation rules and identifying root causes of failures. The app's functionalities include running new validations, reviewing ad hoc and scheduled validation results, analyzing validation failures at the line-item level, navigating validation rule details, and re-running selected validation results. First, you can initiate new validations by specifying entities and rule groups with the **New Validation** option. Once validation is completed, each row displays

an overall result using a status such as **Success**, which indicates that all validation rules were passed, except those at the **Information** control level. **Success with Warnings** means all error-level rules were passed, but at least one warning-level rule failed. On the other hand, **Failed** signifies at least one **Error**-level rule was not met. The **Technical Error** status occurs when a technical issue prevents completion. The overall result is accompanied by a comments status, determined by the **Comments Required** setting of the rules within the rule group. In addition, you can review the error message via the **View Log** option. Also, you can navigate to other apps using the drilldown options.

> **Tip**
>
> When initiating new validation for profit centers or cost centers, currency types are automatically determined based on the company codes to which these entities are assigned.

View Import Jobs for Memo Records 2.0 (F7513)

Application Type: Transactional

This app allows you to view jobs for the Import Memo Records 2.0 app (F6124) and monitor their execution. To use the app, you must carry out the Customizing activities under **Financial Supply Chain Management • Cash and Liquidity Management • Cash Management • Memo Records 2.0**, including **Defining Memo Record Types**, assigning them to predefined categories, and specifying number ranges for memo records and imports. Optionally, field status groups can be created for different memo record types. You can search for import jobs, check their status, planned start time, creation details, and other relevant information. To initiate an import job, you must use the Import Memo Records 2.0 app.

Work List for Conflicted Contracts (F4271)

Application Type: Transactional

This app allows you to manage and resolve conflicts arising from manual changes to performance obligations (POBs) and attributes that conflict with operational system updates. Manual changes can be made to correct attributes without altering the operational document, Customizing, or Business Rule Framework plus (BRF+) rules. However, conflicts may arise if the backend operational system requests further change the POBs. The app allows you to review conflicted contracts, edit attributes, and choose whether to retain manual changes or adopt values from operational documents for all POBs in a contract. If both the system and manual changes modify the same attribute, the contract is flagged as **Pending Review** in the Work List for Conflicted Contracts app (F4271). In addition, you can resolve conflicts by choosing either the operational document value or the manual change value for all POBs in the contract. Selecting manual values allows you to specify a business change reason, while choosing operational values resolves conflicts without altering the contract. Conflicts typically occur when manual changes are overwritten by operational system changes. The app provides tools to review and resolve these conflicts effectively. This ensures you can efficiently resolve discrepancies and maintain data integrity in revenue accounting and reporting (RAR).

Workflow Change Requests – In Approval (F7166)

Application Type: Transactional

As part of bank account management (BAM), this app offers a centralized location for managing submitted change requests awaiting approval, covering updates to master data, organizational structures, and other master data. Thus, the app enables a workflow for opening, modifying, closing, and reopening bank accounts, as well as maintaining payment approvers across multiple accounts. You can approve, reject, claim, release, suspend, and forward tasks, while sorting, grouping, and searching for specific requests. The app's features include task-specific details, attachment support, and entering comments, along with options to send emails containing direct links to relevant information. Moreover, the app also integrates with SAP Jam, enabling you to share comments within your company's collaboration groups.

Write Off Receivables (F6728)

Application Type: Transactional

This app helps accounts receivable accountants to streamline uncollectible accounts receivable processes. Thus, the app helps businesses efficiently remove bad debt and maintain accurate financial reporting. You can filter and identify overdue receivables, select items for write-off, and automatically post adjustments to predefined general ledger accounts. In addition, the app supports reason code management, simulation and validation of journal entries, and attachment of supporting documents for audit purposes. As a prerequisite, you must define reason codes for write-off in the Customizing activity. While it only processes normal open receivable items (excluding special general ledger indicators), the app helps with full write-offs rather than partial adjustments. To ensure proper handling, any unrelated customer payments (for example, missing reference information) should first be cleared using the Clear Incoming Payments app (F0773) before initiating a write-off request with this app. You can use simulate posting entries prior to final posting to prevent errors.

Year-To-Date Balances (W0177)

Application Type: Web Dynpro

This app allows the display of balances for selected dimensions of Universal Journal entries (table ACDOCA) within a specific ledger and ledger fiscal year in the general ledger. You can define any start and end date within the fiscal year to calculate balances, with default settings showing balances for the leading ledger (**OL**) up to the previous fiscal period in company code currency. In addition, you can search year-to-date balances with search parameters such as fiscal period, company code, general ledger account, and posting date with the flexibility of adding dimensions and measures through the **Navigation Panel**. Also, the app allows year-to-date balance display; balance splitting by company code, segment, profit center, cost center, function area, business area, supplier, or customer; and hierarchical views for accounts, cost centers, and profit centers. Additional currency options can be selected in the **Navigation Panel** under **Key Figures**.

You also can export results to Microsoft Excel or as a PDF and navigate to related apps such as Display Line Items in General Ledger (F2217), Manage G/L Account Master Data (Version 2) (F0731A), and Display G/L Account Balances (New) (F0707A), depending on your authorizations. The app utilizes CDS view C_GLAccountYTDBalanceQ to display year-to-date balances.

Tip

To display the data for time characteristics correctly, you should complete necessary analytics configuration steps in SAP S/4HANA, as highlighted in SAP Note 2289865.

Transaction Code

The corresponding transaction in SAP GUI is Transaction FAGLB03.

Chapter 2
Controlling

In this chapter, we will explore various SAP Fiori applications from the controlling functionality, presenting them in alphabetical order for ease of reference. With the apps organized in this way, you gain a structured overview of the features and functionalities of the apps available for the controlling subfunctions, making it easier to navigate and locate specific apps as needed. Each app's entry will include an overview of the app's primary purpose, its key features, and how it integrates within the broader SAP financial landscape.

Allocation Flow (F4022)

Application Type: Analytical

This app provides a visual flow of senders and receivers in an allocation, displaying the corresponding debit and credit amounts. You can choose from three view types: cost centers, direct activity allocations, and funds (for public sector customers). The app allows filtering by individual allocation cycles and enables tracking of all preceding and succeeding steps within an allocation flow. For example, you can search for allocations using various filter criteria such as view type, cost center, fiscal year, actual/plan, plan category, ledger, account number, allocation type and allocation cycle. The app's interactive features let you examine debit/credit amounts by drilling down to the sender, receiver, or account; open flow steps in related apps; and analyze relationships by allocation cycle, account, or a combined view.

Allocation Results (F4363)

Application Type: Transactional

This app allows you to view allocation run results for both test and live runs, and you can analyze any errors or warnings that occurred during the allocation run. You can search completed runs with various search criteria such as view type, fiscal year, run date, ledger, user name, allocation context, allocation type, actual/plan, and test run. The **Status** column displays whether the status of an allocation run is **Completed, Completed with Warnings**, or **Error**. The app helps you create custom report variants to highlight relevant data, switch between allocation run and cycle views, and reverse allocation runs or cycles for specific periods. In addition, the cycle view provides results for the last six fiscal periods from a defined endpoint. You also can toggle between a standard table view or a network graph view and access results for background allocation runs processed via job scheduling apps.

Tip

When reviewing allocation results using this app, you must ensure that the fiscal period range entered matches the exact fiscal period from and fiscal period to values from the original allocation run in the Run Allocation app (F3548).

Analyze Costs by Work Center/Operation (F3331)

Application Type: Analytical

This app provides detailed cost insights at the work center or operation level, helping you analyze overall production costs. With this app, you can navigate to review orders by work center or operation, assess productivity, and identify high-variance areas for cost analysis. In addition, the app offers detailed cost breakdowns, tracing activities that contribute to actual costs to ensure accuracy and resolve discrepancies. You can search for costs by variance search criteria such as period from, period to, category, display currency, plant, work center, cost center, and activity type. The app displays results in both chart and table format for flexible analysis. Thus, by analyzing variance causes, the app helps optimize production efficiency and supports informed decision making for cost management.

Tip

You can analyze costs by looking into high-variance work centers first. This helps you pinpoint operations with unexpected cost fluctuations due to inefficiencies, thereby refining cost allocations to eliminate non-value-adding activities to improve control.

Analyze Production Costs – Event-Based (F4059)

Application Type: Analytical

This app allows cost accountants to track manufacturing costs in real time by linking cost postings directly to business events, eliminating reliance on period-end processing. Thus, leveraging this app can enhance transparency and accuracy throughout the production cycle. The app's features include real-time posting, a detailed breakdown of planned, standard, and actual costs with variance analysis, and event-based accounting that ties postings to material consumption, labor usage, and overhead allocation. You can filter data by company code currency or group currency. The app also improves profitability analysis by providing granular visibility into cost drivers and integrates with related apps for further analysis.

Tip

As a prerequisite, this app requires the Event-Based Production Cost Posting scope item (3F0) to be activated. Thus, with this app, you can ensure cost postings are recorded based on specific business events such as material consumption, labor usage, and overhead allocation rather than traditional period-end processing. Consult SAP Notes 3316122 and 3316799 to understand on-the-fly target cost calculation for manufacturing orders and event-based variance posting for product cost by order, respectively.

Assign Profitability Segment (F4588)

Application Type: Transactional, Reuse Component

This app allows you to transfer postings to a profitability segment, which is defined by a combination of characteristics such as customer, product, plant, and distribution channel in margin analysis (previously known as account-based profitability analysis). This SAP Fiori–based UI is reusable across multiple transactions, with fields shown based on the business process. The account assignment screen is dependent on the operating concern, meaning that only relevant characteristics are available for maintenance. Profitability analysis characteristics are grouped into four sections: **Enterprise Structure**, **Product and Customer**, **Account Assignment**, and **Other**. Custom characteristics created via the Customs Fields and Logic app (F1481) under the market segment business context appear in the **Other** section. Additional characteristic values are derived automatically based on specified inputs, and characteristic groups determine the display order of attributes in different business transactions.

You can assign a profitability segment in apps such as Mange Direct Activity Allocation (F3697), Reassign Cost and Revenues (F2009), Repost Line Items – Cost Accounting (F5549), Manage Revenue Contracts (F3883), and Manage Manual Reservations (F4839).

Tip

To define the set of characteristics that appear on the assignment screens for profitability segments, as a prerequisite, you must complete the **Characteristic Group** Customizing activity under **Controlling • Profitability Analysis • Flow of Actual Values • Initial Steps • Characteristic Group**.

Transaction Codes

In SAP GUI, you can assign profitability segments using Transactions FB01 or FB50. Transaction K35Z can be used to display profitability segment information.

Change Log – Activity Types (F3866)

Application Type: Transactional

This app provides visibility into changes made to activity type master data, showing what was modified, by whom, and when. In addition, you can filter changes by specific activity types, users, or time periods, ensuring efficient tracking. For example, you can search change logs by activity type, type of change, changed by, and changed on. The results are displayed in a grid format, including fields such as activity type, activity type name, activity type validity, new value, and old value. The app also allows filtering by validity periods, searching based on defined criteria, and exporting results to spreadsheets for further analysis.

Transaction Code

The corresponding transaction in SAP GUI is Transaction SCU3. You also can use program RSSCD100 to read tables CDHDR and CDPOS in order to show detailed change logs.

Change Log – Cost Centers (F3761)

Application Type: Transactional

This app allows you to track changes made to cost center master data. The app provides details of what was modified, who made the change, and when it occurred. In addition, you can filter changes by specific cost centers, users, or time periods, ensuring precise monitoring. You can perform a search using fields such as cost center, type of change, changed by, and changed on. The app provides navigation capabilities and supports exporting results as spreadsheets for further analysis.

Change Log – Internal Orders (F5403)

Application Type: Transactional

This app allows you to track changes made to internal order master data, including what was modified, who made the changes, and when they occurred. The app provides filtering options to view changes for specific internal orders, users, or time periods. For example, you can search using criteria such as controlling area, internal order, changed on, and changed by. Once the chosen filter criteria are executed, the app displays data in two tabs: **General** and **Status**. The **General** tab provides data in a tabular format with the flexibility to add or remove fields. In addition, you can search based on defined criteria and export results as spreadsheets. The app also displays modifications to internal order status and changes related to locking or unlocking orders.

Transaction Code

The same information can be obtained using Transaction KO03 in SAP GUI.

Change Log – Profit Centers (F3810)

Application Type: Transactional

This app allows you to track modifications to profit center master data. The app displays details of what was changed, by whom, and when. In addition, you can filter changes by specific profit centers or users, check if attributes such as the person responsible were updated, and verify whether a profit center was added to or removed from a company code. For example, you can search for changes using filter criteria like profit center, type of change, changed on, and changed by. Once the chosen filter criteria are executed, the app displays data in two tabs: **General** and **Company Code**. Besides audit trail information, you also can view the **Old Value** and **New Value** as a result of the change. In addition, filtered change data can be downloaded as

a Microsoft Excel spreadsheet and shared with colleagues for further analysis.

Change Material Prices (F6489)

Application Type: Transactional

This app allows you to create, analyze, simulate, and post material price changes across inventory, including sales order stock and project stock. Prices can be adjusted for a single reference valuation, then applied to all or selected valuations of the same material. In addition, you also can modify valuations individually and observe percentage-based impacts on price and valuation. Before posting material price changes, we recommend simulating the results to identify errors, warnings, or inconsistencies. You can search for material price changes using search fields such as posting date, company code, plant, and document header text.

Tip

To use this app, you must activate universal parallel accounting business function FINS_PARALLEL_ACCOUNTING_BF. If universal parallel accounting is not activated, then SAP recommends using Transaction MR21 for material price adjustments.

Commitments by Cost Center (F3016)

Application Type: Analytical

Commitments are reserves that represent the financial obligation recorded when a purchase order is recorded in SAP. This app allows you to track commitments posted to individual cost centers and compare them with actual expenses, plans, and budgets. The app provides visibility into commitments, enabling you to analyze them alongside actual, planned, and assigned expenses while drilling down to the cost element level. Actual cost data is sourced from the leading ledger in the Universal Journal (table ACDOCA), whereas commitments are recorded in the extension ledger within the same table. On the other hand, planned costs are derived either from the plan data repository in the Universal Journal (table ACDOCP) or from SAP Business Warehouse (SAP BW), based on system settings.

Tip

As a prerequisite, the app requires setting up extension ledgers in Transaction FINSC_LEDGER.

The app supports only the new commitment management framework, which uses the Universal Journal (table ACDOCA), and does not support classic commitment (table COOI). To understand the differences between the approaches, consult SAP Note 2778793.

The app displays the amounts of key date currency translation based on the current day, ensuring consistent data representation across actual, commitment, and plan data. In other words, all values are translated into the display currency using a single exchange rate for uniform reporting. Currently, only exchange rate type **M** is supported for all data, including plan values. Consult SAP Note 3253859 for details.

Transaction Codes

The corresponding transactions in SAP GUI are Transactions KSB2, S_ALR_87013648, and KSO9.

Commitments by Cost Center – Classic Commitment Management (F4998)

Application Type: Analytical

This analytical app allows tracking of classic commitments posted to individual cost centers and provides comparisons with actual expenses, plans, and budgets. You can view cost center commitments, analyze them against actual,

planned, and assigned costs, and drill down to the cost element level for detailed insights. For instance, you can filter data by various parameters such as ledger, plan category, controlling area, fiscal year, fiscal period, company code, cost center, and chart of accounts. Actual cost data is read from the Universal Journal (table ACDOCA), while classic commitment data is based on table COOI entries. Planned costs are derived either from the planning table (table ACDOCP) or SAP Business Warehouse (SAP BW), depending on system settings.

Tip

The app supports universal parallel accounting as of SAP S/4HANA 2023. Classic planning for cost centers, projects, and internal orders is no longer supported in the universal parallel accounting environment. SAP recommends SAP Analytics Cloud for planning activities.

When reading plan data from SAP BW, SAP recommends using the **Commitments in Global Currency** option for displaying amounts in the app to ensure consistency and accurate comparisons across financial records.

Transaction Codes

The closest corresponding transactions in SAP GUI are Transactions KSB2, S_ALR_87013648, and KSO9.

Compress Financial Plan Data (F6537)

Application Type: Transactional

This app allows you to reduce the number of plan data items stored in planning table ACDOCP, improving planning process performance. You can search for summarized financial plan data with search parameters such as plan category, fiscal year of ledger, fiscal period, ledger, and company code. The app permanently deletes the history of changes, retaining only the latest snapshot, while ensuring all plan data remains consistent. You can analyze compression statistics, including records in data selection, records after compression is simulated, compress ratio, and total ratio, to assess the impact. The app offers analytical flexibility with column customization, sorting, and grouping, allowing you to filter plan categories, ledgers, fiscal years, fiscal periods, and company codes. Compression deletes the selected scope's history irreversibly, ensuring only current plan data remains accessible.

Tip

As a prerequisite to use this app, the **Compressing Financial Plan Data** option for each applicable plan category must be enabled in Customizing activity **Controlling · General Controlling · Planning · Maintain Category for Planning**.

Copy Financial Plan Data (F3396)

Application Type: Transactional

This app allows you to select financial plan data from a source category and copy it to a target category. You can search for existing plan data by entering values in filter fields, with at least one required entry in the plan category. When copying data, any matching entries in the target category for fields like ledger, fiscal year, fiscal period, and company code will be overwritten, while unmatched entries without corresponding source data will be deleted. In addition, the app includes an **Analyze Financial Plan Data** function for reviewing copied data and an **Open Financial Plan Data** option in **Related Apps** to create custom reports for selecting relevant data.

Tip

As a prerequisite to copy plan data to a target plan category, you must enable the **Copying Another Plan Category** indicator in the **Maintain Category for Planning** Customizing

activity. In addition, if a category stores budgets rather than regular plan data, the new plan value must exceed the actual values already posted. The app only allows both source category and target category copying of financial plan data when the application type and usage properties are compatible.

Transaction Code

The corresponding transaction in SAP GUI is Transaction CJ9CS, which is used to copy actual to plan work breakdown structures (WBSs) in Project System.

Correction Request (F0700)

Application Type: Transactional

This app allows you to review correction requests for controlling documents, meaning you can ensure that posted costs are accurate and reflect the area's financial situation. Moreover, the app allows for direct notifications to be sent to requestors, informing them of the scheduled correction date for posted costs for controlling documents. As a result, you get an opportunity to validate posted costs prior to approving corrections, which streamlines the approval process to maintain accuracy in controlling documents.

Cost Center Budget Report (F3871)

Application Type: Analytical

This app allows cost accountants to display budget, actual costs, commitments, and available budget for budget-carrying cost centers, with information filtered by plan category, budget carrying cost center, general ledger account group, company code, or fiscal year. To use the app, a budget availability control profile must be maintained in a Customizing activity, with assigned account groups defining which accounts are checked during actual cost postings and analyzed in budget reports. In addition, you can view budget details across multiple cost centers, switch between visual and compact filters, and navigate to cost center details and line items for deeper insights. For example, line item details include the amounts in the actual, commitments, assigned, and available budget categories. The app also integrates with SAP Collaboration Manager, enabling contextual collaboration through notes, objects, messages, and screenshots within the working environment.

> **Tip**
>
> If the app doesn't show any plan amount, SAP recommends running Transaction OB_FCAL (program FINS_GENERATE_FISCAL_PERIOD) to ensure the company code belongs to a cost center and is linked to a fiscal year variant correctly, as discussed in SAP Notes 3590155, 3310298, and 2936255 (question no. 7).

> **Transaction Codes**
>
> Similar transactions in SAP GUI are Transactions KPZ3 and S_ALR_87013648.

Cost Centers – Actual (F0963)

Application Type: Web Dynpro

This analytical app allows you to report actual data of profitability analysis for a ledger's fiscal year. With this app, you can refine search criteria using variables or filters such as company code, profit center, segment, functional area, segment, and general ledger account hierarchy. The app displays query results in a grid format, and you can export data to Microsoft Excel. In addition, the app's **Navigation Panel** allows you to add dimensions from the **Available Fields** section into rows and columns. The app offers results in multiple tabs: **Data Analysis**, **Graphical Display**,

and **Query Information**. The app displays both real and statistical postings distinguished via the **IsStatistical** fields such as **IsStatistical Cost Ctr**. The value of **x** in these fields indicates statistical postings, and **#** represents real postings.

> **Tip**
>
> If the app displays no values on the screen, you must maintain parameter **CAC** (controlling area) for the user. This applies to all analytical apps.

> **Transaction Codes**
>
> The corresponding transactions in SAP GUI are Transactions KSB1 and S_ALR_87013611.

Cost Centers – Plan/Actuals (W0081)

Application Type: Web Dynpro

This analytical app allows you to report both actual and plan data of a cost center for a ledger, fiscal year, and plan category. Plan categories help you structure financial planning by organizing data into versions to support costing accounting control and analysis. You can create multiple categories for various business requirements. With this app, you can refine search criteria using variables or filters such as posting period, company code, project definition, cost center, profit center, functional area, segment, customer, company code, general ledger account (hierarchy), and general ledger account (node). The app displays query results in a grid format, and you can export data to Microsoft Excel. In addition, the app's **Navigation Panel** function allows you to add additional dimensions from the **Available Fields** section to the rows and columns. The app's features include the user's ability to specify search criteria using variables or filters, display query results in a structured grid format, export results to Microsoft Excel for further analysis, manipulate the query results grid by repositioning available characteristics within the grid, and display read-only details about the underlying query, such as its technical name and last update timestamp. The app supports authorizations to view data at the company code level.

Transaction Code

The corresponding transaction in SAP GUI is Transaction S_ALR_87013611.

Delete Financial Plan Data (F4851)

Application Type: Transactional

This app allows you to delete selected erroneous financial plan data from the database permanently. You can refine deletions using filters such as plan category, ledger, fiscal year, fiscal period, and company code to target specific data. Before deletion, you must evaluate data using the **Analyze Financial Plan Data** function, which helps assess the impact of removing selected records. In addition, you can leverage the **Open Financial Plan Data** feature via **Related Apps** to create custom reports for proper identification of data to be deleted. Thus, the app ensures financial plan data consistency by maintaining only required financial plan data thanks to removing incorrect or obsolete entries. This is typically required as part of cleanup activities during mock load activities or prior to going live.

Tip

As a general recommendation, you must ensure any financial data no longer used is achieved rather than deleted. In addition, deleting budget data from a plan category may impact budget availability control functionality. Thus, this action may prevent actual postings due to changes in the available budget.

Delete Financial Plan Data – With Timestamp (F4074)

Application Type: Transactional

This app allows you to reset financial plan data to a specified timestamp, permanently removing erroneous imports. A timestamp represents an event, such as a plan data import, that modified existing records, and the app displays delta data—differences between previous and last-imported plan data—starting from the selected timestamp. In addition, you can refine search results using plan category, ledger, fiscal year, fiscal period, and company code to target specific erroneous entries. To correct data, you can delete all delta data from a given timestamp or restrict deletions based on further criteria. Once deleted, data cannot be restored. The **Analyze Financial Plan Data** option helps assess affected records before deletion, while the **Open Financial Plan Data** in **Related Apps** feature allows you to create reports for accurate identification of impacted data.

Tip

The app supports deleting plan data from the planning table (table ACDOCP), but not from InfoCube /ERP/SFIN_R01. In addition, the app does not allow deleting random entries with multiple timestamps; instead, the app can only remove all data from a specified timestamp **From** value to the current time, as indicated in SAP Note 2936255 (question no. 16).

Display Actual Costing Result (F6073)

Application Type: Analytical

This app allows you to view the results of the actual costing run, including standard price and actual price per price unit. The app provides detailed cost breakdowns under the **Cost Component Split** tab and allows exporting customized spreadsheets using the **Custom Export** button. You can filter results based on criteria such as company code, ledger, controlling area, cost components, plant, and material, and you can specify a key date for accurate results.

Tip

This app displays only main cost component splits, excluding auxiliary cost component splits. The actual price may differ from the periodic unit price shown in Transaction CKM3 due to the **Revaluate Material** settings in Transaction CKMLCP (Costing Cockpit – Actual Costing). Consult SAP Note 3520383 for further guidance.

Transaction Code

The corresponding transaction in SAP GUI is Transaction CKM3.

Display Cost Rates (F5150)

Application Type: Analytical

This app provides a clear view of cost rates used in financial and production processes for cost accountants. The app allows you to review predefined cost rates applied to materials, activities, and overhead calculations while supporting historical tracking for better financial planning. In addition, the app allows you to select controlling area currency or object-specific currency, integrates with costing sheets for accurate allocations, adjusts the calculation price unit for comparative purposes, and offers filtering and sorting options based on criteria such as cost center or activity type. For instance, you can filter cost rates by ledger, cost center, activity type, from fiscal year period, to fiscal year period, and object category. Thus, the app facilitates tracking accurate cost allocations and transparency in cost accounting.

Transaction Codes

The corresponding transactions in SAP GUI are Transactions KP26 and KSBT.

Display Costing Run – Estimated Cost Jobs (F7568)

Application Type: Transactional

This app provides a comprehensive overview of costing runs carried out via the Manage Costing Run – Estimates Costs app (F1865) in product costing. The app allows you to monitor, verify, and track ongoing or completed jobs using the **Show Application Jobs** button. In addition, the app allows you to refine job listings using filters such as company code, valuation area, material, job status, and execution dates. The app also provides detailed logs and error messages for troubleshooting unsuccessful costing runs.

Display Inventory Price Upload Jobs (F7567)

Application Type: Transactional

This app provides a comprehensive overview of inventory price upload activities carried out via the Upload Material Inventory Prices app (F4006) in product costing. The app allows you to monitor, verify, and track ongoing or completed jobs using the **Show Application Jobs** button. In addition, the app allows you to refine job listings using filters such as company code, valuation area, material, job status, and execution dates. The app also provides detailed logs and error messages for troubleshooting unsuccessful uploads.

Tip

You must use the Upload Material Inventory Prices app (F4006) to create jobs for uploading inventory prices.

Display Line Items – Margin Analysis (F4818)

Application Type: Transactional

This app allows you to view journal entry line items relevant to margin analysis, including true account assignments (account assignment type = **EO**) and attributed account assignments (account assignment type = **EA**), to profitability segments. Attributed profitability segments are used when a transaction is assigned to another controlling object such as a work breakdown structure (WBS) element or an internal order, yet profitability characteristics are still derived for reporting purposes. By default, the account assignment type is set to **EO** on the selection header screen, displaying true account assignments, but you can adjust the selection to view attributed assignments. Clearing the selection condition allows visibility into all journal entry line items, with filtering, sorting, and grouping options available. You can search with various other parameters such as company code, general ledger account, posting date, profit center, journal entry created by, fiscal year period, customer, reference document, and source reference document. The app supports displaying line items by ledger, accessing additional details through line item details, and navigating to related apps for further analysis.

Tip

If characteristics have been added with Transaction KEA5 already, neither referenced nor user-defined characteristics appear by default. You need to enhance relevant CDS views to incorporate those additional characteristics manually. Consult SAP Note 3005736 for further details.

Transaction Codes

The closest transactions in SAP GUI are Transactions KE24 and KE24N.

Display Line Items – Production Accounting (F7288)

Application Type: Analytical

This app provides a detailed view of event-based production cost postings, including journal entry line items, allowing you to filter, sort, and group line items in product costing. The app offers real-time insights into work in process (WIP), production order variances, material consumption, and actual production costs, ensuring accurate financial tracking. You can analyze individual cost components such as material consumption, activity confirmations, and overhead allocations; refine data using filters such as ledger, company code, general ledger account, order ID, fiscal year, business transaction type, reference document, plant, material, and cost center; and drill down into related documents and journal entries for further details. The app supports parallel valuation across multiple ledgers using universal parallel accounting.

Tip

Consult SAP Notes 3316799 and 3316687 for further details on event-based variance posting for product cost by order and WIP posting for product cost by order.

Transaction Codes

The closest SAP GUI transactions are Transactions CO03, KOB1, and CKM3. These transactions depend on period-end processing rather than real-time postings.

Display Material Price Change Documents (F6684)

Application Type: Analytical

This app allows you to view price change documents, which are automatically generated when materials are debited or credited, prices are changed, or a standard cost estimate is released with a new standard price. Besides displaying these documents, the app helps you analyze detailed price changes and access related records. With this app, you can search for material price changes with fields such as ledger, material, plant, document date, created on, fiscal year, fiscal period, and created by.

In addition, price change documents can be viewed within the Manage Material Valuations (F2680) and Manage Journal Entries (F0717A) apps, where you can search by the price change journal entry type.

Transaction Codes

The closest transactions in SAP GUI are Transactions CKMPCD and CKMPCSEARCH.

Display Material Value Chain (F4095)

Application Type: Analytical

This app provides a visual representation of material quantities and values across a product value chain, covering procurement, transfers, production, and sales across organizational units such as plants, company codes, and divisions. You can analyze valuated transactions within a plant over a specific period, including actual costing results with price and exchange rate differences. You can display the material value chain flow via search criteria parameters such as fiscal period from, fiscal period to, material, plant, ledger, currency type, and valuation view. Color-coded nodes display material values, while connectors represent transaction flows. The app's features include switching between graph and table views, grouping nodes by various criteria, customizing displayed values, filtering with semantic tags, and viewing inventory calculations and cost component splits. In addition, you can choose your profit center display settings and refine the number of nodes shown using the **Largest Movements** slider and the **Tracing Steps** filter.

Tip

You can group nodes by valuation class or procurement type, which provides additional information about cost structures.

Transaction Code

The closest transaction in SAP GUI is Transaction CKM3N.

Display Material Value Chain – Estimated Cost (F4898)

Application Type: Analytical

This app visualizes the progression of planned quantities and values for a material along a value chain, using estimated costs from a specified validity date. The app provides a structured view of valuated transactions, procurement alternatives, and related activities through color-coded nodes representing material values, activities, and subcontracting details, while connectors illustrate transaction flows. You can filter the estimated cost of material valuation with parameters such as material, plant, costing variant, valid on, and currency type. In addition, the app supports different costing variants, the ability to select between controlling area currency or object currency, varied node types for distinct information, access to node details in a copyable format, the option to restart the flow from any node, and the display of cost component splits on material cards based on the cost component view.

Tip

The app displays information in a value chain flow format; it doesn't offer a tabular format.

The corresponding transaction in SAP GUI is Transaction CK13N, which also displays data in a tabular format. In addition, Transactions CK86 and S_ALR_87099930 show the results of costing runs.

Display Posting Rules – Event-Based (F5228)

Application Type: Analytical

This app allows you to view posting rules for event-based manufacturing orders, including production and process orders, as well as event-based repetitive manufacturing orders (product cost collectors). The app provides insights into the cost allocation per receiver and the distribution method used. When accessed from the order header in the Change Production Order app (CO02), you can create and edit posting rules for event-based manufacturing orders, though editing rules for product cost collectors are not supported. You can filter event-based posting rules by various filter parameters such as order category, order, and product. The app also displays general order details such as product, posting type, and user activity, while allowing you to maintain and review cost allocation specifics, including distribution methods and percentage or equivalence values for each receiver.

Tip

The app does not support editing posting rules for orders in engineer-to-order scenarios.

Display Settlement Documents (F4597)

Application Type: Transactional

This app allows you to view transaction details generated during the settlement and reversal process for a work breakdown structure (WBS) element, an order, or a service document in

controlling. You can search settlement documents using various search criteria such as WBS element, project, settlement year, created on, created by, document number, cost object type, and display reversal documents. In addition, the app provides a comprehensive history of settlement documents, including cost allocation details and information about when and by whom previous settlements were executed. The app allows you to drill down into the details using multiple tabs: **Settled Values**, **Senders**, **Receivers**, **Journal Entries**, and **Settlement Rules**.

You can access this app before or after running or reversing settlements for a single cost object using the Run Settlement – Actual app (F4568), navigating to it from the details section of the selected cost object. The app's features include displaying settlement and reversal documents, analyzing settlement execution history, and reviewing settled values, senders and receivers, journal entries, and settlement rules.

Transaction Codes

The closest transactions in SAP GUI are Transactions KSB5 and CO03.

Event-Based Revenue Recognition – Projects (F4767)

Application Type: Transactional

This app provides real-time visibility into recognized costs and revenues for project-based transactions. You can search using parameters such as billing element, company code, project definition, recognition key, and project manager name. Cost postings are matched to revenues and reported as expenses, while revenue is posted to an income statement account, with recognition and adjusting entries generated simultaneously. The app is fully integrated with the general ledger. This ensures consistent cost-revenue matching, keeping profit and margin data accurate and up to date. In addition, you can analyze recognition values based on accounting principles, drill down into individual journal entries and source documents, adjust recognized revenue and cost of sales, and enter accruals for anticipated losses or unrealized costs. These features are accomplished with the **Set Parameters**, **Revalue**, **Enter Temporary Adjustments**, **Enter Manual Contract Accruals**, and **Related Apps** functions. Enhanced cost management is supported with additional reporting attributes, such as customer group, enabling greater transparency in the revenue recognition process.

Tip

You can structure projects with appropriate levels of hierarchies to achieve desired revenue and cost recognition based on business requirements.

Event-Based Revenue Recognition – Sales Orders (Version 2) (F2441A)

Application Type: Transactional

This app provides real-time revenue and cost adjustments for sales orders. Thus, the app records costs and revenues for sales orders immediately as transactions occur, offering enhanced performance compared to its predecessor app, Event-Based Revenue Recognition – Sales Orders (F2441). The app ensures that cost postings align with revenues, categorizing them as expenses, while revenue is posted to an income statement account. Recognition and adjusting entries are generated simultaneously, maintaining real-time accuracy. The app is fully integrated with the general ledger to ensure that financial data remains synchronized, keeping reported profits and margins up to date while enabling you to access income statements and cost-of-sales reports. In addition, you can analyze recognized values based on accounting principles, drill down from recognition journal entries to cost or revenue postings, and adjust recognized revenue or cost of sales. The app also supports revaluation, allowing recalculations when planned costs change, with simulated values highlighted before posting. In addition, with this app, you can make temporary adjustments

manually, though they are overwritten by subsequent revaluations or periodic revenue recognition runs. Based on business needs, the revenue recognition key can be modified under specific conditions, either before postings occur or when assignment rules or account mappings remain consistent.

Event-Based Revenue Recognition – Service Documents (F6007)

Application Type: Transactional

In SAP, *event-based revenue recognition* refers to costs and revenues recognized in real time as they occur, fulfilling the matching principle of accounting standards. Conversely, in a conventional periodic revenue recognition process, costs and revenue are matched after running settlements and other periodic adjustments during period-end closing.

As the name implies, this app immediately recognizes costs and revenues from service contracts and service orders as transactions occur. The app ensures cost postings match revenues, reported as expenses, while revenues are posted to an income statement account. Meanwhile, recognition and adjusting entries are generated simultaneously, with full general ledger integration (data stored in table ACDOCA) to maintain up-to-date profit and margin reporting. In addition, you can analyze recognition values, drill down into journal entries, manually adjust recognized revenue and costs, and obtain enhanced cost management insights via attributes such as customer group. The **Revalue** function recalculates recognition values when planned costs change, and you must trigger an update for these changes to take effect. Temporary adjustments allow manual modifications but are overwritten by the next revaluation or periodic revenue recognition run. You also can change the revenue recognition key under specific conditions. However, manual changes to the recognition key revert if the service document is modified, requiring configuration updates for permanent adjustments.

> **Tip**
>
> The app displays plan amounts only when read from plan values from sources **310** (ongoing cost and revenues of service order) or **500** (cost and revenue planning of all objects) from table ACDOCP. Their availability depends on the company code, accounting principle, and revenue recognition key assigned to the selected service document.

Event-Based Solution Monitor – Product Costing (F5133)

Application Type: Transactional

This app provides cost accountants with real-time tracking and analysis of event-based postings. Thus, the app ensures financial accuracy and transparency in cost flows. The app allows you to monitor cost postings tied to business events, such as goods issues, receipts, and operation confirmations, offering immediate visibility into cost movements without waiting for period-end processing. For example, you can search for postprocess event-based postings using parameters such as fiscal period, company code, plant, order, order type, product, and posted orders. The app's functionalities include real-time monitoring of postings, error detection with corrective actions, detailed cost analysis at the transaction level, seamless integration with production accounting to validate work in process (WIP) and variance calculations, and audit support to comply with accounting standards.

Financial Plan Data Logs (F5144)

Application Type: Transactional

This app provides a comprehensive overview of application logs for financial plan data. Thus, the app allows you to quickly identify potential errors. You can view log details, filter logs by severity, search for specific message texts, and display message details.

The app is not available as a standalone tile; it can only be accessed from related apps, including Import Financial Plan Data (F1711), Copy Financial Plan Data (F3396), Delete Financial Plan Data – with Timestamp (F4074), Delete Financial Plan Data (F4851), and Import Statistical Key Figure Plan Data (F3779).

> **Tip**
>
> You can filter logs by severity to help prioritize critical errors prior to looking into warning messages.

Financial Plan Data Report (W0163)

Application Type: Web Dynpro

This app allows reporting on plan data across all accounting-relevant entities. You can specify search criteria using variables or filters to refine queries, display results in a structured grid format, and export data to a spreadsheet for further analysis. For example, you can search with plan category, company code, cost center, general ledger fiscal year, ledger, and posting period. The query results grid allows manipulation by adjusting available characteristics within the result layout. In addition, you can access read-only details about the underlying query, including its technical name and last update timestamp.

Functional Areas – Actual (F1583A)

Application Type: Web Dynpro

This app allows you to report actual data of functional areas for a ledger and ledger fiscal year. With this app, you can refine search criteria using variables such as planning category, fiscal period, functional area, company code, material group, customer group, and general ledger account hierarchy. In addition, the app displays query results in a grid format, and you can export data to Microsoft Excel. The app's **Navigation Panel** allows you to add dimensions from the **Available Fields** section into the rows and columns. In addition, the app displays both real and statistical postings distinguished via the **IsStatistical** fields, such as **IsStatistical WBS Elem** and **IsStatistical SO Item**. The value of **x** in these fields indicates statistical postings, and **#** represents real postings. The app displays amounts in transaction, company code, and global currencies. Thus, the app offers analytical insights into actual sales order data.

Transaction Code

There is no directly comparable SAP GUI transaction, but you can use Transaction S_ALR_87011806 to see the profit and loss (P&L) by functional area.

Functional Areas – Plan/Actuals (W0077)

Application Type: Web Dynpro

This app provides detailed comparisons of planned and actual costs of functional areas for a specified ledger, functional area, plan category, and ledger fiscal year. The app allows you to filter search criteria, display results in grid or graphical formats using the **Graphical Display** tab, and export data to Microsoft Excel. For example, you can search for functional area plan cost and actual cost using various parameters such as company code, profit center, customer group, material group, general ledger (account hierarchy), and general ledger account node. In addition, you can add additional dimensions to columns via the **Navigation Panel**. Thus, with this app, cost accountants can regularly review cost variances between planned and actual amounts of functional areas to support management reporting.

Transaction Codes

There is no directly comparable SAP GUI transaction, but you can use Transactions S_ALR_87011806 (Profit and Loss by Functional Area) and S_ALR_87013599 (Cost Elements: Breakdown by FuncArea).

Import Allocation Data – Values (F5596)

Application Type: Transactional

This app allows you to upload values such as amounts, rates, or portions to allocation segments. The app allows you to download value combinations into a Microsoft Excel template for modification, which can then be uploaded back into the app. After uploading, you can validate the data and apply changes to allocation segments. You can search for uploaded data with parameters such as file name, user ID, uploaded on date, and status. The app also provides functionality to download data to an Excel file for troubleshooting. In addition, the app supports most allocation contexts and types, except for top-down distribution. Furthermore, the app provides features such as viewing uploaded files, validating data, applying updates, filtering error lines, and accessing detailed statistics on uploaded sheets, such as row counts and error occurrences.

Import Financial Plan Data (F1711)

Application Type: Transactional

This app allows you to upload financial plan data from a CSV file, either online or via background processing. It supports external planning systems like Microsoft Excel, allowing seamless data integration. After importing, you can compare plan data with actual values by using analytical apps for management accounting, such as Profit and Loss – Plan/Actual (F1710) or Market Segments Plan/Actual (W0078). Nearly all fields from table ACDOCP can be included, and the app automatically validates all values before import.

Tip

This app is typically used to upload small volumes of data. For a larger-volume dataset, the recommended approach to use tools such as SAP Analytics Cloud.

Import Statistical Key Figures Plan Data (F3779)

Application Type: Transactional

A statistical key figure represents nonmonetary data related to organizational units and serves as reference data for universal allocation. This app allows you to upload statistical key figure plan data from a CSV file. Thus, the app facilitates external planning in tools like Microsoft Excel.

Key users can generate reports by using the Custom Analytical Query (F1572) app by leveraging CDS view I_FinStatisticalKeyFigureItem. The generated report allows you to analyze data uploaded plan data. In addition, the uploaded CSV file is validated to ensure all required fields are included prior to processing data.

Inspect Revenue Recognition Postings (F4008)

Application Type: Analytical

This app allows cost accountants to analyze how revenue recognition transactions are related to a project or work breakdown structure (WBS) element impact balance sheet and income statement accounts. The app displays the accounts in a T-account format, providing a clear overview

of recognized costs and revenues. You can select transaction documents to examine their financial impact, view debit and credit entries for revenue recognition postings, and display entries as T accounts or in a table format under the **Accounting Impact** section. For instance, you can search for postings with parameters such as account assignment object, ledger, currency type, posting type, document range, and opening balance. You can track recognized costs and revenues across balance sheet accounts and income statements accounts visually. Clicking an entry highlights corresponding transactions in other accounts, with actual postings marked in blue and recognized entries in red.

Transaction Code

The corresponding transaction in SAP GUI is Transaction REV_REC_MON.

Internal Orders – Plan/Actuals (W0076)

Application Type: Web Dynpro

This app provides detailed comparisons of planned and actual costs of internal orders for a specified order, ledger fiscal year, and plan category. The app allows you to filter search criteria, display results in grid or graphical formats using the **Graphical Display** tab, and export data to Microsoft Excel. For example, you can search for internal order plan and actuals using various parameters such as company code, profit center, functional area, segment, order, general ledger (account hierarchy), and general ledger account node. In addition, you can add additional dimensions to columns via the **Navigation Panel**. Thus, with this app, cost accountants can regularly review cost variances between planned and actual amounts of internal orders to support management reporting.

Transaction Codes

The corresponding transactions in SAP GUI are Transactions KOB1 and S_ALR_87014004.

Inventory Balance Sheet Valuations (F3343)

Application Type: Analytical

This app displays valuation results for inventory such as the lowest market price or first in, first out (FIFO) price, ensuring accurate financial reporting for balance sheet purposes. The app helps businesses assess inventory values based on predefined valuation area and plant. The app displays inventory balance valuations in both chart and table view formats for flexible analysis. As a leading practice, you verify inventory valuation regularly to comply with accounting standards. Thus, the app supports valuation to comply with compliance with accounting standards.

Tip

As a prerequisite, this app requires universal parallel accounting business function FINS_PARALLEL_ACCOUNTING_BF to be enabled.

Transaction Codes

The closest transactions in SAP GUI are Transactions MRY3, MRN0, MRN1, MRN2, MRN3, MRND, and MRNP.

Manage Activity Type Groups (F1027)

Application Type: Web Dynpro

This app allows you to search for, create, display, and edit activity type groups. The app displays activity types in a hierarchical tree. You can jump to different nodes using the **Levels** option. The app also allows validation of duplicate subgroups or activity types within a group and ensures all master data in the controlling area is properly included. You can display the where-used list, delete activity type groups with no references, and compare two groups for common attributes. You can search for a specific group

using the **Find** option. Thus, the app allows you to streamline management of activity type groups.

Transaction Codes

The corresponding transactions in SAP GUI are Transactions KLH1, KLH2, and KLH3.

Manage Activity Types (Version 2) (F1605A)

Application Type: Transactional

In controlling, the *activity type* classifies the activities performed within a company by one or more cost centers. These act as tracing factors for cost allocation, ensuring that costs related to services or production activities are correctly assigned to the receiving cost centers, orders, or processes. With this app, you can display, create, and manage activity type master data. You can search for activity types using various parameters such as activity type, activity unit, valid on, allocation cost element, and cost center categories. The app supports navigation to related apps, including the Where-Used List – Activity Types app (F3867), via the **Where Used** option, thus enabling you to track dependencies. In addition, the app links can be accessed through the **More Links** feature for expanded functionality.

Transaction Codes

The corresponding transactions in SAP GUI are Transactions KL01, KL02, and KL03.

Manage Allocation Tags (F4523)

Application Type: Transactional

In SAP, *allocation tags* are semantic labels assigned to allocation cycles and segments to help categorize and identify them easily. This app allows you to create and assign allocation tags to allocation cycles and segments. Thus, the app facilitates grouping and identification efficiently. These semantic tags can be applied across different allocation contexts and types, providing a structured way to track and label all elements within an allocation process. You can search for allocation tags with various filter criteria, such as allocation tag, created by, and changed by. Details are displayed in the **Tag View** and **Structured View** tabs. You can use the **Assign**, **Create**, **Edit**, and **Delete** buttons to manage allocation tags. In addition, the app offers a **Note** function to insert free text describing the purpose of the semantic tag.

> **Tip**
>
> As a general practice, you must assign a meaningful semantic tag to allocation cycles and segments to identify cost categories with relevant tags for easy cost tracking.

Manage Allocations (F3338)

Application Type: Transactional

This app allows you to manage allocation cycles, ensuring accurate distribution of values and quantities in cost accounting. You can search for specific allocation cycles or filter them by various criteria such as company code, valid to, and cycle description. The app supports cost centers, profit centers, and margin analysis allocation contexts, with allocation types including distribution, overhead allocation, intercompany allocation, and top-down distribution. You can view, create, edit, copy, or delete allocation cycles and segments; lock segments to prevent execution; and conduct validation checks.

The app allows direct navigation to the Run Allocations app (F3548) for execution and supports top-down distribution templates from Customizing. You also can upload allocation cycle and segment definitions using a preformatted XLSX template.

> **Tip**
>
> The **Cumulative** option allows you to aggregate allocation values over multiple periods or multiple runs for a continuous distribution of costs or revenues. This is helpful for tracking long-term allocations. The app does not support the **Cumulative** option prior to SAP S/4HANA 2022.

Manage COGS Splits (F5274)

Application Type: Analytical

When a goods delivery is posted, the total cost of the product is recorded in the cost of goods sold (COGS) account, derived as the source account in configuration settings. This action creates an original journal entry. If COGS splitting is configured, the total cost is split based on cost components of material, and the respective amounts are posted to one or more target accounts. Journal entries are reflected in a separate document. You can use this app to analyze these postings and gain insights into cost distribution. The app's features include displaying journal entries related to COGS; filtering postings by parameters such as ledger, journal entry, fiscal year, fiscal period, type of postings, product or company code; downloading data for further analysis; and accessing COGS split settings. In addition, with the type of postings filter criterion, you can display the COGS split journal entry or the original journal entry. The split results are displayed in four sections: **Split Basis**, **Split Components**, **Split Results**, and **Messages**. You also can navigate to relevant apps.

> **Tip**
>
> As a prerequisite, you must set up **Define Accounts for Splitting COGS** in Customizing menu path **Financial Accounting · General Ledger · Periodic Processing · Integration · Materials Management**.

If multiple cost components are mapped to the same target account, precise fixed amount calculations across all currencies become impractical. As a result, the app displays fixed amounts in the group currency.

Consult SAP Note 2399030 to understand COGS split functionality in different SAP releases since its inception.

Manage Cost Center Budgets (F4307)

Application Type: Transactional

This app allows cost accountants to transfer, supplement, and return cost center budgets. Cost center budget management is an important function to see budget versus actual costs within the same company code. The app supports transferring budget between one or multiple cost centers, with detailed visibility into the sender and receiver item information. In addition, you can supplement budgets, viewing relevant supplement details, and return budgets with access to return item information. Furthermore, the app provides general budget transaction details for all adjustments, supporting precise financial tracking.

Tip

The app supports uploading supporting documentation for budget transfers. As a prerequisite to support this function, you must complete the required Customizing activities in cross-application component and harmonized document management, as recommended in SAP Note 3495989.

Manage Cost Center Groups (F1024)

Application Type: Web Dynpro

This app allows you to search for, create, display, and edit cost center groups. You can create a new cost center group from scratch or use an existing cost center group as a reference. The app allows verification of duplicate subgroups or cost centers within the same group and ensures all master data in the controlling area is accounted for. In addition, you can display the where-used list for a cost center group, delete groups not found in where-used lists, and compare two cost center groups for common attributes. Furthermore, the app supports efficient cost center group management by streamlining data validation and structural oversight.

Transaction Codes

The corresponding transactions in SAP GUI are Transactions KSH1, KSH2, and KSH3.

Manage Cost Centers (Version 2) (F1443A)

Application Type: Transactional

This app allows you to create and edit cost center master data while managing validity periods. The app allows displaying existing cost centers with detailed information, creating new entries, and modifying master data and time-based attributes. You can add or delete validity periods as needed and analyze dependencies by accessing the Where-Used List – Cost Centers app (F3549). You can search cost centers with search criteria such as controlling area, cost center category, standard hierarchy node, editing status, cost center, valid on, and company code. In addition, the app facilities navigation to related apps for comprehensive cost center management.

Tip

For a user to appear in the **User Responsible** field of the app, their record must exist in business partner table BUT200. Thus, you must maintain relevant business partner information via Transaction BP.

Consult SAP Note 3553296 to display joint venture fields for workarounds to enable those fields.

Manage Cost Element Groups (F5373)

Application Type: Transactional

This app allows you to create, edit, and display cost element groups, allowing you to group cost elements that share similar characteristics to facilitate reporting. The app allows you to view cost element groups by chart of accounts and cost element group, node type, and controlling area. The app offers filtering options to display top nodes only, but it also can display end nodes or all nodes using the node type options. In addition, you can copy existing cost element groups to streamline data organization and ensure consistency in financial reporting.

Manage Cost Rates – Actual (F3422)

Application Type: Transactional

This app allows cost accountants and finance controllers to define and maintain actual cost rates for cost center activities to facilitate accurate financial tracking and cost allocation. The app is particularly useful for businesses implementing actual costing, where rates are based on real expenses rather than standard estimates. In addition, the app allows ledger-specific cost rate management, supports intercompany allocations across company codes, and integrates with scope item 33Q (Actual Costing) to align cost rates with transactions. Furthermore, you can manually adjust rates to correct discrepancies from the Actual Cost Rate Calculation – Cost Centers app (KSII). Thus, the app helps cost accountants and financial controllers maintain precise cost structures.

Manage Cost Rates – Plan (F3162)

Application Type: Transactional

This app allows you to plan cost rates for cost center activities within manufacturing to support accurate activity allocations to production orders and material costing. In addition, the app supports multiledger financial planning and facilitates intercompany cost allocations across company codes.

Manage Cost Rates – Services (F3161)

Application Type: Transactional

The app allows you to define cost rates for activity allocations across cost centers, projects, profitability segments, service order items, and sales order items with a controlling object. Beyond cost centers and activity types, the app also supports additional criteria such as service cost levels and individual employees, making it valuable for professional service providers and applicable to manufacturers in service-based scenarios. In addition, the app facilitates intercompany allocations. Thus, the app allows you to define cost rates at multiple levels, such as service cost levels and individual employee rates, to ensure precise allocation across projects and profitability segments.

Tip

To use this app, you must activate the universal parallel accounting and intercompany process enhancements business functions (FINS_PARALLEL_ACCOUNTING_BF and FINS_CO_ICO_PROC_ENH_101, respectively).

Manage Costing Runs – Estimated Costs (F1865)

Application Type: Transactional

As part of standard costing, to determine the material cost, a costing run is performed that calculates and updates the cost of multiple materials simultaneously. This app streamlines costing runs for mass material costing, allowing you to cost, mark, and release materials across multiple plants or companies simultaneously. The app provides a clear overview of costing runs, displaying key details such as the costing version, costing variant, costing run status, and job status. You can create new runs from scratch or by first copying existing ones, with the option to set recurring runs at specified intervals to reduce the manual workload. In addition, the

app allows filtering by status, costing variant, costing run date, costing date from, and valuation variant; batch deletion of multiple costing runs; and navigation to selection lists for efficient cost management.

Tip

The app requires that dependent apps or navigation targets be enabled in order to function seamlessly, as noted in SAP Note 3390035.

Transaction Code

The corresponding transaction in SAP GUI is Transaction CK40N.

Manage Cycle Run Groups (F4935)

Application Type: Transactional

This app allows you to group allocation cycles for parallel processing. Thus, the app reduces execution time significantly compared to sequential processing. Although cycles within a group are processed in a sequential order, multiple cycle run groups can run concurrently. The app's features include creating and managing cycle run groups, assigning allocation cycles to groups, editing existing groups, adding text notes to individual run groups, and tracking the last execution time of a cycle run group. For instance, you can track an allocation cycle with parameters such as cycle run group, allocation cycle, allocation type, ledger, and actual/plan. The app displays cycle run group results in two tabs: **Cycle Run Groups** and **Group Assignments**.

Manage Direct Activity Allocation (F3697)

Application Type: Transactional

Direct activity allocation involves measuring, recording, and allocating business activities based on actual activity consumption. This approach allows cost centers to distribute activity costs such as labor hours, machine usage, or service efforts directly to receiving cost objects such as production orders, internal orders, or projects. This app allows you to display, create, copy, repost, and reverse direct activity allocations while supporting the splitting of controlling documents with large line items to avoid input limits. The app provides various search criteria, such as editing status, entered on, activity type, sender cost center, sender activity type, and cost center. In addition, you can view allocation details, including journal entries, and filter allocations based on reference documents, status, or posting type.

The app provides direct navigation to the Allocation Flow app (F4022) for visualizing document links. You also can download templates for allocation uploads, assign profitability segments, and post activity allocations using cost rates either calculated with the Actual Cost Rate Calculation app (KSII) or maintained manually with the Manage Cost Rates – Actual app (F3422).

Tip

For a large volume of splitting of controlling documents, SAP recommends using batch processing or reducing a posting into smaller groups in order to split controlling documents into multiple entries to prevent input line limitations. Consult SAP Note 3474106 for further details.

Transaction Code

The corresponding transaction in SAP GUI is Transaction KB21N.

Manage Event-Based Posting Errors – Product Costing (F5132)

Application Type: Transactional

This transactional app allows you to identify, analyze, and resolve errors in event-based postings for product costing. The app facilitates

tracking failed postings, simulating corrections, and ensuring accurate cost allocation. The app's functionalities include listing posting errors, analyzing source transactions, simulating corrections before application, manually deleting erroneous postings, and reprocessing unsettled orders to maintain proper valuation of work in process (WIP) or variance calculations. The app provides search filters using parameters such as fiscal period, company code, plant, order, order type, and product. Posted orders are displayed in a tabular format that includes fields such as order, product, ledger, order status, current period WIP amounts in company code currency, and period reserve amount in company code currency. Thus, the app ensures that cost accountants can maintain financial accuracy in event-based production cost postings.

Manage Internal Order Groups (F1026)

Application Type: Web Dynpro

This app allows you to search for, create, display, and edit internal order groups. You can create a new internal order group from scratch or use an existing internal order group as a reference. The app also allows verification of duplicate subgroups or internal orders within the same group and ensures all master data in the controlling area is properly included. You can display the where-used list, delete internal order groups without references, and compare two groups for common attributes. Thus, the app streamlines management of internal order groups.

Transaction Codes

The corresponding transactions in SAP GUI are Transactions KOH1, KOH2, and KOH3.

Manage Internal Order (Web) (F1022)

Application Type: Web Dynpro

This app allows you to search for internal orders and manage their master data by creating, displaying, and editing records. The app supports investment orders by allowing direct creation of asset under construction (AUC) master data and enhances preliminary settlement functionality through an improved settlement type. In addition, you can create new internal orders from scratch or based on existing ones as a reference, track all master data changes via change documents, and enter multilingual texts on a single screen. The app also facilitates comprehensive internal order management while streamlining settlement processes.

Tip

As a prerequisite, the create, edit, and delete functions of this app rely on the Web Dynpro–based Manage Internal Order app (F1604), which must be available for these operations to work. For AUCs to generate automatically, the investment profile configuration must be enabled with the **Manage AuC** indicator.

Transaction Codes

The corresponding transactions in SAP GUI are Transactions KO01, KO02, and KO03.

Manage Internal Orders (F1604)

Application Type: Transactional

This transactional app allows you to search for internal orders and manage internal order master data by creating, displaying, and editing entries. With this app, you can filter, edit, display, and copy internal orders. In the header of the app, you can filter data by various dimensions, such as company code, profit center, functional area, object class, and object type. In addition, the app supports investment orders for direct asset under construction (AUC) master data creation and allows enhancements to the preliminary settlement type. You can track master data changes, create multilingual texts, and generate AUCs through investment orders.

Transaction Codes

The corresponding transactions in SAP GUI are Transactions KO01, KO02, and KO03.

Manage Internal Orders (Version 2) (F1604A)

Application Type: Transactional

This app allows you to create, view, and edit internal order master data while providing navigation to related apps for order management. You can search for internal orders, check system status, and track changes via the Change Log – Internal Orders app (F5403). For instance, you can search by various dimensions, such as internal order, controlling area, company code, order type, responsible cost center, status description, lock, created by, and user responsible. In addition, you can analyze dependencies using the Where-Used List – Internal Orders app (F5402). However, the app does not support creating or managing model or investment orders, nor does it allow UI adaptation or other customization operations, which must be performed using the Manage Internal Orders app (F1604).

Manage Material Valuations (F2680)

Application Type: Analytical

This app provides an overview of material valuation data, including sales order stocks, with search functions to filter materials by company code or plant. The app displays quantities, prices, and total values by currency type and period while offering access to price history and standard cost estimates. In addition, you can analyze material prices, modify inventory values for activities including sales order and project stock, and debit or credit materials accordingly. The app also supports planned price changes, allowing future price adjustments with optional currency translation and threshold-based release mechanisms.

> **Tip**
>
> Material valuation data in this app is displayed according to authorization restrictions set for the plant and ledger, while cost estimates follow authorization limits defined for the company code. Thus, the company code restrictions operate independently from valuation area authorizations, meaning that both must be managed separately for consistent access control from security. However, for transactional purposes, SAP recommends you have restrictive parameters in place when viewing material valuation data to prevent performance issues. On the other hand, for analytical purposes, the Material Valuations Balance Summary app (F1422) is recommended for displaying inventory details of large datasets efficiently.

> **Transaction Code**
>
> The corresponding SAP GUI transaction is Transaction MR21.

Manage Profit Center Groups (F0764)

Application Type: Web Dynpro

This app allows you to efficiently manage profit center groups by enabling tasks such as search, creation, display, and editing of group structures. Using this app, you can create different versions of a profit center group for various validity dates by copying it to an inactive version, though versioning is not available for the standard hierarchy. In addition, the app provides functionality to check for duplicate subgroups, verify master data inclusion within controlling areas, and compare attributes between two profit center groups. You also can delete unused profit center groups, view where-used lists, select standard hierarchies, and adjust master data for accurate financial organization. Profit center and profit center groups are master data, so these objects are not transported. As a

prerequisite, you need to create profit center groups in every client prior to commencing business activity.

> **Tip**
>
> The standard hierarchy is required, which includes all profit centers within a controlling area, representing the organizational structure in profit center accounting. When creating profit centers within the standard hierarchy, statuses cannot be modified in the detail area; instead, you must mark objects for activation or flag inactive versions for deletion in the selection area. This approach facilitates collective editing of profit centers, ensuring a streamlined process and consistency within the hierarchy.

> **Transaction Codes**
>
> The corresponding transactions in SAP GUI are Transactions KCH1, KCH2, and KCH3.

Manage Profit Centers (Version 2) (F3516)

Application Type: Transactional

This app allows you to create, edit, display, and delete profit center master data. You can search for profit centers, view master data, and manage validity periods. Profit centers can be created directly, created via copying, or split, with the ability to assign company codes to specific centers. You can search for profit centers with various filter criteria such as profit center, valid from, valid to, and company code. Deletion is possible only if no master data or business transactions reference the profit center; otherwise, it can be locked before archiving.

In addition, the app provides navigation to the Where-Used List – Profit Centers app (F3751) for reviewing dependencies and the Change Log – Profit Centers app (F3810) for tracking modifications.

Tip

As a leading practice, SAP recommends that you check the **Where-Used List – Profit Centers** option so that you can understand the dependencies prior to making changes to the profit center.

Transaction Codes

The corresponding transactions in SAP GUI are Transactions KE51, KE52, and KE53.

Manage Real-Time Revenue Recognition Issues (F4101)

Application Type: Transactional

This app allows cost accountants to identify and resolve errors encountered during the automatic revenue recognition process. During this process, issues may arise due to various reasons, such as missing master data, unmaintained costs, incorrect configuration, or a closed fiscal period. In addition, with this app, you can view event-based revenue recognition errors; filter by project definition, sales document service document, or provider contract; and access a summary of all errors. Once corrections are made, you can reprocess failed revenue recognition postings to clear them from the error list, ensuring a successful period-end run. The app also allows direct navigation to related revenue recognition apps and supports error list exports to Microsoft Excel.

Tip

Depending on the root cause, the errors displayed in this app may stem not only from revenue recognition but also from integration processes with other components, such as financial accounting or logistics.

Manage Revenue Recognition Issues – Projects (F4100)

Application Type: Transactional

This app allows cost accountants to identify and correct errors that occurred during the execution of the Run Revenue Recognition – Projects app (F4277). The app provides a list of impacted projects and work breakdown structure (WBS) elements based on filter criteria, allowing you to analyze issues, view error messages, and access detailed recognition values. In addition, you can perform simulation runs, reprocess WBS elements to ensure accurate revenue recognition, and delete log items that match the search criteria. The simulation run provides a list of records with the **Success** or **Error** statuses. In addition, you can use various app features, such as **Reprocess, Reprocess and Skip Errors**, and **Show Error Summary**, to process records.

The app offers direct navigation to related apps such as Manage Real-Time Revenue Recognition Issues (F4101) and Run Revenue Recognition – Projects (F4277). Thus, the app helps you to ensure that errors are resolved effectively.

Tip

Consult SAP Note 3041610 to learn about the behavior of the **Reprocess, Reprocess and Skip Errors**, and **Show Error Summary** buttons in this app.

Manage Revenue Recognition Issues – Sales Orders (F4185)

Application Type: Transactional

This app allows you to correct errors from the execution of the Run Revenue Recognition – Sales Orders app (F4276). The app displays sales order items with errors based on filter criteria and provides detailed error messages for your review. You also can reprocess sales order items to ensure accurate revenue recognition or perform a simulation run to preview results. In addition, you can use various app features, such as **Reprocess**, **Reprocess and Skip Errors**, and **Show Error Summary**, to process records.

The app offers direct navigation to the Manage Real-Time Revenue Recognition Issues (F4101) and Run Revenue Recognition – Sales Orders (F4276) apps. Thus, the app streamlines error resolution and financial accuracy.

Manage Revenue Recognition Issues – Service Documents (F4186)

Application Type: Transactional

This app allows cost accountants to correct errors that occurred during the execution of the Run Revenue Recognition – Service Documents app (F4278). The app offers features such as reprocessing service document items to ensure accurate revenue recognition values, performing simulation runs to preview results, and navigating directly to the Manage Real-Time Revenue Recognition Issues (F4101) and Run Revenue Recognition – Service Documents (F4278) apps for further adjustments.

Manage Settlement Rules – Internal Orders (F5695)

Application Type: Transactional

This app allows cost accountants to view and modify settlement rules for internal orders. The app allows you to display existing settlement rules, manage associated distribution rules, and edit settlement parameters for sender objects. You can search for existing settlement rules using search criteria such as internal order, distribute rule created, order type, status, created by, changed by, company code, and controlling area. In addition, you can create new distribution rules to reflect accurate cost allocation within internal order settlements. You also can navigate to other relevant apps.

> **Tip**
>
> This app supports settlement rules for actual settlement, and it does not support planned settlement.

> **Transaction Codes**
>
> The corresponding transactions in SAP GUI that are used to manage settlement rules for internal orders are Transactions KO02 and KOB5.

Manage Statistical Key Figure Values (F3915)

Application Type: Transactional

Statistical key figures (SKFs) are nonmonetary values used in cost allocation and reporting in controlling. They act as tracing factors for distributing costs across cost centers based on measurable characteristics, such as employee headcount, square meters occupied, or utility consumption. This app allows you to display and manage SKF values. The app allows you to create, copy, and reverse the posting of SKFs via the **Copy**, **Reverse**, and **Create** features. SKFs can be searched with various parameters such as status, posting type, and created by. In addition, the app supports splitting large controlling documents into multiple entries to avoid input line limitations. You can filter postings by document, posting date, status, and posting type to facilitate data retrieval and further analysis. The app also allows you to post negative quantities in the **Statistical Quantity** column. Finally, the app provides detailed views of existing postings, streamlining the management and reporting processes.

> **Transaction Code**
>
> The corresponding transaction in SAP GUI is Transaction KB31N.

2

Manage Statistical Key Figures (Version 2) (F1603A)

Application Type: Transactional

Statistical key figures (SKFs) in SAP represent nonmonetary data such as the number of employees or square meters occupied, and they are used for cost allocation and reporting. In controlling, they serve as allocation bases to distribute costs across cost centers, profit centers, or orders. This app allows you to display and manage SKFs, including creating new ones or copying, editing, and deleting existing entries. You can search for an SKF ID, name, or category; view master data; and analyze usage in the Where-Used List – Statistical Key Figures app (F4078). In addition, the app provides navigation to related applications, with options to adjust visible links via the **More Links** feature.

Transaction Codes

The corresponding transactions in SAP GUI are Transactions KK01, KK02, and KK03.

Market Segments – Actual (F0960)

Application Type: Web Dynpro

This analytical app allows you to report actual data of profitability analysis for a fiscal year. With this app, you can refine search criteria using variables or filters such as posting period, company code, functional area, segment, profit center, company code, and general ledger account hierarchy. The app displays query results in a grid format and you can export data to Microsoft Excel. In addition, the app's **Navigation Panel** allows you to add dimensions from the **Available Fields** section into rows and columns.

The app also displays both real and statistical postings distinguished via the **IsStatistical** fields, such as **IsStatistical Order** and **IsStatistical SO Item**. The value of **x** in these fields indicates statistical postings, and **#** represents real postings.

Transaction Codes

There is no direct dimensional reporting in SAP GUI. However, Transactions FAGLL03 or FAGLL03H can be used to get similar results.

Market Segments – Plan/Actuals (W0078)

Application Type: Web Dynpro

This analytical app allows you to report both actual and plan data of profitability analysis for a fiscal year. With this app, you can refine search criteria using variables or filters such as posting period, plan category, company code, functional area, segment, profit center, customer group, material group, company code, and general ledger account hierarchy. The app displays query results in a grid format, and you can export data to Microsoft Excel. In addition, the app's **Navigation Panel** function allows you to add dimensions from the **Available Fields** section into rows and columns. The app's features include the ability to specify search criteria using variables or filters, display query results in a structured grid format, export results to Microsoft Excel for further analysis, manipulate the query results grid by repositioning available characteristics within the grid, and display read-only details about the underlying query, such as its technical name and last update timestamp.

Transaction Codes

There is no direct multidimensional reporting in SAP GUI. However, Transactions KE30, KEDR, KE24, and KE5Z can be used to get similar results.

Material Inventory Values – Balance Summary (F1422)

Application Type: Web Dynpro

This app allows you to obtain insights into material inventory quantities and values for a reporting key date and ledger, offering detailed analysis based on role-specific access. In addition, you can view inventory values and quantities sorted by company code, general ledger account, material group, profit center, and segment. The balance summary includes key figures such as inventory amounts in up to three currencies and dimensions such as plant and valuation area. The app allows filtering by various criteria such as company code, material, and plant; grouping dimensions for analysis; and visualizing results through tables, graphics, or combined formats. Moreover, you can save selections as variants and modify displayed columns with an **ID and Description** from the **Display** dropdowns for an enhanced view of the result set. Thus, the app helps you analyze inventory quantities and values efficiently, enabling you to view inventory balances, see key figures, and use filtering capabilities to refine your reporting.

Material Inventory Values – Line Items (F1423)

Application Type: Web Dynpro

This app allows you to access detailed insights into material inventory values and quantities down to the line item level. The **Navigation Panel** allows you to add dimensions and measures into rows and columns. The app displays journal entries that impact inventory values within a specific reporting key date range, along with source document dates, proposed posting dates, and actual posting dates. In addition, you can view inventory details sorted by company

code, general ledger account, material group, business transaction type, posting date, and document number. The app also supports in-depth analysis with dimensions such as plant, valuation area, and key figures, including inventory quantities and amounts in up to three currencies. The app's features include filtering dimensions by various search criteria such as product group, company code, and plant; grouping them for analysis; visualizing the results in different formats; saving your selections as variants; and modifying displayed IDs or descriptions for an enhanced view of the result set.

Transaction Codes

The closest transactions in SAP GUI are Transactions MB51, MB5B, and MB52.

Material Inventory Values – Rounding Differences (F1440)

Application Type: Web Dynpro

This app helps you analyze rounding differences between actual inventory postings and calculated inventory values for a select period and company code. These differences occur when price and unit values cannot be precisely represented in the minimum currency unit (e.g., EUR with two decimal places). With this app, you can take corrective actions, such as adjusting future material prices for rounding differences. The app also allows you to view inventory values and quantities sorted by company code, general ledger account, material, fiscal year, ledger fiscal period, and plant. In addition, the app supports detailed analysis with dimensions such as key figures and currency amounts, filtering by various criteria, and visualizing results in tables or graphics. You also can group dimensions for analysis, save selections as variants, and modify displayed IDs or descriptions for customization.

Tip

For instance, if a total inventory value of USD 10.00 for three identical widgets cannot be evenly distributed with two decimals, the system assigns the first two widgets valued at 3.33 each, while the third receives the remainder—that is, 3.34.

Monitor Material Prices (F7245)

Application Type: Transactional

This app allows you to analyze material prices across all valuations using two distinct views. One view provides a detailed price comparison between two selected valuations, allowing for precise analysis of price variations. The other view presents material prices over time across all valuations (historical trend view), offering a broader perspective on pricing trends. You can monitor material prices by searching with various fields such as view type, valuation to compare 1, valuation to compare 2, material, price type, company code, plant, from fiscal year period, to fiscal year period, price difference (amount), and price difference (%).

Tip

This app requires activation of the universal parallel accounting business function FINS_PARALLEL_ACCOUNTING_BF. If universal parallel accounting is not activated, SAP recommends using the Manage Material Valuations app (F2680) for material price analysis.

Transaction Codes

The corresponding transactions in SAP GUI are Transactions CKM3N, MR21, MM03, MB51, and CKMPCD.

My Spend (Version 2) (F0366)

Application Type: Transactional

With this transactional app, you can monitor departmental and project budgets and spending in real time. It consolidates financial data into an intuitive interface, providing clear visualizations. The app offers you the chance to review budget and spending summaries by department, project, or expense type; analyze detailed expense data such as actual, committed, total, and budgeted amounts with variance insights; and examine trends across different cost elements. In addition, the app supports dimensional drilldowns to corresponding accounts, offers flexible display modes between charts and tables, and facilitates alerts to responsible personnel. Moreover, you can save spending details in favorites for quick access. The app supports integration with SAP Jam, which allows you to post comments or send emails directly from the app.

Tip

The app does not support statistical internal orders or work breakdown structure (WBS) elements. In addition, in the table view, only the first 100 cost centers, internal orders, and cost elements are displayed if more are used. Configuration in Transactions FPB_MAINTAIN_PERS_S and FPB_MAINTAIN_PERS_M is restricted to groups only.

Transaction Code

Transaction S_ALR_87013620 can be used to display postings for primary and secondary cost elements in SAP GUI.

My Unusual Items (F0368)

Application Type: Transactional

This app allows you to monitor potentially disputable expense items across departments and projects in controlling. The app identifies and collects these items based on predefined rules and patterns and then consolidates them into an inbox for your review. In addition, you can filter disputable expense items using fields such as item name, document number, account, department, currency, or exceeding amount; view detailed information; and either confirm or dispute the items with comments while assigning responsibility. The app also tracks item status, allows you to save disputable expenses as tiles, and integrates with SAP Jam for posting comments and email communication.

Order Costs Details – Event-Based (F4254)

Application Type: Analytical

This app provides a detailed cost comparison for event-based manufacturing orders. Thus, you get insights into actual, plan, target, and control costs at both the order header and item levels. You can search for order costs using various search criteria such as fiscal period from, fiscal period to, plan category, company code, currency, order status, plant, order type, product, and order. The app supports parallel ledger valuations across multiple currencies in a legal view and a group view, using the ledger as a filter, therefore supporting common accounting standards. In addition, you can analyze total variance and split variance categories at the general ledger account, cost component, and business transaction levels. When integrated with the event-based production cost posting scenario (scope item 3F0), the app delivers real-time cost

transparency. The app also provides a full breakdown of production cost postings for production orders (e.g., order type YBM1) and process orders (e.g., order type YBM2), including coproducts and by-products. You can adjust the date range to analyze costs for open and closed orders across fiscal periods and access event-based order costs from the launchpad, using the Analyze Production Costs – Event-Based app (F4059) or the Work in Process – Event-Based app (F3498).

Tip

Using this app, you can perform a comprehensive cost comparison by adjusting the date range to include both open and closed orders across different fiscal periods.

P&L – Plan/Actual/Committed (W0170)

Application Type: Web Dynpro

This app provides a comprehensive comparison of planned, actual, and committed financial values across various dimensions. Using this app, you can search plan, actual, and commitment data using the ledger, fiscal year, plan category, key date, fiscal year, fiscal period, company code, cost center, profit center, segment, general ledger account hierarchy, and general ledger account (node) criteria. In addition, you can analyze deviations between planned and actual amounts, which facilitates root-cause analysis. Furthermore, the app supports drilldown capabilities and has a Microsoft Excel file download option for detailed analysis. The app uses core CDS view C_PnLPlnActlCmtmtJrnlEntrItmQ.

Transaction Codes

The closest transactions in SAP GUI are Transactions S_ALR_87013611 and S_ALR_87013558.

Periodic Allocation Results (F5854)

Application Type: Transactional

This app allows you to visualize cost allocations within a selected cycle using a network graph, detailing the amounts transferred from individual senders (e.g., cost centers) to receivers (e.g., orders). Thus, the app allows you to review the allocation cycle settings before executing the process. The system transforms allocation data into a recursive hierarchy, where nodes represent senders and receivers. In addition, you can create graphs for posted or simulated cycles, select controlling area parameters, and enter version **000**. Large cycles may require additional processing time, indicated by the **Triggered** status, with options to **Cancel** or **Restart** failed transformations. Completed graphs allow customization of displayed hierarchy levels and fields, while the **Calculate** function helps estimate sender and receiver counts.

You can launch the Visualized Periodic Allocation Result app (F5855) for detailed analysis for a cycle directly from the **Graph** list.

Transaction Codes

There is no direct SAP GUI transaction for this app. Once allocation-related Transactions KSU5, KSV5, and KEU5 are executed, the results can be viewed using a transaction such as Transactions FAGLL03 or S_ALR_87013611.

Plan Cost Centers on Periods (F1581)

Application Type: Web Dynpro

This app allows you to enter and calculate plan data for cost centers across different posting periods. The app's **Navigation Panel** allows you to add dimensions from the **Available Fields** section into rows and columns. You can filter values based on dimensions such as fiscal year, category, company code, general ledger account hierarchy, and general ledger account/account

group. In addition, you can input data at the lowest level, such as cost center, period, and general ledger account, allowing the app to aggregate values, or at the highest level, like the cost center or fiscal year, with automatic disaggregation to periods and accounts. In addition, the app supports intermediate-level entries, with the app adjusting values as needed. Furthermore, it facilitates aggregation and disaggregation for various characteristics to enhance planning accuracy.

Tip

The app only supports the global currency for cost center period planning. If you experience errors when switching currencies, SAP recommends using SAP Analytics Cloud as an alternative solution. Consult SAP Note 3481143 for further details.

Transaction Code

The corresponding transaction in SAP GUI is Transaction KP06.

Postprocess Event-Based Posting – Overhead Calculation (F5763)

Application Type: Transactional

This app allows you to review and reprocess event-based overhead errors. After resolving errors, you can simulate processing across all ledgers of the same object before reposting. Successfully fixed errors will no longer appear in the result list. The app's functionalities include displaying overhead errors for a given period, reposting corrected entries, and simulating processing outcomes. In addition, you can view journal entries after reposting and navigating to the related Schedule Product Costing Jobs app (F3683) for mass processing of overhead calculations. The app is essential for production cost accountants and overhead cost accountants, particularly in cases of errors during goods movement, confirmation processes, cost center validity issues, or blocked postings. The app supports both postprocessing for production and process orders as well as product cost collectors. Thus, the app allows you to streamline postprocess event-based postings in product costing.

Postprocess Event-Based Posting – Product Costing (F3669)

Application Type: Transactional

In SAP product costing, *event-based posting* refers to an approach in which production costs are recorded in real time as business events occur, rather than waiting for period-end processing. This app allows you to identify and resolve event-based posting exceptions in production orders, process orders, or product cost collectors with event-based-processing key business events such as **RSEBW** (event-based work in process [WIP] and variance posting) or **PCCEB** (product cost collector manual settlement) for a specified company code and fiscal period. The app is delivered with job templates that facilitate the handling of production orders, process orders, and the product cost collector. In addition, using these templates, you can view posting exceptions, simulate resolutions, revaluate orders, and trigger postings to correct errors caused by issues such as incorrect master data, account configuration failures, or additional costs from journal entries and allocations.

The app also provides visibility into WIP values, total variance, actual costs, and unsettled balances for affected orders. You can export data, simulate corrections before actual posting, and resolve exceptions systematically. Posted orders are moved to the **Posted Production Orders/Process Orders** or **Product Cost Collectors to Post** sections, with the ability to review journal entries via the Display Line Items – Production Accounting app (F7288). Furthermore, you can schedule background jobs for large-scale processing and leverage manual order revaluation and variance reversal through the Schedule Product Costing Jobs app (F3683).

> **Tip**
>
> For orders or product cost collectors with settlement type **P Settled Manually** (event-based processing key **ORDMS** or **PCCMS**), SAP recommends using the **Manual Settlement: Event-Based Order and PCC** job template to process settlements efficiently.
>
> For product cost collections, the app displays period amounts only.

Product and Service Margins (W0164)

Application Type: Web Dynpro

This app allows for comprehensive analysis of revenue, costs, and contribution margin across multiple dimensions for a financial statement version and ledger in margin analysis. You can analyze various KPIs using dimensions such company code, fiscal year, fiscal period, object type, profit center, solution order, service document, service contract, sales order, project definition, product sold group, product sold, customer group, and customer. Important revenue KPIs include **Billed Revenue, Recognized Revenue, Revenue Adjustment, Deferred Revenue**, and **Accrued Revenue**. Cost-related KPIs include **Actual Costs, Recognized Cost of Sales (COS), COS Adjustment, Deferred COS**, and **Accrued COS**. Margin analysis is supported through the **Recognized Margin** KPI, while additional insights are available via **Reserves**. A typical use case of this app is performing margin analysis across different market segments, such as customer or product sold.

> **Tip**
>
> The object type determines which objects or account assignment types are displayed in the report, with values such as **EO** (profitability segment), **SV** (service order), **SC** (service contract), **PR** (project), and **OR** (maintenance-centric service orders). Removing any of these object types excludes their corresponding datasets from the report.

In addition, for account assignment type **OR**, the app considers line items in calculations only if the journal entry is assigned to a maintenance-centric service order (**MCSO**).

Transaction Code

The closest transaction in SAP GUI is Transaction KE30.

Product Cost Collector Details – Event-Based (F6965)

Application Type: Analytical

This app provides a comprehensive breakdown of event-based cost postings for repetitive manufacturing orders (order category **05** – product cost collectors). The app allows you to compare actual and target costs, supporting parallel ledgers in multiple currencies. You can view product cost collector details using search criteria such as ledger, product cost collector, product, plant, and production process. In addition, you can analyze costs cumulatively or within specific fiscal periods, ensuring precise cost tracking.

The app can be accessed from related apps such as Analyze Production Costs – Event-Based (F4059) and Work in Process – Event-Based (F3498). Additional features include detailed general ledger account cost breakdowns, real-time updates based on material consumption and activity confirmation, and a direct link to the Display Posting Rules – Event-Based app (F5228) for deeper analysis.

Product Profitability with Production Variances (W0182)

Application Type: Web Dynpro

This app allows the analysis of contribution margins for individual products, incorporating related product details and available profitability characteristics for a financial statement ver-

sion. You can search for variances using filter criteria such as ledger, company code, fiscal year, fiscal period, customer group, product sold group, sales document, and profit center. In addition, you can drill down into fixed and variable costs using the standard cost component split and report on product variances, billed quantity, and margin per unit. The app also includes line items, which are included in calculations only if the journal entry is assigned to a profitability segment and the product sold field is populated. The app presents amounts in the global currency, with an option to view selected measures in the company code currency by navigating to the **Navigation Panel**, expanding the **Measures** node, and adding the relevant measure to the data grid, such as billed revenue amount in company code currency (measures). The selected measures will then be displayed in the company code currency. Thus, the app's features include analyzing product profitability by drilling down into fixed and variable costs based on the standard cost component split, displaying profitability characteristics, and reviewing measures such as billed revenue and billed quantity.

Tip

Certain measures, such as margin per unit, cannot be added in the app directly. SAP recommends using cube I_PROFITABILITYCUB via the Custom Analytical Queries app (F1572) for this, as suggested in SAP Note 3452289.

Transaction Codes

The closest transactions in SAP GUI are Transactions KE30 and KE24.

Production Cost Analysis (F1780)

Application Type: Analytical

This app allows cost accountants to analyze production costs for non-event-based manufacturing orders and product cost collectors using the period-end production cost posting solution (scope item BEI). The app provides detailed cost breakdowns by account and business transaction, facilitating cost comparisons between budgeted and actual amounts at the order level. In addition, you can track variances, identify root causes, and implement cost-saving measures such as alternative resources or process adjustments. The app supports cost component structures, highlights exceptional orders, and allows both plan costs and order standard costs. SAP's predelivered cost component structure **Y1** for standard costing is not available in this app. However, **YP** is supported with this app for reporting and serves as the cost component structure.

Tip

This app does not support orders in the event-based approach.

Transaction Codes

The corresponding transaction in SAP GUI is Transaction KKBC_ORD, which supports the **Y1** standard costing for order-related costs. The closest other transactions are Transactions CO03, KOB1, CKM3N, and S_ALR_87013127.

Production Cost Overview – Event-Based (F6248)

Application Type: Analytical

This app provides real-time tracking of production costs by linking cost postings to business events such as goods issues, goods receipts, and operation confirmations. This event-based approach ensures immediate visibility into work in process (WIP), variances, and actual production costs, eliminating the need for manual month-end calculations. You can search for product costs with various search parameters

such as fiscal period from, fiscal period to, ledger, company code, currency, plant, order type, and order. In addition, you can drill down into cost postings, analyze production variances, and review WIP capitalization, ensuring accurate and efficient cost management. Thus, the app allows you to perform detailed cost analysis in real time.

Tip

This app is fully integrated with universal parallel accounting and the Universal Journal (table ACDOCA). For further information, consult SAP Note 3315610.

Profit and Loss – Actuals (F0958)

Application Type: Web Dynpro

The profit and loss (P&L) statement summarizes a company's revenues, expenses, and net profit or loss over a specific period to determine profitability. This analytical app allows you to report the actual P&L statement across relevant entities for a fiscal year. With this app, you can refine search criteria using variables or filters such as posting period, company code, general ledger account hierarchy, and general ledger account/account group. The app displays query results in a grid format, and you can export data to Microsoft Excel. The app's **Navigation Panel** allows you to add dimensions from the **Available Fields** section into rows and columns. In addition, the app displays both real and statistical postings distinguished via the **IsStatistical** fields, such as **IsStatistical Order** and **IsStatistical SO Item**. The value of **x** in these fields indicates statistical postings, and **#** represents real postings.

Tip

Consult SAP Note 2809662 to learn more about key date behavior in analytical apps.

2

Transaction Codes

The corresponding transactions in SAP GUI are Transactions F.01, S_ALR_87012284, KE5Z, and S_E38_98000088.

Profit and Loss – Plan/Actual (F1710)

Application Type: Web Dynpro

This analytical app displays both actual and plan data in a profit and loss (P&L) statement format across relevant entities for a fiscal year. With this app, you can refine search criteria using variables or filters such as category, fiscal year, functional area, segment, posting period, company code, general ledger account hierarchy, and general ledger account/account group. Category is an important dimension in this app; it's used to differentiate various types of plan data within financial and management accounting processes. Thus, it helps classify planning versions for budgeting, forecasting, and financial analysis. With this app, you can display multiple plan categories for different requirements. In addition, the app displays query results in a grid format, and you can export data to Microsoft Excel. The app's **Navigation Panel** allows you to add dimensions from the **Available Fields** section into rows and columns. In addition, the app displays both real and statistical postings distinguished via the **IsStatistical** fields, such as **IsStatistical Order** and **IsStatistical SO Item**. The value of **x** in these fields indicates statistical postings, and **#** represents real postings.

Transaction Codes

The corresponding transactions in SAP GUI are Transactions F.01, KE5Z, and S_E38_98000088.

Profit Centers – Actual (F0959)

Application Type: Web Dynpro

This analytical app allows you to report actual data of a profit center for a fiscal year. With this app, you can refine search criteria using variables or filters such as posting period, profit center, company code, and profit center hierarchy. The app displays query results in a grid format, and you can export data to Microsoft Excel. In addition, the app's **Navigation Panel** allows you to add dimensions from the **Available Fields** section into rows and columns. The app also displays both real and statistical postings, distinguished via the **IsStatistical** fields, such as **IsStatistical Order** and **IsStatistical SO Item**. The value of **x** in these fields indicates statistical postings, and **#** represents real postings.

Transaction Codes

The corresponding transactions in SAP GUI are Transactions F.01, S_ALR_87012284, KE5Z, and S_E38_98000088.

Profit Centers – Plan/Actuals (W0079)

Application Type: Web Dynpro

This analytical app allows you to report both actual and plan data of profit centers for a ledger, fiscal year, and plan category. With this app, you can refine search criteria using variables or filters such as posting period, company code, functional area, segment, profit center, customer group, material group, company code, and general ledger account hierarchy. The app displays query results in a grid format, and you can export data to Microsoft Excel. In addition, the app's **Navigation Panel** function allows you to add additional dimensions from the **Available Fields** section into rows and columns. The app's features include the ability to specify search criteria using variables or filters, display query results in a structured grid format, export results to Microsoft Excel for further analysis, manipulate the query results grid by repositioning available characteristics within the grid, and display read-only details about the underlying query, such as its technical name and last update time stamp. The app supports authorizations to view data at the company code level.

Transaction Codes

The corresponding transactions in SAP GUI are Transactions F.01, S_ALR_87012284, KE5Z, and S_E38_98000088.

Projects – Actual (F0961)

Application Type: Web Dynpro

This analytical app allows you to report the actual data of projects for a fiscal year. With this app, you can refine search criteria using variables such as fiscal year, posting period, projects, work breakdown structure (WBS), company code, and general ledger account hierarchy. In addition, the app displays query results in a grid format, and you can export data to Microsoft Excel. The app's **Navigation Panel** also allows you to add dimensions from the **Available Fields** section into rows and columns. The app displays both real and statistical postings, distinguished via the **IsStatistical** fields, such as **IsStatistical WBS Elem** and **IsStatistical SO Item**. The value of **x** in these fields indicates statistical postings, and **#** represents real postings.

Transaction Codes

The corresponding transactions in SAP GUI are Transactions CJI3, CJ74, S_ALR_87013558, and S_ALR_87013557.

Projects – Plan/Actual (W0080)

Application Type: Web Dynpro

This analytical app allows you to report both actual and plan data of project work breakdown structure (WBS) elements for a ledger, ledger fiscal year, and plan category. With this app, you can refine search criteria using variables or filters such as posting period, company code, project definition, WBS, profit center, segment, customer, company code, general ledger account (hierarchy), and general ledger account (node). The app displays query results in a grid format, and you can export data to Microsoft Excel. In addition, the app's **Navigation Panel** function allows you to add additional dimensions from the **Available Fields** section into rows and columns. The app's features include the ability to specify search criteria using variables or filters, display query results in a structured grid format, export results to Microsoft Excel for further analysis, manipulate the query results grid by repositioning available characteristics within the grid, and display read-only details about the underlying query, such as its technical name and last update time stamp. The app supports authorizations to view data at the company code level.

Transaction Codes

The corresponding transactions in SAP GUI are Transactions CJI3, CJI4, S_ALR_87013558, and S_ALR_87013557.

Realignment Results – Profitability Analysis (F2549)

Application Type: Transactional

Realignment is a process in margin analysis that adjusts characteristic values for profitability segments in previously posted data. This app allows you to analyze adjustments made to posted profitability reporting dimensions by the Run Realignment – Profitability Analysis app (KEND). These adjustments ensure that organizational changes are reflected in profitability characteristics, incorporate master data corrections, or enrich profitability data with previously unavailable information. With this app, you can search realignment results using various search criteria, such as ledger, fiscal year, company code, posting date, profit center, functional area, realignment date, material group, and other dimensions. You can review profitability characteristics before and after realignment, analyze results through visual graphs, and inspect individual journal entries for deeper insights. Thus, the app facilitates identifying trends and ensures that organizational changes, master data corrections, and enriched profitability information are accurately reflected in margin analysis. The app also facilitates drilling down to review individual journal entries and provides clarity into how profitability reporting dimensions have been adjusted.

Transaction Code

The corresponding transaction in SAP GUI is Transaction KEND.

Reassign Costs and Revenues (F2009)

Application Type: Transactional

This app allows you to reassign costs and revenues when allocations are incorrect, ensuring accurate financial tracking. It supports splitting controlling documents with large line items into multiple separate documents during posting to prevent input line limits. You can create, copy, and reverse cost allocations while filtering by reference document, posting type, or sender cost center. In addition, the app allows you to specify multiple accounts for allocation, streamlining financial adjustments per business needs.

Tip

This app does not support ledger-specific values for cost and revenue reposting, meaning that the same value is reposted across all ledgers. To manually post ledger-specific values, SAP recommends using the Post General Journal Entries app (F0718).

You must carefully review the sender and receiver cost centers before posting to ensure accurate allocations and prevent unintended adjustments across all ledgers later.

Transaction Codes

The similar transactions in SAP GUI are Transactions KB11N and KB61.

Repost Line Items – Cost Accounting (F5549)

Application Type: Transactional

This app allows you to correct posting errors for costs and revenues by reposting specific line items for cost objects such as cost centers, orders, and work breakdown structure (WBS) elements from cost accounting documents (controlling documents). The app allows you to view reposted line items and reverse them when necessary. You can search for line items using various search criteria such as editing status, controlling area, reference document, journal entry, journal entry item, posting date, created by, status, posting type, and account for allocation. The app displays results in a tabular format with the flexibility to add or remove columns. The **Status** column displays whether the line item is **Posted**, **Partially Reposted**, or ended in an **Error**. In addition, you can perform a reversal posting on the sender object, updating debit/credit indicators immediately for sender and receiver assignments. Furthermore, you can select and process line items for reposting, reallocating amounts from the initial account assignment object to one or more new controlling objects. The app features collective processing, reposting primary cost postings to multiple objects, adjusting portions of posted line items, or assigning entries to different account objects. Once validated, all modifications are updated at once, maintaining consistency. The reference document number links the original entry to the adjustment for accurate recordkeeping.

Tip

You must ensure that the total reposted amount does not exceed the overall value of the sender object's line items if multiple controlling objects are involved.

Transaction Code

The corresponding transaction in SAP GUI is Transaction KB61,

Reverse Event-Based Revenue Recognition (F6638)

Application Type: Transactional

This app allows for the reversal of documents generated in event-based revenue recognition for sales documents, service documents, and projects. You can schedule reversal jobs for each ledger, process individual document reversals or mass changes, and reverse documents linked to specific transactions. The job can be executed as a simulation run to review impacts before posting or as an update run for final processing. In addition, you can select data for reversal using parameters such as posting date, company code, ledger, journal entry, and project definition.

Tip

In general, event-based revenue recognition documents are always posted on the last day of the period. Thus, when specifying the posting date, you must ensure it matches the period's final day; otherwise, the system will not find the relevant documents for processing.

Run Allocations (F3548)

Application Type: Transactional

This app allows cost accountants or finance controllers to perform tasks such as running allocation cycles, performing test runs, or performing actual runs for existing allocation cycles and reviewing the completed results. You can access the app from the Manage Allocations app (F3338) for direct worklist navigation or from the SAP Fiori launchpad to view completed runs. In addition, the app facilitates initiating multiple allocation cycles simultaneously, referencing specific sender data periods for top-down distributions, and cumulating sender data from reference periods. For instance, you can initiate a live run using parameters such as run name, journal entry type, fiscal period from, and fiscal period to. Furthermore, once the allocation run is executed, you can analyze the results, check for errors or warnings, view detailed sender-receiver allocations, reverse completed runs, and modify cycles in the Manage Allocations app. The app also allows you to drag and drop allocation cycles to reorder them within a run.

Tip

The app supports defining multiple cycles in a group for execution. If you want to execute a group of cycles, as a prerequisite, you must maintain and activate the universal allocation cycle hierarchy using the Manage Global Hierarchies app (F2918) first.

Run Overhead Calculation – Actual (F4857)

Application Type: Transactional

This app allows you to calculate overhead calculation for a single project, work breakdown structure (WBS) element, order, or service document. You can filter using parameters such as WBS element, project, company code, and cost object type. This app is generally used during period-end closing to determine overhead for individual cost objects. The allocation is based on primary cost elements, with detailed overhead values displayed for analysis. It is particularly useful for testing overhead calculations following error corrections in mass processing before executing the final run. Upon completion of a run, the app displays results with statistics such as overstand calculated, not relevant, inappropriate status, and error, along with line-item details such as sender, receiver cost element, amount in global currency, and amount in company code currency. Thus, the app's functionalities include performing test runs without impacting actual data, executing update runs that affect data, analyzing overhead results—including calculated values, error messages, and journal entries—reviewing overhead allocations per cost object based on the assigned costing sheet, and reversing overhead calculation runs when needed.

Run Overhead Calculation – Production Orders – Actual app (CO43) with the production accountant business role.

- In addition, by default, all intercompany posting transactions performed through this app are permitted. To restrict intercompany postings entirely or allow them only for specific sender and receiver company code combinations, you must perform Customizing activity **Block Intercompany Postings**.

- Ensure necessary Customizing activities such as costing sheet and intercompany postings are correctly set up before executing the app.

Run Revenue Recognition – Projects (F4277)

Application Type: Transactional

This app allows you to periodically recalculate revenue recognition values for all work breakdown structure (WBS) elements relevant to event-based revenue recognition, where costs and revenues from project-based transactions are recognized in real time. However, a periodic run may be necessary after adjustments to planned costs or deferrals made in the Event-Based Revenue Recognition – Projects app (F4767). You can either recalculate specific WBS elements using the **Revalue** function or run a periodic revenue recognition process. The app supports simulation runs to preview results. After execution of the run, the app selects relevant objects based on the revenue recognition key, compares newly calculated values with previously posted ones, and then generates adjustment postings for discrepancies.

Tip

Consult SAP Note 3403377 to understand the default retention period for closing activities in this app.

Transaction Codes

The corresponding transactions in SAP GUI are Transactions KKA2, CJ88, and CJ8N.

Run Revenue Recognition – Sales Orders (F4276)

Application Type: Transactional

This app allows you to periodically recalculate revenue recognition values for all sales order items relevant to event-based revenue recognition, where costs and revenues are recorded in real time. However, a periodic run may sometimes be necessary—for instance, after manual adjustments are made in the Revenue Recognition (Event-Based) – Sales Orders (Version 2) app (F2441A). You can either use the **Revalue** function to recalculate specific items or execute a periodic revenue recognition run for all sales documents in this app. Upon execution, the app selects relevant objects based on the revenue recognition key by comparing newly calculated values with previously posted ones. Then the app creates adjustment postings if discrepancies are found, and it supports simulation runs to preview the results.

Transaction Code

The corresponding transaction in SAP GUI is Transaction REV_REC_PEC_SD.

Run Revenue Recognition – Service Documents (F4278)

Application Type: Transactional

This app allows you to periodically recalculate revenue recognition values for all service documents relevant to event-based revenue recognition, where costs and revenues are recorded in real time. A periodic run may be required after manual adjustments are posted in the Event-Based Revenue Recognition – Service Documents app (F6007). You can either use the **Revalue** function to recalculate specific items or execute a periodic revenue recognition run. During the run, the app selects relevant objects based on the revenue recognition key, compares newly calculated values with previously posted ones, generates adjustment postings for detected discrepancies, and supports simulation runs to preview results before execution.

Run Settlement – Actual (F4568)

Application Type: Transactional

This app allows you to settle actual costs from a single project, work breakdown structure (WBS) element (with or without hierarchy), order, or service document to specified receivers. This app is typically used during period-end closing to test or execute settlements. The app is useful when new costs are added to a previously settled project or when errors occur in mass processing, enabling you to test corrections before applying them to actual data. In addition, to display data, you can filter data with various parameters such as cost object type, WBS element, project, and company code. The entry screen also features **Run**, **Test Run**, **Reverse**, and **Test Reverse** buttons to facilitate various settlement processing options. For a settlement run, the app offers various parameters such as settlement period, posting date, posting period, asset value date, fiscal year, and value date. The app's features include **Test Run** and **Run** (update run); settlement reversal; analysis of results, including senders, receivers, journal entries, and error messages; and a summary of previous settlements along with the ledger-based settlement history. You also can navigate to the Display Settlement Documents app (F4597) via **Cost Object Details** or smart links to analyze current or past transaction details.

Transaction Codes

The app is similar to Transactions CJ88, VA88, KO88, KO8GH, and CJ8GH in SAP GUI.

Sales Orders – Actual (F1582A)

Application Type: Web Dynpro

This app allows you to report actual sales order data for a ledger fiscal year. With this app, you can refine search criteria using variables such as posting period, functional area, company code, segment, and general ledger account hierarchy. In addition, the app displays query results in a grid format, and you can export data to Microsoft Excel. The app's **Navigation Panel** allows you to add dimensions from the **Available Fields** section into rows and columns. The app also displays both real and statistical postings, distinguished via the **IsStatistical** fields, such as **IsStatistical WBS Elem** and **IsStatistical SO Item**. The value of **x** in these fields indicates statistical postings, and **#** represents real postings. The app displays amounts in transaction, company code, and global currencies. Thus, the app offers analytical insights into actual sales order data.

Schedule Division Accounting Jobs (F5288)

Application Type: Transactional

This app allows you to automate division-level accounting tasks by using predefined templates, such as **Universal Allocation Run – Profit Centers**. The app provides scheduling options for executing periodic activities efficiently. Thus, the app allows you to create, modify, and execute scheduled jobs that are relevant to cost accounting.

Let's compare the Schedule Overhead Accounting Jobs app (F3767) and the Schedule Division Accounting Jobs app (F5288) apps:

- **Schedule Overhead Accounting Jobs**
  This app facilitates scheduling jobs for cost center–related activities. It's available from SAP S/4HANA 2022 and beyond.

- **Schedule Division Accounting Jobs**
  This app focuses on profit center processes. It's available from SAP S/4HANA 2021 and beyond.

Schedule Inventory Accounting Jobs (F2292)

Application Type: Transactional

This app allows you to run background jobs for inventory accounting applications, including inventory value adjustments, balance sheet valuation for the lowest value determination, and cost estimate marking and release. Jobs are scheduled using predefined templates, enabling efficient execution of periodic inventory-related activities. A few predelivered templates are available—such as **Adjust Balance Sheet Accounts – Delta Posting**, which is used to adjust for the lower of either the cost or market valuation; **Adjust Balance Sheet Accounts – Price Change**, which is used to adjust the price based on valuation alternatives; and **Adjust Inventory Values**, which is used to fix inconsistencies in inventory. The selection screen includes scheduling options such as **Start Immediately** or **Start Date** and **Start Time** options. In addition, the app provides various parameters for selection such as company code, valuation area, material, and price control.

Tip

In general, you can use this app for recurring scheduling of key inventory tasks, such as balance sheet adjustments and inventory value corrections.

Transaction Code

The corresponding transaction in SAP GUI is Transaction MR23, which can be used for adjusting inventory valuation manually.

Schedule Overhead Accounting Jobs (F3767)

Application Type: Transactional

With this app, cost accountants can process large volumes of overhead accounting data using background scheduling jobs. The app allows you to schedule overhead allocation, cost rate calculation, and settlement jobs, along with other overhead accounting activities, via predefined job templates such as **Actual Cost Rate Calculation**, **Actual Overhead Calculation: Orders**, **Actual Settlement: Orders**, and **Universal Allocation Run – Cost Centers**. Thus, the app helps optimize performance while reducing loading times. In addition, you can define start dates and recurrence patterns, set dynamic valu-

ation key dates, and configure output controls for tax adjustments like nondeductible input tax and cross-border transactions as part of scheduling tax jobs. You can search for jobs using filter criteria such as the status, from, and to fields. The app's posting parameters allow specifying objects such as cost centers, projects, or work breakdown structure (WBS) elements, along with execution types such as test, actual, or reversal runs. Once the execution is completed, the app displays results in a grid format with key information such as status, log, results, and steps. Furthermore, you can customize output lists to control VAT and input tax line-item displays. To support parallel accounting, the app processes overhead calculations, allocations, and settlements per defined ledger, with job templates supporting multiledger execution. The app also supports parallel processing when handling large volumes of overhead calculations, allocations, or settlements.

Tip

By default, the app processes all intercompany posting transactions. However, you can configure restrictions by either blocking all intercompany postings or limiting postings to specific sender-receiver company code combinations using the **Block Intercompany Postings** Customizing activity under **Controlling · Cost Center Accounting · Actual Postings · Intercompany Postings**.

Schedule Product Costing Jobs (F3683)

Application Type: Transactional

This app allows you to automate and manage event-based production cost postings in product costing. The app provides predefined job templates to handle large volumes of orders, thereby reducing the manual workload. The predelivered job templates to choose from are listed in the **Template Selection** option for both the event-based production costing (scope item 3FO) and period-end production costing (scope item BEI) scenarios. Relevant templates for scope item 3FO include postprocess event-based posting for product costing, postprocess event-based posting for overhead costing, and event-based posting: work in process (WIP) quantity posting for actual costing revaluation. For scope item BEI, available templates include actual overhead calculation for production or process orders and product cost collectors, preliminary settlement for coproducts, WIP calculation, variance calculation by both lot and period, and actual settlement for production or process orders and product cost collectors. In addition, the app offers exception handling capabilities, including event-based posting corrections, WIP quantity postings, overhead calculations, and variance adjustments.

Schedule Sales Accounting Jobs (F1483)

Application Type: Transactional

This app allows you to schedule background jobs for overhead cost allocation to profitability segments based on reference values, percentages, or fixed amounts. You can schedule actual overhead allocations, generate controlling documents for a selected period, define execution timing to either start immediately or start at a selected date and time, test the job before final execution, and monitor its status with log details for error tracking. The app's job templates provide predefined configurations that streamline the scheduling of background jobs for sales accounting processes.

Tip

When scheduling allocation runs for allocation cycles created within the margin analysis allocation context, SAP recommends limiting cycles in a single job to improve performance. Consult SAP Note 3549133 to learn more.

Service Actuals (FIS_FPM_OVP_IPSRVO4)

Application Type: Web Dynpro

This Web Dynpro–based app provides a comprehensive overview of actual service-related costs, such as repairs and maintenance in product costing. The app's features include searching for service transactions using parameters such as company code, fiscal year, service document, general ledger account, service document type, service document, and service document item; displaying results for further analysis; exporting data to Microsoft Excel; and navigating individual transaction details. Thus, the app allows you to monitor service order actual expenses for detailed analysis and reconciliation.

Service – Plan/Actuals (W0199)

Application Type: Web Dynpro

This app allows cost accountants to view planned and actual charges on service orders in margin analysis. The app shows details for service orders, including service order items, and for plant maintenance orders. With this app, you can search with various parameters such as company code, fiscal year, ledger, plan category, fiscal year, general ledger account hierarchy, general ledger account, service document type, service document, and service document item. You can analyze deviations if plan amounts differ from actuals for a specific period. Actual data is determined by the selected ledger in the filters reading from table ACDOCA, while planned data is based on standard or virtual plan categories read from plan data table ACDOCP. Planned costs are linked to plant maintenance orders, whereas planned revenues are associated with service orders with advanced execution. For plant maintenance orders, ongoing costs are calculated within existing plan categories for maintenance orders, while for service orders with advanced execution, revenues are recorded within plan categories for service documents. In addition, for service contracts with hierarchy and plant maintenance orders with subsequent documents, ongoing service contract revenues are calculated and maintained within plan categories for service documents. The app also allows you to navigate to related apps by choosing specific line items.

Transaction Codes

The closest transactions in SAP GUI are Transactions IW39, S_ALR_87014004, and S_ALR_87013019.

Set Financial Plan Data to Zero (F4850)

Application Type: Transactional

This app allows you to set financial plan values to zero without deleting them from the database, using delta records to correct outdated data while preserving accuracy. In addition, you can refine your selections by filtering based on plan category, ledger, fiscal period, and company code. The app's features include setting plan values to zero, undoing changes via the Delete Financial Plan Data – with Timestamp app (F4074), reviewing user actions through the **Show Log** function, and analyzing affected data using the **Analyze Financial Plan Data** function. Furthermore, you can navigate to the Applications Log app (F1487) via **Related Apps** to view actions performed by all users and create custom reports using the **Open Financial Plan Data** function. Thus, the app ensures data consistency by maintaining financial plan data via cleaning up incorrect or obsolete entries.

**Set Report Relevancy –
Set-Based Hierarchies (F5295)**

Application Type: Transactional

In SAP, *set-based hierarchies* are structured groupings of master data, such as cost centers or

profit centers, that are organized into hierarchical levels for reporting and allocation purposes. This app allows you to create new set-based hierarchies, such as cost center and profit center groups based on report-relevant requirements. In addition, you can display or update set-based hierarchies using various search criteria such as set class, organizational unit, and set name, and you can include subnodes. The app displays set report relevancy details in a tabular format with fields including set class, organizational unit, set name, description, report relevant, and auto replicate. In addition, to ensure these hierarchies are available for reporting, you must trigger replication whenever they are created or updated. Without replication, changes will not be incorporated into impacted hierarchies. When adding a new root node item, it is essential to mark it with the **Report Relevant** option to enable replication and reporting usage. The app's functionalities include marking new hierarchy sets as **Report Relevant**, enabling the **Auto Replicate** feature, and setting report relevance at the subnode level.

Tip

The Set Report Relevancy – Set-Based Hierarchies app (F5295) is used for making changes to hierarchies to be report relevant and appear in reports, while the Replicate Runtime Hierarchies app (F1478) ensures that changes to those hierarchies are properly updated and available for reporting.

Transaction Code

The corresponding transaction in SAP GUI is Transaction HRY_REPRELEV.

Statistical Key Figures – Actuals (W0075)

Application Type: Transactional

This app provides an overview of actual quantities posted on statistical key figures (SKFs) for a specified cost center and fiscal period. For the app to display actuals, you must first create SKFs using the Manage Statistical Key Figures (Version 2) app (F1603A) and ensure quantities are posted. The app allows you to filter search criteria, display results in grid or graphical formats using the **Graphical Display** tab, and export data to Microsoft Excel. For example, you can search by various parameters such as fiscal year, SKFs, cost center, and fiscal period. Graphical displays require a unique quantity unit per SKF. In addition, you can adjust the query results grid by modifying field placements, opting to display quantities either in a single column or by separating fixed and sum values into distinct columns. By default, SKFs appear in one column, but additional columns can be added via the **Navigation Panel** by filtering for key figures under columns. The app uses CDS view C_StatisticalKeyFigureItem to display actuals.

> **Tip**
>
> For the app to function correctly, you must set up user parameter **CAC** for the controlling area to derive the fiscal year variant correctly.

> **Transaction Codes**
>
> The corresponding transaction in SAP GUI is Transaction KB31N.

Upload Material Inventory Prices (F4006)

Application Type: Transactional

This app allows for mass updates to material inventory prices by allowing you to download predelivered templates or export existing prices for a specified company code, its plants, and its currency type. You can modify the data and upload updated prices. The app's features include currency conversion, previewing changes before posting, comparing new and current values side by side, and accessing price-change documents directly from the posting log. Upon uploading

material inventory prices, the app displays data grouped in three tabs: **All Items**, **Errors**, and **Warnings**. Price changes for a material are displayed with its current **Status**, and the app enables automatic currency conversion.

> **Tip**
>
> The app is limited to 500 entries for material inventory price uploads. Alternatively, you can use SAP GUI Transaction MR21 or use BAPI BAPI_MATVAL_PRICE_CHANGE, as described in SAP Note 278356.

> **Transaction Code**
>
> The equivalent SAP GUI transaction is Transaction MR21.

Where-Used List – Activity Types (F3867)

Application Type: Transactional

This app allows you to find references to a specific activity type across various objects, including cost centers, receiving orders, work breakdown structure (WBS) elements, activity type groups, and production-related elements such as routings and work centers. The app finds references in both master data and transactional data. In addition, you can specify one or more activity types and view all associated objects, with direct navigation to relevant apps for further exploration. For example, you can search for references to an activity type using parameters such as activity type, used in, and valid on. The app provides comprehensive used-in results, covering cost rate maintenance, line items, receiving orders, production routings, work centers, and dynamic item processor (DIP) profiles, ensuring detailed visibility and efficient tracking of activity type references. For example, receiving cost centers (in line items) include results such as activity type, activity name, cost center, cost center name, journal entry, posting date, quantity, amount in transaction currency, and so on. Thus, the app helps you identify all instances in which a specific activity type is referenced across various business objects for tracking and analysis purposes.

Where-Used List – Cost Centers (F3549)

Application Type: Transactional

This app allows you to identify all master data and transactional data references to specific cost centers used across various objects, including activity types, cost allocation cycles, segments, budget-related cost centers, groups, and hierarchies. Using this app, you can search for one or multiple cost centers and view all associated objects, with the ability to preview relevant details and navigate to other relevant apps for further exploration. The app allows you to search with various filter criteria such as cost center, used to, and valid on. For example, for cost allocation segments, you can display information with fields such as cost center, cycle, segment ID, segment name, allocation type, cost center role, sender rule, and receiver rule. Thus, the app helps you to analyze dependencies to understand cost center usage.

Where-Used List – Internal Orders (F5402)

Application Type: Transactional

This app allows you to locate all references to a specific internal order across various objects, including cost allocation cycles, cost allocation segments, fixed assets, and groups. The app finds references in both master data and transactional data. You can find references with various search criteria, such as controlling area, internal order, used in, and valid in. The app provides comprehensive used-in results—for example, covering cost allocation cycles, cost allocation segments, fixed assets, and groups areas for detailed visibility and efficient tracking. Thus, the app allows you to identify

instances in which an internal order appears, preview related objects, and seamlessly navigate to relevant apps for further analysis.

Tip

When navigating from other apps (inbound), such as Manage Internal Orders (Version 2) (F3893), via the smart link in an internal order, only specific fields—such as internal order, controlling area, and used in—are transferred to the filters. A smart link displayed in an internal order field shows a hyperlink that triggers a popup to see a list of SAP Fiori apps and fact sheets. In other words, each order number appears as a hyperlink. Clicking the hyperlink opens a popover that lists all related transactional apps and fact sheets for the selected internal order. On the other hand, when navigating to other apps (outbound) using a smart link in an internal order, such as Manage Internal Orders (Version 2) (F3893), the app transfers the internal order, controlling area, and company code. Consult SAP Note 3122767 for further details.

Where-Used List – Profit Centers (F3751)

Application Type: Transactional

This app allows you to find business objects and transactional data associated with one or more profit centers, enabling searches across allocation cycles, cost centers, profit center groups, global accounting hierarchies, and sales order items. You can refine the searches using profit center assignments, object categories, and reference documents to obtain precise results. For instance, you can search for the usage of profit centers using various search criteria such as profit center, used in, valid on, and segment. Upon search completion, the app displays information in columns such as object type, numbers, and matches. In addition, you can seamlessly navigate to related applications for deeper analysis and necessary adjustments.

Transaction Code

The corresponding transaction in SAP GUI is Transaction 1KE4 in profit center accounting.

Where-Used List – Statistical Key Figures (F4078)

Application Type: Transactional

Statistical key figures (SKFs) are nonmonetary values used in controlling to allocate costs or track the measurable characteristics of a cost center. This app allows you to identify all occurrences of a specific SKF across SKF groups, cost allocation cycles, segments, and other relevant objects. The app finds references in both master data and transactional data. In addition, you can specify one or more SKFs to view associated objects and explore detailed information. For example, you can search with parameters such as SKF, used in, and valid on. The app displays used-in details for SKFs with the object type, number of objects, and matched categories. For cost allocation cycles, the app displays line items with the SKF, cycle, cycle description, allocation type, ledger, start date, and end date. Furthermore, the interactive links in the app enable navigation to related apps for further analysis.

Transaction Code

The corresponding transaction in SAP GUI is Transaction KK03.

Work in Process – Event-Based (F3498)

Application Type: Analytical

Work in process (WIP) represents the difference between the debit and credit of a production

order, process order, or product cost collector that has not been fully delivered in product costing. In SAP, WIP is dynamically updated as a result of production events such as goods issues, activity confirmations, and goods receipts, improving cost accuracy and reducing manual adjustments. This app provides real-time visibility into WIP and production cost data through event-based, real-time postings without relying on period-end processing for the production orders, process orders, or product cost collectors at both total and individual order levels on a specified key date. This app provides WIP data for these orders on any specified date, with filtering options such as plant and product group. In addition, the app provides insights into recent trends, including WIP changes compared to the previous month and over recent days. The app also displays the planned output quantity, completed quantity to date, and estimated completion cost per order. In addition, it generates WIP quantity documents for valuating WIP at actual prices in actual costing. With this app, you can view various dimensions such as WIP by plant, by product group, and by profit center, both in chart and table views. In addition, the app supports parallel ledger tracking, linking WIP calculations directly to production events. Thus, the app allows variance analysis for early cost deviation detection and integrates seamlessly with financial accounting for compliance and reporting.

Tip

To display WIP data, you must enter the **Key Date** value as a period end date, as WIP postings are generated and settled at period close. The **Key Date** value is the cutoff date to pull all WIP, overhead, and variance posting data.

Consult SAP Note 3316687 for troubleshooting tips for the product cost by order scenario.

Chapter 3
Sales and Distribution

In this chapter, we'll detail all the most useful and relevant SAP Fiori apps available in the sales and distribution space. These apps cover all processes related to presales, such as inquiries and quotations, through the sales cycle, including various incarnations of sales documents (such as sales orders, credit memo requests, debit memo requests, sales orders without charge, customer returns, and contracts), business partners, pricing, and other master data. The distribution phase also will be covered, including the outbound delivery and goods issue process. Where possible, the equivalent or comparable SAP GUI transaction code has been noted. As with other chapters, apps are ordered alphabetically for ease of reference.

Aggregated Business Process Activities (F1810)

Application Type: Analytical

This useful app allows order-to-cash (OTC) managers to get an aggregated overview of the sales order processes for their organization. Key figures are provided for determining the health of the OTC processes, and these can be viewed in graphical format. In addition, a useful feature of the app is the ability to display the processes in a process flow chart, thus allowing you easily to identify issues with a specific given process step.

Analyze and Resolve Blocked Documents – Trade Compliance (F5445)

Application Type: Transactional

This app is part of SAP S/4HANA for international trade. Using this functionality, documents such as sales orders and deliveries can be unblocked due to missing legal regulations, or embargo blocks. The home screen for the app allows you to search for documents by filters such as block type, company code, and sales

document category. The results can be seen in graphical format, for which the type of graph can be changed, as well as in tabular format.

> **Tip**
>
> There are three methods for resolving the trade compliance blocks. Additional licenses can be added, which are assigned automatically to the documents or can be assigned manually to release the documents; additional classifications can be added manually in the document or added to the product in order trigger resolution of the block; or the block can be manually released in the case of embargo blocks.

Analyze Delivery Logs (F0870A)

Application Type: Transactional

Each collective delivery run generates a log in SAP, which details the shipping point, document numbers and items, warehouse, gross weight, and volume. In addition, any error message and suggested remedial activities are recorded in the log too. All this information is available for interrogation in this app. Filter options in the app include log number, created by, created on, shipping point, and status. Clicking one of the logs from the list opens a more granular screen where you can see details of any error messages including message type, message class, message number, and message text.

Navigation to the Create Outbound Deliveries – From Sales Orders app (FO869A) is supported.

Analyze Delivery Performance – Shipped as Planned (F2878A)

Application Type: Analytical

Delivery performance can be monitored using this app. If planned goods issue dates are not observed, you can drill down into the relevant documents, such as sales orders, to analyze the issues. The performance is visualized in the app in graphical format and can be sorted and personalized in the usual manner by using data filters for the retrieval of the data, and settings for the display of the graph and the associated table below.

Navigation to outbound deliveries, document flows, and relevant master data objects is supported, as well as SAP Fiori apps like Customer Master FactSheets (FO046A), Outbound Delivery (S/4HANA) (FO233A), and Product (F2773).

Analyze Detailed Statement Royalties (W0171)

Application Type: Analytical

Condition contract settlement documents for royalty purposes can be viewed using this app. The data is retrieved from the associated sales documents. The selection criteria can be used to determine which data is retrieved. The **Navigation Panel** on the left side of the app shows important information that can be pulled into the display. The display itself can be viewed in raw data format or in graphical format. Navigation to the relevant condition contract is

supported via the Condition Contract (Version 2) app (F3594A).

Analyze Detailed Statement Sales Commission (W0147)

Application Type: Analytical

Condition contracts exist for the payment of sales commissions in an organization, and this app can be used to view the itemized volume for these condition contracts. The selection criteria can be used to determine which data is retrieved. The **Navigation Panel** on the left side of the app shows important information that can be pulled into the display. The app uses the standard CDS view C_DetStmntCommsnQry to retrieve the data. The display itself can be viewed in raw data format or in graphical format. Navigation to the relevant condition contract is supported.

Analyze Detailed Statement Sales Rebate (W0149)

Application Type: Analytical

This app provides a detailed breakdown of business volume for selected contract settlement documents. It focuses on settlement document billing and sales orders as the basis for calculating business volume, utilizing the CDS view C_DetStmntSlsRbteQr. The **Navigation Panel** on the left side of the app shows important information that can be pulled into the display. The display itself can be viewed in raw data format or in graphical format. Navigation to the relevant condition contract is supported.

Application Log for Classification Master Data (F4888)

Application Type: Transactional

This app is part of SAP S/4HANA for international trade. External data providers can supply

SAP clients with information related to commodity code classification data. This information can be displayed using this app. When you view the data in the app, it is classed as inactive data. The data can be activated for use in the product classification of commodity codes. Before activation, the data is run through validation and consistency checks.

Application Log for Product Allocation (F6296)

Application Type: Transactional

Using this app, you can view application logs for product allocation in SAP. Filters exist to retrieve the logs, including severity, date range, category, subcategory, and external reference. From within the app, it is easy to be able to see if any errors have occurred in the product allocation process.

The app can be called directly from the Manage Product Allocation Planning Data app (F2121).

Application Logs – Condition Contract (F7087)

Application Type: Transactional

This app allows efficient monitoring and troubleshooting of condition contract application logs, helping you quickly identify and assess potential issues. Key features of the app enable you to view logs related to contract maintenance and settlement; filter logs by severity, time period, and category; search log texts; and access detailed log information with severity-based filtering.

Navigation to the relevant condition contract is supported.

Application Logs – Settlement Management Document (F7088)

Application Type: Transactional

This app allows efficient monitoring and troubleshooting of settlement management application logs, helping you quickly identify and assess potential issues. Key features of the app enable you to view logs related to contract maintenance and settlement; filter logs by severity, time period, and category; search log texts; and access detailed log information with severity-based filtering.

Assign Product to Product Allocation (F2120)

Application Type: Transactional

Product allocation objects can be created using the Configure Product Allocation app (F2119), and then product allocation sequences can be created using the Manage Product Allocation Sequences app (F2474). Once this is complete, you can use this app to assign plants and materials to the product allocation sequences. These assigned combinations of plants and materials then become relevant for availability checks in sales documents and stock transport orders. You also can assign validity dates for the assignment of plant and materials to control fluctuations in supply and demand.

Navigation to the Monitor Allocation Sequence Changes app (F7143) is supported.

Assign Sold-to Parties (F3894)

Application Type: Transactional

This app relates to the assignment of sold-to parties in sales automated sales processing. Within the app, you can assign a sold-to party to an Electronic Data Interchange (EDI) order or schedule that comes in by allocating that sold-to party to a given supplier number. Based on these assignments, the system can automatically determine the correct sold-to party. It is possible to create, edit, display, copy, and delete the assignments using this app. Another key feature of the app is that you can display changes to assignments.

Transaction Code

A similar transaction in SAP GUI is Transaction VOE3, which allows you to add partner functions for inbound sales IDocs.

Billing Document (F1901)

Application Type: Fact Sheet

This app shows an overview of the billing document in question. It can be used during the creation of billing documents as well as editing and displaying billing documents. From within the app, you also can access the process flow for the document. From within the process flow, you can navigate into each other document (e.g., a delivery or a sales order).

The app is accessed through transactional apps such as Create Billing Documents (F0798) or Manage Billing Documents (F0797).

Tip

You can preview the output documentation of a billing document by using the **Preview** option at the bottom of the screen. Similarly, there are options to **Edit**, **Cancel**, or **Post** the billing document from the same area.

Billing Document Request (F2337)

Application Type: Fact sheet

Billing document requests (BDRs) are SAP objects that contain an itemized list of data for

the purposes of billing, drawn from a single or multiple sources. These BDRs are never for credit items; they're only related to debit items. This app shows an overview of the BDR, including all the individual items contained therein, as well as pricing, finance, and business partner data.

From within the app, you can navigate to the individual items as well as the business partners (sold-to, bill-to, and payer).

Business Process Activities (F1609)

Application Type: Analytical

This useful app gives order-to-cash (OTC) managers a good overview of the sales order processes for their organization. Key figures are provided for determining the health of the OTC processes, and these can be viewed in graphical format. In addition, a useful feature of the app is the ability to display the processes in a process flow chart, thus allowing you to easily identify issues with a specific process step. Useful activities within the app include tracking by validity date against delivery or billing blocked orders; monitoring the number of rejected sales orders; monitoring lead times against orders such as order to delivery time, order to goods issue time, order to billing time, etc.; and monitoring changes that have been made to business-critical fields within a sales order. It is possible to choose the chart type (e.g., line chart, bar chart, donut chart, pie chart) that is displayed against the KPIs.

Classify Products – Commodity Codes (F2151)

Application Type: Transactional

As part of the SAP S/4HANA for international trade functionality, this app can be used to classify materials to a specific commodity code, using a specific numbering scheme. From within the search features of the app, you can find unclassified products based on product number, product description, material type, material group, and division. The app presents a list of products based upon the selection criteria used, and you can select single or multiple entries to classify them.

Commodity code classifications can be set using validity dates, and remarks can be added during the process.

Tip

Classification of commodity codes in SAP ERP 6.0 is different from classification in SAP S/4HANA. SAP ERP 6.0 allows you to adjust classification of commodity codes directly in the material master data record using Transaction MM02. SAP S/4HANA only allows this through the SAP Fiori classification process listed here. However, in an environment where SAP S/4HANA and SAP ERP 6.0 exist side by side, IDoc processes can be used for the distribution of SAP S/4HANA commodity codes to SAP ERP 6.0 materials via IDoc type MATMAS.

Classify Products – Customs Tariff Numbers (F3146)

Application Type: Transactional

As part of SAP S/4HANA for international trade, this app can be used to classify materials to a specific custom tariff number, using a specific numbering scheme. From within the search features of the app, you can find unclassified products based on product number, product description, material type, material group, and division. The app presents a list of products based upon the selection criteria used, and you can select single or multiple entries to classify them. Custom tariff number classifications can be set using validity dates, and remarks can be added during the process.

Classify Products – Intrastat Service Codes (F2156)

Application Type: Transactional

As part of SAP S/4HANA for international trade, this app can be used to classify materials to a specific Intrastat service code, using a specific numbering scheme. From within the search features of the app, you can find unclassified products based on product number, product description, material type, material group, and division. The app presents a list of products based upon the selection criteria used, and you can select single or multiple entries to classify them. Intrastat service code classifications can be set using validity dates, and remarks can be added during the process.

Classify Products – Legal Control (F2390)

Application Type: Transactional

As part of SAP S/4HANA for international trade, this app can be used to classify materials to a specific legal control process, using a specific numbering scheme. From within the search features of the app, you can find unclassified products based on specific legal control regulations. The app presents a list of products based upon the selection criteria used, and you can select single or multiple entries to classify them to a given control class and control group. Legal control classifications can be set using validity dates, and remarks can be added during the process.

Condition Contract (Version 2) (F3594A)

Application Type: Fact Sheet

This app allows you to analyze in detail all the different types of condition contracts, such as sales and supplier rebate agreements, internal commission agreements, and royalty agreements. All data in the condition contract is displayed in an easy-to-use format, including the relevant business partner, contract value, open accruals amount, activation status, settlement lock status, payment terms, and settlement dates.

Navigation to the condition contract is supported.

Configure Alternative Control (F2698)

Application Type: Transactional

The SAP S/4HANA available-to-promise (ATP) functionality allows you to set controls around sources of supply for demand in sales orders. This is done by activating alternative-based confirmation (ABC). Alternatives are defined in this app through the allocation of value combinations of fields, with assignment to a substitution strategy. Values are assigned to these field combinations that can then trigger a substitution strategy in the ATP process, if required.

Navigation to the Configure Substitution Strategy (F2699) and Manage Characteristic Combinations (F5303) apps is supported.

Configure BOP Segment (F2158)

Application Type: Transactional

As part of the suite of apps related to SAP S/4HANA for advanced available-to-promise (ATP), this app allows you to configure the selection conditions and how to run backorder processing (BOP) for the advanced ATP check. From the app, you can define a set of selection and exclusion conditions in order to automate the rescheduling and redistribution of allocated stock to sales documents, in line with the overall business strategy.

The next step in the process after the BOP segment is configured is to configure the BOP variant and execute the rescheduling according to the rules that have been defined. As such, navigation to the SAP Fiori apps Configure BOP Variant (F2160) and Schedule BOP Run (F2665) is supported. Similarly, you can monitor the BOP run in the related app, Monitor BOP Run (F2159).

Tip

ATP has existed in SAP since the beginning, but advanced ATP is a much newer iteration of the functionality, with extended features. The functionality of advanced ATP relies upon a hierarchical prioritization of customers to determine how available inventory is allocated. This prioritization process is far more flexible and granular than standard ATP. It should be noted, however, that organizations that wish to use advanced ATP must pay an additional software license to SAP. In addition, there are security objects available for advanced ATP that must be assigned to relevant users, in addition to a whole suite of SAP Fiori apps. It is also important to note that advanced ATP is only available as a function by using SAP Fiori; it is not available in SAP GUI.

Configure BOP Variant (F2160)

Application Type: Transactional

As part of the suite of SAP Fiori apps related to advanced available-to-promise (ATP), this app allows you to define the variant for backorder processing (BOP) to automate the rescheduling and redistribution of allocated stock to sales documents, in line with the overall business strategy. Options in the app allow you to display, change, edit and create new variants, assign confirmation strategies, define requirement categories, as well as simulate the BOP run.

Tip

Exceptions can occur during the BOP run. These can be handled neatly by configuring a fallback variant to sweep up the exceptions and process them.

3

Configure Custom BOP Sorting (F2983)

Application Type: Transactional

As part of the suite of SAP Fiori apps related to advanced available-to-promise (ATP), within this app, you can define logic for the requirement prioritization that is handled in the Configure BOP Segment app (F2158). The app supports create, display, edit, and copy functions for the sort sequences. Sort sequences can be searched for by using attributes such as sequence name, sort attribute, created by, and changed by.

Tip

Use this app to define sort orders for back-order processing (BOP) that cannot be defined using simple alphanumeric logic. For example, if it is important that the customer order for prioritization is first customer number 321, then 123, then 322, this sort order can be manually added in this app.

Configure Order Fulfillment Responsibilities (F2246)

Application Type: Transactional

This app is where you can assign responsibility for specific groups of materials to specific users. These users can control the order fulfillment for these materials using the Release for Delivery app (F1786), which can be navigated to directly from within this app. The order fulfillment responsibilities need to be assigned to at least one specific criterion, such as plant or material group, for which materials are assigned. The responsibilities can be assigned for a given period using validity dates.

Configure Product Allocation (F2119)

Application Type: Transactional

Product allocation schemas can be viewed, edited, and created using this app. These objects can be created for sales documents and stock transport orders in order to ensure effective and timely allocation of products to sales orders and stock transport orders. Product allocation schemas can be restricted to a period using validity dates.

Configure Sourcing Profile (F7754)

Application Type: Transactional

This app is used to configure third-party order processing. Standard advanced available-to-promise (ATP) runs define the allocation of stock to sales documents based on specific rules. In the context of third-party order processing, a follow-up strategy can be defined in order to cater for any remaining quantity that has not been confirmed. This follow-up strategy requires a sourcing profile and can be relevant for either direct shipments (drop shipments) where the supplier delivers directly to the customer, or bought-in processing (cross-dock) where the supplier delivers to an internal plant and then the organization delivers to the end customer.

The app allows the strategy to be defined to use alternative-based confirmation (ABC) and provides a business add-in (BAdI) to influence the result of the source determination.

Configure Substitution Strategy (F2699)

Application Type: Transactional

Sales document requirements can't always be confirmed as requested. In this instance, a substitution strategy can be defined for use in consideration of fulfillment of the requirements.

These substitution strategies are assigned to alternative controls in the Configure Alternative Control app (F2698). Controls can be displayed in a list in the app by filtering on fields such as strategy name, substitution item type, created by, and changed by.

Create Billing Documents (F0798)

Application Type: Transactional

This app is exclusively used for the creation of billing documents (invoices and credit memos) directly from sales and distribution document items on the billing due list, which are represented in tabular format from within the app. It is possible to filter the data displayed in the billing due list by using the standard filter bar. It is also possible to create temporary billing documents for review before finalization.

Tip

The **Billing Settings** menu provides useful features when processing billing documents through this app, such as the ability to set the billing date and billing type before billing, create separate billing documents for each item, automatically (or not) post the billing documents to accounting, and display the billing documents after creation.

Transaction Code

The SAP GUI equivalent transaction is Create Billing Documents – VF01, Maintain Billing Due List, accessed via Transaction VF04.

Create Outbound Deliveries – From Sales Orders (F0869A)

Application Type: Transactional

The app provides real-time visibility into delivery operations and logging. Features include the

creation of outbound delivery runs from sales order lists, the ability to display comprehensive logs for sales orders and deliveries, and the ability to track system messages for delivery creation, including successful processes and potential errors. Standard SAP Fiori filters are available for restricting data retrieval such as ship-to party, shipping point, planned creation date, priority, and sales document.

You can navigate directly to the Schedule Delivery Creation app (F2228) to schedule a job to automate the mass creation of deliveries.

Transaction Code

The SAP GUI equivalent app is Transaction VL01N (Create Outbound Deliveries).

Create Preliminary Billing Documents (F2876)

Application Type: Transactional

Sometimes it may be necessary to discuss the contents of a billing document, such as pricing, quantities, and discounts, with clients before finalizing the agreed document. This can be carried out using this app. The billing due list can be read, and preliminary billing documents can be created for this purpose. Note that preliminary credit notes are not able to be created.

Any changes that need to be carried out can be made using Transaction VFP2 (Change Preliminary Billing Documents).

Create Sales Orders – Automatic Extraction (F4920)

Application Type: Transactional

This app can be used to simulate and create sales orders based upon an external purchase order file in PDF format, as unstructured data. The process begins with the upload of a purchase order file (English and German languages are supported). A company code must be referenced in the app before you can upload the file. It is possible to customize a duplication check in the Implementation Guide (IMG) to stop the same file being uploaded twice. Next, assuming that a sales organization is assigned correctly to the company code, and the correct header fields (sales order type, sales area, sold-to party) and item fields (material, quantity, unit of measure) are specified, then a sales order simulation can be carried out, or the creation of a sales order. The data in the request can then be displayed, edited, or deleted from within the app. Finally, change and application log details can be viewed for each sales order request. The change log details the field and value changes, whereas the application log details the processing success or failure measures from the BAdI call.

Credit Memo Request (F1846)

Application Type: Fact Sheet

This app shows an overview of all details related to a credit memo request, in a tabular format. Header-level fields such as sales area, credit type, sold-to, and net value are immediately available, as are item-level fields. A graphical process flow for the document flow is also available. Any credit memo request approvals that are relevant can be seen in this view too (provided the approval workflow has been correctly configured in the system).

Supported drilldown navigation includes the Customer 360° View – Version 2 app (F2187A) and the Manage Credit Memo Requests app (F1989). This app is normally navigated to from the Manage Credit Memo Requests app (F1989).

Customer 360° View (Version 2) (F2187A)

Application Type: Transactional

This app shows a central hub where you can navigate to any number of different apps related to

a specific customer. Master data from the customer can be displayed in the app, such as sales areas. Furthermore, from within the app, a list of documents such as quotations, sales orders, sales contracts, customer returns, credit memo requests, debit memo requests, scheduling agreements, and billing documents, can all be listed. In addition, any attachments and fulfillment issues for the customer can be displayed. The app supports navigation to related apps for all these documents.

This app is the updated version of the predecessor app, Customer 360° View (F2187). The new version was released for SAP S/4HANA 2022 FPS01.

> **Tip**
>
> Use the SAP Fiori enterprise search functionality to search for a customer by name and then navigate directly to the Customer 360° View app to explore all documents and data for that customer.

Customer Master Fact Sheets (F0046A)

Application Type: Fact Sheet

You can use this app to show a full overview of the customer master data. This relates to all information related to name and address, contacts, bank accounts, sales area, and company code.

Navigation is supported to related master data records such as contact records and other related customers via SAP Fiori apps such as Outbound Delivery (S/4HANA) (F0233A), Returns Delivery (S/4HANA) (F0234A), Manage Outbound Deliveries (F0867A), Create Outbound Deliveries – from Sales Orders (F0869A), Analyze Delivery Logs (F0870A), and Analyze Delivery Performance – Shipped as Planned (F2878A).

Customer Return (F1815)

Application Type: Fact Sheet

This app shows an overview of all details related to a customer return. Key information for the return document is shown, including header and item details, business partners, and all related documents in the process flow. Navigation to related data such as business partner master data and other documents within the workflow is supported through drilldown capabilities.

This app is normally navigated to from the Manage Customer Returns and version 2 of the same app (F1708 and F4832), but navigation to this app is supported from various other apps too, including Customer 360° View (F2187 and F2187A).

Customer Returns – Return Rate (F4092)

Application Type: Analytical

Analysis of the returns rate for sales orders can be analyzed using this app, according to a list of standard KPIs. This data can be shown both in tabular and graphical formats, according to the selected settings. There are nine standard KPIs used within the app for analysis of returns rate. These KPIs are normally focused on two key dimensions: value and quantity. The KPIs include percentage of returns as a factor of incoming sales orders, amount due for delivery or billing, and number of sales orders referenced by customer returns.

Tip

By selecting the graphical format and selecting a date range, you can view the monthly rolling trend for returns against sales orders.

Customers Overview (F4645)

Application Type: Analytical

This app provides a centralized hub for all data regarding customers. The overview is made up of cards that represent data related to your selections in the filters area. Each card represents an SAP Fiori app, and the summary data shown on the card is dynamically updated according to the filters. The following cards are available:

- **Frequency of Sales Order**
  Shows the average number of sales orders per month for the last six months. Clicking the card will navigate you to the Manage Sales Orders app (F1873).

- **Incoming Sales Orders**
  Shows the average net value of sales orders per month for the last six months. Clicking the card will navigate you to the Manage Sales Orders app (F1873).

- **Open Quotations**
  Shows the number of quotations that are still open. Clicking the card will navigate you to the Manage Quotations – Version 2 app (F5630).

- **Rejected Quotations**
  Shows the number of partially or fully rejected quotations per month for the last six months. Clicking the card will navigate you to the Manage Sales Quotations – Version 2 app (F5630).

- **Open Sales Orders**
  Shows the number of sales orders that are still open. Clicking the card will navigate you to the Manage Sales Orders app (F1873).

- **Rejected Sales Orders**
  Shows the number of partially or fully rejected sales orders per month for the last six months. Clicking the card will navigate you to the Manage Sales Orders app (F1873).

- **Sales Volume by Sales Area**
  Shows the net sales volume from all sales areas for the customers selected for the current year. Clicking the card will navigate you

to the Sales Volume – Check Open Sales app (F2270).

- **Customer Returns**
  Shows the total number of items returned per month for the last six months. Clicking the card will navigate you to the Customer Returns – Return Rate app (F4092).

- **Sales Volume/Profit Margin**
  Shows the sales volume versus profit margin graphically for the last six quarters. Clicking the card will navigate you to the Sales Volume – Profit Margin app (F2271).

- **Delivery Performance**
  Shows the last six quarters' delivery performance in graphical format. The key figures available are Delivered as Requested (delivery by customer's requested delivery date) and Delivered as Committed (delivery by the committed delivery date).

- **Sales Contract Fulfilment**
  Shows the number of sales contracts that are not completed. Clicking the card will navigate you to the Manage Sales Contracts app (F1851).

- **Customer Contact**
  Shows the contacts for the customers selected in the filter bar. Clicking the card will navigate you to the Customer – 360° View – Version 2 app (F2187A).

- **Quick Actions**
  Shows a list of quick actions for navigating to transactional apps to create sales orders, quotations, and returns.

Debit Memo Request (F1848)

Application Type: Fact Sheet

This app shows an overview of all details related to a debit memo request in a tabular format. Header-level fields such as sales area, document type, sold-to, and net value are immediately available, as are item-level fields. A graphical process flow for the document flow is also available.

Supported drilldown navigation includes the Customer 360° View – Version 2 app (F2187A)

and the Manage Debit Memo Requests app (F1988). This app is normally navigated to from the Manage Debit Memo Requests (F1988) app.

Delivery Performance – Delivered as Requested (F2783)

Application Type: Analytical

Track and improve customer satisfaction by monitoring sales order delivery performance in real-time. This app shows delivery analytics and alerts you to delays or issues, allowing quick access to detailed sales records and customer data. The app measures delivery accuracy by comparing customers' requested delivery dates on their sales orders versus when items actually arrived as denoted by the actual delivery date on the sales orders. In the case of partial deliveries, the latest outbound delivery is considered.

> **Tip**
>
> Personalize your app by selecting your preferred chart type and using visual filters. You also can select the drilldown apps available to you from the options in the table by clicking the sales order number, the sold-to party number, or the material number.

Delivery Schedules – Demand Deviation (F5651)

Application Type: Transactional, Analytical

This app shows changes and fluctuations in demand for specific sales scheduling agreements. It provides an overview of the quantity changes that have taken place over a given period of time. By clicking **Compare** and navigating to the **Demand Deviation** tab, you can view the differences between the schedule line quantities in the newer delivery schedule compared to the older delivery schedule. It is also possible to apply a **Deviation Profile** in order to apply deviation levels (acceptable, warning, and alert) according to absolute thresholds. Deviation profiles are created in the Manage Deviation Profiles app (F4781).

Display Classified Products – International Trade (F3789)

Application Type: Transactional

As part of the SAP S/4HANA for international trade solution, this app allows you to display materials that have been classified using one of the international trade classifications of commodity code, custom tariff number, and legal control. All areas of the classification can be displayed in the app including full change history and remarks created during the processing of the classification.

Navigation to the fact sheet Product app (F2773) is supported.

Display Credit Memos (F7502)

Application Type: Transactional

This app allows logistics service providers to manage invoice reversals efficiently. Carriers can view credit memos submitted to shippers, create credit memos to reverse processed invoices, and post them to financial accounting. The system enforces a strict policy of one credit memo per invoice, streamlining documentation and financial processing. Standard SAP Fiori filters such as credit memo date, terms of payment, sold-to party, and credit memo number are available to restrict the amount of data retrieved.

Display International Trade Classification (F3935)

Application Type: Transactional

As part of the SAP S/4HANA for international trade solution, this app allows you to show all

valid international trade classifications for a given material. This covers the following classifications: commodity codes, custom tariff numbers, custom tariff numbers with end users, legal control, and Intrastat service codes.

Navigation to the Manage Product Master app (F1602) is supported.

> **Tip**
>
> It is possible to navigate directly to this SAP Fiori app by selecting a product from the Manage Product Master Data app (F1602) and choosing the **Open In** option, then selecting the appropriate function.

Display License Assignments – Trade Compliance (F4047)

Application Type: Transactional

As part of the SAP S/4HANA for international trade solution, this app allows you to display details related to specific legal control licenses. The licenses can be searched for using filters such as official license number, legal regulation, license type, and license owner. Once the license is identified, a drilldown option is available to show the sales documents that are assigned to that license.

Navigation to the Manage Licenses app (F2545) is supported.

Display Settlement Dates (F4964)

Application Type: Transactional

This app provides comprehensive management of settlement dates for condition contracts across various business scenarios. You can filter settlement dates using multiple criteria such as date, contract details, organizational units, and status. The app supports tracking settlement dates for customer rebates, supplier rebates, commissions, and royalties through predefined variants. Key capabilities include displaying detailed settlement information, accessing contract specifics, managing settlement documents, and integrating with worklists and financial processing. You can easily filter and view settlement dates, add them to existing worklists, create new worklists, and release settlement documents to finance.

navigation path to the various other "Import" apps as follows:

- Import Sales Orders (F4293)
- Import Sales Scheduling Agreements (F6191)
- Import Sales Quotations (F6381)
- Import Sales Orders Without Charge (F6382)
- Import Sales Contracts (F7066)

Import Sales Contracts (F7066)

Application Type: Transactional

Using this app allows you to create sales contracts in SAP from a Microsoft Excel file. From within the app, you can download the template by clicking the **Download Template** button. The template is opened in Microsoft Excel and shows a field list with all mandatory fields (marked with an asterisk) and all optional fields. The second worksheet is where the document data is entered.

Tip

Prerequisites for a successful data load are that the spreadsheet must be in XLSX format, it must contain data for all the required fields, it must contain at least one document, and it must be smaller than 10 MB and have no more than 1,000 rows of data.

Import Sales Documents (F6192)

Application Type: Transactional

This app allows you to create different types of sales documents in SAP from a Microsoft Excel file. From within the app, you can download the template by clicking the **Download Template** button. The template is opened in Microsoft Excel and shows a field list with all mandatory fields (marked with an asterisk) and all optional fields. The second worksheet is where the document data is entered. The app provides a quick

Import Sales Orders (F4293)

Application Type: Transactional

Using this app allows you to create sales orders in SAP from a Microsoft Excel file. From within the app, you can download the template by clicking the **Download Template** button. The template is opened in Microsoft Excel and shows a field list with all mandatory fields (marked with an asterisk) and all optional fields. The second worksheet is where the document data is entered.

Tip

Prerequisites for a successful data load are that the spreadsheet must be in XLSX format, it must contain data for all the required fields, it must contain at least one document, and it must be smaller than 10 MB and have no more than 1,000 rows of data.

Import Sales Orders Without Charge (F6382)

Application Type: Transactional

Using this app allows you to create sales orders without charge in SAP from a Microsoft Excel file. From within the app, you can download the template by clicking the **Download Template** button. The template is opened in Microsoft Excel and shows a field list with all mandatory fields (marked with an asterisk) and all optional fields. The second worksheet is where the document data is entered.

Tip

Prerequisites for a successful data load are that the spreadsheet must be in XLSX format, it must contain data for all the required fields, it must contain at least one document, and it must be smaller than 10 MB and have no more than 1,000 rows of data.

Tip

Prerequisites for a successful data load are that the spreadsheet must be in XLSX format, it must contain data for all the required fields, it must contain at least one document, and it must be smaller than 10 MB and have no more than 1,000 rows of data.

Import Sales Quotations (F6381)

Application Type: Transactional

Using this app allows you to create sales quotations in SAP from a Microsoft Excel file. From within the app, you can download the template by clicking the **Download Template** button. The template is opened in Microsoft Excel and shows a field list with all mandatory fields (marked with an asterisk) and all optional fields. The second worksheet is where the document data is entered.

Tip

Prerequisites for a successful data load are that the spreadsheet must be in XLSX format, it must contain data for all the required fields, it must contain at least one document, and it must be smaller than 10 MB and have no more than 1,000 rows of data.

Import Sales Scheduling Agreements (F6191)

Application Type: Transactional

Using this app allows you to create sales scheduling agreements in SAP from a Microsoft Excel file. From within the app, you can download the template by clicking the **Download Template** button. The template is opened in Microsoft Excel and shows a field list with all mandatory fields (marked with an asterisk) and all optional fields. The second worksheet is where the document data is entered.

Incoming Sales Orders – Flexible Analysis (F1249)

Application Type: Web Dynpro

This app is used to understand trends for incoming sales orders. The key figures used in the app revolve around the net value of the orders, the quantity, and the number of items that are relevant for billing or delivery, or that have been confirmed for shipping. As is normal with SAP Fiori apps, standard filters are available such as year, month, sales organization, product group, and sold-to party. As with most SAP Fiori apps, selection variants are also supported.

The app is contained as a card within the My Sales Overview app (F2200) and the Sales Management Overview app (F2601).

Invoice List (F2739)

Application Type: Fact Sheet

This app shows an overview of all details related to an invoice list. Header-level fields such as company code, sales organization, payer, billing date, payment terms, and posting status are immediately available, as are item-level fields. The item-level fields in an invoice list document refer to the individual billing documents that make up the invoice list. Some of the item-level fields include item number, billing document number, sold-to party, and net value. You can navigate directly to the billing document from the app by clicking the **Billing Document** number. This opens the Billing Document app (F1901).

This app is normally navigated to from the Manage Invoice Lists app (F2740).

List Incomplete Credit Memo Requests (F6234)

Application Type: Transactional

This app is a variant of the List Incomplete Sales Documents app (F2430). By using this specific app, the sales and distribution document category is prepopulated in the List Incomplete Sales Documents app to show only credit memo requests. The app allows you to search for incomplete credit memo requests according to various filters available and display them as a list. From the list, you can navigate directly to the sales document in question to add missing data.

Tip

Performance issues are common with this app if no filters are added. It is therefore recommended that filters are always added to limit the number of hits.

Transaction Code

The SAP GUI equivalent app is Transaction V.02 (Display Incomplete Orders).

List Incomplete Customer Returns (F6237)

Application Type: Transactional

This app is a variant of the List Incomplete Sales Documents app (F2430). By using this specific app, the sales and distribution document category is prepopulated in the List Incomplete Sales Documents app, to show only customer returns. The app allows you to search for incomplete customer returns according to various filters and display them as a list. From the list, you can

navigate directly to the sales document in question to add missing data.

Tip

Performance issues are common with this app if no filters are added. It's therefore recommended that filters are always added to limit the number of hits.

Transaction Code

The SAP GUI equivalent is Transaction V.02 (Display Incomplete Orders).

List Incomplete Debit Memo Requests (F6235)

Application Type: Transactional

This app is a variant of the List Incomplete Sales Documents app (F2430). By using this app, the sales and distribution document category is prepopulated in the List Incomplete Sales Documents app, to show only debit memo requests. The app allows you to search for incomplete debit memo requests according to various filters and display them as a list. From the list, you can navigate directly to the sales document in question to add missing data.

Tip

Performance issues are common with this app if no filters are added. It is therefore recommended that filters are always added to limit the number of hits.

Transaction Code

The SAP GUI equivalent is Transaction V.02 (Display Incomplete Orders).

List Incomplete Sales Contracts (F6236)

Application Type: Transactional

This app is a variant of the List Incomplete Sales Documents app (F2430). By using this specific app, the sales and distribution document category is prepopulated in the List Incomplete Sales Documents app, to show only sales contracts. The app allows you to search for incomplete sales contracts according to various filters and display them as a list. From the list, you can navigate directly to the sales document in question to add missing data.

Tip

Performance issues are common with this app if no filters are added. It is therefore recommended that filters are always added to limit the number of hits.

Transaction Code

The SAP GUI equivalent is Transaction V.06 (Display Incomplete Contracts).

List Incomplete Sales Documents (F2430)

Application Type: Transactional

This app is the central app that is called by all the other "List Incomplete" apps. From within this app, you can display a list of all the incomplete sales documents for any document category. The app allows you to search for incomplete sales documents according to various filters and display them as a list. From the list, you can navigate directly to the sales document in question to add missing data.

Tip

Performance issues are common with this app if no filters are added. It is therefore recommended that filters are always added to limit the number of hits.

Transaction Code

The SAP GUI equivalent is Transaction V.00 (Display Incomplete SD Documents).

List Incomplete Sales Inquiries (F6232)

Application Type: Transactional

This app is a variant of the List Incomplete Sales Documents app (F2430). By using this specific app, the sales and distribution document category is prepopulated in the List Incomplete Sales Documents app, to show only sales inquiries. The app allows you to search for incomplete sales inquiries according to various filters and display them as a list. From the list, you can navigate directly to the sales document in question to add missing data.

Tip

Performance issues are common with this app if no filters are added. It is therefore recommended that filters are always added to limit the number of hits.

Transaction Code

The SAP GUI equivalent is Transaction V.03 (Display Incomplete Inquiries).

List Incomplete Sales Item Orders Without Charge (F6233)

Application Type: Transactional

This app is a variant of the List Incomplete Sales Documents app (F2430). By using this specific app, the sales and distribution document category is prepopulated in the List Incomplete Sales Documents app, to show only sales orders without charge. The app allows you to search for incomplete sales orders without charge according to various filters and display them as a list. From the list, you can navigate directly to the sales document in question to add missing data.

Tip

Performance issues are common with this app if no filters are added. It is therefore recommended that filters are always added to limit the number of hits.

Transaction Code

The SAP GUI equivalent app is Transaction V.02 (Display Incomplete Orders).

List Incomplete Sales Item Proposals (F6238)

Application Type: Transactional

This app is a variant of the List Incomplete Sales Documents app (F2430). By using this specific app, the sales and distribution document category is prepopulated in the List Incomplete Sales Documents app, to show only sales item proposals. The app allows you to search for incomplete sales item proposals according to various filters and display them as a list. From the list, you can navigate directly to the sales document in question to add missing data.

Tip

Performance issues are common with this app if no filters are added. It is therefore recommended that filters are always added to limit the number of hits.

Transaction Code

The SAP GUI equivalent is Transaction V.02 (Display Incomplete Orders).

List Incomplete Sales Quotations (F6239)

Application Type: Transactional

This app is a variant of the List Incomplete Sales Documents app (F2430). By using this specific app, the sales and distribution document category is prepopulated in the List Incomplete Sales Documents app, to show only sales quotations. The app allows you to search for incomplete sales quotations according to various filters and display them as a list. From the list, you can navigate directly to the sales document in question to add missing data.

Tip

Performance issues are common with this app if no filters are added. It is therefore recommended that filters are always added to limit the number of hits.

Transaction Code

The SAP GUI equivalent app is Transaction V.04 (Display Incomplete Quotations).

List Sales Documents by Object Status (F2714)

Application Type: Transactional

This app allows you to view a list of all sales documents according to a specific system or user status. From the list, you can navigate directly to the status overview page for the selected sales document or sales document item, in a classic SAP GUI application (accessed via a transaction code). For example, if the sales document selected is a sales order, navigation will take you directly to the Display Sales Documents app (Transaction VA03). From there, you can change the status by selecting the appropriate radio button. User statuses are customizable statuses in SAP S/4HANA.

Navigation to all document types is supported, as is navigation to the master data of the sold-to party.

Tip

Performance issues are common with this app if no filters are added. It is therefore recommended that filters are always added to limit the number of hits.

Manage Access Sequences – Substitutions (F5302)

Application Type: Transactional

This app allows you to create and edit alternative controls for the confirmation of requirements in sales orders. This is done by creating a substitution control and assigning a corresponding combination of characteristic values. The app supports the creation of multiple accesses for each combination of characteristic values, reordering the access sequence, and the specification of time periods using validity dates.

Navigation to the related apps Manage Substitution Controls – Products (F4787), Manage

Characteristic Combinations (F5303), and Manage Substitution Controls – Locations (F5312) is supported.

Manage Batches (F2462)

Application Type: Transactional

Batches for materials can be managed in this app. Functions of search, display, edit and create are all supported. Batches can be found using the standard filter criteria, such as material, batch, plant, batch status, and shelf-life expiration date. The results of the search are displayed in a list format, with navigation to the batch record itself showing relevant information such as plant, inspection lots, batch usage, and change history.

Tip

Batch information is also an option in the enterprise search in SAP Fiori. Once the batch information is found and displayed in the list, one of the options in the enterprise search results is the Manage Batches app, which is used to show a more granular level of detail.

Manage Billing Document Requests (F2960)

Application Type: Transactional

Billing document requests (BDRs) are SAP objects that contain an itemized list of data for the purposes of billing, drawn from single or multiple sources. These BDRs are never for credit items; they are only related to debit items.

The app allows you to create, reject, and delete BDRs. Navigation to the following apps is supported from within this app:

- Manage Billing Documents (F0797)
- Create Billing Document (F0798)
- Billing Document Request (F2337)

> **Tip**
>
> Use the **Download Excel Template** and **Upload from Excel** features to create BDRs automatically from an XLSX file.

Manage Billing Documents (F0797)

Application Type: Transactional

All types of SAP billing documents can be managed in this app, including invoices, credit memos, debit memos, and invoice cancellations. Key activities that can be carried out from within the app include displaying and filtering a list of SAP billing documents, canceling billing documents, posting billing documents to accounting, and editing billing documents. It's also possible to highlight multiple billing documents and carry out a split analysis to display why multiple billing documents have been created by SAP instead of a single billing document. The following quick actions are available with a single click:

- Display Billing Document
- Display Split Analysis
- Cancel Billing Document
- Post Billing Document to Accounting

It is also possible to display a preview of the billing document output by clicking the **Preview** button when viewing an individual billing document.

Manage Business Partner Master Data (F3163)

Application Type: Transactional

This is a comprehensive list report app that allows you to manage your business partner master data through the functions of creating, updating, displaying, and copying. Creation is supported by adding a person or an organization. Basic fields can be added using the dialog box, although more details can be added by

using the **Details** option. Updating business partners can be achieved by clicking the **Edit** button once the business partner is located from the search fields. The copy function will copy all data from the source record into the new entry, except the business partner number. Fields should be edited as needed. If you unintentionally close the browser window, the record is saved automatically as a draft and can be retrieved later for completion.

> **Transaction Code**
>
> The SAP GUI equivalent app is Transaction BP (Business Partner).

Manage Characteristic Catalogs (F3829)

Application Type: Transactional

Characteristics, which are used within classes for various master data objects, can be grouped together into characteristics catalogs. With this app, you can add attributes (which can be used as characteristics in process flows such as availability checks) as characteristics to catalogs. For example, to help with defining advanced available-to-promise (ATP) flows, it might be useful to create a new characteristic catalog for geographical region of the ship-to partner. In this example, you might have a new characteristic catalog called "US Regions", then have West USA and East USA as regions within that catalog. It may then be useful to connect the US state codes to this characteristic catalog, thereby enabling use of the "West USA" characteristic catalog to drive results in advanced ATP. Further functions allow you to define new value groups, connect characteristics across multiple catalogs, and define authorizations for characteristic values and groups of values.

This app is most typically utilized in the advanced ATP process. When configuring an advanced ATP process, the SAP standard-delivered set of characteristics is too large to use effectively, and therefore it makes sense to group the relevant characteristics for a process into a characteristic catalog.

Manage Characteristic Combinations (F5303)

Application Type: Transactional

With this app, you can create multiple combinations of characteristics from sales order documents for usage in the advanced available-to-promise (ATP) process. It is possible to set up a combination within the app and assign a name and several attributes as characteristics, such as sales organization, sold-to party, and distribution channel.

Tip

The usage and catalog type in the combination are fixed once the object is saved in the app. In addition, once a characteristic combination is used within an access in advanced ATP, it is not possible to change the combination of characteristics.

Manage Commodity Codes (F2516)

Application Type: Transactional

As part of SAP S/4HANA for international trade, this very simple app allows you to add, edit, and display commodity codes for use in the classification of products. Commodity codes can have a description as well as a validity date range. It is also possible to add a customs unit of measure against the commodity code.

Manage Condition Contracts – Customers (F5413)

Application Type: Transactional

Customer condition contracts can be managed using this app. This includes setting and removing condition contracts and settlement locks. All

details of the customer condition contract can be viewed in the app and navigation is supported to related apps such as Manage Condition Contracts – Sales Rebates (F6739). You also can view the document flow from within the app.

Manage Condition Contracts – External Commissions (F5989)

Application Type: Transactional

External sales commissions are created as condition contracts in SAP, and this app allows you to manage those documents. This includes setting and removing condition contracts and settlement locks. All details of the condition contract can be viewed in the app, and navigation is supported to related apps in order to create and change condition contracts.

Manage Condition Contracts – Internal Commissions (F7263)

Application Type: Transactional

Internal sales commissions are created as condition contracts in SAP, and this app allows you to manage those documents. This includes setting and removing condition contracts and settlement locks. All details of the condition contract can be viewed in the app, and navigation is supported to related apps in order to create and change condition contracts.

Manage Condition Contracts – Royalties (F5414)

Application Type: Transactional

Royalty agreements are created as condition contracts in SAP, and this app allows you to manage those documents. This includes setting and removing condition contracts and settlement locks. All details of the condition contract can be viewed in the app, and navigation is supported to related apps in order to create and change condition contracts.

Manage Condition Contracts – Sales Rebates (F6739)

Application Type: Transactional

Sales rebates are created as condition contracts in SAP, and this app allows you to manage those documents. This includes setting and removing condition contracts and settlement locks. All details of the condition contract can be viewed in the app, and navigation is supported to related apps in order to create and change condition contracts.

Manage Content from Data Provider – Commodity Codes (F3429)

Application Type: Transactional

Some data providers can provide listings of commodity codes for input into another organization's SAP S/4HANA system. This app allows you to review and validate that data before allowing the activation. Data initially resides in SAP S/4HANA from linked external sources and exists as inactive data. You can use this app to interrogate the system and selectively recheck or activate this data. Once the data is activated, it moves from the **Inactive Versions** tab in the app to the **Active Versions** tab. Once the data is activated, it is available for use in classification of commodity codes to materials by using either the Classify Products – Commodity Codes app (F2151) or the Reclassify Products – Commodity Codes app (F2152).

Manage Content from Data Provider – Control Classes (F3582)

Application Type: Transactional

Some data providers can provide listings of control classes for legal control purposes, for input

into another business's SAP S/4HANA system. This app allows you to review and validate that data before allowing the activation. Data initially resides in SAP S/4HANA from linked external sources and exists as inactive data. You can use this app to interrogate the system and selectively recheck or activate the data. Once the data is activated, it moves from the **Inactive Versions** tab in the app to the **Active Versions** tab.

Once the data is activated, it is available for use in classification of control classes to materials by using the Classify Products – Legal Control app (F2390).

Manage Content from Data Provider – Customs Tariff Numbers (F3581)

Application Type: Transactional

Some data providers can provide listings of customs tariff numbers for input into another business's SAP S/4HANA system. This app allows you to review and validate that data before allowing the activation. Data initially resides in SAP S/4HANA from linked external sources and exists as inactive data. You can use this app to interrogate the system and selectively recheck or activate the data. Once the data is activated, it moves from the **Inactive Versions** tab to the **Active Versions** tab in the app.

Once the data is activated, it is available for use in classification of customs tariff numbers to materials by using either the Classify Products – Customs Tariff Numbers app (F3146) or the Classify Products for End-Uses – Customs Tariff Numbers app (F4388).

Manage Control Classes (F2518)

Application Type: Transactional

As part of SAP S/4HANA for international trade, this app can be used to manage the control classes that are used to fulfill legal regulations for the control of goods movements. A control class is defined as a unique code for the classification of goods and requires a license from an authority for controlling goods movement. An example of a control class is the export control classification number (ECCN). Filters can be used in the app to find the control classes based on the numbering scheme, control class, description, and validity dates. It is possible to create, change, display, and delete control classes from the app.

Manage Control Groupings (F2515)

Application Type: Transactional

As part of SAP S/4HANA for international trade, this app can be used to manage the control groupings that are used to group products based on their common legal regulations. It is then possible to assign products to these groupings. This makes maintenance of the legal regulations quicker, as groupings can be checked to determine the license types required. Filters can be used in the app to find the control groupings based on the legal regulation, control grouping, and description. It is possible to create, change, display, and delete control groupings from the app.

Manage Countries under Embargo (F2791)

Application Type: Transactional

Each legal regulation can have countries that are specified as being under embargo and therefore restricted from the supply of goods contained within those legal regulations. The embargo checks can be ring fenced to a time period using validity dates. Filters can be used in the app to find the countries under embargo based on the legal regulation, country, and validity dates. It is possible to create, change, display, and delete countries under embargo from the app.

Manage Credit Memo Requests (Version 2) (F1989A)

Application Type: Transactional

This app displays credit memo requests in a list according to the filter criteria used. The intention is to be able to show all credit memo requests and allow you to navigate to related apps for each record. There are quick actions that can be taken for each record selected: **Reject All Items**, **Set Billing Block**, and **Remove Billing Block**. In addition, you can create a credit memo request from one of the quick actions available. Version 2 of this app offers a cleaner user interface with additional standard options for filters.

Navigation to the Customer 360° View – Version 2 (F2187A) and Credit Memo Request (F1846) apps is supported.

This app is the updated version of the Manage Credit Memo Requests app (F1989), which is still available but has been superseded by this app.

Manage Customer Master Data (F0850A)

Application type: Transactional

This app allows you to manage your customer master data through the business partner functionality. The app uses a whole list of standard CDS views in order to support the correct data retrieval. As part of the functionality, you can create, edit, and copy customer master data. The copy function will copy all fields from the source data, unless the **Copy with Preselection** option is selected, in which case you can copy across only the organizational data (such as company codes and sales areas) that are selected in the dialog box. As with other master data SAP Fiori apps, you can save your changes in draft mode.

Transaction Code

The SAP GUI equivalent is Transaction BP (Business Partner).

Manage Customer Materials (F2499)

Application Type: Transactional

This app allows you to create information regarding materials that are specific to a customer. This has traditionally been called a customer material info record in SAP. As with other "Manage" apps, you can search for customer material records using the filter bar with options such as editing status, customer material, customer, material, sales organization, and distribution channel. There are options to create a new customer material record, copy an existing record, edit an existing record, and delete an existing record.

Tip

Any records that are not fully completed are automatically saved as a draft and can be completed later. Draft records are also available to be used in the **Copy** function.

Manage Customer Returns – Create (F2651)

Application Type: Transactional

You can use this app to create a customer returns order. The app is normally called from within the Manage Customer Returns app (F1708). From within the app, you can search for reference documents such as sales orders or deliveries for a given customer. Once a suitable reference document is established, you can create the return with reference to the document. The app will then step through the various stages of creation of the return. Returns orders created based on advanced returns management (ARM) are also supported.

Tip

What is advanced returns management (ARM) in SAP? ARM is an automated end-to-end process that handles returns from customers, including forwarding documents such as returns to suppliers, credits, return purchase orders, and goods movements, all of which can be automated. ARM is triggered by a special type of return sales order, which activates a new tab in the return sales document, with options for returns inspection, follow-on documents, and credits and refunds. As a result, the entire flow can be finalized in a single step. ARM can be used in combination with the Manage Customer Returns – Create app to create ARM returns documents.

Manage Customer Returns – Edit (F2650)

Application Type: Transactional

This app can be used to edit existing customer returns documents. The app is normally called from within the Manage Customer Returns app (F1708). From within the app, you can search for customer returns documents for a given customer. Once a suitable customer returns document is selected, you can edit the document. The app will then step through the various stages for editing the return.

Returns orders created based on advanced returns management (ARM) are also supported.

Manage Customer Returns – Refund (F2652)

Application Type: Transactional

This app can be used to determine refunds for customer returns documents. The app is normally called from within the Manage Customer

Returns app (F1708). From within the app, you can search for customer returns documents for a given customer. Once a suitable customer returns document is selected, you can edit the document to determine the refund. The app will then step through the various stages to determine the refund.

Returns orders created based on advanced returns management (ARM) are also supported.

Manage Customer Returns (Version 2) (F4832)

Application Type: Transactional

This app displays customer returns in a list according to the filter criteria used. The intention is to be able to show all customer returns and allow you to navigate to related apps for each record. There are quick actions that can be taken for each record selected: You can **Edit**, **Delete**, **Determine Refund**, and **Withdraw Approval Request** for a given returns document. In addition, you can create a customer returns document from one of the quick actions available. Returns orders that are created based on advanced returns management (ARM) are also supported.

This app supersedes the Manage Customer Returns app (F1708) as of SAP S/4HANA 2022, and it offers a cleaner user interface with additional standard options for filters.

Manage Customs Tariff Numbers (F3122)

Application Type: Transactional

As part of SAP S/4HANA for international trade, this app can be used to manage the customs tariff numbers. Filters can be used in the app to find the customs tariff numbers based on the numbering scheme, customs tariff number, description, and validity dates. It is possible to create, change, display, and delete customs tariff numbers from the app.

Manage Debit Memo Requests (F1988)

Application Type: Transactional

This app displays debit memo requests in a list according to the filter criteria used. The intention is to be able to show all debit memo requests and allow you to navigate to related apps for each record. There are quick actions that can be taken for each record selected: **Reject All Items**, **Set Billing Block**, and **Remove Billing Block**. In addition, you can create a debit memo request from one of the quick actions available.

Navigation to the Customer 360° View – Version 2 (F2187A) and Debit Memo Request (F1848) apps is supported.

Manage Delivery Schedule Processing (F3895)

Application Type: Transactional

This app allows you to create, edit, and delete delivery scheduling runs for specific customers and/or unloading points, which have been created via Electronic Data Interchange (EDI). There are filters available based on sold-to customers, unloading point, and editing status.

Tip

It is possible that the customer purchase order and unloading point specified in the originating sales scheduling agreement may differ from the inbound delivery schedule. From within this app, you can specify whether this should be checked and updated.

Manage Deviation Profiles (F4781)

Application Type: Transactional

Delivery schedules can differ markedly from actual order quantities. This app allows you to set thresholds for the deviations in order to flag them with key statuses: **Alert**, **Warning**, or **Acceptable**. The thresholds can be aligned to deviations in order quantities as well as deviations in cumulative received order quantities. In the app, you can create deviation profiles by setting percentage numbers to the thresholds for each status, as well as search and filter on deviation profiles and edit and delete deviation profiles. The deviation profiles are used in the Delivery Schedules – Demand Deviations app (F5651).

Tip

Deviations are recorded in the app using color coding (red = alert, amber = warning, gray = acceptable).

Manage Documents – Trade Compliance (F2826)

Application Type: Transactional

As part of SAP S/4HANA for international trade, this app is used to manage all documents related to legal controls, embargoes, and SAP Watch List Screening. From the app, you can use legal control functionality, which allows you to check the status of the documents; view the attributes assigned to the legal control such as control class, control grouping, customs tariff number, commodity code and license; and adjust the classification and license assignment to process the trade compliance document. You also can work with embargoes by checking the document status and processing embargo blocks. The app also displays the status of documents from SAP Watch List Screening. Filters can be used in the app to find documents based on legal regulation, company code, plant, document category, document number, document date, and partner country.

Manage Duplicate Sales Documents (F3245)

Application Type: Analytical

This app allows you to search and remedy duplicated sales documents such as inquiries, quotations, sales orders, contract, credit memo requests, debit memo requests, and returns. Search criteria include creation period, sales document, sales document category, sales organization, sold-to party, created by, and status. Duplicates are identified by sales document category, sales document type, sales area, sold-to party, net value, currency, creation date, and customer reference.

It is also possible to compare sales documents based on key parameters: created by, requested delivery date, rejection status, overall status, and material. The app allows displaying the duplication results in graphical format whereby you can see the results for each period.

Navigation to all related apps to change the sales documents is supported. These apps are SAP GUI transactions such as Change Sales Orders (VA02), Change Sales Inquiries (VA12), Change Sales Quotations (VA22), and Change Contract (VA42).

Manage Exclusions – Locations (F5315)

Application Type: Transactional

This app can be used as part of the overall advanced available-to-promise (ATP) process. As part of that process, substitutions can be configured using the Configure Substitution Strategy app (F2699). However, it may be required that some locations are excluded from the substitution strategy for a given period of time. For example, if a plant had a defined shutdown period, this plant can be excluded using this app. With this app, you can use the functions of **Create**, **Edit**, and **Delete** to manage exclusions for locations.

This app is one of five that can be used to manage location substitutions in advanced ATP. The other apps are as follows:

- Manage Substitution Reasons – Locations (F5313)
- Manage Substitution Groups – Locations (F5311)
- Manage Substitution Controls – Locations (F5312)
- Manage Substitutions – Locations (F5314)

Manage Exclusions – Products (F4786)

Application Type: Transactional

This app can be used as part of the overall advanced available-to-promise (ATP) process. As part of that process, substitutions can be configured using the Configure Substitution Strategy app (F2699). However, it may be required that some products are excluded from the substitution strategy for a given period of time. For example, if a product is known to be out of stock for a given month, then it can be excluded from substitution processes for that month using this app. With the app, you can use the **Create**, **Edit**, and **Delete** functions to manage exclusions for products.

This app is one of five that can be used to manage product substitutions in advanced ATP. The other apps are as follows:

- Manage Substitution Reasons – Products (F4789)
- Manage Substitution Groups – Products (F4788)
- Manage Substitution Controls – Products (F4787)
- Manage Substitutions – Products (F4785)

Manage Intrastat Service Codes (F2517)

Application Type: Transactional

As part of SAP S/4HANA for international trade, this app can be used to manage Intrastat service codes. Filters can be used in the app to find the Intrastat service codes based on the numbering scheme, Intrastat service code, description, and validity dates. It is possible to create, change, display, and delete Intrastat service codes from the app.

Manage Invoice Lists (F2740)

Application Type: Transactional

This app displays all invoice lists according to your filter criteria. Invoice lists can be created, displayed, filtered, sorted, grouped, and canceled. The list, as with any SAP Fiori list report app, can be exported to a Microsoft Excel spreadsheet. It is also possible to navigate to the individual billing documents that make up the invoice list.

Transaction Code

The SAP GUI equivalent app is Transaction VF23 (Display Invoice List).

Manage JIT Customer Data (F3011)

Application Type: Transactional

Just-in-time (JIT) supply is a process whereby materials and components are delivered or produced to fulfill demand only when they are needed. The benefit of following this approach is that it minimizes inventory holdings and therefore costs, and it reduces waste. This app helps you manage JIT customer data, including delivery locations and status information. Designed for master data specialists in JIT supply, it allows you to read essential JIT customer information such as sales organizations and distribution channels. It is possible to update customer supply areas and unloading points, map external customer statuses, and handle call processing details all in one place. The intention here is to facilitate the movement of inventory for JIT purposes, to the right location at the right time for the right customers. For example, some customers pay require special production processes that require a certain assembly location to be determined.

> **Tip**
>
> Customer information for JIT data is relevant for sales and distribution purposes, and therefore a JIT-relevant customer is always defined for a specific sales area.

Manage JIT Customer Supply Control (F3010)

Application Type: Transactional

This app allows you to combine data such as just-in-time (JIT) customer data, customer supply area, plant, and component group material, in order to create and manage customer supply controls. Supply controls are specifically intended to allow you to control the business processing of JIT calls from customers. The controls can be activated at multiple levels including the customer, customer/plant, customer supply area (combination of customer, plant, and customer supply area), and component group material (combination of customer, plant, customer supply area, and component group material).

Manage Licenses (F2545)

Application Type: Transactional

As part of SAP S/4HANA for international trade, this app can be used to manage licenses for the exporting and importing of products that require approval. Regional authorities can specify that the movement of certain goods must be controlled via licensing methods. These licenses can restrict goods movement to metrics such as quantity quotas, or maximum value of goods. License requirements are normally defined based upon export control classification numbers (ECCNs) or commodity codes. With this app, you can create, display, change, and delete licenses for legal regulations. Licenses can be

searched for by using filters such as legal regulation, license category, license type, license number, official license number, licenses owner, status, and remark.

Manage Outbound Deliveries (F0867A)

Application Type: Transactional

With this well-used app, you can manage your outbound deliveries. You can post goods issue, reverse the post, edit deliveries (depending on the goods issue status), and perform picking. To edit deliveries, the goods issue status can only be **Not Yet Processed** (**A**). Delivery items that have the status **Completed** (**C**) or **Partially Processed** (**B**) cannot be edited.

Navigation to related apps such as Customer Master FactSheets (F0046A), Outbound Delivery (S/4HANA) (F0233A), and Pick Outbound Delivery (F0868) is supported.

> **Tip**
>
> Goods issues and reversal of goods issues can be carried out for multiple documents at once by selecting each line and then clicking the appropriate quick action.

Manage Preliminary Billing Document Workflows (F4274)

Application Type: Transactional

This app is used by configuration experts to define workflows for the approval of preliminary billing documents. You can create a workflow with individual steps for approval. For example, if a preliminary billing document has a net value of greater than $100,000, then it can be routed to a specific approver. The approvers will receive the notification to approve in their SAP Fiori notifications area as well as the My Inbox app (F0862).

> **Tip**
>
> Multiple approvers can be determined via a single workflow if several conditions are met simultaneously. For example, if a condition exists that the document will need to be approved by approver 1 if the net value is over $10,000 and approver 2 if the net value is over $20,000, and the document net value is $25,000, then both conditions are met, and both approvers will receive the approval request.

Manage Preliminary Billing Documents (F2875)

Application Type: Transactional

A preliminary billing document is used during customer negotiations, before the details of the billing are confirmed. As soon as the agreement with the customer has been established, the preliminary billing document can be output, but it is not posted to accounting. These documents are often also called proforma billing documents. They can be finalized by creating another subsequent standard billing document with reference to the sales document or delivery in the usual fashion.

Navigation to the Manage Billing Documents (FO797) and Preliminary Billing Document (F2874) apps is supported.

> **Transaction Code**
>
> The SAP GUI equivalent app is Transaction VFP2 (Change Preliminary Billing Documents).

Manage Price Fixation Options (F4268)

Application Type: Transactional

Floating prices (prices that move with market relevant attributes such as exchange rates) in sales documents and purchasing documents can be replaced by an agreed upon fixed price

using price fixation options. This app can be used to create price fixation options, display an overview of the options, show the status of the options, as well as define code that can be run to calculate the overall used quantity to which a fixed price can be assigned.

> **Tip**
>
> If the currencies on the price fixation option matches the target currency, then the fields **Currency from**, **Currency to**, and **Fixed Exchange Rate** are automatically hidden.

Manage Prices – Sales (F4111)

Application Type: Transactional

This app can be used to search for, create, edit, and delete sales pricing condition records. Standard filter criteria for searching includes condition type, key combination, validity dates, customer, sales area, and product. Quick actions include **Create**, **Edit**, **Copy**, and **Delete**. It is also possible to import condition records from a template (which can be downloaded from a quick action in the app). Similarly, you can download the conditions in the results into a Microsoft Excel format.

Navigation to the Manage Subscription Product-Specific Data (F3560) and Manage Tax Rates – Sales (F6972) apps is supported.

> **Transaction Code**
>
> The SAP GUI equivalent app is Transaction VK13 (Display Sales Prices).

Manage Product Allocation Planning Data (F2121)

Application Type: Transactional

Product allocation objects can be created using the Configure Product Allocation app (F2119), and then product allocation sequences can be created using the Manage Product Allocation Sequences app (F2474). Once this is complete, you can maintain characteristic value combinations and planned allocation quantities for the time periods of the characteristic value combinations. These assigned combinations then become relevant for availability checks in sales documents and stock transport orders.

Navigation to the Application Log for Product Allocation (F6296), Monitor Allocation Object Changes (F714), and Monitor Allocation Value Combination Changes (F7144) apps is supported.

Tip

External files of these characteristic value combinations can be uploaded into the app in CSV or XLSX format.

Manage Product Allocation Sequences (F2474)

Application Type: Transactional

This app is part of the advanced available-to-promise (ATP) suite of SAP Fiori apps. Issues in supply arising from all inventory being allocated to a single customer can be avoided by using this app to define product allocation sequences. The app can be used to pull together combinations of product allocation objects as factors in product allocation for confirming quantities during availability checks. For example, during the COVID-19 pandemic, a supplier of personal protective equipment such as masks may have wanted to allocate a certain quantity of this inventory to public healthcare providers within a certain country, to avoid inventory shortfalls in key sectors due to excessive high demand. The **Create**, **Edit**, **Delete**, and **Display** functions are available for product allocation sequences.

Navigation to the Monitor Allocation Sequence Changes app (F7143) is supported.

Manage Route Version Proposals (F6022)

Application Type: Transactional

Once a proposal profile is assigned to the route at header level, you can use this app to create master data based on the route version proposals. You can create, check, block, unblock, accept, delete, and export route version proposals. In addition, you can add or update and delete new stops within the route version proposal.

Tip

Use the **Download Template** button to fill in data in a Microsoft Excel spreadsheet for the creation of route version proposals.

Manage Sales Contracts (Version 2) (F5987)

Application Type: Transactional

This app allows you to list, filter, create, change, and display sales contracts in SAP. Standard filters are available such as sales contract number, sold-to party, customer reference, contract expiration data, and contract document date.

This app is an updated version of the Manage Sales Contracts app (F1851), with additional functionality available. In addition to the **Create Contract** and **Reject All Items** quick actions, with this updated app, you can set or remove the billing block, set or remove a rejection reason, update pricing, and delete the document. You also can add attachments from within the app.

Navigation to the updated Customer – 360° View app (F2187A) is supported.

Transaction Code

The SAP GUI equivalent app is Transaction VA43 (Display Sales Contracts).

Manage Sales Document Workflows (F3014)

Application Type: Transactional

This app is used by configuration experts to define workflows for the approval of sales documents. You can create a workflow with individual steps for approval. For example, if a sales document has a net value of greater than $100,000, then it can be routed to a specific approver. The approvers will receive the notification to approve in their SAP Fiori notifications area as well as the My Inbox app (F0862).

Navigation to the Manage Workflows app (F2190) is supported.

Tip

Multiple approvers can be determined via a single workflow if several conditions are met simultaneously. For example, if a condition exists that the document will need to be approved by approver 1 if the net value is over $10,000 and approver 2 if the net value is over $20,000, and the document net value is $25,000, then both conditions are met, and both approvers will receive the approval request.

Manage Sales Documents with Customer-Expected Price (F2713)

Application Type: Transactional

Inbound sales orders, normally through Electronic Data Interchange (EDI), specify the customer-expected price. This app allows you to list all the sales documents that show discrepancies between the net price in the sales order and the customer-expected price. Options in the app include the ability to accept or decline the customer-expected price, or to reject the item altogether. Acceptance of the customer-expected price allows you to change the net price and progress the document for further processing. If the customer-expected price is declined, then

the net value is respected, and the document is released for further processing. These options are available for multiple documents at a time as well as for single processing.

Navigation to SAP GUI Transaction VA03 (Display Sales Orders) is supported.

Manage Sales Inquiries (F2370)

Application Type: Transactional

This app displays sales inquiries in a list according to the filter criteria used, including sales inquiry, sold-to party, customer reference, overall status, and document date. The intention is to be able to show all sales inquiries and allow you to navigate to related apps for each record. There are quick actions that can be taken for each record selected: **Reject All Items** or **Create Inquiry**. Clicking the sales inquiry number brings up a dialog box where you can select related SAP GUI transactions to navigate to, such as Create Inquiry (VA11), Change Inquiry (VA12), and Display Inquiry (VA13).

Navigation to the Customer 360° View – Version 2 (F2187A) app is supported, and clicking the individual sales inquiry line will launch the factsheet Sales Inquiry app (F2369).

Transaction Code

The SAP GUI equivalent app is Transaction VA13 (Display Sales Inquiry).

Manage Sales Item Proposals (F2583)

Application Type: Transactional

This app displays sales item proposals in a list according to the filter criteria used. The intention is to be able to show all sales item proposals and allow you to navigate to related apps for each record. Standard SAP Fiori filters are available for fields such as sales item proposal number, description, and validity dates. You can use quick actions to delete one or multiple documents from the list or create new sales item proposals. Clicking the sales item proposal number brings up a dialog box where you can select related SAP GUI transactions to navigate to, such as Create Sales Item Proposal (VA51), Change Sales Item Proposal (VA52), and Display Sales Item Proposal (VA53).

Transaction Code

The SAP GUI equivalent app is Transaction VA53 (Display Sales Item Proposal).

Manage Sales Orders (Version 2) (F3893)

Application Type: Transactional

This heavily used app displays sales orders in a list according to the filter criteria used, including sales order, sold-to party, customer reference, requested delivery date, overall status, and document date. The intention is to be able to show all sales orders and allow you to navigate to related apps for each record. There are quick actions that can be taken for each selected record: **Create Sales Order, Reject All Items, Set Delivery Block, Remove Delivery Block, Set Billing Block**, and **Remove Billing Block**. Delivery blocks and billing blocks are useful in sales orders for putting the order header or order item on hold for delivery or billing. For example, there may be an ongoing dispute with the customer, which means that the items should not be delivered; in this case, a temporary delivery block can be added until the dispute is resolved. Normally, to add a delivery or billing block, it would be necessary to open the sales document and navigate to the field or the item field and enter the block. However, with this app, the quick actions button allows you to populate the block from the list. The app is also intelligent enough to recognize when a billing block or delivery block cannot be added (e.g., if the line item has been completely processed) and, on these occasions, the quick action button for delivery block and billing block would be grayed out. In the case of creating sales orders, once the **Create Sales Order** or **Create with Reference** action has been initiated, you can add all header, item, and schedule line information into the new sales order from within the app.

This app is an updated version of the Manage Sales Orders (F1873) app. All the same functionality is available in this app, with the added ability to create, change or display individual sales orders directly in the app, without having to navigate to related apps.

Navigation to the Customer 360° View – Version 2 (F2187A), Manage Sales Orders – (F1873), Sales Volume – Check Open Sales (F2270), and Display Credit Exposure (F4826) apps is supported.

Transaction Code

The SAP GUI equivalent app is Transaction VA03 (Display Sales Order).

Manage Sales Orders Without Charge (F2305)

Application Type: Transactional

This app displays sales orders without charge in a list according to the filter criteria used, including sales order without charge, sold-to party, customer reference, required delivery date, overall status, and document date. The intention is to be able to show all sales orders without charge and allow you to navigate to related apps for each record. There are quick actions that can be taken for each record selected: **Create Sales Orders Without Charge, Reject All Items, Set Delivery Block, Remove Delivery Block**, and **Set Order Reason**. All these quick actions (except the creation function) can be carried out for multiple documents in the list. It is also possible to extract the list to Microsoft Excel.

Navigation to the Customer 360° View – Version 2 app (F2187A) and SAP GUI Transaction VA03 (Display Sales Order) is supported.

Transaction Code

The SAP GUI equivalent app is Transaction VA03 (Display Sales Order).

Manage Sales Plans (F2512)

Application Type: Transactional

Sales plans exist to set targets for a sales team within a defined period. With this app, different types of sales plans can be created: value plans or quantity plans. Value plans are sales plans based on a certain target monetary value, which can be aligned either to incoming sales orders (which represents the total net value of sales order items that are relevant for billing) or sales volume (which represents the total net value of the invoiced sales in a given period). Quantity plans are sales plans based on a certain quantity of sales orders (incoming sales orders) or a total quantity of sales volume (quantity of invoiced items). To set the sales plans, certain data must be added, such as the name and description of the sales plan, name, and description of the version (standard versus "stretch" version), planned from and to dates, sales plan type (value or quantity), currency, and dimensions. The dimensions can be set to provide more granularity, for example, specific products or customers or a combination of the two.

Once the sales plans are set, the targets can be compared with actual performance by using the Sales Performance – Plan/Actual app (F2941). Filters in the Manage Sales Plans app include sales plan ID, plan description, planned by, and currency. There are quick actions that can be taken for each record selected: **Copy to New Version**, **Release**, **Reopen**, **Delete**, and **Create**. Once you create a sales plan, you can assign it to a sales team, so every member of that team can see it.

Tip

It is possible to create a sales plan as a value plan (with value-based targets based on either the total net value of incoming sales orders or the total invoiced sales), or a quantity plan (total incoming sales orders quantity of items, or total quantity of invoice of debit items).

Manage Sales Price Workflows (F5525)

Application Type: Transactional

This app is used by configuration experts to define workflows for the approval of sales prices. You can create a workflow with steps for approval. For example, if a sales price for a given geographical region represented by a sales area must be approved, then it can be routed to a specific approver. The approvers will receive the notification to approve in their SAP Fiori notifications area as well as the My Inbox app (F0862). New workflows are configured by adding the following steps:

1. Define the properties of the workflow, such as name and validity dates.
2. Add conditions for starting the workflow.
3. Create the steps in sequence. Note here that conditions that are set and assigned to individual steps are ignored. The conditions are only valid for starting the workflow.

The app can be used not only to create new sales price workflows, but also to copy existing workflows, display details of existing workflows, and delete existing workflows.

Manage Sales Quotation (Version 2) (F5630)

Application Type: Transactional

This app displays sales quotations in a list according to the filter criteria used, including sales quotation, sold-to party, customer reference, overall status, valid-to, and document date. The intention is to be able to show all sales quotations and allow you to navigate to related apps for each record. There are quick actions that can be taken for each record selected: **Create Quotation**, **Extend Validity**, and **Reject All Items**. Clicking the sales quotation number brings up a dialog box where you can select related SAP GUI transactions to navigate to, such as Transactions VA21 (Create Sales Quotation), VA22 (Change Sales Quotation), and VA23 (Display Sales Quotation). Once the **Create** or **Create with Reference** has been initiated, you can add all header and line item information into the new sales quotation from within the app.

This app is an updated version of the Manage Sales Quotations (F1852) app. All the same functionality is available in this app, with the added ability to create, change or display individual sales quotations directly in the app, without having to navigate to related SAP Fiori apps.

Navigation to the Sales Volume – Check Open Sales (F2270), Display Credit Exposure (F4826), and Customer 360° View – Version 2 (F2187A) apps is supported.

Transaction Code

The SAP GUI equivalent app is Transaction VA23 (Display Sales Quotations).

Manage Sales Scheduling Agreements (F3515)

Application Type: Transactional

This app displays sales scheduling agreements in a list according to the filter criteria used. The intention is to be able to show all sales scheduling agreements and allow you to navigate to related apps for each record. Standard SAP Fiori filters are available to filter by data such as sales scheduling agreement number, sold-to party, sales organization, customer reference, created by, and last changed by. It is possible to create a sales scheduling agreement by using the quick action **Create Scheduling Agreements**. Clicking the sales scheduling agreement number brings up a dialog box where you can select related SAP GUI transactions to navigate to, such as Create Sales Scheduling Agreement (VA31), Change

Sales Scheduling Agreement (VA32), and Display Sales Scheduling Agreement (VA33).

Navigation to the Customer 360° View – Version 2 app (F2187A) and SAP GUI Transactions VA31 (Create Sales Scheduling Agreements) and VA33 (Display Sales Scheduling Agreements) is supported.

Transaction Code

The SAP GUI equivalent app is Transaction VA33 (Display Sales Scheduling Agreements).

Manage Scheduling Worklists – Settlement Management (F5856)

Application Type: Transactional

Settlement management must be coordinated and managed by individuals within a business effectively. To achieve this, you can create scheduling worklists. This app allows you to create scheduling worklists and manage and monitor your scheduling worklists in a detailed way. These scheduling patterns can be applied to both individual items and groups of worklists that share common attributes. Scheduling worklists can be found using filters in the app such as editing status, scheduling worklist ID, scheduling worklist template, scheduling status, last run status, job execution date and time, and scheduling worklist run category.

Manage Settlement Documents (F3254)

Application Type: Transactional

Settlement documents, as part of the settlement management function, can be hard to keep control of in such a complex process. This app allows you to get a holistic overview of each settlement document, including the condition contract and bill-to party it is assigned to and the value of the document. Filters can be used to search for settlement documents, such as settlement document number, settlement document type, bill-to party, invoicing party, posting status, condition contract, posting date, and process category.

Manage Substitution Controls – Locations (F5312)

Application Type: Transactional

As part of the advanced available-to-promise (ATP) functions in SAP S/4HANA, predefined groups of substitutions can be created to manage substitutions in the availability check. A substitution control is used in order to consume predefined groups of substitution. The app can be used to create, edit, and delete a control, create language specific descriptions, and assign and remove the assignment of a control to a substitution group.

This app is one of five that can be used to manage location substitutions in advanced ATP. The other apps are as follows:

- Manage Substitution Reasons – Locations (F5313)
- Manage Substitution Groups – Locations (F5311)
- Manage Substitutions – Locations (F5314)
- Manage Exclusions – Locations (F5315)

Manage Substitution Controls – Products (F4787)

Application Type: Transactional

As part of advanced available-to-promise (ATP) functions in SAP S/4HANA, predefined groups of substitutions can be created to manage substitutions in the availability check. A substitution control is used in order to consume predefined groups of substitution. The app can be used to create, edit, and delete a control, create language specific descriptions, and assign and remove the assignment of a control to a substitution group.

This app is one of five that can be used to manage product substitutions in advanced ATP

functions in SAP S/4HANA. The other apps are as follows:

- Manage Substitution Reasons – Products (F4789)
- Manage Substitution Groups – Products (F4788)
- Manage Substitutions – Products (F4785)
- Manage Exclusions – Products (F4786)

Manage Substitution Groups – Locations (F5311)

Application Type: Transactional

When plant and storage location substitutions are configured, substitution groups for the objects can be managed using this app. For example, if plants have the same business functions, they can be grouped together for substitution purposes into a substitution group. Storage locations can be grouped together in the same way.

This app is one of five that can be used to manage location substitutions in advanced available-to-promise (ATP). The other apps are as follows:

- Manage Substitution Reasons – Locations (F5313)
- Manage Substitution Controls – Locations (F5312)
- Manage Substitutions – Locations (F5314)
- Manage Exclusions – Locations (F5315)

Manage Substitution Groups – Products (F4788)

Application Type: Transactional

When product substitutions are configured, substitution groups for the objects can be managed using this app. For example, if substitute products have the same business functions, they can be grouped together for substitution purposes into a substitution group.

This app is one of five that can be used to manage location substitutions in advanced available-to-promise (ATP) functions in SAP S/4HANA. The other apps are as follows:

- Manage Substitution Reasons – Products (F4789)
- Manage Substitution Controls – Products (F4787)
- Manage Substitutions – Products (F4785)
- Manage Exclusions – Products (F4786)

Manage Substitution Reasons – Locations (F5313)

Application Type: Transactional

When plant and storage location substitutions are created, it is possible to use this app to create and manage reasons for substitutions. When you create the reason, you must state which location type it is being used for (plant or storage location).

This app is one of five that can be used to manage location substitutions in advanced available-to-promise (ATP). The other apps are as follows:

- Manage Substitution Groups – Locations (F5311)
- Manage Substitution Controls – Locations (F5312)
- Manage Substitutions – Locations (F5314)
- Manage Exclusions – Locations (F5315)

Manage Substitution Reasons – Products (F4789)

Application Type: Transactional

When product substitutions are created, you can use this app to create and manage substitution reasons, which help explain why products are substituted in or excluded from a substitution. The app can be used to create, edit, or delete reasons, and descriptions can be added in multiple languages.

This app is one of five that can be used to manage location substitutions in advanced available-to-promise (ATP) functions in SAP S/4HANA. The other apps are as follows:

- Manage Substitution Groups – Products (F4788)
- Manage Substitution Controls – Products (F4787)
- Manage Substitutions – Products (F4785)
- Manage Exclusions – Products (F4786)

Tip

Reason names must be unique and can have no more than 20 characters. The description can be up to 60 characters.

Manage Substitutions – Locations (F5314)

Application Type: Transactional

Substitution location can be used in the availability check to substitute an alternative plant or storage location in an out-of-stock situation. Multiple plant or storage location substitutions can be created and maintained.

This app is one of five that can be used to manage location substitutions in advanced available-to-promise (ATP) functions in SAP S/4HANA. The other apps are as follows:

- Manage Substitution Reasons – Locations (F5313)
- Manage Substitution Groups – Locations (F5311)
- Manage Substitution Controls – Locations (F5312)
- Manage Exclusions – Locations (F5315)

Manage Substitutions – Products (F4785)

Application Type: Transactional

Substitution products can be used in the availability check to substitute an alternative product in an out-of-stock situation. Multiple product substitutions can be created and maintained.

This app is one of five that can be used to manage location substitutions in advanced available-to-promise (ATP) functions in SAP S/4HANA. The other apps are as follows:

- Manage Substitution Reasons – Products (F4789)
- Manage Substitution Groups – Products (F4788)
- Manage Substitution Controls – Products (F4787)
- Manage Exclusions – Products (F4786)

Manage Tax Rates – Sales (F6972)

Application Type: Transactional

This app can be used to search for, create, edit, and delete sales pricing condition records related to taxation. Standard filter criteria for searching includes condition type, key combination, validity dates, tax code, withholding tax code, departure country, customer tax classification, and material tax classification. Quick actions include **Create**, **Edit**, **Copy**, and **Delete**. It is also possible to import condition records from a template (which can be downloaded from a quick action in the app). Similarly, you can download the conditions in the results into a Microsoft Excel format.

Navigation to the Manage Prices – Sales app (F4111) is supported.

Manage Trading Contracts (F6649)

Application Type: Transactional

This app is part of the global trade management functionality in SAP S/4HANA, and it allows you to search, display, edit, and create trading contracts. A *trading contract* in SAP is a document that allows legal contracts to be defined and stored between customers and vendors. In this way, sales orders and purchase orders can be maintained in exactly the same transaction or app. Trading contracts are, by definition, driven by sales order documents. It is possible to integrate sales and purchasing together by creating trading contracts that automate the creation of purchase orders from the originating sales order. Options are available for filtering by header and item data according to the business partners involved, validity dates, and trading contract number. You also can sort and personalize the lists, as well as export the data in spreadsheet format.

These display, edit, and create functions are supported by navigation targets to SAP GUI Transactions WB21 (Create Trading Contract), WB22 (Change Trading Contract), and WB23 (Display Trading Contract).

Manage Workflows – Condition Contracts (F3680)

Application Type: Transactional

This app allows you to configure automatic workflows for condition contracts so that they can be released. The options for the approver, when receiving the notification, are to approve the condition contract, reject, or send back for rework. The workflow can be aligned to one of two processes. Firstly, there is the option to release the condition contract completely. Secondly, there is the option to release the condition contract for settlement purposes. Exception handling also can be defined in this app to determine what happens after a condition contract workflow is rejected. The app can be used to activate and deactivate workflows, display details of existing workflows, copy existing workflows, or delete existing workflows.

> **Tip**
>
> When a document has been approved or rejected, you can notify users by email that this action has been carried out. SAP provides standard email templates for achieving this: for releasing condition contracts, use email template WCB_CC_RELEASED_EMAIL; for rejected condition contracts for release, use WCB_CC_REJECTED_EMAIL; for releasing condition contracts for settlement, use WCB_CCS_RELEASED_EMAIL; and for rejected condition contracts for settlement, use WCB_CCS_REJECTED_EMAIL.

Manage Workflows – Settlement Documents (F3681)

Application Type: Transactional

Workflows for the approval of the release of settlement documents can be configured using this app. The options for the approver, when receiving the notification, are to approve the settlement document, reject, or send back for rework. Approved documents are released to accounting. Exception handling also can be defined in this app to determine what happens after a settlement document workflow is rejected. The app can be used to activate and deactivate workflows, display details of existing workflows, copy existing workflows, or delete existing workflows.

Mass Change of Credit Memo Requests (F5281)

Application Type: Transactional

This app is used to make mass changes to the sales documents that relate to sales document category **K** (credit memo request). This sales document category is defaulted into the Mass Change of Sales Documents app (F5091) by using this app. Once the mass change is initiated, it runs as a background job.

The initial step is to search for the credit memo requests that need to be changed by using the standard SAP filters available, such as sales order number, sales area, customer reference, overall delivery status, material number, sold-to party, and ship-to party. From the list, select the documents you want to change using the mass change feature, and then drop down the **Change** quick action to reveal the options for amendments. The options for changing depend upon the sales document category selected (for this app, sales document category **K**). If the background job fails, you can repeat the changes for the individual item that failed or schedule a new job altogether.

> **Tip**
>
> It is possible to define jobs to run in parallel, for performance purposes, by configuring this via the following menu path: **SPRO • Sales and Distribution • Sales • App-Specific Settings • Apps for Sales Documents • Mass Change of Sales Documents • Define Parallel Processing for Mass Change of Sales Documents**.

Mass Change of Customer Returns (F5279)

Application Type: Transactional

This app is used to make mass changes to the sales documents that relate to sales document category **H** (returns). This sales document category is defaulted into the Mass Change of Sales Documents app (F5091) by using this app. Once the mass change is initiated, it runs as a background job.

The initial step is to search for the customer returns that need to be changed by using the standard SAP filters available, such as sales order number, sales area, customer reference, overall delivery status, material number, sold-to party, and ship-to party. From the list, select the documents you want to change using the mass change feature, and then drop down the **Change** quick action to reveal the options for amendments. The options for changing depend upon the sales document category selected (for this app, sales document category **H**). If the background job fails, you can repeat the changes for the individual item that failed or schedule a new job altogether.

Tip

It is possible to define jobs to run in parallel, for performance purposes, by configuring this via the following menu path: **SPRO • Sales and Distribution • Sales • App-Specific Settings • Apps for Sales Documents • Mass Change of Sales Documents • Define Parallel Processing for Mass Change of Sales Documents**.

Tip

It is possible to define jobs to run in parallel, for performance purposes, by configuring this via the following menu path: **SPRO • Sales and Distribution • Sales • App-Specific Settings • Apps for Sales Documents • Mass Change of Sales Documents • Define Parallel Processing for Mass Change of Sales Documents**.

Mass Change of Debit Memo Requests (F5280)

Application Type: Transactional

This app is used to make mass changes to the sales documents that relate to sales document category **L** (debit memo requests). This sales document category is defaulted into the Mass Change of Sales Documents app (F5091) by using this app. Once the mass change is initiated, it runs as a background job.

The initial step is to search for the debit memo requests that need to be changed by using the standard SAP filters available, such as sales order number, sales area, customer reference, overall delivery status, material number, sold-to party, and ship-to party. From the list, select the documents you want to change using the mass change feature, and then drop down the **Change** quick action to reveal the options for amendments. The options for changing depend upon the sales document category selected (for this app, sales document category **L**). If the background job fails, you can repeat the changes for the individual item that failed or schedule a new job altogether.

Mass Change of Sales Contracts (F5275)

Application Type: Transactional

This app is used to make mass changes to the sales documents that relate to sales document category **G** (contracts). This sales document category is defaulted into the Mass Change of Sales Documents app (F5091) by using this app. Once the mass change is initiated, it runs as a background job.

The initial step is to search for the sales contracts that need to be changed by using the standard SAP filters available, such as sales order number, sales area, customer reference, overall delivery status, material number, sold-to party, and ship-to party. From the list, select the documents you want to change using the mass change feature, and then drop down the **Change** quick action to reveal the options for amendments. The options for changing depend upon the sales document category selected (for this app, sales document category **G**). If the background job fails, you can repeat the changes for the individual item that failed or schedule a new job altogether.

Tip

It is possible to define jobs to run in parallel, for performance purposes, by configuring this via the following menu path: **SPRO • Sales and Distribution • Sales • App-Specific Settings • Apps for Sales Documents • Mass Change of Sales Documents • Define Parallel Processing for Mass Change of Sales Documents**.

3

Mass Change of Sales Documents (F5091)

Application Type: Transactional

This app is used to make mass changes to any type of sales document. Once the mass change is initiated, it runs as a background job.

The initial step is to search for the sales documents that need to be changed by using the standard SAP filters available, such as sales order number, sales area, customer reference, overall delivery status, material number, sold-to party, and ship-to party. From the list, select the documents you want to change using the mass change feature, and then drop down the **Change** quick action to reveal the options for amendments. The options for changing depend upon the sales document category selected. If the background job fails, you can repeat the changes for the individual item that failed or schedule a new job altogether.

sales area, customer reference, overall delivery status, material number, sold-to party, and ship-to party. From the list, select the documents you want to change using the mass change feature, and then drop down the **Change** quick action to reveal the options for amendments. The options for changing depend upon the sales document category selected (for this app, sales document category **C**). If the background job fails, you can repeat the changes for the individual item that failed or schedule a new job altogether.

> **Tip**
>
> It is possible to define jobs to run in parallel, for performance purposes, by configuring this via the following menu path: **SPRO • Sales and Distribution • Sales • App-Specific Settings • Apps for Sales Documents • Mass Change of Sales Documents • Define Parallel Processing for Mass Change of Sales Documents**.

> **Tip**
>
> It is possible to define jobs to run in parallel, for performance purposes, by configuring this via the following menu path: **SPRO • Sales and Distribution • Sales • App-Specific Settings • Apps for Sales Documents • Mass Change of Sales Documents • Define Parallel Processing for Mass Change of Sales Documents**.

Mass Change of Sales Orders Without Charge (F5336)

Application Type: Transactional

This app is used to make mass changes to the sales documents that relate to sales document category **I** (sales orders without charge). This sales document category is defaulted into the Mass Change of Sales Documents app (F5091) by using this app. Once the mass change is initiated, it runs as a background job.

The initial step is to search for the sales orders without charge that need to be changed by using the standard SAP filters available, such as sales order number, sales area, customer reference, overall delivery status, material number, sold-to party, and ship-to party. From the list, select the documents you want to change using the mass change feature, and then drop down the **Change** quick action to reveal the options for amendments. The options for changing depend upon the sales document category selected (for this app, sales document category I). If the background job fails, you can repeat the changes for

Mass Change of Sales Orders (F5211)

Application Type: Transactional

This app is used to make mass changes to the sales documents that relate to sales document category **C** (sales orders). This sales document category is defaulted into the Mass Change of Sales Documents app (F5091) by using this app. Once the mass change is initiated, it runs as a background job.

The initial step is to search for the sales orders that need to be changed by using the standard SAP filters available, such as sales order number, the individual item that failed or schedule a new job altogether.

> **Tip**
>
> It is possible to define jobs to run in parallel, for performance purposes, by configuring this via the following menu path: **SPRO** • **Sales and Distribution** • **Sales** • **App-Specific Settings** • **Apps for Sales Documents** • **Mass Change of Sales Documents** • **Define Parallel Processing for Mass Change of Sales Documents**.

Mass Change of Sales Quotations (F5277)

Application Type: Transactional

This app is used to make mass changes to the sales documents that relate to sales document category **B** (quotations). This sales document category is defaulted into the Mass Change of Sales Documents app (F5091) by using this app. Once the mass change is initiated, it runs as a background job.

The initial step is to search for the quotations that need to be changed by using the standard SAP filters available, such as sales order number, sales area, customer reference, overall delivery status, material number, sold-to party, and ship-to party. From the list, select the documents you want to change using the mass change feature, and then drop down the **Change** quick action to reveal the options for changes. The options depend on the sales document category selected (for this app, sales document category **B**). If the background job fails, you can repeat the changes for the individual item that failed or schedule a new job altogether.

> **Tip**
>
> It is possible to define jobs to run in parallel, for performance purposes, by configuring this via the following menu path: **SPRO** • **Sales and Distribution** • **Sales** • **App-Specific Settings** •

Apps for Sales Documents • Mass Change of Sales Documents • Define Parallel Processing for Mass Change of Sales Documents.

Mass Change of Sales Scheduling Agreements (F5278)

Application Type: Transactional

This app is used to make mass changes to the sales documents that relate to sales document category **E** (scheduling agreements). This sales document category is defaulted into the Mass Change of Sales Documents app (F5091) by using this app. Once the mass change is initiated, it runs as a background job.

The initial step is to search for the sales scheduling agreements that need to be changed by using the standard SAP filters available, such as sales order number, sales area, customer reference, overall delivery status, material number, sold-to party, and ship-to party. From the list, select the documents you want to change using the mass change feature, and then use the **Change** quick action dropdown to reveal the options for changes. The options depend on the sales document category selected (for this app, sales document category **E**). If the background job fails, you can repeat the changes for the individual item that failed or schedule a new job altogether.

> **Tip**
>
> It is possible to define jobs to run in parallel, for performance purposes, by configuring this via the following menu path: **SPRO** • **Sales and Distribution** • **Sales** • **App-Specific Settings** • **Apps for Sales Documents** • **Mass Change of Sales Documents** • **Define Parallel Processing for Mass Change of Sales Documents**.

Mass Changes Across Sales Document Categories (F5335)

Application Type: Transactional

This app is used to make mass changes to the multiple sales documents across different sales document categories. Once the mass change is initiated, it runs as a background job. This app is interchangeable with the Mass Change of Sales Documents (F5091) app. The only clear difference here is that this app is used to specify documents by document category type (e.g., sales inquiries = **A**, sales quotations = **B**, sales orders = **C**, etc.)

The initial step is to search for the sales scheduling agreements that need to be changed by using the standard SAP filters available, such as sales order number, sales area, customer reference, overall delivery status, material number, sold-to party, and ship-to party. From the list, select the documents you want to change using the mass change feature, and then drop down the **Change** quick action to reveal the options for amendments. The options for changing depend upon the sales document category selected. If the background job fails, you can repeat the changes for the individual item that failed or schedule a new job altogether.

Tip

It is possible to define jobs to run in parallel, for performance purposes, by configuring this via the following menu path: **SPRO • Sales and Distribution • Sales • App-Specific Settings • Apps for Sales Documents • Mass Change of Sales Documents • Define Parallel Processing for Mass Change of Sales Documents**.

Monitor Allocation Object Changes (F7142)

Application Type: Analytical

As part of the SAP S/4HANA functions for advanced available-to-promise (ATP), this app,
which was introduced with SAP S/4HANA 2023, allows you to display all change documents for product allocation objects. A product allocation object is a mechanism used to control and limit the availability of certain products or product groups, especially in high-demand or supply-constrained scenarios. It allows companies to allocate limited quantities of products to specific customers, regions, or channels, thus ensuring fair distribution and protecting priority commitments. Examples of product allocation object criteria are customer, customer group, sales organization, time periods (daily, weekly, monthly), and product hierarchy. Combinations of these attributes can be used to drive allocation of inventory at the advanced ATP stage. Changes to header-level details (e.g., period type or quantity unit), changes to characteristic value combination levels (e.g., characteristic values), and changes on the period level (e.g., quantities) can all be displayed in the app.

Monitor Allocation Sequence Changes (F7143)

Application Type: Analytical

As part of the SAP S/4HANA functions for advanced available-to-promise (ATP), this app, which was introduced with SAP S/4HANA 2023, allows you to display all change documents for product allocation sequences. The product allocation sequence is used to define the order in which SAP checks product allocation objects. The sequence is checked step by step until available allocation quantities are found. Changes to product allocation sequence master data, such as header data, group data like group type, constraint data like backward/forward consumption or allocation rate, and production/location assignment data like material or validities, can all be displayed in the app.

Monitor Allocation Value Combination Changes (F7144)

Application Type: Analytical

As part of the SAP S/4HANA functions for advanced available-to-promise (ATP), this app, which was introduced with SAP S/4HANA 2023, allows you to display all change documents for product allocation characteristic value combinations. Product allocation characteristic value combinations are the rows or entries in a product allocation object that determine how much of a product is available for a specific combination of attributes. Changes to product allocation characteristic value combination level (which is based on characteristics defined in the allocation object, such as customer group, distribution channel, product, etc.) as well as changes at period level like quantities, can all be displayed in the app.

Monitor BOP Run (F2159)

Application Type: Transactional

The Configure BOP Variant app (F2160) is used to create backorder processing (BOP) runs. This app follows from the Configure BOP Variant app, in that you can monitor the results of the BOP run at different levels. Typically, an administrator would check the results of the BOP run to ensure that they are in line with the organization's strategy for BOP. If the results do not comply, then you can change the variant and re-execute the run. The BOP run also can be rerun in simulation mode. Additional features include displaying the substituted storage location if alternative-based confirmation (ABC) is active.

Tip

If the results are not as expected, the Release for Delivery app (F1786) can be used on a document-by-document basis to change the confirmed quantities manually.

Monitor Condition Contracts – Customers (F2954)

Application Type: Transactional

SAP S/4HANA has replaced the rebates functionality with condition contracts for customers. With this app, you can get a comprehensive overview of the condition contracts that are in use. Search filters are available for fields such as customer number, contract type, and sales organization. Filters can be displayed in standard list format or in graphical format. Individual condition contracts can be changed using the quick actions **Release** and **Lock for Settlement**. In addition, you can create new condition contracts by selecting the **Create** button.

Monitor Condition Contracts – External Commissions (F3480)

Application Type: Transactional

With this app, you can get a comprehensive overview of the condition contracts that are used for external sales commission agreements. Search filters are available for fields such as supplier number, contract type, and purchasing organization. Filters can be displayed in standard list format or in graphical format. Individual condition contracts can be changed using the quick actions **Release** and **Lock for Settlement**. In addition, you can create new condition contracts by selecting the **Create** button.

Monitor Condition Contracts – Internal Commissions (F7264)

Application Type: Transactional

With this app, you can get a comprehensive overview of the condition contracts that are used for internal sales commission agreements. Search filters are available for fields such as personnel number, contract type, and sales organization. Filters can be displayed in standard list format or in graphical format. Individual condition contracts can be changed using the **Release** and **Lock for Settlement** quick actions. In addition, you can create new condition contracts by selecting the **Create** button.

Monitor Condition Contracts – Royalties (F4578)

Application Type: Transactional

With this app, you can get a comprehensive overview of the condition contracts that are used for royalties administration. Search filters are available for fields such as supplier number, contract type, and purchasing organization. Filters can be displayed in standard list format or in graphical format. Individual condition contracts can be changed using the quick actions **Release** and **Lock for Settlement**. In addition, you can create new condition contracts by selecting the **Create** button.

Monitor Condition Contracts – Sales Rebates (F6883)

Application Type: Transactional

With this app, you can get a comprehensive overview of the condition contracts that are used for sales rebates. Search filters are available for fields such as customer number, contract type, and sales organization. Filters can be displayed in standard list format or in graphical format. Individual condition contracts can be changed using the quick actions **Release** and **Lock for Settlement**. In addition, you can create new condition contracts by selecting the **Create** button.

Monitor Customer JIT Calls – Component Groups (F4749)

Application Type: Analytical

This app allows you to monitor the usage of just-in-time (JIT) component groups according to

their internal processing status. The internal processing status of a JIT call indicates what stage of the processing lifecycle the call has reached within the SAP system and how far the system has gone in creating and processing the relevant documents (e.g. delivery schedules, delivery notes, goods issues). Possible statuses vary according to the configuration settings in the SAP system, but typical values include JIT call created but not yet processed (**01**), delivery schedule or delivery document created (**02**), outbound delivery has been picked or packed (**03**), goods issue posted (**04**), JIT call completed (fully processed; **05**), and error (**99**).

Search filters are available for fields such as internal processing status, plant, and ship-to party. The filters can be displayed in a standard list format or in graphical format. The results can be displayed in graphical only format, tabular format only, or both together.

Monitor Customer JIT Calls Components (F4748)

Application Type: Analytical

This app allows you to monitor the usage of just-in-time (JIT) components according to their internal processing status. The internal processing status of a JIT call indicates what stage of the processing lifecycle the call has reached within the SAP system and how far the system has gone in creating and processing the relevant documents (e.g. delivery schedules, delivery notes, goods issues). Possible statuses vary according to the configuration settings in the SAP system, but typical values include JIT call created but not yet processed (**01**), delivery schedule or delivery document created (**02**), outbound delivery has been picked or packed (**03**), goods issue posted (**04**), JIT call completed (fully processed; **05**), and error (**99**).

Search filters are available for fields such as internal processing status, component material, plant, and ship-to party. The filters can be displayed in a standard list format or in graphical format. The results can be displayed in graphical only format, tabular format only, or both together.

Monitor Mass Changes of Sales Documents (F5210)

Application Type: Transactional

Using the Mass Change of Sales Documents app (F5091), you can update multiple sales documents at one time. The changes are scheduled in a background job. This app can be used to monitor the status of the background job, which can be **In Progress**, **Completed**, or **Failed**, as well as repeating the change for selected individual documents or by scheduling an entirely new job. Search functions are available for fields such as job status, created by, and start date and time.

Navigation to the entire sales suite of "Manage" SAP Fiori apps is supported, such as Manage Sales Orders (F1873).

Tip

It is possible to mark failed jobs as **Completed**, but once this is done, it is not possible to change the job status or any other details.

Monitor Product Allocation Characteristic Value Combinations (F3476)

Application Type: Transactional, Analytical

As part of the SAP S/4HANA functions for advanced available-to-promise (ATP), this app allows you to monitor the quantity and consumption data for the product allocation characteristic value combinations that belong to a product allocation object. Typically, when you are faced with errors in the product allocation functions, this is the best place to start. An example of an error could be that the allocation quantity is zero. This normally means that allocation quantities have not yet been maintained

for the relevant product allocation characteristic value combination. In addition, sometimes product allocation characteristic value combinations can overlap causing ambiguity and incorrect results. This app can be used to identify such cases and change so that they are distinct.

Navigation to related apps such as Configure Product Allocation (F2119) and Manage Product Allocation Planning Data (F2121) is supported.

Monitor Product Allocation Order Items (F3477)

Application Type: Transactional, Analytical

This app allows you to monitor the usage of the product allocation feature against sales orders and stock transport orders. The results can be viewed in a table format or graphically according to their availability confirmation check status. Typically, when you are faced with errors in the product allocation functions related to order items, this is the best place to start. Order item errors in product allocation include such errors as a situation where the allocation has already been fully consumed; this is normally fixed by increasing the allocation quantity, prioritizing orders in a different way, or retriggering backorder processing to redistribute the available stock.

Navigation to related apps such as Configure Product Allocation (F2119) and Manage Product Allocation Planning Data (F2121) is supported.

Monitor Product Allocation Periods (F3475)

Application Type: Transactional, Analytical

This app allows you to monitor the usage of the product allocation feature a specified product allocation period range. The results can be viewed in a table format or graphically according to their availability confirmation check status. Typically, when you are faced with errors in the product allocation functions related to a given period, this is the best place to start. Typical errors here are related to an incorrect assignment of the allocation period time bucket. The fix would be to review the allocation period and correct the date ranges.

Navigation to related apps such as Configure Product Allocation (F2119) and Manage Product Allocation Planning Data (F2121) is supported.

Monitor Route Data (F4558)

Application Type: Transactional

With this app, you can monitor the usage of route data and where the route ID currently is in its lifecycle. Search functions exist for fields such as route ID, route lifecycle status, settlement status, mobile data status, route type, route category, departure date, processing date, license plate, and driver. The app is useful in that you can not only track all details of the route ID, but also analyze error messages, enable and disable manual settlement for a route, or cancel a route.

Monitor Settlement Documents (F4165)

Application Type: Transactional

As part of the suite of products for settlement management in SAP S/4HANA, this app can be used to show an overview of all types of settlement documents within a given time frame. Various drilldowns are available to interrogate the data for the settlement document, the bill-to party, and the condition contract. Search functions exist for fields such as settlement document, condition contract, and process category (which can be used to filter for settlement documents that are related to a specific condition contract process such as customer rebates or supplier rebates). Results can be shown in tabular format or in graphical format.

Monitor Value Chains (F4854)

Application Type: Transactional

The value chain in SAP is the end-to-end process and activities required in order to create, deliver, and support a product or service. This app can be used specifically to track the effectiveness of the following value chain processes:

- **Advanced intercompany sales**
  Intercompany sales are triggered when the selling company within an organization is different from the delivering company within an organization. This is triggered by the company code that is assigned to the sales organization versus the company code assigned to the delivering plant. Advanced intercompany sales is a richer version of traditional intercompany sales, with automation in the creation of the purchase order to the affiliate company and then a subsequent automation of the second intercompany sales order to sell from the affiliate company to the selling company. The process triggers a movement of the stock into valuated stock in transit and then out of valuated stock in transit when the selling company posts goods issue.

- **Advanced intercompany stock transfer**
  Advanced intercompany stock transfer is similar to advanced intercompany sales, except the intention here is to move stock from the affiliate company to the selling company. The same process flow applies.

- **Sell-from-stock with valuated stock in transit**
  It is possible sell from valuated stock in transit. In this process, the first goods movement type from the outbound delivery is to move the stock from unrestricted to valuated stock in transit. At this point, the stock still belongs to the selling company. According to the external transfer of control date, the goods issue from stock in transit is posted, meaning that the stock no longer belongs to the selling company.

Usefully, the app allows you to display the entire process as a flow diagram at either the header or item level. From within the app, you can reprocess documents that are in error, cancel documents, and navigate to related apps.

My Inbox – Approve Credit Memo Requests (F5042)

Application Type: Transactional

The app allows you to review and act on credit memo request approvals directly from your inbox. It provides all relevant details of the credit memo request, such as all involved business partners, items and full item details, quantities, prices, and process flow, allowing approvers to make quick and informed decisions. You can release, reject, request rework, forward, or suspend the credit memo requests, ensuring efficient handling of sales requests and maintaining smooth operational workflows.

My Inbox – Approve Customer Returns (F5053)

Application Type: Transactional

The app allows you to review and act on customer returns directly from your inbox. It provides all relevant details of the customer return, such as all involved business partners, items and full item details, quantities, prices, and process flow, allowing approvers to make quick and informed decisions. You can release, reject, request rework, forward, or suspend the customer returns, ensuring efficient handling of sales requests and maintaining smooth operational workflows.

My Inbox – Approve Debit Memo Requests (F7438)

Application Type: Transactional

The app allows you to review and act on debit memo request approvals directly from your

inbox. It provides all relevant details of the debit memo request, such as all involved business partners, items and full item details, quantities, prices, and process flow, allowing approvers to make quick and informed decisions. You can release, reject, request rework, forward, or suspend the debit memo requests, ensuring efficient handling of sales requests and maintaining smooth operational workflows.

My Inbox – Approve Preliminary Billing Documents (F5043)

Application Type: Transactional

The app allows you to review and act on preliminary billing document approvals directly from your inbox. It provides all relevant details of the preliminary billing document, such as all involved business partners, items and full item details, quantities, prices, and process flow, allowing approvers to make quick and informed decisions. You can release, reject, request rework, forward, or suspend the preliminary billing documents, ensuring efficient handling of sales requests and maintaining smooth operational workflows.

My Inbox – Approve Sales Orders (F5041)

Application Type: Transactional

The app allows you to review and act on sales order approvals directly from your inbox. It provides all relevant details of the sales order request, such as all involved business partners, items and full item details, quantities, prices, and process flow, allowing approvers to make quick and informed decisions. You can release, reject, request rework, forward, or suspend the sales order requests, ensuring efficient handling of sales requests and maintaining smooth operational workflows.

My Inbox – Approve Sales Orders Without Charge (F5432)

Application Type: Transactional

The app allows you to review and act on sales orders without charge approvals directly from your inbox. It provides all relevant details of the sales orders without charge, such as all involved business partners, items and full item details, quantities, prices, and process flow, allowing approvers to make quick and informed decisions. You can release, reject, request rework, forward, or suspend the sales order without charge, ensuring efficient handling of sales requests and maintaining smooth operational workflows.

My Inbox – Approve Sales Quotations (F5044)

Application Type: Transactional

The app allows you to review and act on sales quotations approvals directly from your inbox. It provides all relevant details of the sales quotation request, such as all involved business partners, items and full item details, quantities, prices, and process flow, allowing approvers to make quick and informed decisions. You can release, reject, request rework, forward, or suspend the sales quotation requests, ensuring efficient handling of sales requests and maintaining smooth operational workflows.

My Inbox Sales Documents (F6012)

Application Type: Transactional

The app allows you to review and act on all different types of sales document approvals directly from your inbox. It provides all relevant details of the sales document request, such as all involved business partners, items and full item details, quantities, prices, and process flow, allowing approvers to make quick and informed

decisions. You can release, reject, request rework, forward, or suspend the sales document, ensuring efficient handling of sales requests and maintaining smooth operational workflows.

Further information on the sales documents is retrieved by using the My Inbox – Approve Sales Orders (F5041), My Inbox – Approve Credit Memo Requests (F5042), My Inbox – Approve Sales Quotations (F5044), My Inbox – Approve Customer Returns (F5053), My Inbox – Approve Sales Orders Without Charge (F5432), and My Inbox – Approve Debit Memo Requests (F7438) apps.

My Sales Overview (F2200)

Application Type: Analytical

This hugely popular app can act as a central hub for sales functions. Actionable cards, which relate to separate SAP Fiori apps, are grouped together into a dashboard format. Standard cards that are displayed include **Open Sales Quotations**, **Open Sales Order**, **Blocked Credit Memo Requests**, **Customer Returns**, **Customer Contacts**, **Sales Order Fulfilment**, **Sales Quotation Pipeline**, and **Incoming Sales Orders**, and quick actions such as **Create Sales Order**, **Create Sales Quotation**, and **Create Credit Memo Request**. Most of these cards show dynamic graphical content on the front of the card, some of which can be changed by using dropdown options on the card (such as time period within the **Incoming Sales Orders** card). The cards visible by default are set according to the roles assigned to you. The screen placement of each card can be modified by simply dragging the cards around the screen. Search fields are available in the filter bar, to restrict the content shown on the cards. These fields include display currency, sold-to party, sales area, sales office, and sales group. These filters can be defaulted in from your default values in your SAP Fiori user settings.

There are multiple apps available as navigation targets, as each card represents a target SAP Fiori app, including all the "Manage" sales apps such as Manage Sales Orders (F1873), as well as standard SAP GUI transactions such as Create Sales Orders (VA01).

> **Tip**
>
> In your user menu, select **Manage Cards** to select which cards to show. In addition, it is important to note that users who have multiple roles assigned to them may experience performance issues with this app if too much data is retrieved. To mitigate this, limit the number of cards to show to only the ones that are most important for your role.

Order-to-Cash Performance (F2005)

Application Type: Analytical

This very useful app allows order-to-cash (OTC) managers to keep track of all the OTC processes within their organization. KPIs are used to give an overall health check of the OTC processes. The KPIs are organized into graphical illustrations of the company's OTC performance and can be time bound using date ranges. Tasks such as detecting changes to critical fields and tracking delivery and billing blocks are possible. In addition, an extremely useful feature is to view the lead times of the sales process—for example:

- The length of time taken between ordering and invoicing
- The length of time taken between ordering and delivery (order readiness)
- The length of time taken between ordering and goods issue
- The length of time taken between goods issue and invoicing

All these KPIs can be visualized in a bar chart, along with quantity statistics such as the value of the orders in each category, number of orders created, number of orders rejected, number of delivery blocks applied, and number of billing blocks applied. As with all analytical apps, the types of charts used can be changed from within the app.

Navigation to related apps such as Business Process Activities (F1609), Aggregated Business Process Activities (F1810), Order to Cash Performance Monitor – Time Series (F2006), and Order to Cash Performance – Overview (F2242) is supported.

Order-to-Cash Performance – Overview (F2242)

Application Type: Analytical

This app shows very similar data to the Order-to-Cash Performance app (F2005), but the difference here is that the data is organized into a standard SAP overview page. This overview page has cards representing navigation to related apps. Each card shows dynamic data on the front representing an overview of the health of the organization's order-to-cash (OTC) processes. The cards available are as follows:

- **Sales Orders per Month** (represented as a line graph)
- **Sales Order Items per Month** (represented as a line graph)
- **Deliveries, Pickings, Goods Issues per Month** (represented as a column chart)
- **Invoices per Month** (represented as a line graph)
- **Sales Order Totals** (represented as a bar chart)
- **Sakes Order Item Totals** (represented as a bar chart)
- **Delivery Totals** (represented as a bar chart)
- **Billing Lead Time** (represented as a bar chart)
- **Sales Order Lead Time** (represented as a bar chart)
- **Delivery Lead Time** (represented as a bar chart)

3

Order-to-Cash Performance Monitor – Time Series (F2006)

Application Type: Analytical

This very useful app allows order-to-cash (OTC) managers to keep track of all the OTC processes within their organization. It is almost identical to the Order-to-Cash Performance (F2005) app with the exception that the KPIs are, by default, time series defined for the last 365 days. KPIs are used to give an overall health check of the OTC processes. The KPIs are organized into graphical illustrations of the company's OTC performance and can be time bound using date ranges. Tasks such as detecting changes to critical fields and tracking delivery and billing blocks are possible. In addition, an extremely useful feature is to view the lead times of the sales process—for example:

- The length of time taken between ordering and invoicing
- The length of time taken between ordering and delivery (order readiness)
- The length of time taken between ordering and goods issue
- The length of time taken between goods issue and invoicing

All these KPIs can be visualized in a bar chart, along with quantity statistics such as the value of the orders in each category, number of orders created, number of orders rejected, number of delivery blocks applied, and number of billing blocks applied. As with all analytical apps, the types of charts used can be changed from within the app.

Navigation to related apps such as Business Process Activities (F1609), Aggregated Business Process Activities (F1810), Order to Cash Performance Monitor – Time Series (F2006), and Order to Cash Performance – Overview (F2242) is supported.

Outbound Delivery (S/4HANA) (F0233A)

Application Type: Fact Sheet

This app shows general information related to the outbound delivery, such as ship-to party, shipping point, delivery date, and individual items and their quantity requirements and plant and storage location. As with all SAP Fiori fact sheet apps, this app can be accessed from other apps. The key usage is to access the app from an enterprise search result for outbound deliveries, but the app also can be accessed from related apps such as Manage Outbound Deliveries (F0867A), Pick Outbound Delivery (F0868), Analyze Delivery Logs (F0870A), Customer 360° View – Version 2 (F2187A), Analyze Delivery Performance – Shipped as Planned (F2878A), Settle Route Data (F4242), Check-Out/In Differences (F4741), and Route Document Flow (F4742).

It is possible to navigate directly to related apps such as Material Master (F0338A), Sales Order (S/4HANA) (F1814), and Customer Master Fact-Sheets (F0046A).

Transaction Code

The SAP GUI equivalent app is Transaction VL03N (Display Outbound Delivery).

Overview of JIT Supply to Customer (F3933)

Application Type: Analytical

This app offers you a way to get a full overview of just-in-time (JIT) calls for a specific customer. The content is displayed in the app in a set of panels with graphical or tabular information displayed on each panel. The screen placement of each panel can be modified by simply dragging the panels around the screen. Standard panels include JIT calls by status, alerts (where the status of the call can be changed in a dropdown box in the panel), ship-to party summary, and JIT customer summary. Search functions include searching by ship-to party and sales area.

Navigation to the Manage Customer JIT Call (F3008) and Manage JIT Customer data (F3011) apps is supported.

Predicted Delivery Delay (F3408)

Application Type: Analytical

This app is used to predict potential delays in deliveries for open sales documents. The app can show the predicted delay to the planned delivery creation date, the requested delivery date, the predicted delay of delivery creation, the predicted delay of delivery processing, and the overall predicted delivery delay for sales document items. The results of the predictions are displayed as one of the following, where there are two objects to be predicted:

- **Late**: If one of the predictions is late and the other is early, the late status is prioritized and the overall status is set to late.
- **Early**: If one of the predictions is early and the other is on time, the early status is prioritized and the overall status is set to early.
- **On Time**: If one of the predictions is on-time and the other is not predicted, the on-time status is prioritized and the overall status is set to on time.
- **Not Predicted**: If both predictions are not predicted, the overall status is also not predicted. This happens when no training has taken place for either of the attributes used.

You can search for open sales documents by fields such as sales document, sold-to party, sales organization, material, plant, required delivery date, and delivery block. The results are displayed in tabular and graphical formats.

Tip

The app uses machine learning tools. A prerequisite for using this app is that all scenarios are created in the Intelligent Scenarios app (F4469). The model created must be trained using the Intelligent Scenario Management app (F4470).

Preliminary Billing Document (F2874)

Application Type: Fact Sheet

Sometimes it may be necessary to discuss the contents of a billing document, such as pricing, quantities, and discounts, with clients before finalizing the agreed document. This app can be used to display an overview of the preliminary billing documents.

Any changes that need to be carried out can be made using Transaction VFP2 (Change Preliminary Billing Documents). The app is called from the SAP Fiori apps Customer 360° View – Version 2 (F2187A), Manage Preliminary Billing Documents (F2875), and Create Preliminary Billing Documents (F2876).

Transaction Code

The SAP GUI equivalent app is Transaction VFP3 (Display Preliminary Billing Document).

Pricing Elements – Flexible Analysis (F4572)

Application Type: Web Dynpro

This app allows you to analyze pricing in sales documents according to condition types that are used. The results can show all sales documents that use a specific condition type in a condition time period, counting back from the number of days from the current date. The app allows you to see all the pricing elements that have an influence over the sales volume for a given period. For example, it may be beneficial to display the sales documents and condition types for a specific quarter, to identify which discounts have been applied, which taxes have been charged and how much revenue has been generated through delivery charges. The amount displayed is "normalized" for credit memos by multiplying the value by minus 1. As with all the flexible analysis Web Dynpro SAP Fiori apps, available fields and columns can be changed. Search features are available in fields such as exchange rate type, currency, condition type, sales area, sales office, and date. The minimum entries in the selection are number of days from current date, exchange rate type, and display currency. It is recommended, for performance reasons, to use additional features to narrow down the search.

Product Allocation Overview (F3474)

Application Type: Analytical

This SAP Fiori overview page gives you the ability to see a complete overview of the product allocation situation for a given set of product allocation objects, periods, and characteristic value combinations. The overview page has selected cards to navigate to the related apps such as **Overloaded Periods**, **Underloaded Periods**, **Highloaded Periods**, **Characteristic Value Combinations**, and **Product Allocation Order Items**. Most of these cards show dynamic graphical content on the front of the card, some of which can be changed by using dropdown options on the card (such as **Load** within the **Highloaded Periods** card). The cards visible by default are set according to the roles assigned to you. The screen placement of each card can be modified by simply dragging the cards around the screen.

> **Tip**
>
> In your user menu, select **Manage Cards** to select which cards to show. In addition, it is important to note that users who have multiple roles assigned to them may experience performance issues with this app if too much data is retrieved. To mitigate this, limit the number of cards to show to only the ones that are most important for your role.

Quotation Conversion Rates (F1904)

Application Type: Analytical

This app allows sales managers to track the conversion rate of quotations to sales orders. In addition, the app employs machine learning functionality to predict future conversion rates. The quotation conversion rates are displayed in a tabular form but also can be displayed graphically. The app uses CDS view C_SLSQTANCONVERSIONRATEQ to retrieve the relevant data. To show the prediction rates, you must click the **Show Mini Charts** button at the top right of the screen; mini charts are hidden by default.

Follow-up actions are available by navigating to related apps such as Manage Sales Quotations (F1852).

> **Tip**
>
> A prerequisite for using predictive values is to enable prediction in the Intelligent Scenario Management app (F4470).

Reclassify Products – Commodity Codes (F2152)

Application Type: Transactional

This app is part of SAP S/4HANA for international trade. Using this functionality, commodity codes can be reassigned to classified products. It is also possible to assign commodity codes with validity dates to products, as well as add remarks during the process. Search functions are available for fields such as numbering scheme, valid on date, commodity code, product, product description, product type, product group, and division.

Reclassify Products – Customs Tariff Numbers (F3147)

Application Type: Transactional

This app is part of SAP S/4HANA for international trade. Using this functionality, customs tariff numbers can be reassigned to classified products. It is also possible to assign customs tariff numbers with validity dates to products, as well as add remarks during the process. Search functions are available for fields such as numbering scheme, valid on date, customs tariff number, product, product description, product type, product group, and division.

Reclassify Products – Intrastat Service Codes (F2155)

Application Type: Transactional

This app is part of SAP S/4HANA for international trade. Using this functionality, Intrastat service codes can be reassigned to classified products. It is also possible to assign Intrastat service codes with validity dates to products, as well as add remarks during the process. Search functions are available for fields such as numbering scheme, valid on date, Intrastat service code, product, product description, product type, product group, and division.

Reclassify Products – Legal Control (F2391)

Application Type: Transactional

This app is part of SAP S/4HANA for international trade. Using this functionality, control classes and control groupings can be reassigned to classified products. It is also possible to assign control classes and control groupings with validity dates to products, as well as add remarks

during the process. Search functions are available for fields such as legal regulation, valid on date, control class, control grouping, product, product description, product type, product group, commodity code, and division.

Release for Delivery (F1786)

Application Type: Transactional

This app allows you to view the overall delivery and availability situation for a given set of sales parameters such as material, created on date range, material group, and sales document number. The app shows sales documents, along with the situation regarding required material availability date assigned to the sales order versus shipping dates and confirmed quantities. Using the app, order fulfilment managers can redistribute available quantities of inventory between selected sales documents. It is also possible to use the app to display only the documents related to materials that certain teams are responsible for.

Resolve Blocked Documents – Trade Compliance (F2792)

Application Type: Transactional

This app is part of SAP S/4HANA for international trade. Using this functionality, documents such as sales orders and deliveries can be blocked due to missing legal regulations, or embargo blocks. The home screen for the app allows you to search for documents by filters such as block type, company code, and sales document category.

Although this app is still available, it is deprecated for the current release. We recommend using the Analyze and Resolve Blocked Documents – Trade Compliance (S/4HANA) app (F5445).

Resolve Payment Card Issues (F2758)

Application Type: Transactional

This app is used to schedule background jobs that can process sales documents that have payment cards associated with them, which have not yet been authorized. The documents are selected according to the status of the payment card, and the validity date of the payment. The background job can be scheduled with a job template that has previously been created, a job name, and organizational data for the selection, such as sales organization, distribution channel, division, sales office, and sales group.

Resolve Payment Card Issues – Reauthorizations (F4793)

Application Type: Transactional

Payment card processing has an authorization horizon, and when sales orders or deliveries are created, the material availability date or the billing date may be outside the authorization horizon. In addition, the available-to-promise (ATP) check may not return a date. In these cases, payment cards can only be preauthorized. This app is used to search for sales documents (orders and deliveries) that require reauthorization of their associated payment cards.

Tip

If multiple documents are found that need reauthorizing, then they can be processed together by selecting them using the checkboxes and then clicking the **Reauthorize** button.

Returns Delivery (S/4HANA) (F0234A)

Application Type: Fact Sheet

This app shows an overview of individual returns deliveries, with options to navigate to related apps such as Customer Master Fact-Sheets (F0046A), Material Master (F0338A), and Customer Return (F1815). The document flow for the returns delivery can be displayed as a graphical process flow at the bottom of the screen. The app is normally accessed via the enterprise search when returns deliveries have been selected.

Route Document Flow (F4742)

Application Type: Transactional

Specific routes can be interrogated with this app to show all customer visits that have uncompleted documents, such as goods issue, freight order, and billing document. It is possible to navigate to these documents to complete the process from within the app. The app shows all the documents relevant for a specific route, with information as to whether the documents are completed or not in the **Uncompleted Documents** field. It is possible to click the line to view the status of the documents in a document flow menu tree.

From here, you can click the **Related Apps** button to navigate to the apps Outbound Delivery (F0233A), Manage Billing Documents (F0797), Material Documents Overview (F1077), Return Delivery (F1996), and Track Sales Order Details (F2981).

Route Overview (F4743)

Application Type: Transactional

This overview page is a useful app to show all route-related data. Routes can be filtered using the source location, route type, and departure date. The screen shows all the information related to the selected routes as a series of cards, which can be clicked on to navigate to related apps. The screen placement of each card can be modified by simply dragging the cards around the screen. The cards available are as follows:

- Routes with Settlement Status
- Routes by Lifecycle Status
- Route Settlement Status (as a donut chart)
- Route Lifecycle Stats (as a donut chart)
- Visits with Uncompleted Documents
- Check-In Differences
- Invoice Differences
- Quick Links, such as **Route Document Flow**, **Collected Payments**, **Settle Route Data**, and **Monitor Route Data**

Navigation to the related Settle Route Data (F4242), Monitor Route Data (F4558), Route Document Flow (F4742), and Collected Payments (F4747) apps is supported.

Tip

In your user menu, select **Manage Cards** to select which cards to show. In addition, it is important to note that users who have multiple roles assigned to them may experience performance issues with this app if too much data is retrieved. To mitigate this, limit the number of cards to show to only the ones that are most important for your role.

Route Process Tracking (F4744)

Application Type: Analytical

With this app, you can track the full lifecycle of a route through all documents where the route is relevant. This includes outbound deliveries, freight orders in transportation management (TM), warehouse execution status (where extended warehouse management [EWM] is employed), picking, and goods movement. The filter options are numerous, and the results can be displayed in a table format, with statuses shown at each stage.

Sales Contract Fulfillment Rates (F1905)

Application Type: Analytical

Sales contract value usage can be monitored using this app. The sales contract fulfilment rate is the rate at which the value in your sales contracts is being fulfilled by subsequent documents. Using the app, you can identify sales contracts that are not being utilized to their maximum, as well as identifying sales contracts that have an upcoming renewal date. The data is displayed in chart format, with three standard dimensions in use:

- **Sales Organization**: Shows in bar chart format the contract fulfilment rate percentage for each sales organization
- **Customer**: Shows in bar chart format the contract fulfilment rate percentage for each customer
- **Top 10 Contracts by Target Value**: Shows in bubble chart format the top 10 sales contracts in terms of target value

Sales Contract (S/4HANA) (F2026)

Application Type: Fact Sheet

This app shows an overview of individual sales contracts. Available for display are the sales contract items, as well as the business partners involved in the sales contract. In addition, related sales documents such as sales orders can be viewed. Navigation to related apps such as Customer – 360° View (F2187) and Manage Sales Contracts (F1851) is supported.

Transaction Code

The SAP GUI equivalent app is Transaction VA43 (Display Sales Contract).

Sales Inquiry (F2369)

Application Type: Fact Sheet

This app shows an overview of individual sales inquiries. You can also view the business partners involved in the sales inquiry and related sales documents such as sales orders. Navigation to related apps such as Customer – 360° View (F2187) is supported.

Transaction Code

The SAP GUI equivalent app is Transaction VA13 (Display Sales Inquiry).

Sales Management Overview (F2601)

Application Type: Analytical

This app allows sales managers to see a full overview of the sales cycle in analytical cards, with dynamic data, usually in the form of charts, on each card. The cards can be clicked on to navigate to related apps. The cards visible by default are set according to the roles assigned to you. The screen placement of each card can be modified by simply dragging the cards around the screen. Cards available are as follows:

- Incoming Sales Orders
- Blocked Sales Orders
- Incomplete Sales Documents
- Overdue Sales Orders
- Profit Margin
- Customer Returns
- Backorder Items

You can navigate to related apps: Sales Order Fulfillment Issues (Version 2) (F0029A), Incoming Sales Orders – Flexible Analysis (F1249), Sales Volume – Profit Margin (F2271), List Incomplete Sales Documents (F2430), Sales Order Items – Backorders (F5307), and Customer Returns (W0139).

Tip

In your user menu, select **Manage Cards** to select which cards to show. In addition, it is important to note that users who have multiple roles assigned to them may experience performance issues with this app if too much data is retrieved. To mitigate this, limit the number of cards to show to only the ones that are most important for your role.

Sales Order Fulfillment Issues (Version 2) (F0029A)

Application Type: Transactional, Analytical

This app provides great insight into blocking processes for sales orders. These blocks can be in sales orders themselves (incomplete data, unconfirmed quantities, billing and delivery blocks, etc.), in supply (purchase order not created, manufacturing issue, etc.), in deliveries (incomplete data, shipping block, etc.), and in invoicing (e.g., transfer to accounting not completed). Additional blocks include international trade blocks such as legal control blocks and customer screening and embargo blocks, which can be pulled in from integration with SAP Global Trade Services (SAP GTS).

It is possible to select appropriate sales documents from the list to be able to carry out relevant quick actions. If an item is selected, then only the relevant quick actions are available. For example, if a sales order does not have a delivery block, then the quick action **Remove Delivery Block** is grayed out.

Multiple filters are available for selecting the data, and these filters can be displayed as raw data or in chart format. Navigation to the Track Sales Order Details app (F2981) is supported.

Sales Order Items – Backorders (F5307)

Application Type: Analytical

The intention of this app is to allow you to identify blockages in the sales cycle that are the cause of backorders. The definition of a backorder is an order that is unable to be confirmed or can only be partially confirmed. The confirmation normally fails due to a lack of product availability. This app shows in the format of a chart, the total number of backorders for various dimensions. The dimensions available are as follows:

- By Week
- By Customer Group
- By Customer Classification
- By Sold-to Party
- By Plant
- By Material Group
- By Product Hierarchy
- By Material
- By Sales Order

The app uses the Analytics – Confirmation of Sales Orders CDS view (C_SlsOrdConfAnlytsQry) for retrieval of the data.

Navigation to the Sales Orders – Demand Fulfilment (F2458) and Sales Order Items – Confirmed as Requested (F5308) apps is supported.

Tip

By viewing the data in this app by customer, you have a good opportunity to see which customers are not getting fair service and redistribute available stock to neglected customers.

Sales Order Items – Confirmed as Requested (F5308)

Application Type: Analytical

This app shows the percentage of confirmed sales orders in the last three weeks. The data is shown in the format of a chart. The figures are shown as either:

- Confirmed ratio: This shows the ratio of sales order items that have been confirmed as requested for the full quantity, with no delays.
- Delayed ratio: This shows the ratio of sales order items that have been confirmed as requested for the full quantity, but *not* on the customer requested date.

The app uses the CDS view Analytics – Confirmation of Sales Orders (C_SlsOrdConfAnlytsQry) for retrieval of the data.

Navigation to the Sales Orders – Demand Fulfilment (F2458) and Sales Order Items – Confirmed as Requested (F5308) apps is supported.

Sales Order Without Charge (F2303)

Application Type: Fact Sheet

This app allows you to view all details for a specific sales order without charge in one place. A sales order without charge is a specific category of sales order that defaults in an item category that is not relevant for billing. Typically, these are used to provide free samples to customers, replacements, promotional material, or shipment of goods for warranty reasons. The process flow for a sales order without charge is 1) sales order, 2) delivery, 3) goods issue. No billing document is generated. You can display the sales order items, the business partners involved in the sales order without charge, and related sales documents such as billing documents.

Navigation to related apps such as Customer – 360° View (F2187) and Manage Sales Orders (F2305) is supported. The app is a navigation target in many other apps in the sales cycle.

Transaction Code

The SAP GUI equivalent app is Transaction VA03 (Display Sales Order).

Sales Order (S/4HANA) (F1814)

Application Type: Fact Sheet

This app is one of the most heavily used apps in the sales and distribution space. It allows you to view all details for a specific sales order in one place. You can display the sales order items, the business partners involved in the sales order, and related sales documents such as deliveries. Navigation to related apps such as Customer – 360° View (F2187) and Manage Sales Orders (F1873) is supported. The app is a navigation target in many other apps in the sales cycle.

Tip

The document flow for the sales order can be displayed as a graphical process flow by clicking the **Process Flow** option in the app.

Transaction Code

The SAP GUI equivalent app is Transaction VA03 (Display Sales Order).

Sales Orders – Demand Fulfillment (F2458)

Application Type: Analytical

This app allows you to identify bottlenecks in the sales process, where it may not be possible to deliver a customer's goods to them by the customer required date. Selection filters are available for fields such as week or year, sold-to party, product, plant, delivery status, and sales document type. The detail in the app is displayed graphically with appropriate drilldowns to related apps such as Sales Order Items – Backorders (F5307) and Sales Order Items – Confirmed as requested (F5308)

This app is an analytical app and uses the Analytics – Confirmation of Sales Orders (C_SlsOrd-ConfAnlytsQry) CDS view for retrieval of the data.

Tip

By default, the app will show all sales orders in the system. To avoid performance issues, the filters should be configured appropriately in the Manage KPIs and Reports app (F2814) for the group **Sales Order** and **Check Sales Volume** key figures.

Sales Performance – Plan/Actual (F2941)

Application Type: Analytical

Sales plans can be created for an organization's sales team by using the Manage Sales Plans (app F2512). Once the sales plans have been created, you can use this app to compare actual sales performance against the plan. To be able to use this app, the sales plan must have been created by the user of the app or be shared with a team of which they are a part. Sales plans can be value plans (by net value) or actual plans (by sales volume). These plans are compared in this app with performance by month, quarter, year, or a specific planned period. The visualization of the performance is shown in the app in a graph combining a column chart and a line chart, for plan versus actual respectively, as well as the raw data in a table format showing the planned value or volume against the actual value or volume, with grand totals. This visualization can be changed from within the app.

The app supports navigation to related apps Quotation Conversion Rates (F1904), Sales Contract Fulfillment Rates (F1905), and Sales Volumes – Check Open Sales (F2270).

Sales Quotation (S/4HANA) (F1871)

Application Type: Fact Sheet

This app allows you to view all details for a specific sales quotation in one place. You can display the sales quotation items, business partners involved in the sales quotation, and related sales documents such as sales orders. By clicking the **Process Flow** option in the app, you can view the document flow for the sales quotation in graphical format.

Navigation to related apps such as Customer – 360° View – Version 2 (F2187A) and Manage Sales Quotations (F1852) is supported. The app is a navigation target in many other apps in the sales cycle.

Transaction Code

The SAP GUI equivalent app is Transaction VA23 (Display Sales Quotation).

Sales Scheduling Agreement (F3720)

Application Type: Fact Sheet

This app allows you to view all details for a specific sales scheduling agreement in one place. You can display the sales scheduling agreement items, the business partners involved in the document, and related sales documents such as sales orders. By clicking the **Process Flow** option in the app, you can view the document flow for the sales scheduling agreement in graphical format.

Navigation to related apps such as Customer – 360° View – Version 2 (F2187A) is supported. The app is a navigation target in many other apps in the sales cycle.

Transaction Code

The SAP GUI equivalent app is Transaction VA33 (Display Sales Scheduling Agreement).

Sales Scheduling Agreements – Product Demand (F4508)

Application Type: Analytical

Scheduling agreements offer a valuable insight into upcoming product demand. This is an important app in the sales cycle, where you can monitor the demand from scheduling agreements for specific products within a given time frame. The intention of the app is to get an overview of demand for specific customers to view whether demand is declining or increasing from that customer for specific product lines. The visualization can be in chart format or tabular format, or both. However, it is important to note that if a mix of products is selected with differing units of measure, the results will only be displayed in tabular format. It is possible to drill down into the data to get more granular results based upon sales organization, product, and sold-to party. Selection filters are available for fields such as release type, time scale, delivery date, product, product group, sold-to party, customer group, and scheduling agreement.

Navigation to related apps such as Manage Scheduling Agreements (F3515) and Display Sales Scheduling Agreements (VA33) is supported.

Sales Volume – Check Open Sales (F2270)

Application Type: Analytical

Tracking sales volumes over time is an important function in the role of a sales manager, and this app allows you to see trends easily from month-to-month. The results are displayed in graphical format against the timeline specified or can be displayed as a table. Additional drill-downs are available based on dimensions such as sales organization, customer group, product, or sold-to party.

Navigation to related apps such as Sales Volume – Profit Margin (F2271) and Sales Volume – Credit Memos (F5310) is supported.

This app uses the CDS view Analytics – Sales Volume and Open Sales (C_SalesAnalyticsQry_1).

Tip

By default, the app will show all sales credit memos in the system.

To avoid performance issues, the filters should be configured appropriately in the Manage KPIs and Reports app (F2814) for the group **Sales Volume** and **Check Sales Volume** key figures.

Sales Volume – Credit Memos (F5310)

Application Type: Analytical

Internal errors within an organization usually result in sales credit memos being raised to the customer. With this app, you can monitor the number and frequency of sales credit memos resulting from returns, complaints, and cancellations. The results are displayed in graphical format against the timeline specified or can be displayed as a table. Additional drilldowns are available based on dimensions such as sales office, sales document type, product, or sold-to party.

This app uses the Analytics – Sales Volume (C_SalesVolumeAnalyticsQry) CDS view.

Tip

By default, the app will show all sales credit memos in the system. To avoid performance issues, the filters should be configured appropriately in the Manage KPIs and Reports app (F2814) for the group **Sales Volume** and **Credit Memo** key figures.

Sales Volume – Detailed Analysis (F2235)

Application Type: Analytical

In SAP, sales volume is defined by the total invoiced sales (which have not been canceled) within a specific period. This app shows you a graphical display of sales volume over time for different filter selections, such as date range, sales organization, customer, and material.

Drilldowns are available to show more granular details of the billing documents and items.

Navigation to related apps such as Billing Document (F1901), Sales Volume – Check Open Sales (F2270), Sales Volume – Profit Margin (F2271), and Sales Volume – Credit Mamos (F5310) is supported.

This app uses CDS views C_SalesVolumeAnalysisQuery and C_SalesVolumebyYearQuery.

Sales Volume – Profit Margin (F2271)

Application Type: Analytical

An important part of analyzing organizational performance is measuring profit margin against sales volume. This app allows you to view how well specific sales organizations are performing in terms of profit margins. The SAP Fiori tile shows dynamic data against the top three sales organizations with their profit margins in percentages. Selection fields are available such as month or year, sold-to party, product, and plant. Results of the selection can be visualized in the app in graphical or tabular format. Drilldown options are available to show a further granularity of data according to dimensions such as sales office, sales document, sold-to party, and product. This app uses the CDS view Analytics – Sales Volume (C_SalesVolumeAnalyticsQry).

Navigation to related apps such as Billing Document (F1901), Sales Volume – Check Open Sales (F2270), and Sales Volume – Profit Margin (F2271) is supported.

Tip

By default, the app will show all sales volumes in the system. To avoid performance issues, the filters should be configured appropriately in the Manage KPIs and Reports app (F2814) for the group **Sales Volume** and **Profit Margin** key figures.

Schedule Accruals Reversal – External Agent Contracts (F4146)

Application Type: Transactional

Condition contracts can be set up for the payment of external agent contracts. When these condition contracts become obsolete, they are flagged with the status **Logically Deleted**. Using this app, you can monitor the reversal of open accruals for obsolete condition contract for sales commissions for external agents. It is possible to use this app to perform multiple functions, such as the creation of periodic background jobs to reverse open accruals for obsolete condition contracts, receiving notifications for failed jobs, and monitoring and displaying information about the scheduled jobs.

Schedule Accruals Reversal – Obsolete Customer Condition Contracts (F2201)

Application Type: Transactional

Condition contracts can be set up for payment to customers. When these condition contracts become obsolete, they are flagged with the status **Deleted**. Using this app, you can monitor the reversal of open accruals for obsolete condition contract for customers. It is possible to use this app to perform multiple functions, such as the creation of periodic background jobs to reverse open accruals for obsolete condition contracts, receiving notifications for failed jobs, and monitoring and displaying information about the scheduled jobs.

Schedule Accruals Reversal – Royalty Contracts (F5270)

Application Type: Transactional

Condition contracts can be set up for the payment of royalties. When these condition contracts become obsolete, they are flagged with the status **Logically Deleted**. Using this app, you can monitor the reversal of open accruals for obsolete condition contract for royalty payments. It is possible to use this app to perform multiple functions, such as the creation of periodic background jobs to reverse open accruals for obsolete condition contracts, receiving notifications for failed jobs, and monitoring and displaying information about the scheduled jobs.

Schedule Accruals Reversal – Supplier Contracts (F2204)

Application Type: Transactional

Condition contracts can be set up for payment to suppliers. When these condition contracts become obsolete, they are flagged with the status **Deleted**. Using this app, you can monitor the reversal of open accruals for obsolete condition contract for supplier payments. It is possible to use this app to perform multiple functions, such as the creation of periodic background jobs to reverse open accruals for obsolete condition contracts, receiving notifications for failed jobs, and monitoring and displaying information about the scheduled jobs.

Schedule Accruals Update – Customer Contracts (F2942)

Application Type: Transactional

Condition contracts can be set up for payment to customers. When these condition contracts are updated to change accrual relevant data such as the condition rate, you must perform a retroactive accrual update. It is possible to use this app to perform multiple functions, such as the creation of periodic background jobs to update accruals for changed condition contracts, receiving notifications for failed jobs, and monitoring and displaying information about the scheduled jobs.

Schedule Accruals Update – External Agent Contracts (F4144)

Application Type: Transactional

Condition contracts can be set up for the payment of external agent contracts. When these condition contracts are updated to change accrual relevant data such as the condition rate, you must perform a retroactive accrual update. It is possible to use this app to perform multiple functions, such as the creation of periodic background jobs to update accruals for changed condition contracts, receiving notifications for failed jobs, and monitoring and displaying information about the scheduled jobs.

Schedule Accruals Update – Royalty Contracts (F5269)

Application Type: Transactional

Condition contracts can be set up for the payment of royalties. When these condition contracts are updated to change accrual relevant data such as the condition rate, you must perform a retroactive accrual update. It is possible to use this app to perform multiple functions, such as the creation of periodic background jobs to update accruals for changed condition contracts, receiving notifications for failed jobs, and monitoring and displaying information about the scheduled jobs.

Schedule Billing Creation (F1519)

Application Type: Transactional

This app is used to create background billing jobs for the creation of invoices for sales documents. The input parameters into the app include scheduling options such as job template (select from a dropdown list), job name (free text), the **Start Immediately** flag, start date and time, and recurrence pattern (select as single run or recurring). You also can set billing data, such as billing date from, billing date to, and

billing type; organizational data, such as sales area, shipping point, and sales and distribution document; and customer data such as sold-to party, destination country, and region. It is also possible to copy an existing job to create a similar version. All parameters are copied.

> **Tip**
>
> The job will automatically post all created billing documents to accounting, unless the **Block posting to accounting** checkbox is selected. If this is selected, then the documents must be posted to accounting manually using the Manage Billing Documents (F0797) app or Transaction VA02 (Change Billing Documents).

Schedule Billing Creation for Preliminary Billing Documents (F5137)

Application Type: Transactional

A preliminary billing document is used during customer negotiations, and before the details of the billing are confirmed. As soon as the agreement with the customer has been established, the preliminary billing document can be output but is not posted to accounting. These documents are often also called proforma billing documents. They can be finalized by creating another subsequent standard billing document with reference to the sales document or delivery in the usual fashion. This app is used to create background billing jobs for the creation of invoices for preliminary billing documents. The input parameters into the app include scheduling options such as job template (select from a dropdown list), job name (free text), the **Start Immediately** flag, start date and time, and recurrence pattern (select as single run or recurring). You also can set billing data, such as billing date from, billing date to, billing type, preliminary billing document, and the only finalized documents checkbox; organizational data, such as sales area; and customer data, such as sold-to party and destination country. It is also possible to copy an existing job to create a similar version. All parameters are copied.

Tip

The job will automatically post all created billing documents to accounting, unless the **Block posting to accounting** checkbox is selected. If this is selected, then the documents must be posted to accounting manually using the Manage Billing Documents (F0797) app or the Change Billing Documents (VA02) transaction.

Schedule Billing Output (F1510)

Application Type: Transactional

This app is used to create background jobs to create output documents for billing documents. It is possible to create, display, cancel, copy, and edit background jobs within the app. The input parameters into the app include job details such as job template (select from a dropdown list), job name (free text), the **Start Immediately** flag, and the start date and time. You also can set output data such as output channel (i.e. email, print, etc.) and billing data such as billing document, billing date from, sales organization, distribution channel, division, sold-to party, payer, and destination country.

Tip

There are performance and restriction considerations to be aware of with this app. For example, the selection boxes for sales organization, distribution channel, division, and destination country, all allow multiple selections. However, selecting multiple options will cause an error in the job. Therefore, only specify a single value or no value at all in these fields.

In addition, outputting billing documents that have over 1,000 items on them to print or email will cause errors with Adobe Document Services when it attempts to render to PDF.

Schedule Billing Release (F1518)

Application Type: Transactional

This app is used to create accounting entries for all billing documents that have previously not been passed to financial accounting. The input parameters into the app include scheduling options such as job template (select from a dropdown list) and job name (free text), as well as parameters such as payer, sales organization, billing document, billing type, billing category, created by, and creation date. You can use the app to create accounting entries restricted to documents that have not yet been posted for specific reasons. These reasons are listed in the selection criteria under the **Incomplete due to** section. The available reasons are:

- Pricing error: The document contains at least one pricing error, such as missing mandatory condition.
- Accounting block: The document or customer has been blocked for posting to accounting.
- Error in accounting interface: Normally refers to errors in account determination but also includes attempts to post to a closed accounting period.
- Error in authorization: The user posting the document to accounting did not have sufficient authorization to complete the task.

It is also possible to copy an existing job to create a similar version. All parameters are copied.

Schedule BOP Run (F2665)

Application Type: Transactional

This app is used to create background jobs for backorder processing (BOP) runs. The input parameters into the app include general information such as job template (select from a dropdown list) and job name (free text) and scheduling options such as the **Start Immediately** flag, recurrence pattern (select single run or recurring), and start date and time. You also can select parameters such as creation date, created by, last change date, changed by, variant name (mandatory), delivery group option, and parallel document update profile. It is possible to run the job in simulation mode by selecting the **Simulation Mode** checkbox.

The results can be displayed in the Monitor BOP Run app (F2159).

Schedule Business Partner Screening – Watch List Screening (F8016)

Application Type: Transactional

SAP Watch List Screening is a service offered by SAP Business Technology Platform (SAP BTP), which must be licensed separately. The service screens the customers and suppliers and returns information regarding the viability of doing business with the partner. As part of the international trade suite of apps, this app allows you to schedule a background job to transfer customers and suppliers to SAP Watch List Screening for individual or all business partners with the defined roles created. The input parameters into the app include template selection parameters such as job template (select from a dropdown list) and job name (free text), as well as scheduling options such as the **Start Immediately** flag, recurrence pattern (select single run or recurring), and start date and time. You also can set customer/supplier-specific parameters such as business partner number.

Schedule Content Request to Data Provider – Commodity Codes (F3051)

Application Type: Transactional

Commodity codes can be provided by external content providers and uploaded into SAP for use within the suite of apps for international trade. With this app, you can schedule a background job to retrieve new versions of the list of commodity codes available, for updating SAP. This allows the commodity codes to be kept up to date. The app shows several fields for the data

that has been retrieved: version (number of the data package, assigned by the data provider), revision, version type (initial or delta), data provider ID, consistency status (**Not Checked**, **Not Consistent**, or **Consistent**), application status (**Not Checked**, **Not Activatable**, **Activatable**), application log, and data selection (restrictions by numbering scheme and data provider are available).

The versions provided can then be used in the Manage Content from Data Provider – Commodity Codes app (F3429).

Schedule Content Request to Data Provider – Control Classes (F3568)

Application Type: Transactional

Control classes can be provided by external content providers and uploaded into SAP for use within the suite of apps for international trade. With this app, you can schedule a background job to retrieve new versions of the list of control classes available, for updating SAP. This allows the control classes to be kept up to date. The app shows several fields for the data that has been retrieved: version (number of the data package, assigned by the data provider), revision, version type (initial or delta), data provider ID, consistency status (**Not Checked**, **Not Consistent**, or **Consistent**), application status (**Not Checked**, **Not Activatable**, **Activatable**), application log, and data selection (restrictions by numbering scheme and data provider are available).

The versions provided can then be used in the Manage Content from Data Provider – Commodity Codes app (F3429).

Schedule Content Request to Data Provider – Customs Tariff Numbers (F3569)

Application Type: Transactional

Customs tariff numbers can be provided by external content providers and uploaded into

SAP for use within the suite of apps for international trade. With this app, you can schedule a background job to retrieve new versions of the list of customs tariff numbers available, for updating SAP. This allows the customs tariff numbers to be kept up to date. The app shows several fields for the data that has been retrieved: version (number of the data package, assigned by the data provider), revision, version type (initial or delta), data provider ID, consistency status (**Not Checked**, **Not Consistent**, or **Consistent**), application status (**Not Checked**, **Not Activatable**, **Activatable**), application log, and data selection (restrictions by numbering scheme and data provider are available).

The versions provided can then be used in the Manage Content from Data Provider – Commodity Codes app (F3429).

Schedule Contract Settlement – Customer Contracts (F2202)

Application Type: Transactional

Condition contracts can be set up for payment to customers. With this app, you can create background jobs for the creation of settlement documents for the settlement of customer condition contracts. Multiple selection options are available for the creation of the job, including template selection parameters such as job template (select from a dropdown list) and job name (free text); scheduling options such as the **Start Immediately** flag, recurrence pattern (single run or recurring), and start date and time; settlement control options such as settlement date; and contract selection options such as condition contract number, condition contract type, and condition contract category.

It is possible to use this app to perform multiple functions, such as the creation of periodic background jobs, receiving notifications for failed jobs, and monitoring and displaying information about the scheduled jobs.

Schedule Contract Settlement – External Agent Contracts (F4145)

Application Type: Transactional

Condition contracts can be set up for payment of sales commissions to external agents. With this app, you can create background jobs for the creation of settlement documents for the settlement of condition contracts for sales commissions with external sales agents. Multiple selection options are available for the creation of the job, including template selection parameters such as job template (select from a dropdown list) and job name (free text); scheduling options such as the **Start Immediately** flag, recurrence pattern (single run or recurring), and start date and time; settlement control options such as settlement date; and contract selection options such as condition contract number, condition contract type, and condition contract category.

It is possible to use this app to perform multiple functions, such as the creation of periodic background jobs, receiving notifications for failed jobs, and monitoring and displaying information about the scheduled jobs.

Schedule Contract Settlement – Royalty Contracts (F5271)

Application Type: Transactional

Condition contracts can be set up for payment of royalties. With this app, you can create background jobs for the creation of settlement documents for the settlement of condition contracts for royalties. Multiple selection options are available for the creation of the job, including template selection parameters such as job template (select from a dropdown list) and job name (free text); scheduling options such as the **Start Immediately** flag, recurrence pattern (single run or recurring), and start date and time; settlement control options such as settlement date; and contract selection options such as condition contract number, condition contract type, and condition contract category.

It is possible to use this app to perform multiple functions, such as the creation of periodic background jobs, receiving notifications for failed jobs, and monitoring and displaying information about the scheduled jobs.

Schedule Contract Settlement – Supplier Contracts (F2203)

Application Type: Transactional

Condition contracts can be set up for payment to suppliers. With this app, you can create background jobs for the creation of settlement documents for the settlement of condition contracts to suppliers. Multiple selection options are available for the creation of the job, including template selection parameters such as job template (select from a dropdown list) and job name (free text); scheduling options such as the **Start Immediately** flag, recurrence pattern (single run or recurring), and start date and time; settlement control options such as settlement date; and contract selection options such as condition contract number, condition contract type, and condition contract category.

It is possible to use this app to perform multiple functions, such as the creation of periodic background jobs, receiving notifications for failed jobs, and monitoring and displaying information about the scheduled jobs.

Schedule Creation of Preliminary Billing Documents (F4563)

Application Type: Transactional

A preliminary billing document is used during customer negotiations, and before the details of the billing are confirmed. As soon as the agreement with the customer has been established, the preliminary billing document can be output but is not posted to accounting. These documents are often also called proforma billing documents. They can be finalized by creating another subsequent standard billing document with reference to the sales document or delivery in the usual fashion.

This app is used to create preliminary billing documents from the billing due list. The input parameters in the app include scheduling options such as job template (select from a dropdown list), job name (free text), the **Start Immediately** flag, start date and time, and recurrence pattern (select as single run or recurring), as well as parameters such as billing date from and billing date to, billing type, and preliminary billing document. You also can set organizational data such as sales area, shipping point, and sales and distribution document and customer data such as sold-to party and destination country.

It is also possible to copy an existing job to create a similar version. All parameters are copied.

Schedule Customer Settlement Creation (F5633)

Application Type: Transactional

This app is used to create customer settlements. The input parameters in the app include scheduling options such as job template (select from a dropdown list), job name (free text), the **Start Immediately** flag, start date and time, recurrence pattern (select as single run or recurring), as well as parameters such as document number, posting date, document date, invoicing party, payment recipient, bill-to party, payer, customer, settlement process type, settlement document type, and process category. You also can set organizational data such as purchasing organization, purchasing group, sales organization, distribution channel, division, company code, and company code—customer.

It is possible to use this app to perform multiple functions, such as the creation of periodic background jobs, receiving notifications for failed jobs, and monitoring and displaying information about the scheduled jobs.

353

Schedule Customer Settlement List Creation (F5634)

Application Type: Transactional

This app is used to create customer settlement lists. The input parameters in the app include scheduling options such as job template (select from a dropdown list), job name (free text), the **Start Immediately** flag, start date and time, recurrence pattern (select as single run or recurring), as well as parameters such as document number, posting date, document date, invoicing party, payment recipient, bill-to party, payer, customer, settlement process type, settlement document type, and process category. You also can set organizational data such as purchasing organization, purchasing group, sales organization, distribution channel, division, company code, and company code—customer.

It is possible to use this app to perform multiple functions, such as the creation of periodic background jobs, receiving notifications for failed jobs, and monitoring and displaying information about the scheduled jobs.

Schedule Delivery Creation (F2228)

Application Type: Transactional

This app is used to create background delivery jobs for the creation of outbound deliveries for sales documents and purchase orders. The input parameters in the app differ according to whether purchasing or sales documents are being used for creation. The input parameters for sales orders include scheduling options such as job template (select from a dropdown list), job name (free text), the **Start Immediately** flag, start date and time, and recurrence pattern (select as single run or recurring). You also can provide general delivery data such as shipping/receiving point, delivery creation date, delivery creation date (days), delivery priority, shipping conditions, ship-to party, unloading point, sales document type, sales organization, distribution channel, division, goods issue date, and goods

issue date (days). Sales order data includes sales and distribution document, sold-to party, complete delivery, forwarding agent, sales office, sales group, department, and delivery block, and material includes material, customer material, material group, promotion, plant, and storage location.

It is also possible to copy an existing job to create a similar version. All parameters are copied.

Schedule Document Collection – Settlement Document (F2558)

Application Type: Transactional

With this app, you can create background jobs for the creation of collective settlement documents with reference to two-step condition contracts. Multiple selection options are available for the creation of the job, including template selection parameters such as job template (select from a dropdown list) and job name (free text) and scheduling options such as the **Start Immediately** flag, recurrence pattern (single run or recurring), and start date and time. You also can provide document data options such as document number, posting date, document date, bill-to party, payer, settlement process type, settlement document type, and process category, as well as organizational data such as sales organization, distribution channel, division, and company code.

It is possible to use this app to perform multiple functions, such as the creation of periodic background jobs, receiving notifications for failed jobs, and monitoring and displaying information about the scheduled jobs.

Schedule Document Completion – Settlement Management (F5500)

Application Type: Transactional

With this app, you can create background jobs for the mass completion of settlement management documents. Multiple selection options are available for the creation of the job, including template selection parameters such as job template (select from a dropdown list) and job name (free text), as well as scheduling options such as the **Start Immediately** flag, recurrence pattern (single run or recurring), and start date and time. You also have document data options such as document number, posting date, document date, invoicing party, payment recipient, alternative supplier, bill-to party, payer, settlement process type, settlement document type, and process category. Finally, you also can provide organizational data such as purchasing organization, purchasing group, sales organization, distribution channel, division, and company code.

It is possible to use this app to perform multiple functions, such as the creation of periodic background jobs, receiving notifications for failed jobs, and monitoring and displaying information about the scheduled jobs.

Schedule Document Output – Settlement Management (F2418)

Application Type: Transactional

With this app, you can create background jobs for the mass creation of settlement management output documents. Multiple selection options are available for the creation of the job, including template selection parameters such as job template (select from a dropdown list) and job name (free text), as well as scheduling options such as the **Start Immediately** flag, recurrence pattern (single run or recurring), and start date and time. You also have document data options such as document number, posting date, document date, bill-to party, payer, settlement process type, settlement document type, and document category. Organizational data includes the sales organization and company code.

It is possible to use this app to perform multiple functions, such as the creation of periodic background jobs, receiving notifications for failed jobs, and monitoring and displaying information about the scheduled jobs.

Schedule Document Pricing – Settlement Management (F5502)

Application Type: Transactional

With this app, you can create background jobs for mass pricing of settlement management documents. Multiple selection options are available for the creation of the job, including template selection parameters such as job template (select from a dropdown list) and job name (free text), as well as scheduling options such as the **Start Immediately** flag, recurrence pattern (single run or recurring), and start date and time. Document data options include document number, posting date, document date, invoicing party, payment recipient, alternative supplier, bill-to party, payer, settlement process type, settlement document type, and process category. Organizational data includes purchasing organization, purchasing group, sales organization, distribution channel, division, and company code.

It is possible to use this app to perform multiple functions, such as the creation of periodic background jobs, receiving notifications for failed jobs, and monitoring and displaying information about the scheduled jobs.

Schedule Document Release – Freight Cost Allocation (F6068)

Application Type: Transactional

With this app, you can create background jobs for releasing freight cost allocation documents to accounting. Multiple selection options are available for the creation of the job, including template selection parameters such as job template (select from a dropdown list) and job name (free text), as well as scheduling options such as the **Start Immediately** flag, recurrence pattern (single run or recurring), and start date and time. Document data options include document number, posting date, document date, settlement process type, settlement document type, and process category. Organizational data includes purchasing organization, purchasing group, sales organization, distribution channel, division, and company code.

It is possible to use this app to perform multiple functions, such as the creation of periodic background jobs, receiving notifications for failed jobs, and monitoring and displaying information about the scheduled jobs.

Schedule Document Release – Settlement Management (F2417)

Application Type: Transactional

With this app, you can create background jobs for releasing settlement management documents to accounting. Multiple selection options are available for the creation of the job, including template selection parameters such as job template (select from a dropdown list) and job name (free text), as well as scheduling options such as the **Start Immediately** flag, recurrence pattern (single run or recurring), and start date and time. Document data options include document number, posting date, document date, bill-to party, payer, settlement process type, settlement document type, and process category. Organizational data includes sales organization, distribution channel, division, and company code.

It is possible to use this app to perform multiple functions, such as the creation of periodic background jobs, receiving notifications for failed jobs, and monitoring and displaying information about the scheduled jobs.

Schedule Document Reversal – Settlement Management (F2557)

Application Type: Transactional

With this app, you can create background jobs for the reversal of settlement documents. Multiple selection options are available for the creation of the job, including template selection

parameters such as job template (select from a dropdown list) and job name (free text), as well as scheduling options such as the **Start Immediately** flag, recurrence pattern (single run or recurring), and start date and time. Document data options include document number, posting date, document date, bill-to party, payer, settlement process type, settlement document type, and process category. Organizational data includes sales organization, distribution channel, division, and company code.

It is possible to use this app to perform multiple functions, such as the creation of periodic background jobs, receiving notifications for failed jobs, and monitoring and displaying information about the scheduled jobs.

Schedule Document Revoke Completion – Settlement Management (F5501)

Application Type: Transactional

With this app, you can create background jobs to revoke the completion of settlement management documents. Multiple selection options are available for the creation of the job, including template selection parameters such as job template (select from a dropdown list) and job name (free text), as well as scheduling options such as the **Start Immediately** flag, recurrence pattern (single run or recurring), and start date and time. Document data options include document number, posting date, document date, invoicing party, payment recipient, alternative supplier, bill-to party, payer, settlement process type, settlement document type, and process category. Organizational data includes purchasing organization, purchasing group, sales organization, distribution channel, division, company code, and company code (customer).

It is possible to use this app to perform multiple functions, such as the creation of periodic background jobs, receiving notifications for failed jobs, and monitoring and displaying information about the scheduled jobs.

Schedule Follow-On Document Processing (F6438)

Application Type: Transactional

With this app, you can create background jobs for the creation of follow-on documents for trading contracts such as sales orders and purchase orders. Multiple selection options are available for the creation of the job, including template selection parameters such as job template (select from a dropdown list) and job name (free text), as well as scheduling options such as the **Start Immediately** flag, recurrence pattern (single run or recurring), and start date and time. Document data options include document number, posting date, document date, invoicing party, payment recipient, alternative supplier, bill-to party, payer, settlement process type, settlement document type, and process category. Organizational data includes purchasing organization, purchasing group, sales organization, distribution channel, division, and company code.

It is possible to use this app to perform multiple functions, such as the creation of periodic background jobs, receiving notifications for failed jobs, and monitoring and displaying information about the scheduled jobs.

Schedule Goods Issue for Deliveries (F2259)

Application Type: Transactional

With this app, you can create background jobs for the posting of goods issues for outbound deliveries. Multiple selection options are available for the creation of the job, including template selection parameters such as job template (select from a dropdown list) and job name (free text), as well as scheduling options such as the **Start Immediately** flag, recurrence pattern (single run or recurring), and start date and time. Parameter options include shipping point/receiving point, planned goods movement date, vendor returns material authorization (RMA)

number, customer returns order, material batch number, forwarding agent, ship-to party, route schedule, route, and handling unit.

It is possible to use this app to perform multiple functions, such as the creation of periodic background jobs, receiving notifications for failed jobs, and monitoring and displaying information about the scheduled jobs.

Schedule Message Processing (F5649)

Application Type: Transactional

With this app, you can create background jobs for the processing of output messages for settlement management documents. Some settlement management documents may have the status of **In Preparation** or **Not Processed**. Using this app, the documents with these statuses can be processed, output messages created, and the status changed accordingly. Multiple selection options are available for the creation of the job, including template selection parameters such as job template (select from a dropdown list) and job name (free text), as well as scheduling options such as the **Start Immediately** flag, recurrence pattern (single run or recurring), and start date and time. Message output options include output type, transmission medium, sort order, processing mode, massage language, user name, created on date and processing date, and sort criteria.

It is possible to use this app to perform multiple functions, such as the creation of periodic background jobs, receiving notifications for failed jobs, and monitoring and displaying information about the scheduled jobs.

Schedule Message Redetermination of Settlement Management Documents (F5650)

Application Type: Transactional

With this app, you can create background jobs for the redetermination of output messages for settlement management documents. Multiple selection options are available for the creation of the job, including template selection parameters such as job template (select from a dropdown list) and job name (free text), as well as scheduling options such as the **Start Immediately** flag, recurrence pattern (single run or recurring), and start date and time. Document data options include document number, reference, assignment, payment reference, posting date, document date, created by, created on, invoicing party, payment recipient, alternative supplier, bill-to party, payer, settlement process type, settlement document type, and process category. Organizational data includes purchasing organization, purchasing group, sales organization, distribution channel, division, and company code.

It is possible to use this app to perform multiple functions, such as the creation of periodic background jobs, receiving notifications for failed jobs, and monitoring and displaying information about the scheduled jobs.

Schedule Postprocessing – Watch List Screening (F3052)

Application Type: Transactional

With this app, you can create background jobs for the postprocessing of documents for SAP Watch List Screening integration. This transfer watch list screening decisions from the screening hits UI to the international trade document. Only basic selection options are available for the creation of the job, including template selection parameters such as job template (select from a dropdown list) and job name (free text), as well as scheduling options such as the **Start Immediately** flag, recurrence pattern (single run or recurring), and start date and time.

Schedule Recheck Documents – Trade Compliance (F4285)

Application Type: Transactional

With this app, you can create background jobs to recheck trade compliance documents. In previous versions of the app, prior to SAP S/4HANA 2023, this recheck could only be performed on blocked trade compliance documents. With the 2023 edition, you can perform this recheck for all trade compliance documents, not just blocked documents.

Multiple selection options are available for the creation of the job, including template selection parameters such as job template (select from a dropdown list) and job name (free text), as well as scheduling options such as the **Start Immediately** flag, recurrence pattern (single run or recurring), and start date and time. Parameter options include document date, document category, company code, and document number.

Schedule Reprocessing of Documents – Global Trade Services (F3050)

Application Type: Transactional

SAP Global Trade Services (SAP GTS) can be integrated seamlessly into SAP S/4HANA as part of the international trade functionality. With this app, you can create background jobs to schedule the reprocessing of documents which, for technical reasons, were not transferred to SAP GTS.

Multiple selection options are available for the creation of the job, including template selection parameters such as job template (select from a dropdown list) and job name (free text), as well as scheduling options such as the **Start Immediately** flag, recurrence pattern (single run or recurring), and start date and time. Parameter options include status and from and to dates.

Schedule Resolution of Value Chain Issues (F6868)

Application Type: Transactional

The related Monitor Value Chains app (F4854) allows you to monitor the advanced intercompany sales, advanced intercompany stock transfer, and sell-from-stock with valuated stock in transit processes. This app allows you to resolve business process issues related to these value chains by scheduling a background job. Multiple selection options are available, including template selection parameters such as job template (select from a dropdown list) and job name (free text). Scheduling options are available, such as the **Start Immediately** flag, recurrence pattern (single run or recurring), and start date and time.

It is possible to use this app to perform multiple functions, such as the creation of periodic background jobs, receiving notifications for failed jobs, and monitoring and displaying information about the scheduled jobs.

Schedule Sales Document Output (F2459)

Application Type: Transactional

With this app, you can create background jobs for the scheduling of output messages for sales documents. Multiple selection options are available for the creation of the job, including template selection parameters such as job template (select from a dropdown list) and job name (free text), as well as scheduling options such as the **Start Immediately** flag, recurrence pattern (single run or recurring), and start date and time. Output data options include output channel, receiver ID, receiver role, and output type. Sales document data includes sales document type, sold-to party, document creation date, division, sales document, sales organization, and distribution channel.

It is possible to use this app to perform multiple functions, such as the creation of periodic back-

ground jobs, receiving notifications for failed jobs, and monitoring and displaying information about the scheduled jobs.

Schedule Sales Document Status Update – Global Trade Services (F5721)

Application Type: Transactional

SAP Global Trade Services (SAP GTS) can be integrated seamlessly into SAP S/4HANA as part of the international trade functionality. If the connection between the two systems is broken for any length of time during the trade compliance check of a sales document, this app can be used to re-transfer the SAP GTS status to the SAP S/4HANA documents. Multiple selection options are available for the creation of the job, including template selection parameters such as job template (select from a dropdown list) and job name (free text), as well as scheduling options such as the **Start Immediately** flag, recurrence pattern (single run or recurring), and start date and time. Parameter options are available such as status and from and to dates.

Schedule Service Transaction Status Update – Global Trade Services (F7925)

Application Type: Transactional

SAP Global Trade Services (SAP GTS) can be integrated seamlessly into SAP S/4HANA as part of the international trade functionality. If the connection between the two systems is broken for any length of time during the trade compliance check of a service transaction, this app can be used to re-transfer the SAP GTS status to the SAP S/4HANA service transaction. Multiple selection options are available for the creation of the job, including template selection parameters such as job template (select from a dropdown list) and job name (free text), as well as scheduling options such as the **Start Immediately** flag, recurrence pattern (single run or recurring), and start date and time. Parameter options are available such as status and from and to dates.

Schedule Settlement Document List Creation (F5648)

Application Type: Transactional

With this app, you can create background jobs to create settlement document lists. Multiple selection options are available for the creation of the job, including template selection parameters such as job template (select from a dropdown list) and job name (free text), as well as scheduling options such as the **Start Immediately** flag, recurrence pattern (single run or recurring), and start date and time. Document data options include document, posting date, document date, invoicing party, bill-to party, and payer. Organizational data includes purchasing organization, sales organization, and company code.

It is possible to use this app to perform multiple functions, such as the creation of periodic background jobs, receiving notifications for failed jobs, and monitoring and displaying information about the scheduled jobs.

Schedule Transfer of Bill of Materials – Global Trade Services (F2807)

Application Type: Transactional

SAP Global Trade Services (SAP GTS) can be integrated seamlessly into SAP S/4HANA as part of the international trade functionality. This app can be used to create background jobs to transfer bills of material to SAP GTS. Multiple selection options are available for the creation of the job, including template selection parameters such as job template (select from a dropdown list) and job name (free text), as well as scheduling options such as the **Start Immediately** flag, recurrence pattern (single run or recurring), and start date and time. Parameter options include valid from, description, low-level code, material, and plant.

Schedule Transfer of Billing Documents – Global Trade Services (F3740)

Application Type: Transactional

SAP Global Trade Services (SAP GTS) can be integrated seamlessly into SAP S/4HANA for international trade. This app can be used to create background jobs to transfer billing documents to SAP GTS. Multiple selection options are available for the creation of the job, including template selection parameters such as job template (select from a dropdown list) and job name (free text), as well as scheduling options such as the **Start Immediately** flag, recurrence pattern (single run or recurring), and start date and time. Parameter options include created on, created by, billing type, billing document, sales organization, and distribution channel.

Schedule Transfer of Changed Master Data – Global Trade Services (F6041)

Application Type: Transactional

SAP Global Trade Services (SAP GTS) can be integrated seamlessly into SAP S/4HANA for international trade. This app can be used to create background jobs to transfer changed master data to SAP GTS. Multiple selection options are available for the creation of the job, including template selection parameters such as job template (select from a dropdown list) and job name (free text), as well as scheduling options such as the **Start Immediately** flag, recurrence pattern (single run or recurring), and start date and time. Parameter options include message type (which refers to the type of master data, such as suppliers, customers, BOMs, products, etc.).

Schedule Transfer of Contact Persons – Global Trade Services (F2803)

Application Type: Transactional

SAP Global Trade Services (SAP GTS) can be integrated seamlessly into SAP S/4HANA for international trade. This app can be used to create background jobs to transfer contact persons to SAP GTS. Multiple selection options are available for the creation of the job, including template selection parameters such as job template (select from a dropdown list) and job name (free text), as well as scheduling options such as the **Start Immediately** flag, recurrence pattern (single run or recurring), and start date and time. Parameter options include time of transfer log and contact person.

Schedule Transfer of Customer Product Names – Global Trade Services (F3743)

Application Type: Transactional

SAP Global Trade Services (SAP GTS) can be integrated seamlessly into SAP S/4HANA for international trade. This app can be used to create background jobs to transfer customer product names to SAP GTS. Multiple selection options are available for the creation of the job, including template selection parameters such as job template (select from a dropdown list) and job name (free text), as well as scheduling options such as the **Start Immediately** flag, recurrence pattern (single run or recurring), and start date and time. Parameter options include label, sales organization, distribution channel, created on, created by, customer, material group, and material.

Schedule Transfer of Customers – Global Trade Services (F2806)

Application Type: Transactional

SAP Global Trade Services (SAP GTS) can be integrated seamlessly into SAP S/4HANA for international trade. This app can be used to create background jobs to transfer customers to SAP GTS. Multiple selection options are available for the creation of the job, including template selection parameters such as job template (select from a dropdown list) and job name (free text), as well as scheduling options such as the **Start**

Immediately flag, recurrence pattern (single run or recurring), and start date and time. Parameter options include title of transfer log and customer.

Schedule Transfer of Duty-Paid Stock – Global Trade Services (F5702)

Application Type: Transactional

SAP Global Trade Services (SAP GTS) can be integrated seamlessly into SAP S/4HANA for international trade. This app can be used to create background jobs to transfer inventory in SAP S/4HANA to SAP GTS for the purposes of including the stock in duty paid stock. Duty paid stock refers to imported goods for which all applicable customs duties, taxes and import charges have been fully paid and cleared. SAP GTS needs this information for the provision of import declarations. Multiple selection options are available for the creation of the job, including template selection parameters such as job template (select from a dropdown list) and job name (free text), as well as scheduling options such as the **Start Immediately** flag, recurrence pattern (single run or recurring), and start date and time. Parameter options include customs ID and material.

Schedule Transfer of Inbound Deliveries – Global Trade Services (F4541)

Application Type: Transactional

SAP Global Trade Services (SAP GTS) can be integrated seamlessly into SAP S/4HANA for international trade. This app can be used to create background jobs that transfer inbound deliveries to SAP GTS. Multiple selection options are available for the creation of the job, including template selection parameters such as job template (select from a dropdown list) and job name (free text), as well as scheduling options such as the **Start Immediately** flag, recurrence pattern (single run or recurring), and start date and time. Parameter options include created on, created by, delivery type, and delivery.

Schedule Transfer of Material Documents – Global Trade Services (F3741)

Application Type: Transactional

SAP Global Trade Services (SAP GTS) can be integrated seamlessly into SAP S/4HANA for international trade. This app can be used to create background jobs to transfer material documents to SAP GTS. Multiple selection options are available for the creation of the job, including template selection parameters such as job template (select from a dropdown list) and job name (free text), as well as scheduling options such as the **Start Immediately** flag, recurrence pattern (single run or recurring), and start date and time. Parameter options include document date, posting date, goods movement type, material, material document, material document year, material type, material type, and plant.

Schedule Transfer of Min./Max. Product Prices – Global Trade Services (F3745)

Application Type: Transactional

SAP Global Trade Services (SAP GTS) can be integrated seamlessly into SAP S/4HANA for international trade. This app can be used to create background jobs to transfer minimum/maximum product prices to SAP GTS. Multiple selection options are available for the creation of the job, including template selection parameters such as job template (select from a dropdown list) and job name (free text), as well as scheduling options such as the **Start Immediately** flag, recurrence pattern (single run or recurring), and start date and time. Parameter options include log title, period in number of days, materials, and plants.

Schedule Transfer of Outbound Deliveries – Global Trade Services (F4539)

Application Type: Transactional

SAP Global Trade Services (SAP GTS) can be integrated seamlessly into SAP S/4HANA for international trade. This app can be used to create background jobs to transfer outbound deliveries to SAP GTS. Multiple selection options are available for the creation of the job, including template selection parameters such as job template (select from a dropdown list) and job name (free text), as well as scheduling options such as the **Start Immediately** flag, recurrence pattern (single run or recurring), and start date and time. Parameter options include created on, created by, delivery type, delivery, sales office, sales organization, and shipping point.

Schedule Transfer of Product Attributes – Global Trade Services (F7231)

Application Type: Transactional

SAP Global Trade Services (SAP GTS) can be integrated seamlessly into SAP S/4HANA for international trade. This app can be used to create background jobs to transfer product attributes for special customs procedures to SAP GTS. Special customs procedures include declarations such as inward processing, customs warehousing and duty suspension. Attributes from the material master such as tariff code (HS code), origin of goods, valuation, material type, customs procedure code, and weight, volume, and packaging are all examples of product attributes that are required for fulfilling the special customs procedures. Multiple selection options are available for the creation of the job, including template selection parameters such as job template (select from a dropdown list) and job name (free text), as well as scheduling options such as the **Start Immediately** flag, recurrence pattern (single run or recurring), and start date and time. Parameter options include customs ID, material, plant, storage location, batch, valuation area, and valuation type.

Schedule Transfer of Product Prices – Global Trade Services (F3746)

Application Type: Transactional

SAP Global Trade Services (SAP GTS) can be integrated seamlessly into SAP S/4HANA for international trade. This app can be used to create background jobs to transfer product prices to SAP GTS. Multiple selection options are available for the creation of the job, including template selection parameters such as job template (select from a dropdown list) and job name (free text), as well as scheduling options such as the **Start Immediately** flag, recurrence pattern (single run or recurring), and start date and time. Parameter options include log title and material.

Schedule Transfer of Products – Global Trade Services (F2805)

Application Type: Transactional

SAP Global Trade Services (SAP GTS) can be integrated seamlessly into SAP S/4HANA for international trade. This app can be used to create background jobs to transfer products to SAP GTS. Multiple selection options are available for the creation of the job, including template selection parameters such as job template (select from a dropdown list) and job name (free text), as well as scheduling options such as the **Start Immediately** flag, recurrence pattern (single run or recurring), and start date and time. Parameter options include title of transfer log and material.

Schedule Transfer of Sales Documents – Global Trade Services (F4540)

Application Type: Transactional

SAP Global Trade Services (SAP GTS) can be integrated seamlessly into SAP S/4HANA for

international trade. This app can be used to create background jobs to transfer sales documents to SAP GTS. Multiple selection options are available for the creation of the job, including template selection parameters such as job template (select from a dropdown list) and job name (free text), as well as scheduling options such as the **Start Immediately** flag, recurrence pattern (single run or recurring), and start date and time. Parameter options include sales document type, created on, created by, division, sales document, sales office, sales group, sales organization, and distribution channel.

Schedule Transfer of Service Transactions – Global Trade Services (F7926)

Application Type: Transactional

SAP Global Trade Services (SAP GTS) can be integrated seamlessly into SAP S/4HANA for international trade. This app can be used to create background jobs to transfer service transactions to SAP GTS. Multiple selection options are available for the creation of the job, including template selection parameters such as job template (select from a dropdown list) and job name (free text), as well as scheduling options such as the **Start Immediately** flag, recurrence pattern (single run or recurring), and start date and time. Parameter options include business transaction category, transaction ID, created at, created by, and transaction type.

Schedule Unplanned Contract Settlement – Customer Contracts (F6792)

Application Type: Transactional

In SAP S/4HANA settlement management, when a condition contract is created, it is normal to create a settlement date. However, contracts that have no settlement date are referred to as *unplanned*. This app can be used to create background jobs to create settlement documents for the unplanned settlement of customer condition contracts. Multiple selection options are available for the creation of the job, including template selection parameters such as job template (select from a dropdown list) and job name (free text), as well as scheduling options such as the **Start Immediately** flag, recurrence pattern (single run or recurring), and start date and time. Parameter options include settlement date, settlement date type, reference settlement date, customer, condition contract, contract process variant, condition contract type, semantic type, condition contract category, company code, purchasing organization, purchasing group, sales organization, distribution channel, division, and status.

It is possible to use this app to perform multiple functions, such as the creation of periodic background jobs, receiving notifications for failed jobs, and monitoring and displaying information about the scheduled jobs.

Schedule Unplanned Contract Settlement – External Agent Contracts (F6796)

Application Type: Transactional

In SAP S/4HANA settlement management, when a condition contract is created, it is normal to create a settlement date. However, contracts that have no settlement date are referred to as *unplanned*. This app can be used to create background jobs to create settlement documents for the unplanned settlement of external agent condition contracts. Multiple selection options are available for the creation of the job, including template selection parameters such as job template (select from a dropdown list) and job name (free text), as well as scheduling options such as the **Start Immediately** flag, recurrence pattern (single run or recurring), and start date and time. Parameter options include settlement date, settlement date type, reference settlement date, supplier, condition contract, contract process variant, condition contract type, condition

contract category, company code, purchasing organization, purchasing group, and status.

It is possible to use this app to perform multiple functions, such as the creation of periodic background jobs, receiving notifications for failed jobs, and monitoring and displaying information about the scheduled jobs.

Schedule Unplanned Contract Settlement – Royalty Contracts (F6795)

Application Type: Transactional

In SAP S/4HANA settlement management, when a condition contract is created, it is normal to create a settlement date. However, contracts that have no settlement date are referred to as *unplanned*. This app can be used to create background jobs to create settlement documents for the unplanned settlement of royalty condition contracts. Multiple selection options are available for the creation of the job, including template selection parameters such as job template (select from a dropdown list) and job name (free text), as well as scheduling options such as the **Start Immediately** flag, recurrence pattern (single run or recurring), and start date and time. Parameter options include settlement date, settlement date type, reference settlement date, supplier, condition contract, contract process variant, condition contract type, condition contract category, company code, purchasing organization, purchasing group, and status.

It is possible to use this app to perform multiple functions, such as the creation of periodic background jobs, receiving notifications for failed jobs, and monitoring and displaying information about the scheduled jobs.

Schedule Unplanned Contract Settlement – Sales Rebates (F6797)

Application Type: Transactional

In SAP S/4HANA settlement management, when a condition contract is created, it is normal to create a settlement date. However, contracts that have no settlement date are referred to as *unplanned*. This app can be used to create background jobs to create settlement documents for the unplanned settlement of sales rebate condition contracts. Multiple selection options are available for the creation of the job, including template selection parameters such as job template (select from a dropdown list) and job name (free text), as well as scheduling options such as the **Start Immediately** flag, recurrence pattern (single run or recurring), and start date and time. Parameter options include settlement date, settlement date type, reference settlement date, customer, condition contract, contract process variant, condition contract type, condition contract category, company code, sales organization, distribution channel, division, and status.

It is possible to use this app to perform multiple functions, such as the creation of periodic background jobs, receiving notifications for failed jobs, and monitoring and displaying information about the scheduled jobs.

Schedule Update Settlement Calendar – Condition Contracts (F6871)

Application Type: Transactional

In SAP S/4HANA settlement management, when a condition contract is created, it is normal to create a settlement date. The settlement date refers to the date on which the system determines and executes the settlement of accrued amounts such as rebates, commissions, or bonuses, for a customer, supplier, or partner. It is the date used to trigger the creation of settlement documents such as credit memos and debit memos. This app can be used to create background jobs to update the settlement calendar in condition contracts. Multiple selection options are available for the creation of the job, including template selection parameters such as job template (select from a dropdown list) and job name (free text), as well as scheduling options such as the **Start Immediately** flag,

recurrence pattern (single run or recurring), and start date and time. Parameter options include condition contract type, semantic type, condition contract category, condition contract, contract process variant, reference, assignment, created by, created on, supplier, customer, and status.

It is possible to use this app to perform multiple functions, such as the creation of periodic background jobs, receiving notifications for failed jobs, and monitoring and displaying information about the scheduled jobs.

Schedule Update Settlement Calendar – External Agent Contracts (F6875)

Application Type: Transactional

In SAP S/4HANA settlement management, when a condition contract is created, it is normal to create a settlement date. The settlement date refers to the date on which the system determines and executes the settlement of accrued amounts such as rebates, commissions, or bonuses, for a customer, supplier, or partner. It is the date used to trigger the creation of settlement documents such as credit memos and debit memos. This app can be used to create background jobs to update the settlement calendar in external agent condition contracts. Multiple selection options are available for the creation of the job, including template selection parameters such as job template (select from a dropdown list) and job name (free text), as well as scheduling options such as the **Start Immediately** flag, recurrence pattern (single run or recurring), and start date and time. Parameter options include condition contract type, condition contract category, condition contract, contract process variant, reference, assignment, created by, created on, supplier, customer, and status.

It is possible to use this app to perform multiple functions, such as the creation of periodic background jobs, receiving notifications for failed jobs, and monitoring and displaying information about the scheduled jobs.

Schedule Update Settlement Calendar – Royalty Contracts (F6874)

Application Type: Transactional

In SAP S/4HANA settlement management, when a condition contract is created, it is normal to create a settlement date. The settlement date refers to the date on which the system determines and executes the settlement of accrued amounts such as rebates, commissions, or bonuses, for a customer, supplier, or partner. It is the date used to trigger the creation of settlement documents such as credit memos and debit memos. This app can be used to create background jobs to update the settlement calendar in royalty condition contracts. Multiple selection options are available for the creation of the job, including template selection parameters such as job template (select from a dropdown list) and job name (free text), as well as scheduling options such as the **Start Immediately** flag, recurrence pattern (single run or recurring), and start date and time. Parameter options include condition contract type, condition contract category, condition contract, contract process variant, reference, assignment, created by, created on, supplier, customer, and status.

It is possible to use this app to perform multiple functions, such as the creation of periodic background jobs, receiving notifications for failed jobs, and monitoring and displaying information about the scheduled jobs.

Schedule Update Settlement Calendar – Sales Rebates (F6872)

Application Type: Transactional

In SAP S/4HANA settlement management, when a condition contract is created, it is normal to create a settlement date. The settlement date refers to the date on which the system determines and executes the settlement of accrued amounts such as rebates, commissions, or bonuses, for a customer, supplier, or partner. It

is the date used to trigger the creation of settlement documents such as credit memos and debit memos. This app can be used to create background jobs to update the settlement calendar in sales rebate condition contracts. Multiple selection options are available for the creation of the job, including template selection parameters such as job template (select from a dropdown list) and job name (free text), as well as scheduling options such as the **Start Immediately** flag, recurrence pattern (single run or recurring), and start date and time. Parameter options include condition contract type, condition contract category, condition contract, contract process variant, reference, assignment, created by, created on, supplier, customer, and status.

It is possible to use this app to perform multiple functions, such as the creation of periodic background jobs, receiving notifications for failed jobs, and monitoring and displaying information about the scheduled jobs.

Scheduling Worklist Jobs – Condition Contract Settlement (F7138)

Application Type: Transactional

With this app, you can monitor any scheduled jobs for the settlement of condition contracts. The list of jobs is displayed in tabular format and can be searched using a free search against the job name. Additional filters include status and date from-to. It is possible, using the quick actions icons, to create, copy, view details, cancel, and restart the jobs by selecting from the list.

Select Dispatches and Customer Returns – Intrastat Declaration (F2507)

Application Type: Transactional

Intrastat declarations are a key part of SAP S/4HANA for international trade. With this app, you can schedule background jobs to read dispatches and customer returns based on SAP billing documents in order to create Intrastat declarations. New jobs can be created or copied from existing jobs. Existing jobs also can be canceled. Multiple selection options are available for the creation of the job, including template selection parameters such as job template (select from a dropdown list) and job name (free text), as well as scheduling options such as the **Start Immediately** flag, recurrence pattern (single run or recurring), and start date and time. Parameter options include provider of info, declaration year, and declaration month.

Select Receipts and Returns to Supplier – Intrastat Declaration (F2508)

Application Type: Transactional

Intrastat declarations are a key part of SAP S/4HANA for international trade. With this app, you can schedule background jobs to read receipts and returns to suppliers based on SAP billing documents and purchase orders in order to create Intrastat declarations. New jobs can be created or copied from existing jobs. Existing jobs also can be canceled. Multiple selection options are available for the creation of the job, including template selection parameters such as job template (select from a dropdown list) and job name (free text), as well as scheduling options such as the **Start Immediately** flag, recurrence pattern (single run or recurring), and start date and time. Parameter options include provider of info, declaration year, and declaration month.

Settle Route Data (F4242)

Application Type: Transactional

Routes in SAP S/4HANA can be displayed using this app, with various features enabled in order to update the route. The full route and settlement information can be displayed in the app, with a process flow shown between each step. Each step in the process flow can be selected to see more information related to the route. The following features are available:

- Add customer documents to the route
- Assign inventory groups
- Record payments
- Identify differences in the check out and check in process and rectify them
- Identify and change products loaded and unloaded during stops
- Check whether tolerances have been exceeded by using tolerance groups
- Settle routes in foreground or schedule in background

Navigation to various related apps is supported, such as Outbound Delivery (S/4HANA) (FO233A), Returns Delivery (S/4HANA) (FO234A), Manage Billing Documents (FO797), Material Documents Overview (F1077), and Sales Order (S/4HANA) (F1814).

Settlement Document (Version 2) (F3658A)

Application Type: Fact Sheet

This centralized fact sheet app allows you to review all details related to any kind of settlement document in a single hub. The intention is to allow you to determine any related follow-up actions that need to be taken regarding settlement documents. A full graphical process flow is available for the settlement document under the **Document Flow** tab of the fact sheet. Account assignment for the document can be seen by selecting the **Account Assignment** menu option.

Track Sales Order Details (F2981)

Application Type: Transactional

This app is extremely useful in understanding the flow and all the details for a specific sales order. There is an additional level of granularity, focused on a single sales order, compared with the related app, Track Sales Orders (S/4HANA) (F2577). The document flow is shown graphically in a tree structure, with additional features such as the ability to view outstanding issues against the documents. Only the order, order without charge, and debit memo request document categories are supported for this app. More details for each document can be shown by selecting the required node from the structure.

A whole selection of related apps are supported through navigation from the Track Sales Order Details app, such as Sales Order Fulfillment Issues (Version 2) (F0029A), Track Sales Orders (S/4HANA) (F2577), and Predicted Delivery Delay (F3408).

Tip

In the SAP ERP 6.0 world, a document flow would only show the document number, status, and date of the related documents. The Track Sales Order Details app can show you the same, but in a graphical process flow, and in addition, the full details for all the related documents are visible at a click of a button together on a single screen. For this reason, it is recommended to use this app for tracking specific sales orders rather than the traditional document flow attached to sales documents in SAP GUI transactions.

Transaction Code

The equivalent transaction code in SAP GUI is Transaction VA03 (Display Sales Order).

Track Sales Orders (S/4HANA) (F2577)

Application Type: Analytical

This very useful app provides a centralized hub where you can trace the progress of a specific sales order from a list of sales orders generated from filter options such as sales document, sold-to party, customer reference, requested delivery

date, overall status, and document date. The ensuing list shows the sales documents with their process phases (such as in order, in delivery, in invoice, or in accounting), displayed as a single text, but also as a series of ticks or exclamation marks against each phase. Clicking a specific line will take you to the details for that document to show any outstanding issues, billing or payment details, attachments, business partners involved in the document, and the graphical display of the process flow.

Navigation to the related Track Sales Order Details app (F2981) is supported in order to view much more information related to the sales order and its related documents.

Transaction Code

The equivalent transaction code in SAP GUI is Transaction VA03 (Display Sales Order).

Trade Compliance Document (F5977)

Application Type: Fact Sheet

As with all fact sheet apps, the enterprise search in the SAP Fiori launchpad can be used to open this app. Filters in the dropdown list in enterprise search, next to the search input fields, can be used to narrow down your search parameters. Filters such as sales orders, sales contracts, purchase orders, company code, plant, and partner country and region are available in the enterprise search filter bar. Once a document is selected and the fact sheet app is opened, all information related to the trade compliance document can be viewed in one place, including document number, overall status, legal control status, embargo status, SAP Watch List Screening status, and business partners.

Navigation to the related Manage Documents – Trade Compliance app (F2826) is supported.

Chapter 4
Inventory and Warehouse Management

In this chapter, we'll detail all the most useful and relevant SAP Fiori apps available in the inventory and warehouse management space. This relates to all processes between the creation of an outbound or inbound delivery and the shipping of the goods. It also covers warehouse movements between locations, but it does not cover manufacturing or assembly processes, which are covered in the chapters on materials management and production planning. Warehouse management in SAP covers logistics warehouse management, as well as stockroom management and extended warehouse management (EWM). Outbound processes related to inventory management are also covered. This chapter will also touch upon the key apps available in the transportation management (TM) functionality in SAP S/4HANA. Many of these processes are covered by SAP GUI transactions that have adopted the SAP Fiori look and feel. This chapter doesn't cover those; instead, it concentrates on the SAP Fiori apps that are not covered by existing SAP GUI transactions.

Advanced Shipping and Receiving Cockpit (F7366)

Application Type: Transactional

This app allows you to monitor freight orders and loading/unloading points in the warehouse. The app is flexible enough to handle deliveries generated in extended warehouse management (EWM) and in inventory management. Multiple selection options to determine the data to display are available, including freight order, warehouse number, loading/unloading point, warehouse door, warehouse door status, loading/unloading completed, carrier, transportation planning on/at, execution status, and lifecycle status. You can click one of the listed freight orders to navigate to the related Web Dynpro app, Load or Unload Freight Orders (/SCMTMS/ASR()).

Allocation – ALP (F2745)

Application Type: Analytical

This app provides KPIs in an analytical list page (ALP). The list of KPIs shows the weight, volume, and value of the transportation allocation. Each unit of measure can show total and average utilization. The KPIs are displayed in a chart, with options to change the type of chart displayed, and the data is displayed for the current and the previous month so that you can identify key trends. The data can be viewed by carrier, allowing you to identify the total weight, volume, and container utilization for each carrier.

Allocation Compliance (F2673)

Application Type: Analytical

This overview page offers an alternative view of the transport allocation in a single hub. The list of KPIs shows the weight, volume, and value of the transportation allocation. The data is retrieved by selecting one of the following cards from the overview page:

- **Tradelane Weight Load Analysis**
- **Tradelane Volume Load Analysis**
- **Tradelane Container Load Analysis**
- **Transportation Mode Category Analysis**
- **Overall Weight Utilization Distribution**
- **Overall Container Utilization Distribution**

Technically, this page retrieves data using CDS view `C_TranspAllocComplianceQ`.

Analyze Delivery Logs (F0870A)

Application Type: Transactional

Collective background jobs for the creation of deliveries (outbound and inbound) can be analyzed using this app. Selection options such as log number, created by, created on, shipping point, and log status are available. A list of background jobs and log numbers is displayed in a table format. Clicking an entry will take you to a more detailed screen, on which you can see the details of the collected messages and the full list of delivery numbers.

Analyze Stock in Date Range (F6185)

Application Type: Transactional

This app can be used by inventory managers to analyze goods movements for specific materials or ranges of materials within a given date range. There are standard SAP use cases provided for filters, as follows:

- Standard: Displays the most vital stock quantity and value fields for the material
- Stock quantity: Displays net receipt quantity and net goods issue quantity
- Supplier consignment stock: Displays the supplier consignment (**K**) special stock type

It's also possible to create your own filter view. The results are organized into two separate page views: the **Main** page and the **Detail** page. The **Main** page displays a list of all the materials with stock counts and an option to navigate directly to the detail page. The **Detail** page displays a microchart showing all relevant stock details for the material selected.

You can also define background jobs for inventory movement from this app.

> **Tip**
>
> Authorizations are key for this app and could lead to zeros being displayed for certain attributes in the results table. These attributes can be hidden, or the authorizations can be analyzed to assess what is missing from the user record.

Analyze Warehouse Task (F5123)

Application Type: Analytical

This app can be used to analyze all tasks assigned as part of extended warehouse management (EWM). These tasks include picks, putaways, internal goods movements, posting changes, goods receipt postings, and goods issue postings. You can create warehouse tasks with reference to a warehouse request. You can also create warehouse tasks without a reference—for example, for internal goods movements.

Analyzer Warehouse Order (F5122)

Application Type: Analytical

Warehouse tasks have specific KPIs assigned to them, and this app allows you to view these KPIs in more detail. Navigation to drilldown reports is supported. Various selection filters are available such as warehouse order queue, warehouse, warehouse order, activity area, and warehouse order status. The results are displayed in a table, with variants available for configuration. Each warehouse order has a separate row, and users can see the status, activity area, queue, warehouse order creation rule, and process type. This analytical overview app is relevant for the warehouse clerk role in SAP Fiori and was introduced in SAP S/4HANA 2021.

Application Log for Just-In-Time Supply to Customer (F4961)

Application Type: Transactional

Just-in-time (JIT) call processing jobs create application logs in SAP. This app can be used by shipping specialists to identify issues to resolve in a centralized hub. Multiple options for selecting the data to display are available, including severity, from and to dates, category, subcategory, and external reference.

Tip

When creating a JIT log, add a sales order ID as an external reference. As a result, you can filter all logs related to that sales order.

Application Logs for Manufacturing Logistics (F5189)

Application Type: Transactional

Manufacturing logistics processes create processing jobs with application logs in SAP. This app can be used by warehouse clerks to identify issues to resolve in a centralized hub. Multiple options for selecting the data to display are available, including severity, from and to dates, category, subcategory, and external reference. The data is displayed in a table along with the description of the subcategory it refers to, such as logistical reevaluation, route, warehouse task, warehouse request, warehouse order, or tour.

Block Bulk Bins from Putaway (F5485)

Application Type: Transactional

This app allows warehouse clerks to block the putaway of goods into storage bins in the bulk area. This exercise can be done in the foreground (immediate) or in the background as a scheduled job. With the app, scheduled jobs can be created as copies of existing jobs. In addition, jobs can be viewed, monitored, or canceled. Multiple selection options are available for the creation of the job, including template selection parameters such as job template (select from dropdown list) and job name (free text), as well as scheduling options such as the **Start Immediately** flag, recurrence pattern (single run or recurring), and start date and time. Parameter options related to the putaway task are available.

Business Share (F2674)

Application Type: Analytical

This overview page related to transportation management (TM) in SAP S/4HANA allows managers to get real-time insight into the KPIs required to manage transportation of products by carriers to customers. The intention is to provide a centralized hub for all insight into a business's dealings with carriers. The overview page is organized into the following cards, which provide individual KPIs for analysis:

- **Weight Fulfillment Ratio**
  This card shows how carriers have performed against their allotted weight for transportation (e.g., they may have shipped 151,250 kg out of a total allocation of 208,210 kg). This fulfillment ratio can also be shown as a percentage.

- **Volume Fulfillment Ratio**
  This card shows how carriers have performed against their allotted volume for transportation (e.g., they may have shipped 1000 m^3 out of a total allocation of 2500 m^3). This fulfillment ratio can also be shown as a percentage.

- **Road Fulfillment Ratio**
  This card shows, in graphical format, the volume of goods transported by road for the actual booked quantity against the target quantity.
- **Sea Fulfillment Ratio**
  This card shows, in graphical format, the volume of goods transported by sea for the actual booked quantity against the target quantity.
- **Air Fulfillment Ratio**
  This card shows, in graphical format, the volume of goods transported by air for the actual booked quantity against the target quantity.
- **Trade Lane Fulfillment**
  This card shows how each trade lane performs against its allotted share. The units can be weight or volume.

You can navigate from the app to the Business Share – ALP app (F2746), where you can get analytical information related to your carrier business.

Business Share – ALP (F2746)

Application Type: Analytical

This very useful SAP Fiori app gives you a much more detailed level of information regarding carrier business. Information available in the Business Share app (F2674) can be displayed in this app as well, with graphical and table overviews for each category. Navigation between this app and related apps in the transportation management (TM) function are supported, such as Freight Booking Execution (F2677) and other freight-related apps.

Change Inbound Delivery (F1706)

Application Type: Transactional

With this app, you can change an inbound delivery to carry out tasks such as changing the delivery quantity, deleting items, entering batch

information, creating or changing handling units, and posting goods receipt. It is also possible to upload and delete attachments, such as photographs or documents in various formats.

You can navigate to the related Create Inbound Delivery (F1705) and Process Warehouse Tasks (F4595) apps.

Tip

When a delivery arrives at the goods-in location, use this app to cross-reference the delivery note from the driver against the expected delivery and make changes in SAP accordingly. It's also possible to add items to an open inbound delivery, in case the delivery from the driver contains items belonging to a different purchase order.

Transaction Code

The associated transaction in SAP GUI is Transaction VL32N (Change Inbound Delivery).

Change Inbound Delivery – Production (F5162)

Application Type: Transactional

Inbound deliveries can be from external sources but also can be related to items arriving in the warehouse location from the production lines. This app allows you to compare the inbound delivery in SAP with the physical goods arriving from production. The available tasks include changing the delivery quantity, deleting delivery items, entering batch information, creating or changing handling units, and posting or reversing goods receipts. It is also possible to upload and delete attachments, such as photographs or documents in various formats.

You can navigate to the related Change Inbound Delivery (F1706) and Process Warehouse Tasks (F4595) apps.

Check-Out/In Differences (F4741)

Application Type: Transactional

Routes have planned quantities that can be tracked against actual quantities by using this app. The app tracks planned and actual check-out quantities for the route against expected and actual check-in quantities. This is useful for analyzing the efficiency of the execution of the route.

You can navigate to the related Outbound Delivery (FO233A), Return Delivery (F1996), and Track Sales Order Details (F2981) apps.

Classify Products – Commodity Codes for Transportation (F7151)

Application Type: Transactional

As part of the SAP S/4HANA for international trade functionality, this app can be used to classify materials to a specific commodity code, using a specific numbering scheme. From within the search features of the app, you can find unclassified products based on product number, product description, material type, material group, and division. The app presents a list of products based upon the selection criteria used, and you can select single or multiple entries to classify them. Commodity code classifications can be set using validity dates, and remarks can be added during the process.

Configure Activity Attributes (F6439)

Application Type: Transactional

With this app, you can use business process scheduling (BPS) flexibly to configure the durations of logistics activities and define the necessary working times for scheduling them. These durations and working times are based on attributes in the sales documents. Available activities in the app include the following:

- Creation of accesses that contain characteristic value combinations based on scheduling activity

- Assigning values to each characteristic value combination
- Changing the sequencing priority of the access
- Changing the status of the accesses to one of the available statuses: active, inactive, or obsolete
- Controlling the use of characteristic value combinations based on authorization groups

> **Tip**
>
> After manually changing any activity attributes, it is recommended to run rescheduling for the sales documents in question.

Configure Alternative Determination (F5354)

Application Type: Transactional

As a part of the advanced available-to-promise (ATP) solution, this app allows you to define alternative determinations for sales order requirements. It is possible to define rating attributes and hard constraints within the app. Templates can also be defined that contain preconfigured alternative determinations for confirming requirements fully, on time, or earlier than on time but with as large a quantity as is possible. Rating attributes to drive the alternative determination, such as minimum delay, minimum number of plants, and minimum number of storage locations, can be configured, as can hard constraints to limit the alternative determination, such as maximum confirmation ratio.

Configure Supply Protection (F4569)

Application Type: Transactional

This useful app can be used to protect the quantities of a given material in a plant against other demand. There are two approaches to providing this protection:

- Horizontal protection: Complete groups are protected against each other.
- Vertical protection: Priority is assigned to groups, so with vertical protection, every group is only protected against demands originating from groups with a lower priority.

Quantities that are protected are defined by a combination of a material and a plant, with other characteristics available as well. Supply protection objects also can be defined, such as sales orders, stock transport orders, and so on. In addition, protection can be defined by validity dates. Supply protection integrates natively with backorder processing and product allocation.

Confirm Freight Orders (F5694)

Application Type: Transactional

This app can be used to respond to freight orders sent by the shipper. In this app, you can review freight orders that require confirmation and view their details, such as source and destination, transportation, cargo, dangerous goods, driver, stops, notes, and attachments. You can also update the freight order with the driver and registration, and initiate changes to the confirmed end date/time, price limit, or currency.

Count Physical Inventory (F3340)

Application Type: Transactional

Many physical inventory counting processes are still carried out using paper. With this app, you can count the results of the physical inventory documents after the warehouse operator has manually recorded the count results on the physical inventory documents. The app allows you to make various changes to physical inventory documents, such as activating/deactivating the documents, printing documents and defining the layout of the printouts, assigning products and handling units as well as serial numbers, and assigning a count reference to the items.

Create Inbound Delivery (F1705)

Application Type: Transactional

With this app, you can create an inbound delivery to carry out tasks such as adding the delivery quantity, entering batch information, creating handling units, and posting goods receipt. It is also possible to upload attachments, such as photographs or documents in various formats.

Navigation to the related Change Inbound Delivery app (F1706) is supported.

Transaction Code

The associated transaction in SAP GUI is Transaction VL31N (Create Inbound Delivery).

Create JIT Delivery Schedule (In Background) (F5001)

Application Type: Transactional

With this app, just-in-time (JIT) delivery schedules can be created with reference to scheduling agreements as a background job. The schedule is based on JIT calls received from customers. Multiple selection options are available for the creation of the job, including template selection parameters such as job template (select from dropdown list) and job name (free text), as well as scheduling options such as the **Start Immediately** flag, recurrence pattern (single run or recurring), and start date and time. Parameter options related to the JIT delivery schedule include till specific date, till date, call components, external status, plant, and sold-to party.

Create Outbound Deliveries from Sequenced JIT Calls (F3897)

Application Type: Transactional

With this app, you can create outbound deliveries from sequenced just-in-time (JIT) calls on a package unit, according to specific shipping

4

points. It is possible to split the delivery document based upon the criteria maintained in the Manage JIT Customer Data app (F3011). Selection criteria for finding package groups include the just-in-sequence (JIS) packaging unit ID. Often used in the automotive industry, JIS comes into play when parts need to arrive not only just in time, but also in the correct sequence required for the efficient operation of the production line. Other selection criteria include component group material, ship-to party, unloading point, shipping point, requirement date and time, planned shipping date and time, and planned delivery date. From the list of package groups displayed, you can create outbound deliveries or view logs.

Create Outbound Deliveries from Summarized JIT Calls (F3922)

Application Type: Transactional

With this app, you can create outbound deliveries from summarized just-in-time (JIT) calls on component groups. Selection criteria for finding package groups include ship-to party, shipping point, planned creation date, delivery date and time, planned goods issue date and time, unloading point, external JIT call number, and requirement date and time. From the list of package groups displayed, you can create outbound deliveries or view logs.

Create Physical Inventory Documents (F3197)

Application Type: Transactional

Inventory managers can use this app to create physical inventory documents for materials or groups of materials. This is part of the cycle count preparation, and the app assists with finding materials to be counted. Selection options for filtering include the plant, storage location, and not-counted-since date. The list of materials

in the table is sorted by stock type (unrestricted stock, blocked stock, quality inspection stock, etc.). There are options to export the data to a spreadsheet or to attach documents.

Transaction Code

The associated transaction in SAP GUI is Transaction MI01 (Create Physical Inventory Document).

Display Freight Quotations (F5553)

Application Type: Transactional

Accepting freight requests for quotation (RFQs) creates freight quotations in SAP to send to the shipper. These quotations can be viewed using this app. The app displays all the freight quotations that have been awarded or rejected by the shipper as well as those that are still in pending status. You can view more information for the freight quotation as well, such as its source, destination transportation, cargo, dangerous goods, transportation stops, notes, and attachments.

Display Serial Numbers (F5147)

Application Type: Analytical

Employees such as warehouse clerks and inventory managers can use this app to display a list of serial numbers associated with a specific material. The list displayed is fully up to date according to the last goods movement for the serial number. Frequently, this app is accessed from the related Stock – Single Material (F1076) and Stock – Multiple Materials (F1595) apps.

Transaction Code

The associated transaction in SAP GUI is Transaction IQ09 (Display Material Serial Number).

Expected Goods Receipt (F3298)

Application Type: Transactional

This app allows you to create background jobs for the transfer of expected goods receipts for purchase orders and production orders to extended warehouse management (EWM). You also can view the log for all jobs and cancel or restart jobs. Multiple selection options are available for the creation of a job, including template selection parameters such as job template (select from dropdown list) and job name (free text), as well as scheduling options such as the **Start Immediately** flag, recurrence pattern (single run or recurring), and start date and time. Parameter options related to the just-in-time (JIT) delivery schedule include the warehouse number, selection start date, selection end date, supplier, purchasing document type, confirmation control, material, purchasing document, production order type, and production order.

Explain Scheduling Results (F4976)

Application Type: Transactional

This app is part of the transportation management (TM) suite of apps and is used to analyze scheduling results. The results are displayed in a table format with information such as the planning profile, scheduling settings, planning horizon, scheduling direction, scheduling strategy, status, and action. Each item can be selected to open more detailed information, such as solution steps. Each activity in the detailed information area can be viewed chronologically in a Gantt chart.

Freight Booking Execution (F2677)

Application Type: Analytical

This overview page holds individual cards that show information related to transportation management (TM) freight bookings. The following cards are available:

- **Lifecycle Status Analysis**
  This card shows the number of bookings, grouped by lifecycle status. The statuses available are draft (status 00), new (status 01), and in process (status 02).
- **Freight Bookings in Transit**
  This card shows the number of freight bookings that are currently in transit. Drilldown options here allow you to view details of the freight bookings.
- **Carrier On-Time Performance**
  This card shows the percentage of freight bookings that are delivered on time, split by carrier. Note that a freight booking is only classed as on time if all of its events are classed as on time.
- **Location Delay Analysis**
  This card shows the number of delayed freight bookings for each location, with an analysis of which events are delayed.
- **Execution Block Analysis**
  This card shows all freight bookings that have an execution block, grouped by execution block reason.
- **Delayed Freight Bookings in Transit**
  This card shows all freight bookings that are delayed in transit. The delay here represents the difference between the actual and expected dates. Drilldown options are available to view details of the freight bookings.
- **Location On-Time Performance**
  This card displays, as a percentage, the on-time freight bookings for each location. The delays also are displayed with numbers (e.g., 10/12 events on time).
- **Planning Block Analysis**
  This card shows all freight bookings with an associated planning block, grouped by planning block reason.
- **Blocking Discrepancies**
  This card shows freight bookings with discrepancy types associated with them, such as delivery of item or delay.
- **Activity On-Time Performance**
  This card shows the on-time performance for individual events within the freight booking,

such as loading and unloading, thus allowing you to identify bottlenecks in the process.

- **Unexpected Events**
  This card allows you to monitor unexpected events related to freight bookings, such as accidents or delays.

- **Non-Blocking Discrepancies**
  This card shows freight bookings with discrepancy types associated with them that did not result in a block.

- **Carrier Delay Analysis**
  This card shows an analysis of the delays in freight bookings for each carrier. This can be used to evaluate carrier performance in order to identify areas of concern.

The app uses two CDS views for the retrieval of data: C_FRTBKGEXECUTIONSTSQ_CDS and C_FRTBK-GEXECUTIONQUERY_CDS.

Freight Booking Execution – ALP (F2747)

Application Type: Analytical

This very useful app gives you a much more detailed level of information about your carriers. Information available in the Freight Booking Execution app (F2677) can be displayed in this app as well, with graphical and table overviews for each category. Navigation between this app and related apps in the transportation management (TM) function are supported, such as Freight Booking Execution (F2677) and other freight-related apps.

Freight Booking Execution Status – ALP (F2748)

Application Type: Analytical

This app allows you to view transports that are blocked in planning or in execution, grouped together by the block reason. The information is provided in both graphical and tabular format. Selection options for filtering include event on-time ratio by carrier, event on-time ratio by location, event on-time ratio by event, and

number of events by no block. For the listed freight bookings, you can perform quick actions by selecting **Change Freight Booking**.

Navigation to the Freight Booking Execution app (F2677) is supported.

Freight Booking Quantities – ALP (F2749)

Application Type: Analytical

This app allows you to view transport quantities in freight booking, grouped together by various selection options. The information is provided in both graphical and tabular format. Selection options for filtering include the carrier and the transportation category. For the listed freight bookings, you can perform quick actions by selecting **Change Freight Booking**.

Navigation to the Freight Booking: Quantity-Driven Operational Business app (F2678) is supported.

Freight Booking: Quantity-Driven Operational Business (F2678)

Application Type: Analytical

This overview page holds individual cards that show information related to quantities and volumes in transportation management (TM) freight bookings. The following cards are available:

- **Gross Weight Distribution**
  This card shows the change in gross weight across all weeks in the current and previous months for all completed freight bookings in a graphical format.

- **Gross Volume Distribution**
  This card shows the change in gross volume distribution across all weeks in the current and previous months for all completed freight bookings in a graphical format.

- **Quantity Distribution**
  This card shows the change in quantity distribution across all weeks in the current and previous months for all completed freight booking, in a graphical format.

- **Transportation Mode Category Analysis**
  This card displays a chart with the total number of freight bookings for each transportation mode category, by gross weight and gross volume.

- **Logistics Network Analysis**
  This card shows geographical locations with the total weight for completed freight bookings.

- **Carrier Quantity Analysis**
  This card shows completed freight bookings by carrier, with gross weight, gross volume, and container count.

- **Shipper Quantity Analysis**
  This card shows completed freight bookings by shipper, with gross weight, gross volume, and container count.

- **Ship-to Party Quantity Analysis**
  This card shows completed freight bookings by ship-to party, with gross weight, gross volume, and container count.

- **Utilization Distribution**
  This card shows the lowest percentage utilization by carriers for completed freight booking.

- **Quantity Analysis**
  This card shows the transportation mode category (road, air, sea, etc.) alongside the gross volume and gross weight for completed freight bookings.

- **Distance Analysis**
  This card shows the transportation mode category (road, air, sea, etc.) alongside the total distance and duration for completed freight bookings.

Freight Cost Allocation Document (F6215)

Application Type: Fact Sheet

This app provides a centralized hub where you can view all freight cost allocations in one place. The details in the app allow you to view account assignment information for the freight costs, navigate to pertinent related apps such as

Manage Freight Cost Allocation Documents (F5514), and get more detailed information about freight cost allocation documents, such as pricing information and the document flow.

Freight Order Execution (F2675)

Application Type: Analytical

This overview page holds individual cards that show information related to transportation management (TM) freight orders. The following cards are available:

- **Lifecycle Status Analysis**
  This card shows the number of orders grouped by lifecycle status: draft (status 00), new (status 01), or in process (status 02).

- **Execution Block Analysis**
  This card shows all freight orders with an execution block, grouped by execution block reason.

- **Planning Block Analysis**
  This card shows all freight orders with a planning block, grouped by planning block reason.

- **Delayed Freight Orders in Transit**
  This card shows all freight orders that are delayed in transit. The delay here represents the difference between the actual and expected dates. Drilldown options are available to view the details of the freight orders.

- **Blocking Discrepancies**
  This card shows freight orders with discrepancy types associated with them, such as delivery of item or delay.

- **Non-Blocking Discrepancies**
  This card shows freight orders with discrepancy types associated with them that did not result in a block.

- **Carrier On-Time Performance**
  This card shows the percentage of freight orders delivered on time, split by carrier. Note that a freight order is only classed as on time if all of its events are classed as on time.

- **Location On-Time Performance**
  This card displays the on-time freight orders

for each location as a percentage. The delays also are displayed with numbers (e.g., 10/12 events on-time).

- **Activity On-Time Performance**
  This card shows the on-time performance for individual events within the freight order, such as loading and unloading, allowing you to identify bottlenecks in the process.

- **Unexpected Events**
  This card allows you to monitor unexpected events related to freight orders, such as accidents or delays.

The app uses two CDS views to retrieve the data: `C_FrtOrdExecutionStsQ` and `C_FrtOrdExecution-Query`.

Freight Order Execution – ALP (F2750)

Application Type: Analytical

This app allows you to see a detailed view of freight order execution, grouped together by various selection options. The information is presented in both graphical and tabular format. Selection options for filtering include event on-time ratio by carrier, event on-time ratio by location, event on-time ratio by event, and number of events by no block. For the listed freight bookings, you can perform quick actions by selecting **Change Freight Booking**.

Freight Order Execution Status – ALP (F2751)

Application Type: Analytical

This app allows you to view freight orders that are blocked in planning or in execution, grouped together by block reason. The information is presented in both graphical and tabular format. Selection options for filtering include the number of freight orders by lifecycle and number of planning blocked freight orders by transportation management (TM) category. For the listed freight bookings, you can perform quick actions by selecting **Change Freight Order**.

Freight Order Quantities – ALP (F2752)

Application Type: Analytical

This app allows you to view freight order quantities, grouped together by various filter options. The information is presented in graphical and tabular format. Selection options for filtering include gross weight by carrier and number of blocked freight orders by transportation management (TM) category. For the listed freight bookings, you can perform quick actions by selecting **Change Freight Order**.

Freight Order: Quantity-Driven Operational Business (F2676)

Application Type: Analytical

This overview page holds individual cards showing information related to quantity analysis for transportation management (TM) freight orders. The following cards are available:

- **Gross Weight Distribution**
  This card shows the change in gross weight across all weeks in the current and previous months for all completed freight orders in graphical format.

- **Gross Volume Distribution**
  This card shows the change in gross volume distribution across all weeks in the current and previous months for all completed freight orders in graphical format.

- **Quantity Distribution**
  This card shows the change in quantity distribution across all weeks in the current and previous months for all completed freight orders in graphical format.

- **Transportation Mode Category Analysis**
  This card displays a chart with the total number of freight orders for each transportation mode category, by gross weight and gross volume.

- **Logistics Network Analysis**
  This card shows geographical locations with the total weight for completed freight orders.

- **Carrier Quantity Analysis**
  This card shows completed freight orders by carrier, with gross weight, gross volume, and container count.

- **Shipper Quantity Analysis**
  This card shows completed freight bookings by shipper, with gross weight, gross volume, and container count.

- **Ship-to Party Quantity Analysis**
  This card shows completed freight orders by ship-to party, with gross weight, gross volume, and container count.

- **Utilization Distribution**
  This card shows the lowest percentage utilization by carriers for completed freight orders.

- **Quantity Analysis**
  This card shows the transportation mode category (road, air, sea, etc.) alongside the gross volume and gross weight for completed freight orders.

- **Distance Analysis**
  This card shows the transportation mode category (road, air, sea, etc.) alongside the total distance and duration for completed freight orders.

Navigation to the Freight Order Quantities – ALP app (F2752) is supported.

Inbound Delivery (S/4HANA) (F0232A)

Application Type: Fact Sheet

This app offers information about inbound deliveries. For example, header data like gross weight and net weight, volume, number of packages, document date, planned goods movement date, and actual goods movement date are available on the initial screen, as are the various statuses: putaway status, putaway confirmation status, warehouse management status, and overall goods movement status. Additional information related to item-level data and business partners is also available, as are process flows.

This app, like many fact sheet apps, is best accessed from the enterprise search results. The app provides a detailed overview of an inbound delivery, and you can navigate to several related apps, such as Material Master (F0338A), Purchase Order (Version 2) (F0348A), and Supplier (F0354).

Transaction Code

The associated transaction in SAP GUI is Transaction VL33N (Display Inbound Delivery).

Inventory KPI Analysis (F3749)

Application Type: Analytical

Inventory analysts can use this app to monitor an organization's KPIs effectively, thus ensuring accurate forecasting. The standard KPI filters are as follows:

- **Stock changes**
  Displays stock value changes for the selected materials in the currency of the company code over a given time period.

- **Consumption changes**
  The *consumption value* is the result of the consumption multiplied by the current price. This option displays the changes in consumption for materials in the currency of the company code over a given time period.

- **Inventory aging changes**
  This option can be defined as a percentage relationship between the average stock quantity and consumption quantity for the selected materials over a given time period.

- **Inventory turnover changes**
  This option can be defined as a relationship between the average stock quantity and consumption quantity for the selected materials over a given time period.

- **Range of coverage changes**
  This option can be defined as a relationship between the current stock quantity and the relationship between consumption and a specified number of days for the selected materials over a given time period.

Manage Charges for Freight Orders (F7104)

Application Type: Transactional

Carriers can use this app to view, update, and approve freight charges. The freight orders in flight can be viewed in the app, with the option to list them by status. From the list of freight orders generated per your selection criteria, you can view information such as transportation details, source and destination details, cargo details, transportation stages, charges, notes, and attachments. You also can edit a freight order to add or update a charge. As this app is normally used by carriers, the function for collaboration with carriers must be configured first. This needs to be done in the Maintain Collaboration Users app (F4911).

Manage Commodity Codes for Transportation (F7150)

Application Type: Transactional

As part of the SAP S/4HANA for international trade functionality, this very simple app allows you to add, edit, and display commodity codes for use in transportation. Commodity codes can have a description as well as a validity date range. You also can add a custom unit of measure against the commodity code.

Manage Execution of Freight Orders (F6013)

Application Type: Transactional

This app is available to carriers or service providers to monitor and manage freight orders that have been sent by a shipper and are in the process of execution. You can search using a selection of parameters and display a list in tabular format of all freight orders that currently have the *in execution* status. You also can view details of transportation stops as well as report

a planned or unplanned event during a transportation stop. As this app is normally used by carriers, the function for collaboration with carriers must be configured first. This needs to be done in the Maintain Collaboration Users app (F4911).

Manage Freight Agreement RFQ Masters (F5395)

Application Type: Transactional

Shipper transportation requirements are often detailed in freight agreement request for quotation (RFQ) documents. The RFQ document is used in the selection process to select a carrier or carriers to fulfill transportation requirements. It is possible to use this app to create new freight agreement RFQs as well as edit and display existing freight agreement RFQs. It is also possible to create a freight agreement RFQ using an industry-standard template.

> **Tip**
>
> Templates for freight agreement RFQs can be customized in the IMG, using the path **Transportation Management • Master Data • Agreement RFQs and Quotations • Define Freight Agreement RFQ Master Templates**. From within this IMG activity, templates can be created with RFQ items as standard and commodity codes can be assigned.

Manage Freight Agreement RFQs (F4916)

Application Type: Transactional

This app can be used to monitor and negotiate rates with shippers. The carriers themselves can access this app to respond to freight agreement requests for quotation (RFQs). It is possible to use a spreadsheet to enter rate details and attach files to the RFQ, and it is also possible to reject the RFQ and thus choose not to participate. As this app is normally used by carriers, the function for collaboration with carriers must be configured first. This needs to be done in the Maintain Collaboration Users app (F4911).

Manage Freight Cost Allocation Documents (F5514)

Application Type: Transactional

This app is used to manage freight cost allocation documents. You can release the documents using the **Release** quick action in the app. The data is viewed in tabular format, and filters such as editing status, freight cost allocation document, settlement document type, posting status, sales organization, purchasing organization, and posting date can be applied. The details screen for each freight cost allocation document can be accessed by clicking the document's line from the table view. In the details, you can view the administration data, journal entry data, pricing elements, and items, as well as a visualization of the document flow.

Manage Freight Invoices (F7106)

Application Type: Transactional

Carriers can use this app to manage the freight invoices that have been sent to a shipper. The status of each invoice determines what actions are available for it, as follows:

- **Submitted**
  Submitted by the shipper; you can create a credit memo to reverse the invoice.
- **Not Submitted**
  Created but not submitted by the shipper; you can edit all areas in full and submit.
- **For Approval**
  Invoices awaiting approval for changes proposed by the shipper; you can view and edit the invoice, accept the shipper's proposal, or edit and propose new changes.

- **Rejected/Canceled**
  Canceled or rejected internally or by the shipper; you can view details.

As this app is normally used by carriers, the function for collaboration with carriers must be configured first. This needs to be done in the Maintain Collaboration Users app (F4911).

Manage Freight Orders for Invoicing (F7732)

Application Type: Transactional

A carrier can use this app to manage the creation and amendment of invoices for freight orders. The results of searches are displayed in tabular format, and fields such as source location, destination location, and means of transport can be used as filters. The results can be viewed by status (**Not Invoiced**, **Partially Invoiced**, or **Completely Invoiced**). New invoices can be raised using this app for orders in the **Not Invoiced** or **Partially Invoiced** status.

As this app is normally used by carriers, the function for collaboration with carriers must be configured first. This needs to be done in the Maintain Collaboration Users app (F4911).

Manage Freight RFQs (F5552)

Application Type: Transactional

This app can be used to manage and respond to requests for quotation (RFQs) from shippers. The carriers themselves can access this app to manage their response. It is possible to accept the RFQ without making any changes; propose changes to the price, currency, or validity dates; or reject the RFQ and thus choose not to participate. As this app is normally used by carriers, the function for collaboration with carriers must be configured first. This needs to be done in the Maintain Collaboration Users app (F4911).

Manage Loading Lanes (F6681)

Application Type: Transactional

This app is used by warehousing clerks to manage loading lanes and assign them to a given route. Selection options include route ID, warehouse, route type, lifecycle status, start storage type, and route parameter group. From the selected criteria, a list of routes is generated, from which loading lanes can be assigned. You also can view changes that have been made in the app.

Manage Packing Group Specification for Sequenced JIT Calls (F3896)

Application Type: Transactional

Multiple just-in-time (JIT) calls can be packed together. This app can be used to manage the packing specifications for JIT calls. Each component group can be packed in different ways depending upon a variety of attributes, such as customer or destination. Packaging specifications can be created, edited, and deleted using this app. Selection options can be used to bring up a full list of packaging specifications that includes data such as the description, packaging materials, number of slots, and status.

Manage Physical Stock – Products (F5340)

Application Type: Transactional

This app is a useful way of identifying inventory in storage bins. You can search via a variety of selection fields, such as warehouse number, product, storage bin, handling unit, and stock type, to bring up a list of products that meet your criteria. From within the app, you also can use the **Consume** and **Scrap** quick actions to post unplanned goods issues. This can be done for a single stock item or for several at once.

Manage Reload Requests (F4745)

Application Type: Transactional

Products that are to be loaded as part of customer deliveries can be amended to add products or deliveries via reload requests. The reload requests list can be generated by using fields such as editing status, route ID, reload location, and created on. Detailed information is shown when you click one of the resulting reload IDs, such as information about the sales area, route information such as stops and locations, administrative data such as dates created and changed, and information about products in the reload ID and any logs. You can use this app to view, edit, or delete reload requests.

Manage Resource Groups (F4243)

Application Type: Transactional

This app is used by warehousing clerks to manage resource groups within warehouses. A *resource group* is a selection of resources, such as vehicles, that can be assigned together in a group. You can use the app to make settings for the resource group that are relevant to actions such as processing using radio frequency (RF) guns. You also can use the app to create new resource groups, edit existing resource groups, or delete existing resource groups.

Manage Route Groups (F4533)

Application Type: Transactional

This app can be used to create groups of routes for a specific warehouse. Route groups can be displayed, edited, created, or deleted from within the app. Descriptions, languages, assigned routes, and assigned users can all be managed in the app. Routes also can be unassigned. Selection options such as route group ID, warehouse, and route ID are available for filtering the results, which appear in tabular format.

Manage Route Parameter Groups (F5958)

Application Type: Transactional

Route parameter groups are groups of specific parameters that can be assigned to multiple routes. This app is used to manage the assignment of parameters to a route parameter group. Selection options such as route parameter group, warehouse number, and cycle selection parameter are available for filtering the results, which appear in tabular format.

Manage Route Version Proposals (F6022)

Application Type: Transactional

With this app, you can create condition records for the management of routes for route determination. There are several functions available in the app—for example, create new route version proposals, check existing route version proposals, block/unblock/accept route version proposals, and update route version proposals. Selection options are available for filtering the data, based upon fields such as route version proposal ID, warehouse number, route ID, valid from, proposal is blocked, and lifecycle status.

Manage Routes (F4390)

Application Type: Transactional

With this app, you can create routes with start and end points and defined stops as well as define route versions. There are several functions available in the app—for example, create new routes with route versions and stops; copy routes and route versions; create new versions for an existing route; block, deactivate, or delete active route versions; and add attachments to routes. Selection options are available for filtering the data, based upon fields such as route ID, warehouse, route type, lifecycle status, start storage type, and route parameter group.

Manage Selection Variants (F5143)

Application Type: Transactional

As part of the extended warehouse management (EWM) suite of apps, this app is used to define and maintain selection variants for the execution of warehouse tasks. The selection variant has predefined field values that are used during the warehouse process for execution of tasks such as wave preparation execution. Selection options for filters are available for fields such as editing status, name, status, and type.

Manage Stops (F4532)

Application Type: Transactional

Stops are assigned to route versions, and this app allows you to maintain those stops through functions such as creating, copying, and deleting transfer points. Single stops or stop groups can be managed in this way. Maintenance options in the app are listed as quick action buttons as follows: **Edit**, **Copy**, **New Transfer Point**, **New Storage Bin Stop**, **New PSA Group**, and **Delete**. You can change the route header data by selecting a route ID from the list or change a version by selecting **Route Version**. You also can view the where-used list to see which routes stops are used in.

Manage Tours (F4600)

Application Type: Transactional

Warehouse operators can manage and maintain tours using this app. The app can be used not only to create tours for use in warehouse management but also to assign loading employees to a tour. Tours can be listed in the app with their assigned route, planned loading start time, and planned departure time, as well as resource and driver. Handling unit data also is available with statuses, as well as rack data, which is available as a microchart embedded in the table.

Manage Users (F4244)

Application Type: Transactional

Radio frequency (RF) frameworks are used often in warehouse management to conduct tasks in the warehouse processes. This app is used to define and maintain users for working within the RF framework. Activity profiles, such as picking and loading, are assigned to users, as are route group IDs. This app allows this user data to be created, edited, copied, and deleted.

Manage Warehouse Documents (F5341)

Application Type: Transactional

As part of the extended warehouse management (EWM) suite of SAP Fiori apps, this app is used to search for, display, and edit warehouse documents (i.e., documents for goods movements like receipts and issues). Key features of the app include the ability to view full details of the warehouse documents, including the serial numbers of the items; change the documents to enter a reason for movement or a short text; and cancel warehouse documents (or individual items on the document) for which the status is **Confirmed** or **Partially Canceled**. Selection filters are available for fields such as warehouse number, warehouse document, warehouse process category, product, storage bin, created by, created on, account assignment category, and account assignment object.

Monitor Freight Cost Allocation Document Items (F5513)

Application Type: Transactional

Freight cost allocation document items can be searched for and analyzed effectively using this app. These freight cost allocation document items can be displayed in graphical format and listed by plant or by product, with their associated value. The selection items can be viewed graphically or in raw format. You also can check the overall status of the freight cost allocation

document item and navigate to the corresponding document to perform any necessary follow-up tasks.

Monitor Periodic Supply Protection Maintenance (F5503)

Application Type: Transactional

As part of the advanced available-to-promise (ATP) solution, this app can be used to check and review the log messages for the automatic activation of supply protection items. Selection objects are available to refine and filter the data retrieved, such as severity, from date and to date, category, subcategory, and external reference.

Navigation to the related Configure Supply Protection app (F4569) is supported.

My Tours (F4553)

Application Type: Transactional

This app is a very useful way for a warehouse operative to get a visual map of and guidance for the unloading process defined within a specific single tour. The app provides all the detailed information necessary to view the status of not-yet-completed tours, along with the details of the handling units yet to be unloaded. Although this is a display-only app, it is extremely useful for understanding the full process of the specific tour, seeing the loading positions for each handling unit, reviewing the planned departure and arrival dates, and examining route and stop information.

Overview Inventory Management (F2769)

Application type: Analytical

This app provides a snapshot of the most important and timely information and tasks relevant to the business case. Presented through a set of actionable cards, it allows users to quickly identify priorities, make faster decisions, and take immediate action. The following cards are available:

■ **Recent Material Documents**
Provides detailed information for each selected item. It also offers the option to reverse a material document if needed, helping you manage and correct inventory transactions efficiently. It redirects to the Material Documents Overview app (F1077) when the card is selected.

■ **Overdue Materials – GR Blocked Stock**
Displays overdue materials for which a goods receipt (GR) has been posted into the nonvaluated GR blocked stock, helping you identify items that may require further action or investigation. Redirects to the Overdue Materials – GR Blocked Stock app (F2347).

■ **Stock Value by Stock Type**
Provides an overview of material stock values, organized by different stock types. Redirects to the Stock – Multiple Materials app (F1595).

■ **Stock Value by Special Stock Type**
Gives an overview of your material stock values, categorized by special stock types.

■ **Warehouse Throughput History**
Provides a clear view of all goods movements within a company. This allows you to explore various dimensions and key figures and lets you drill down to specific material document items to find detailed information quickly and easily. Redirects to the Goods Movement Analysis app (W0055).

■ **Monitor Purchase Order Items**
With the view switch, you can easily toggle between viewing overdue items and items that are currently awaiting approval, allowing for better tracking and prioritization of tasks. Redirects to the Monitor Purchase Order Items app (F2358).

■ **Overdue Materials – Stock in Transit**
Presents the top 10 stock transport order items experiencing the longest delays for which the goods receipt process is still pending. This helps identify critical bottlenecks in stock transfers that require immediate attention. Redirects to the Overdue Materials – Stock in Transit app (F2139).

Overview Inventory Processing (F2416)
Application type: Analytical

This app provides a centralized and real-time snapshot of key inventory-related information and tasks, displayed through a set of actionable cards. These cards highlight the most relevant and urgent matters, allowing you to focus on what needs immediate attention, make faster decisions, and take prompt action. You can filter the entire overview page by parameters such as plant or storage location, which dynamically updates the displayed cards to reflect only relevant data, including overdue materials, outbound deliveries, and warehouse throughput. Selecting a card header typically navigates to the corresponding app, while clicking an item lets you drill down into specific details. The following cards are available:

■ **Recent Inventory Counts**
Gives detailed item-level information about specific material to help you make informed decisions for different physical inventory strategies, like annual, continuous, or cycle counting.

■ **Overdue Materials – Stock in Transit**
Presents the top 10 stock transport order items experiencing the longest delays for which the goods receipt process is still pending. This helps identify critical bottlenecks in stock transfers that require immediate attention.

■ **Inbound Delivery List**
Displays both completed and open inbound deliveries, offering a comprehensive list resulting from collective delivery processing. This is used to access key information related to general shipping activities and helps track the status of inbound goods.

- **Outbound Delivery List**

  Provides a list of outbound deliveries, including those that are completed and those still open. This reflects the outcome of collective delivery processing and serves as a source of information for monitoring and managing general shipping activities.

- **Overdue Materials – GR Blocked Stock**

  Displays overdue materials for which a goods receipt has been posted into the nonvaluated goods receipt–blocked stock, helping you identify items that may require further action or investigation.

- **Monitor Purchase Order Items**

  With the view switch, you can easily toggle between viewing overdue items and items that are currently awaiting approval, allowing for better tracking and prioritization of tasks.

- **Recent Material Documents**

  Displays a list of material document items and provides detailed information for each selected item. You can also reverse a material document if needed, helping you manage and correct inventory transactions efficiently.

- **Warehouse Throughput History**

  Provides a clear view of all goods movements within a company. This allows you to explore various dimensions and key figures and lets you drill down to specific material document items to find detailed information quickly and easily.

Tip

If the cards are not displayed by default, make sure that the **Plant** key field is populated in order to display the SAP Fiori cards with metrics from the system.

Pack Outbound Deliveries (F3193)

Application Type: Transactional

The follow-up process to picking is often packing, and this app can be used by warehouse operators to pack outbound deliveries. The functions available in the app are reasonably comprehensive, including packing using advanced or basic packing modes, creating and changing shipping handling units, packing in the background, excluding exceptions from packing, and using favorite packaging materials. You can also connect externally to a scale to transfer the recorded weight to the app. Packing can be handled in the app for a handling unit, a storage bin, or an outbound delivery order.

Transaction Code

The associated transaction in SAP GUI is Transaction VL02N (Change Outbound Delivery).

Pack Warehouse Stock (F3738)

Application Type: Transactional

As part of the extended warehouse management (EWM) suite of SAP Fiori apps, stock in the warehouse that exists in a specific handling unit or storage bin can be transferred to another handling unit or storage bin using this app. With this app, you can create and change handling units, pack in the background, exclude exceptions from packing, and use favorite packaging materials. You can also connect externally to a scale to transfer the recorded weight to the app. Packing can be handled in the app for either a handling unit or a storage bin.

Pick by Cart (F2793)

Application Type: Transactional

This app can be used by a warehouse operator to pick multiple orders in a single trip. This is useful for improving the productivity and efficiency of the warehouse operations. Handling units can be displayed in graphical format, showing their logical positions on a pick cart. The process follows three steps:

1. Preparation: Operators place handling units in the correct logical positions in the pick cart.

2. Picking: The warehouse operator picks the products for the destination handling units. The tasks in the picking step are grouped together by logical attributes such as storage bin.

3. Unloading: Operators unload the handling units at their destinations. Again, this step can contain multiple handling units based upon logical attributes such as a common destination bin.

Pick Outbound Delivery (F0868)

Application Type: Transactional

Picking is the process of removing goods from storage bins and transferring them to a staging area in preparation for final loading and shipment. This app, used in inventory-managed warehouses, allows you to update the picking quantity for individual items on the outbound delivery. Once the picking is complete (and the picking status is updated to **Completely Processed**), you can then post the goods issue.

Transaction Code

The associated transaction in SAP GUI is Transaction VL02N (Change Outbound Delivery).

Planograph (F3434)

Application Type: Analytical

This app allows warehouse clerks to see a full overview of storage locations used to store explosive materials. The app covers the full lifecycle of the explosive materials in the warehouse, including viewing all stock in the storage location, viewing the compatibility status of the materials, performing quality inspections, printing the grid view of the *planograph* (the visual layout of the materials in the location), and saving your selections in a custom SAP Fiori tile.

Post Goods Receipt for Inbound Delivery (F2502)

Application Type: Transactional

This app allows you to post goods receipts with reference to an inbound delivery, making it an essential tool for receiving specialists in daily warehouse operations. It provides a list of relevant inbound deliveries from suppliers, supporting stock procurement, direct consumption (with single or multiple account assignments), and vendor consignment stock. The app displays account assignment details but doesn't recalculate quantities in partial receipts. It also enables shelf-life checks when required fields are maintained in the material master and supports the entry of production or expiration dates. You can record reasons for goods movements and manage batch creation and editing if the appropriate roles are assigned. For materials with split valuation, the app can assign or display valuation types, though consistency across documents requires manual input beforehand. In subcontracting scenarios, it provides a component view showing related materials and supports adjustments to component quantities or units. It also handles serial numbers at both the header and component levels. Barcode scanning is supported by using internal or external devices, with options for simple or GS1-standard barcodes, and manual entry is available. The app can print slips and labels, display detailed storage-level data, and adjust stock values using the current material price regardless of the reporting date.

Post Goods Receipt for Process Order (F6352)

Application Type: Transactional

This app allows you to post goods receipts specifically tied to process orders. On the initial screen, you can either enter a production order manually or use the scan function to retrieve an order. The app includes multiple tabs for easier navigation. The **General Information** tab captures key fields such as printing options, notes, delivery note, document date, and post-ing date. The **Items** tab displays details like material, open quantity, plant, storage location, and stock type (e.g., unrestricted use, quality inspection, or blocked). A **Split Item (+)** function within the **Distribution** column allows you to split a material and enter the necessary details for each split. The **Attachments** tab enables uploading supporting documents or adding links. This app simplifies the goods receipt process by listing all relevant process orders and associated materials ready for posting.

Post Goods Receipt for Production Order (F3110)

Application Type: Transactional

This app allows you to post goods receipts spe-cifically tied to production orders. On the initial screen, you can either enter a production order manually or use the scan function to retrieve an order. The app includes multiple tabs for easier navigation. The **General Information** tab cap-tures key fields such as printing options, notes, delivery note, document date, and posting date. The **Items** tab displays details like material, open quantity, plant, storage location, and stock type (e.g., unrestricted use, quality inspection, or blocked). A **Split Item (+)** function within the **Dis-tribution** column allows users to split a material and enter the necessary details for each split. The **Attachments** tab enables uploading sup-porting documents or adding links. This app simplifies the goods receipt process by listing all relevant production orders and associated materials ready for posting.

Post Goods Receipt for Purchasing Document (F0843)

Application Type: Transactional

This app allows you to post goods receipts with reference to various types of purchasing docu-ments, such as purchase orders or scheduling

agreements. When goods are delivered for a spe-cific purchasing document, referencing it during the goods receipt process is essential to ensure that all relevant departments are aligned. This linkage automatically updates inventory levels and triggers the necessary financial postings as soon as the receipt is posted. The app includes several functional tabs. The **General Information** tab captures key details such as delivery note, document date, posting date, and print options (no print, individual slip, individual slip with inspection text, or collective slip). The **Items** tab provides a searchable list of materials, display-ing information such as open quantity, deliv-ered quantity, plant, storage location, and stock type (e.g., unrestricted use, quality inspection, blocked, or goods receipt blocked stock). The **Attachments** tab allows you to upload relevant documents or include helpful links to support the receipt process.

> **Tip**
>
> The app supports the scanning functionality through the **Scan** option on the initial screen of the app.

Post Goods Receipt Without Reference (F3244)

Application Type: Transactional

This app allows you to post goods receipts with-out needing a prior purchase order or document reference, making it especially valuable for warehouse clerks handling day-to-day opera-tions. It streamlines the process of recording incoming goods directly into the system, help-ing maintain accurate and up-to-date inventory records. The app is organized into several tabs. The **General Information** tab includes fields such as printing options (no print, individual slip, individual slip with inspection text, or collective slip), note, delivery note, document date, and posting date. The **Items** tab lets you enter item details like the ID, material, quantity, plant, and storage location, with options to create, copy, or delete entries. The **Attachments** tab allows you to upload supporting documents or insert relevant links, ensuring complete and traceable goods receipt entries.

Process E-Commerce Returns (F1955)

Application Type: Transactional

This app is related to the processing of customer returns for decentralized extended warehouse management (EWM) systems, which can be connected to SAP S/4HANA or SAP ERP 6.0. From within the app, it is possible to identify returns deliveries and manage the returns through the end-to-end returns process. Handling units, product barcodes in GS1 format, EAN and GTIN numbers, serial-managed products, and batch-managed products are all handled in the app.

Tip

Be aware that the app has some limitations. For example, you cannot use it to process returns for items that are bundled together in a bill of materials (BOM).

Process Warehouse Tasks (F4595)

Application Type: Transactional

As part of the extended warehouse management (EWM) suite of SAP Fiori apps, individual warehouse tasks such as picking, putaway, and internal movements can be managed using this app. Single or multiple warehouse tasks can be processed simultaneously. Selection filters are available based on fields such as warehouse number, warehouse order, warehouse task, warehouse task status, and warehouse process category. The **Confirm**, **Cancel**, and **Print** quick actions are available.

Navigation to the related Manage Product Master app (F1602) is supported.

Process Warehouse Tasks – Internal Movements (F4289)

Application Type: Transactional

As part of the extended warehouse management (EWM) suite of SAP Fiori apps, individual warehouse internal movements can be managed using this app. Single or multiple internal movements can be processed simultaneously. Selection filters are available based on fields such as warehouse number, warehouse order, warehouse task, and warehouse task status. The **Confirm**, **Cancel**, and **Print** quick actions are available.

Navigation to the related Manage Product Master app (F1602) is supported.

Process Warehouse Tasks – Picking (F3880)

Application Type: Transactional

As part of the extended warehouse management (EWM) suite of SAP Fiori apps, individual warehouse picking can be managed using this app. Single or multiple picks can be processed simultaneously. Selection filters are available based on fields such as warehouse number, warehouse order, warehouse task, warehouse task status, warehouse process category, and product or handling unit warehouse task. The **Confirm**, **Cancel**, and **Print** quick actions are available.

Navigation to the related Manage Product Master app (F1602) is supported.

Process Warehouse Tasks – Putaway (F4150)

Application Type: Transactional

As part of the extended warehouse management (EWM) suite of SAP Fiori apps, individual warehouse putaway can be managed using this app. Single or multiple putaways can be processed simultaneously. Selection filters are available based on fields such as warehouse number, warehouse order, warehouse task, warehouse task status, warehouse process category, and product or handling unit warehouse task. The **Confirm**, **Cancel**, and **Print** quick actions are available.

Navigation to the related Manage Product Master app (F1602) is supported.

Review Scheduling Result – SAPGUI Adapter (F7275)

Application Type: Transactional

As part of the advanced available-to-promise (ATP) solution, this app can be used to review the results of the scheduling process for sales orders processed using Transactions VA01 and VA02 in SAP GUI. Features of the app include the ability to display scheduled dates and activities, together with total and net durations, and display determined attributes and values, together with the sources of the determination. You can also show nonworking days and times, as well as any information regarding backward or forward scheduling that has determined the results.

Run Outbound Process – Deliveries (F1704)

Application Type: Transactional

As part of the extended warehouse management (EWM) suite of SAP Fiori apps, this app provides a way to monitor the entire outbound delivery process. The overview initial screen offers a visualization of all the deliveries scheduled for the date range specified. Selection options are available based on fields such as warehouse number, carrier, departure date/time for route, and the **Include Completed Deliveries** checkbox. Quick actions are available to carry out the following tasks: print loading list, create tasks, goods issue, reverse goods issue, and edit.

Navigation to the Process Warehouse Tasks app (F4595) is supported.

> **Tip**
>
> This app is best used for small and medium warehouses. It is not optimized for outbound processes that use waves.

Run Outbound Process – Production (F5164)

Application Type: Transactional

As part of the extended warehouse management (EWM) suite of SAP Fiori apps, this app provides a way to monitor the entire outbound delivery process for outbound deliveries for production. The initial overview screen offers a visualization of all the deliveries scheduled for the date range specified. Selection options are available based on fields such as warehouse number, carrier, departure date/time for route, and the **Include Completed Deliveries** checkbox. Quick actions are available to carry out the following tasks: print loading list, create tasks, goods issue, reverse goods issue, and edit.

Navigation to the Process Warehouse Tasks app (F4595) is supported.

Run Outbound Process – Transportation Unit (F1703)

Application Type: Transactional

As part of the extended warehouse management (EWM) suite of SAP Fiori apps, this app provides a way to monitor the entire outbound delivery process for transportation units. The overview initial screen offers a visualization of all the transportation units scheduled for the date range specified. Selection options are available based on fields such as carrier and planned departure date. The following quick action options are available: **Create Tasks**, **Goods Issue**, and **Edit**.

Tip

This app is best used for small and medium warehouses. This is because it is not optimized for outbound processes which use waves.

Scenario Builder (F3214)

Application Type: Transactional

This app allows transportation managers to create transportation management (TM) data quickly for the purposes of testing and training. Scenario templates can be created that include all the scenario data required in order to execute a TM process. These scenarios can then be used over and over again for creation of test scripts and training sessions.

Schedule Goods Receipt for Inbound Deliveries (F2776)

Application Type: Transactional

This app can be used to create background jobs to create goods receipts for inbound deliveries. Multiple selection options are available for the creation of the job, including template selection parameters such as job template (select from dropdown list) and description (free text), as well as scheduling options such as the **Start Immediately** flag, recurrence pattern (single run or recurring), and start date and time. Parameter options include shipping/receiving point, purchasing document, item, created on, created by, warehouse number, storage location, supplier, external delivery ID, and inbound delivery.

You can use this app to perform a number of functions, such as the creation of periodic background jobs, receiving notifications for failed jobs, and monitoring and displaying information about scheduled jobs.

Schedule Inbound Delivery Creation (F2798)

Application Type: Transactional

This app can be used to create background jobs that can create inbound deliveries for purchase orders. Multiple selection options are available for the creation of the job, including template selection parameters such as job template (select from dropdown list) and description (free text), as well as scheduling options such as the **Start Immediately** flag, recurrence pattern (single run or recurring), and start date and time. Parameter options include the document number, item number, plant, storage location, warehouse number, and delivery date.

You can use this app to perform a number of functions, such as the creation of periodic background jobs, receiving notifications for failed jobs, and monitoring and displaying information about scheduled jobs.

Schedule Outbound Delivery Creation for Sequenced JIT Calls (F4962)

Application Type: Transactional

This app can be used to create background jobs to create outbound deliveries for sequenced just-in-time (JIT) calls from customers. Multiple selection options are available for the creation of the job, including template selection parameters such as job template (select from dropdown list) and description (free text) as well as scheduling options such as the **Start Immediately** flag, recurrence pattern (single run or recurring), and start date and time. Parameter options include component group material, customer supply area, distribution channel, delivery creation date, division, internal processing status, sales organization, shipping point, ship-to party, planned shipping date from and to, planned shipping time from and to, requirement date from and to, and requirement time from and to.

You can use this app to perform a number of functions, such as the creation of periodic background jobs, receiving notifications for failed jobs, and monitoring and displaying information about scheduled jobs.

Schedule Outbound Delivery Creation for Summarized JIT Calls (F4963)

Application Type: Transactional

This app can be used to create background jobs that can create outbound deliveries for summarized just-in-time (JIT) calls from customers. Multiple selection options are available for the creation of the job, including template selection parameters such as job template (select from dropdown list) and description (free text) as well as scheduling options such as the **Start Immediately** flag, recurrence pattern (single run or recurring), and start date and time. Parameter options include component group material, customer supply area, distribution channel, delivery creation date, division, internal processing status, sales organization, shipping point, ship-to party, planned shipping date from and to, planned shipping time from and to, requirement date from and to, and requirement time from and to.

You can use this app to perform a number of functions, such as the creation of periodic background jobs, receiving notifications for failed jobs, and monitoring and displaying information about scheduled jobs.

Schedule Tour Creation (F5166)

Application Type: Transactional

This app can be used to create background jobs that can create tours of scheduled routes. Multiple selection options are available for the creation of the job, including template selection parameters such as job template (select from dropdown list) and description (free text) as well as scheduling options such as the **Start Immediately** flag, recurrence pattern (single run or

recurring), and start date and time. Parameter options include warehouse number, selection variant, route ID, route type, route parameter group, and calculation start date.

You can use this app to perform a number of functions, such as the creation of periodic background jobs, receiving notifications for failed jobs, and monitoring and displaying information about scheduled jobs.

Schedule Wave Preparation (F5355)

Application Type: Transactional

This app can be used to create background jobs to create wave preparations. Multiple selection options are available for the creation of the job, including template selection parameters such as job template (select from dropdown list) and description (free text) as well as scheduling options such as the **Start Immediately** flag, recurrence pattern (single run or recurring), and start date and time. Parameter options include warehouse number, warehouse document category, document type, delivery priority, route ID, selection variant, provision date (begin), provision time (begin), provision date (end), and provision time (end).

You can use this app to perform a number of functions, such as the creation of periodic background jobs, receiving notifications for failed jobs, and monitoring and displaying information about scheduled jobs.

Stock – Multiple Materials (F1595)

Application Type: Analytical

This app provides a comprehensive view of material stocks across various plants and storage locations, helping inventory managers track and manage inventory efficiently. You can filter data by material, plant, storage location, base unit, and reporting date, with the option to adapt filters as needed. The app displays key details such as material ID, description, plant, special stock type, and stock levels (unrestricted, quality inspection, and blocked). Selecting a material ID hyperlink enables navigation to related apps, including the Product app (F2773), which offers in-depth product details like type, group, category, unit of measure, related plants, purchasing data, and sales order information. Stock data can be presented in tabular format, and additional insights include stock values, storage location data, and carbon footprint metrics, supporting organizations focused on sustainability through integration with SAP Sustainability Footprint Management. The app also features advanced search capabilities to filter for active materials or those without deletion indicators, and it calculates stock values using current or historical prices. You can export results to spreadsheets, send material details via email, and collaborate using Microsoft Teams, making the app a powerful tool for real-time inventory analysis and strategic decision-making.

Stock Reporting Overview (F6266)

Application Type: Transactional

This app enables a real-time view of stock reports, making it easier to monitor inventory levels across different materials and locations. You can drill down into individual reports for detailed insights into inventory status and history and navigate directly to related material documents for a deeper review of specific stock movements. The app supports customizable filters such as stock report number, report year, supplier, and processing date through the **Adapt Filters** option. Additional functions include deleting, duplicating, triggering output, copying stock reports, and adjusting settings for a more personalized experience.

Stock – Single Material (F1076)

Application Type: Analytical

This app provides a detailed and interactive view of material stock across various plants and

storage locations, helping inventory managers effectively monitor and analyze stock levels. You can view stock by plant or storage location and filter by reporting date. It displays various stock types, including unrestricted use, blocked, quality inspection, restricted use, returns, stock in transit, tied empties, transfer stock (both plant and storage location), and valuated goods receipt blocked stock. The app supports both tabular and chart formats, offering a clear comparison of stock type quantities. Additional features include barcode scanning for accurate data capture and the ability to pin the header for easier navigation. The app also allows you to use the **Open In** option to jump to related apps for deeper stock analysis and operational tasks (more than 50 possible apps, as listed in **Define Link List**).

You can navigate to the following apps: Manage Stock (F1062), Transfer Stock – In Plant (F1061), Transfer Stock – Cross-Plant (F1957), Material Document Overview (F1077), Stock – Multiple Materials (F1595), and Display Serial Numbers (F5147).

Tendering (F2679)

Application Type: Analytical

This app can give a tendering manager KPIs to manage the tendering business. The overview page will give the tendering manager KPIs for the current and previous month. The following cards are available on the overview page:

- **Acceptance Analysis**
  Review the proportion of freight request for quotations (RFQs) for which a response has been received from a carrier.

- **Discrepant Selected Carrier**
  Review carriers deviating from the tendering-awarded carrier.

- **Tendering Cycle Time Analysis**
  Review the average cycle time by carrier.

- **Response Acceptance Comparison**
  Review the number of positive and negative responses from carriers to RFQs.

- **Tendering Ratio**
  Review which tendering types are the most successful for each carrier.
- **Open, Overdue Requests**
  Review number of open and overdue RFQs.
- **Open, Non-Overdue Requests**
  Review number of open and non-overdue RFQs.
- **Tendering Peer-to-Peer Cycle Analysis**
  Review the number of peer-to-peer requests to which a carrier has responded with a quotation.

The data is retrieved from SAP using CDS view C_TenderingQuery.

Tendering – ALP (F2753)

Application Type: Analytical

This app can give a tendering manager KPIs in order to manage the tendering business. Cycle times and carriers can be used in the filter criteria, and the results are shown in graphical as well as tabular format. A **Change Tendering** quick action is available for the selected documents.

Navigation to the Tendering overview page (F2679) is available.

Transportation Costs (F2672)

Application Type: Analytical

As part of the transportation management (TM) functionality in SAP S/4HANA, this app can give a transportation manager KPIs in order to manage their costs. The overview page will give the manager KPIs for the current and previous month. The following cards are available on the overview page:

- **Trade Lane Weight Load Analysis**
  Review the weights listed by carrier in order to identify the best trade lane.
- **Trade Lane Volume Load Analysis**
  View aggregation of current, minimum, and maximum volumes transported by carriers.

- **Trade Lane Container Load Analysis**
  View aggregation of current, minimum, and maximum container counts transported by carriers.
- **Transportation Mode Category Analysis**
  View the weight, volume, and quantities for each transportation mode category (air, sea, land).
- **Overall Weight Utilization Distribution**
  View the total weight each carrier transports by month.
- **Overall Container Utilization Distribution**
  View the total container count each carrier transports by month.

The data is retrieved from SAP using CDS view C_TranspAllocComplianceQ.

Transportation Costs – ALP (F2754)

Application Type: Analytical

This analytical list page gives transportation managers an overview of all the transportation costs in a single view. The data is structured in tabular and graphical format and shows useful information like costs per week; invoiced amount per week; and metrics split by carrier, such as gross weight per week, volume per week, containers per week, stage distance traveled per week, and destination countries. The page relies upon data extracted from two CDS views: C_FreightSettlementCostQ and C_TranspOrdInvcgBlockStsQ.

Transportation Invoice Blocked – ALP (F2755)

Application Type: Analytical

As part of the transportation management (TM) function in SAP, this app can give a transportation manager KPIs to manage transportation orders that are blocked for invoicing. The data is structured in tabular and graphical format and shows useful information such as a summary graph of costs per week and a table showing all documents that are blocked, the block reason and category, the transportation management category (sea, rail, air, etc.), the document number and document category, and the cost amount. The view uses CDS view C_TranspOrd-InvcgBlockStsQ to retrieve the data.

Warehouse KPIs – Operations (F4024)

Application Type: Analytical

As a part of the extended warehouse management (EWM) suite in SAP, warehouse clerks can use this app to get an overview of relevant warehouse operations. The following cards are available on the overview page:

- **ODO Items by GI Status**
  Displays the outbound delivery order items grouped by goods issue status (not relevant, completely processed, partially processed, open).

- **ODO Items Without GI by Ship-To Party**
  Displays the outbound delivery order items that have not yet progressed to goods issue, grouped by ship-to party.

- **ODO Items Without GI by Planned GI Time**
  Displays the outbound delivery order items that have not yet progressed to goods issue, grouped by planned goods issue time.

- **ODO Items Without Pick Warehouse Tasks by Planned GI Time**
  Displays the outbound delivery order items that have not yet progressed to picking, grouped by planned goods issue time.

- **Open Warehouse Tasks by Warehouse Process Type**
  Shows the open warehouse tasks (picking, packing, etc.), grouped by warehouse process type.

- **Open Warehouse Tasks by Activity Area**
  Shows the open warehouse tasks (picking, packing, etc.), grouped by activity type.

- **Open Warehouse Tasks (Overdue) by Overdue Time in Hours**
  Shows the open and overdue warehouse tasks (picking, packing, etc.), displayed in chronological order by overdue time in hours.

- **Open Pick Warehouse Tasks by Activity Area**
  Shows the open picking warehouse tasks, grouped by activity area.

Selection options are available, such as warehouse, activity area, warehouse process type, creation date of warehouse task, type of outbound delivery order, planned goods issue date, and actual goods issue date.

You can navigate to related apps such as Run Outbound Process – Deliveries (F1704), Process Warehouse Tasks – Picking (F3880), and Process Warehouse Tasks – Putaway (F4150).

Warehouse Outbound Delivery Orders (F4969)

Application Type: Analytical

As a part of the extended warehouse management (EWM) suite in SAP, warehouse clerks can use this app to view a more granular level of detail around outbound delivery orders. Selection options such as warehouse, warehouse order queue, and warehouse order status are available.

Navigation to related apps such as Run Outbound Process – Deliveries (F1704) is supported.

Chapter 5
Production Planning and Manufacturing

In this chapter, we'll explore the SAP Fiori apps for the production planning and manufacturing process area. This area encompasses a variety of functions across the board, such as demand planning, material requirements planning (MRP), production engineering and operations, capacity requirements planning, predictive material and resource planning (pMRP), Kanban, and many more. Where applicable, we've noted a comparable SAP GUI transaction code. Apps are listed in alphabetical order.

Analyze Change Impact (F2664)

Application Type: Transactional

The app allows business users, particularly production engineers, to assess the potential impact of changes on manufacturing objects, such as bills of materials (BOMs)—including engineering BOMs (EBOMs) and manufacturing BOMs (MBOMs)—and shop floor routings. This is particularly valuable when product design changes trigger updates to the process plan, BOMs, or production orders. By using this app, you can determine which objects may be affected by the changes and decide whether to put production orders on hold or add the affected objects to the change record for later action. The app provides a detailed analysis of the impact, identifying related and impacted objects, and allows you to drill down into these objects to check if they are directly impacted by the change. It highlights any net changes between the current and previous versions of BOMs, pinpointing deleted, modified, or added components. You can navigate to other apps, such as the Display Planning Routing app (MSFR13), to gather further details about material or operation changes.

More importantly, for unitized BOMs and routings, the app also identifies affected units and impacted shop floor routings. The analysis results can be viewed in a list format, which consolidates all impacted objects, or in a network view, which visualizes the relationships between these objects. This approach allows you to understand the scope of changes quickly and take immediate actions as necessary, such as adding impacted objects to the change record or performing follow-up tasks through related apps.

> **Tip**
>
> This app is part of the production engineering and operations (PEO) process area but can be accessed using the Manage Change Records app (F2097), which falls under the application component PLM-CR (Change Record). Furthermore, this functionality can be set up in Customizing using menu path **SPRO • Production • Manufacturing for Production Engineering and Operations • Production Engineering • Change Impact Analysis**.

Analyze PIR Quality (Deprecated) (F1943)

Application Type: Analytical, Web Dynpro

Planned independent requirements (PIRs) represent future demand forecasts for materials, and this app allows you to analyze these requirements for accuracy and alignment with business needs at the plant level. The app has a navigation panel and several tabs: **Data Analysis**, **Graphical Display**, and **Query Information**. Based on the available fields (e.g., material, plant, MRP controller) in the navigation panel, you can design rows and columns with results immediately displayed in the **Data Analysis** tab. You have numerous options to display these results, such as filter, sort, display descriptions or IDs only, and more. The app can be used to identify discrepancies, perform quality assurance in forecasting, align forecasts with sales and operations planning, monitor changes over time, and support new production introductions.

You can navigate to other apps, such as Stock – Single Material (F1076), Stock – Multiple Materials (1595), Manage Sales Orders (F1873), and Dead Stock Analysis (F2899), among many others.

This app was deprecated as of the SAP S/4HANA 2022 FPS02 release. The successor app is the Maintain PIRs app (F3445), for the SAP S/4HANA 2023 FPS03 release.

Application Jobs for Manage Buffer Levels (F5074)

Application Type: Transactional

This app allows you to manage application jobs for buffer levels. The app can be accessed through the Manage Buffer Levels app (2706) using the **Logs** button. The app lists application jobs for buffer levels, detailing information such as the job status, results, steps, job name, and so on. You can create, copy, cancel, view job details, and even restart failed jobs within the app. When creating a new job, you will need to specify the job template ID and job name, as well as define scheduling options such as the job's recurrence pattern and parameter settings like the proposal run suspension start and end dates, material, material group, procurement type, adjustment mode, and more, depending on your preferences.

Assign Work (F3435)

Application Type: Transactional

This app lists all existing operation activities for production and process orders at the plant level. You can toggle between unassigned and all operations in the system using the **All** and **Unassigned** options and see the status. You can adapt the filters to suit your needs with options including material, scheduled start date, quality assurance status, plant, work center, and so on. You can orchestrate the actual execution of production order operations and activities. Operators can access their tasks in work center work queues, or supervisors can explicitly distribute work to their team members by moving tasks to personal work queues. When assigning production operators to operation activities, the system displays a dialog box of all operators that have a corresponding manufacturing user defined in the system and whose default work center is identical to the one defined for the operation activity. Furthermore, a production supervisor has the option to change the time for an operation activity using the **Edit Target Time** button.

Transaction Code

This app is closely linked to Transaction MPE_AS_BUILT (Multilevel As-Built Report) in SAP GUI.

Buffer Positioning (F3282)

Application Type: Transactional

This app is intuitive in that it allows you to list and identify products that require buffering at the plant level. Products are listed in **Upstream** and **Downstream** sections, in which you can choose a product and view all the contextual information and all possible actions that can be performed on that product by navigating to related apps. You have the flexibility to filter results based on the selection criteria, including MRP type, product, product group, plant, sales organization, decoupled lead time (DLT) indicator, and supplier. You can also display products based on their positioning by switching between the **Positioning-Internal** and **Positioning-External** tabs.

Tip

You can switch to the **Network Graph View** to display a graphical view of the full product structure to easily identify decoupling points. By using network graphs, you also can view the longest path, product flow, bill of materials (BOM), and BOM usage.

Buyoff Timeline (F4792)

Application Type: Transactional, Reuse Component

The app helps you facilitate the electronic approval of manufacturing activities on the shop floor. It aims to eliminate paper documentation while maintaining the same level of recordkeeping for manufacturing tasks. The app allows you to trigger a *buyoff cycle*, a process in which personnel on the shop floor validate the completion of specific tasks associated with a product at a work center. Buyoffs are essential in ensuring high-quality standards are maintained; they serve as an official way to approve and document the completion of key activities.

Capacity Evaluation (F6798)

Application Type: Transactional

This app helps you see how busy different work centers in a plant are. You can group similar work centers together and check how much they are being used. The app's visual tools and analytics make it easier to understand and manage capacity, helping production planners improve efficiency and make better decisions. For example, a production planner at a car factory can check if the welding stations are overloaded. Using this app, the planner can group all welding work centers and see that one station is at, say, 95% capacity while another is at 45% capacity. Based on this, they can redistribute work to balance the load and avoid delays. You can switch between chart and table views in the app, and results can be displayed based on the selection criteria. This app is assigned to the production planner role (SAP_BR_PRODN_PLNR).

Capacity Scheduling Board (F3951)

Application Type: Transactional

This app allows you to analyze the schedule of operations and to dispatch, reschedule, or deallocate individual bottleneck operations on pacemaker work centers in a graphical chart. To fully utilize the app, you need to define an area of responsibility and industry type. You can adapt filters as required; for example, business users can choose to display work centers that have a dispatched (**DSPT**) or not dispatched (**DALL**) status, or even choose to display work centers based on the evaluation horizon. You can use the work center dropdown functionality to display pacemaker work centers or all work centers as required. At the top of the screen, you can select the **Set Strategy** function to set strategies used for rescheduling orders, including the planning mode, scheduling mode, and direction. Here you can choose the scheduling control—find slot (finite), infinite scheduling, find slot (finite), or insert operation (finite)—and the direction can either be backward or forward. You can even choose to display order details by selecting the **Order Details** button at the top of the initial screen, which will showcase more information about the order.

The durations of the operations are displayed as bars on the time axis, considering the working and nonworking times of corresponding work centers. The operations to be evaluated can be selected using filters such as horizon, work center, work center group, and plant. Operations that match all filter criteria are displayed in green and are referred to as *filtered operations*. All other operations are displayed in gray and are referred to as *unfiltered operations*. In addition to this color coding, the operations are displayed with different color intensities and with additional graphical elements depending on the order type and operation status. You can display either the bottleneck operations of pacemaker work centers or the operations of all work centers in the defined area of responsibility.

> **Tip**
>
> We recommend using the **Zoom** function to enlarge the contents of the board for easier navigation.

Capacity Scheduling Table (F3770)

Application Type: Transactional

This app allows you to filter and manage orders, focusing on bottleneck operations and ensuring the dispatch and deallocation of operations to pacemaker work centers. This helps prevent bottlenecks and delays and ensures timely production and sales order fulfillment. You can filter orders by work center, product, horizon, and status. The app provides detailed information about the material to be produced, production version, pacemaker work center, required capacity for bottleneck operations, operation dates, and more. You also can control important scheduling parameters, such as the planning mode and scheduling strategy, and use visual indicators (colors) to determine if the work center capacity is sufficient. The app supports mass dispatching and deallocation, realignment of scheduled orders, and rescheduling operations to meet production needs. You can navigate to the graphical **Schedule Overview** area to view operations on a time axis, facilitating the adjustment of scheduling conflicts. The app is designed for optimizing work center capacity, especially for dispatching orders to pacemaker work centers, using features like midpoint scheduling and rescheduling capabilities.

Check Major Assembly Projects (F3769)

Application Type: Transactional

This app helps you to maintain the consistency of a project created for a major assembly (i.e., complex technical projects for products such as trains, power stations, ships, aircraft, and more). It enables you to detect issues of the major assembly bills of materials (BOMs) and the networks needed to assemble a certain model and unit. You can filter based on the set selection criteria: project definition, background processing status, unit, and so on. Furthermore, the list can be filtered based on issue type—for example, parameter effectivity issues, network issues, network activity not yet released, or BOM explosion issues. Listed projects can show the associated error message, processing status, and project definition.

Within the app, you can navigate to Transaction CJ2ON (Project Builder), an SAP GUI app.

Check Material Coverage (F0251)

Application Type: Transactional

The app helps you monitor and resolve material shortages. It gives detailed insights into material requirements, stock levels, and shortages while providing actionable solutions to address these issues. The app includes a **Stock/Requirements List** tab, which you can use to view a breakdown of material stock, including current and future demands. You can visualize stock availability and identify shortages that may disrupt production. It also has a **Material Info** tab for more information about the materials and a **Notes** tab to maintain a record of decision-making.

The app helps you handle shortages by providing proposed solutions. These solutions can be simulated to preview their impact and applied based on suitability. The system considers various segments such as net requirements or individual customer requirements. Furthermore, you also can perform MRP runs to explore the stock/requirements list, check for shortages, and identify potential disruptions. You can then quickly contact suppliers to request adjustments to by using the Manage Change Requests app (F3406). More importantly, the system rates proposed solutions on a scale of 1 to 3, indicating the effectiveness; for instance, three stars means the issue is fully resolved, while two stars means the issue is partially resolved.

You can navigate to the following apps for more detailed analysis and processing: Display Process Order (COR3), Process Order Object Page (F2263), Monitor Stock/Requirements List (MD04), and Product (F2773).

> **Tip**
>
> SAP reiterates that the system cannot propose solutions for materials produced in house. Instead, this app is primarily used to support externally procured materials.

Convert Planned Orders (F4171)

Application Type: Transactional

This app lists planned orders for conversion. The **Action** dropdown function allows you to convert planned orders to process orders or production orders. You also can make use of the collective conversion functions for production orders, process orders, and purchase requisitions. You can make use of the selection criteria to display planned orders as required by using the material and planned order type (consignment order, individual customer order, stock order, project order, reservation, or planned independent requirement [PIR]). It is also important to note that by selecting a specific planned order, you can see contextual information about the planned order under the **General Information** tab.

By selecting the material hyperlink in the **Planned Order** screen, you can navigate to the Compliance Information app (F3226) and the Manage Product Master app (F1602). Similarly, clicking the planned order hyperlink will redirect you to the Display Planned Order: Stock Order app (MD13). The app uses CDS view C_CONVERTPLANNEDORDER to retrieve data.

Create MRP Change Requests (F5416)

Application Type: Transactional

The app lists purchase schedule lines, with options to see the associated material, supplier, delivery date open quantity, proposed date, and proposed quantity based on your selection criteria. The selection criteria include the purchase order, material, supplier, and request status (e.g., answered, new, none, requested). For the purchase schedule lines, you can create change, requests which can be sent to suppliers using an external system to request changes. For example, business users can request the delivery date of the purchase order schedule line. This enables communication between suppliers and the business by making sure that material availability is ensured and bottlenecks are prevented before-hand. Based on the feedback from the suppliers, you can either apply or discard change requests using the **Apply Change Request** or **Discard Change Request** buttons.

Create Optimal Orders for Shipment (F1700)

Application Type: Transactional

The app allows you to efficiently plan and optimize material orders. Through analyzing material shortage definitions, optimal order dates, and vendor data, you can maximize transportation capacity, reduce delivery costs, and fast-track procurement processes. In other words, the app allows you to view and manage materials that need to be ordered based on business needs for timely replenishments. The resulting material list can be filtered based on filters such as the ABC indicator (classification of materials by value and movement), material and material group, and so on. The app displays a list of materials with the following details:

- Proposed quantity: Recommended order quantity for each material
- Replenishment date: The date materials should be restocked
- Stock availability (next 21 days): A chart showing stock projections over the next three weeks

The app supports creating purchase orders for one or multiple materials from the same vendor. For instance, you can place an order directly for a selected material or navigate to the **Order Several Materials** screen to bundle materials in one order.

Choosing the material hyperlink at the material list level on the initial screen displays a popup

showcasing the contextual information for the selected material, such as the current stock (real-time inventory data) and MRP data. After opening a selected material, you can navigate to the Check Material Coverage app (F0251) or Product app (F2773).

Create Supply Application Job (DD) (F5186)

Application Type: Transactional, Reuse Component

This app is used to view logs for scheduled jobs related to the creation of supply orders in demand-driven replenishment. It allows you to track the results of jobs that automatically create supply orders to meet demand-driven replenishment needs. These logs help monitor the status of the supply orders generated from the demand-driven replenishment process, ensuring timely replenishment and a smoother production or sales order fulfillment process.

Define Flexible Constraints (F4449)

Application Type: Transactional

This app lists material constraints showcasing, for example, constraint ID, material, supplier, plant, and so on. You can use filters such as material, plant, period type (week, month), constraint category (material, production line, stock transfer and supplier), and more. You also can add, copy, and delete material constraints as required. Furthermore, constraints can be added to single materials or multiple materials. For example, by selecting the **Add** function, you can add constraint quantities to a material for defined period. These constraints are considered during simulations, helping to restrict demand over specified time periods. The app also supports managing constraints for in-house materials, supplier-procured materials, stock transfer materials from other plants, and production line materials. You also can manage constraints in draft mode before finalizing them.

The app is closely tied to the Process pMRP Capacity Simulations app (F3952) and the Multi-Level Simulation View app (F4157).

> **Tip**
>
> When adding a material constraint in the initial screen, choosing **Multiple Materials** will allow you to add a material group at the plant level, which ultimately lists all the materials belonging to this material group.

Demand-Driven Replenishment (F3276)

Application Type: Transactional

This app gives an overview of buffer information organized by planning priority, helping you ensure timely replenishment and avoid stock shortages that could lead to delays in production or sales orders. Key features of the app include the ability to define filters to narrow down the products displayed and save predefined filter settings as a variant for easy access. You can view the bill of materials (BOM) tree structure up to the next buffered material and identify products with shortage alerts. The app highlights buffers that are below the reorder point, marked with yellow or red for low stock priority, and displays the proposed quantity needed to refill the buffer to the maximum stock level to avoid shortages. You can create supply orders based on the proposed quantity, and view stock projections through time-dependent buffer level charts. The app also allows you to manage demand and supply by adjusting the

order quantity, date, or source of supply as needed.

Demand-Driven Replenishment Scope for Products (DD) (F6436)

Application Type: Transactional

This app allows you to set the demand-driven scope for products in bulk. This app is used within the context of application jobs, where you can manage the status of jobs (failed, canceled, finished, scheduled, etc.) and set criteria like date ranges. When creating a new application job, there are three steps: template selection, scheduling options, and parameter configuration. During the template selection step, you must choose a job template. The app allows you to create, copy, view details, cancel, or restart the application job based on your use cases.

> **Tip**
>
> You can use the SAP_SCM_DD_SCOPE_PRODUCT (DD Scope for Products) job template, corresponding to the SAP_SCM_DD_SCOPE_PRODUCT job template catalog entry.
>
> Also, note that when configuring the job parameters, you can choose to maintain or not maintain logs.

Detect MRP Situations (F3853)

Application Type: Transactional

This app helps you handle MRP situations by bringing critical issues to your attention via situation handling. When creating a new application job, you are guided through three steps: selecting a template, choosing scheduling options, and defining parameters. On the template selection screen, you must choose a job template to proceed. The app allows you to filter by various selection criteria, such as status (e.g., failed, canceled, finished, scheduled) and date range, to refine your search for relevant issues.

You can utilize various options, such as **Create**, **Copy**, **Details**, **Cancel**, and **Restart**, to manage the application jobs. By using this feature, you can leverage situation handling in the Manage Material Coverage app (FO251A) to ensure that important MRP issues are promptly identified and addressed.

Display MRP Key Figures (F1426)

Application Type: Transactional

This app displays the low-level code steps taken during MRP. Selecting a specific low-level code step will display contextual information such as the low-level code description, consumed time, and table statistics. The app lists a log that includes a user name, date and time stamp, and a run status and a fatal error text (if any). Selecting a log will display contextual information about that log, which includes the user name for which MRP was performed, showcasing the status of the run and the number of materials that could have failed in the **Materials Failed** field. You can choose to view the chart by using the base unit of measure, log name, or start time stamp. In the **Run History** tab, you can display in a chart the number of materials failed, materials planned, and elapsed time for a specific MRP run. Furthermore, the app displays all the MRP Live steps.

Display MRP Master Data Issues (F1425A)

Application Type: Transactional

The app lists materials with MRP master data issues at the issue list level. For instance, you can filter to include materials, MRP controller, issue category, accepted (yes, no), and so on. The app allows you to display all the master data

issues that would have occurred during an MRP Live planning run. You can accept or revoke the acceptance of issues as required by using the **Accept** or **Revoke Acceptance** button.

Through the material hyperlink in the **Information** tab for a specific selection, you can navigate to the Compliance Information app (F3226) and the Manage Product Master Data app (F1602).

This app is the successor app to Display MRP Master Data (F1425), which was deprecated as of the SAP S/4HANA 1909 FPS01 release.

Tip

You can define the link list on the popup screen that is displayed after selecting the material hyperlink in the **Information** tab by choosing the **More Links** option. This presents a sizeable number of apps to choose from, which is important as it allows you to select relevant apps to jump to each time the Display MRP Master Data Issues app (F1425A) is in use for further processing and analysis.

Display Operation Activity Assignments (F7967)

Application Type: Transactional

This app allows you to visualize all the planned assignments for an operation activity. For instance, you can view components, work instructions, and buyoffs. Furthermore, assigned teams for operation activities also can be seen by navigating to the Assign Work app (F3435). The app also provides insight into the teams and operators who have been assigned to perform the work on a given operation activity through the Assign Work app. Although this app is not available as a tile on the launchpad, it can be accessed via other apps (like the Assign Work app).

Embedded Work Instruction (F3511)

Application Type: Transactional, Reuse Component

This app allows you to create and manage version-controlled routings, which define the sequence of operations necessary to produce a product. Accessible through the Rich Text Editor app (F2818) or Transaction MSFR4 (Manage Shop Floor Routings) in SAP GUI, the app allows you to specify the sequence of operations independent of time, ensuring flexibility and consistency across production cycles. It includes detailed information on work center resources, specifying where the work will be performed and the resources consumed, along with technical specifications such as standard times and capacity. The app also supports the inclusion of work instructions for each operation, guiding production workers on how to execute tasks efficiently. You also can integrate detailed operational activities to further clarify the production process. Once a routing version is released, it becomes part of the production execution and changes are no longer possible, ensuring that the defined sequence and instructions are followed strictly throughout production.

Excess Component Consumption (F2171)

Application Type: Analytical

This app helps you compare the expected and actual amount of material wasted during production. The selection criteria include the start and end performance period as key entries, and you can add extra filters such as MRP controller, material, batch, and so on. Business users can identify the biggest causes of excess material usage and find ways to reduce waste. The app works for materials with confirmed orders, and it helps prevent too much waste, which might lead to shortages. It provides insights and trends. The following KPIs are available:

- **Component Scrap Deviation by Material (Column Chart)**
  Displays the excess component consumption as a percentage and component scrap % for the top three materials with the highest deviations.

- **Component Scrap by Deviation Amount (Bubble Chart)**
  Displays the excess consumption cost deviation, which can help a business to identify the items with the greatest financial loss.

- **Component Consumption Quantity (Tabular View)**
  Displays the component consumption quantity in a table, with columns showing different parameters (based on table settings) such as production supervisor, material, withdrawn quantity, requirement quantity, original quantity, and so on.

- **Component Scrap % (Tabular View)**
  Displays details about component scrap in a table, showing details such as the excess component %, component scrap %, component scrap in %, planned scrap in %, and so on based on the table settings.

- **All Dimensions and Measures (Tabular View)**
  Displays a sizable number of parameters in a table that can help you visualize important information in one go—for instance, information about the withdrawn amount for a particular reporting period, storage location for the affected material(s), and so on.

> **Tip**
>
> You have the flexibility to change the measures and dimensions (x and y axes) for these KPIs in the chart settings. It is also important to note that the overall excess component consumption is displayed as a percentage at the top of the initial screen (validate).
>
> Based on need, you also can select the **Open Full Screen** icon in either the table or chart view to clearly visualize the different KPIs in a full-screen view.

Find Standard Texts (F2553)

Application Type: Transactional

This app allows you to search and use prewritten standard texts instead of writing everything from scratch. These texts can be used for work instructions in production processes, making it easier and faster to create clear guidelines while avoiding duplication. On the initial screen of the app, you can select from a variety of filter criteria such as keywords, category, plant, usage type (**Only Copy**, **Reference**, **Reference or Copy**), status (**In Progress**, **Released**, etc.), and the plant level.

You also can add new standard texts using the **Add** button. Standard texts are quite useful in a variety of business contexts. For instance, imagine that a factory supervisor needs to create work instructions for assembling washing machines. Instead of writing the same safety guidelines and quality checks repeatedly, they can use this app to find and insert prewritten standard texts, which ultimately saves time and effort and makes certain that consistency is maintained across all work instructions.

Selecting the **Add** button will redirect users to the Manage Standard Texts app (F2554) for further processing.

Hand Over Purchase Requisitions (F4795)

Application Type: Transactional

For a particular area of responsibility, this app allows you to view all the existing purchase requests that are generated during an MRP run. This app allows you to check, review, and make quick edits (if necessary). Once everything is ready, in a single step, the purchase requests can be sent to the purchasing team for further processing. Via filters, you can list purchase requisitions based on the material, supplier, or status description (which in this case could be requests that have been handed over or those still pending a handover). Using various buttons, you can hand over requests, edit them, and even delete purchase requisitions.

Historical Buffer Performance (F7147)

Application Type: Analytical

This app helps process owners analyze past buffer stock performance by identifying unusual stock levels (outliers) over the past month. You can filter the resulting line items based on product, plant, MRP area, MRP controller, and so on. The filter criteria include the red and yellow zone parameters: low red, low yellow, high yellow, high red, and above high red. This allows a business to check if stock levels were too low or too high compared to the expected buffer limits (red and yellow zones). For example, say that a production planner checks buffer stock levels for engine parts in a specific plant. The app shows that in the previous month, the stock level dropped below the low red threshold several times, meaning there was a risk of running out. With such an insight, the production planner can adjust buffer levels to prevent shortages in the future.

Kanban Board (F4630)

Application Type: Transactional

The Kanban Board app helps process owners track Kanban containers, showcasing how materials are moving through production. It gives a clear view of work progress, material use, and shortages, helping to spot missing parts and bottlenecks in various areas. The board uses a different set of colors to show container statuses and issues. The app is also used to send Kanban signals when more materials are needed. For instance, a production supervisor at a car assembly plant checks the Kanban Board and sees that seat belts in the assembly line are marked in red, meaning they are running low. They use the board to trigger a Kanban signal, automatically requesting a new batch from the warehouse to prevent delays.

Kanban Control Cycle Analysis (F5235)

Application Type: Analytical

This app helps you analyze and track Kanban control cycles for materials in production. You can filter data by plant, supply area, material, control cycle, and so on to view detailed insights. The app lets you zoom in, switch views, and display data in a table, a chart, or both at the same time for better visualization. For example, say a production planner at a food packaging plant wants to check how often cardboard boxes are replenished in a specific production area. With this app, they can filter by plant and material, view the data in a chart to spot trends, and use the table for detailed numbers, which obviously helps adjust the Kanban cycle to ensure a steady supply of boxes.

You can choose to view the Kanban control cycles using the following KPIs or dimensions: **Plant of Supply Area, Purchasing Organization, Release Date of Kanban Cycle, Replenishment Type, Storage Location of Supply Area, Storing Position, Supplier, Base Unit of Measure, Kanban Control Cycle Category, Lifecycle Status of Kanban Control Cycle, Lock Date of Kanban Control Cycle, Material, Maximum Number of Empty Containers Exceeded**, and **Trigger Group**.

Legacy Integration App for Work Instruction (F2926)

Application Type: Transactional, Reuse Component

This app allows you to create and manage version-controlled routings in the production process. These routings are not time dependent, meaning that they remain stable and consistent over time. A routing version outlines the sequence of individual operations necessary to produce a product, ensuring that each step in the production process is followed accurately. The routing version contains crucial information, such as the work centers where the operations will be performed, the resources consumed in the work centers, and the technical specifications needed for the operations. These technical specifications include important elements like standard times, capacity requirements, and work instructions for operators. Operation activities also can be used within the routing to describe the production process in more detail, providing a clearer understanding of the production steps. Once a routing version is released, it becomes locked and cannot be changed, ensuring consistency and accuracy during production execution. This ensures that the same set of instructions is followed throughout the manufacturing process, providing reliability and efficiency.

Maintain MRP Controllers (F1386)

Application Type: Transactional

This app lets users add, edit, or remove MRP controllers at the plant level. Each material used in MRP is linked to an MRP controller in the **MRP 1** tab (**MRP Procedure** section) in the material master. The *MRP controller* is the person or team responsible for making sure the material is available when needed. If required, users can modify the name of an existing MRP controller or delete a controller, provided that no material assignments exist for it. This helps maintain an efficient and accurate assignment of MRP controllers to materials, ensuring that each material is properly managed and monitored for availability.

Maintain PIRs (F3445)

Application Type: Transactional

This app allows you to create, update, and track planned independent requirements (PIRs) for materials at the plant level. PIRs help plan production or procurement of materials—for example, before customer orders arrive, ensuring stock is available when suppliers have long lead times. You can filter the PIR data by material, plant, accuracy, MRP area, and so on or even use time-based filters (e.g., period indicator—

monthly or weekly) to track planning over different periods. You can view accuracy levels for the current and previous period to see how well-planned quantities match actual demand as well as create, upload, or edit PIRs to improve forecasting and negotiate better rates.

This is the successor app to the Analyze PIR Quality app (F1943), which was deprecated as of SAP S/4HANA 2022 FPS02.

Tip

The accuracy filter helps evaluate how well PIRs align with actual material consumption, as follows:

- Less than 50%
- Between 50% and 75%
- Between 76% and 100%
- Between 101% and 125%
- Greater than 125%

Maintain Time-Dependent Stock Levels (F5726)

Application Type: Transactional

The app allows you to set and adjust safety stock levels for materials at different times. Instead of keeping a fixed amount of safety stock, you can define stock levels that change over time based on expected demand, which helps prevent shortages while avoiding excess inventory. You can filter materials by material number, plant, MRP type, and so on. You also can choose a safety stock method (static safety stock or time-dependent safety stock) as well as set safety stock levels for specific time periods to match changing demands. Via the available buttons, you can add, download, or upload stock-level data.

On another note, for the selection criteria at the top of the initial screen, you can add more than one item in the **MRP Safety Stock Method** field. For example, you can add both static safety stock and time-dependent safety stock, which

allows you to combine static and time-dependent stock safety strategies. For example, say that a production planner at a beverage company knows that the demand for bottled water increases in summer. Instead of keeping the same safety stock all year, they can use this app to increase stock levels in summer and reduce them in winter. This ensures that they always have enough supply during peak season without overstocking in low-demand months. This app becomes invaluable for businesses looking to prevent stock shortages, reduce excess inventory costs, and ultimately improve supply chain efficiency.

Tip

To upload or download a template, you can use the **Upload Intervals** dropdown functionality on the initial screen of the app at the material list level.

Manage Action Settings (F2836)

Application Type: Transactional

This app allows you to control and customize actions in a production process. *Actions* are specific tasks that must occur during manufacturing, such as updating the status of an object, sending production confirmations, or blocking an object from further processing. In the app, you can choose an object type such as a machine, operation, production order, buyoff, or bill of materials (BOM) header, and select a plant in which the action will apply. You can then proceed to view a list of all available actions, such as buyoff, hold production, and assemble components. The system can check for action rules, such as requiring a reason code or notes, and check user qualifications (to ensure only trained workers can approve quality inspections). You can assign action handlers, such as attaching specific rules and settings to each action, defining whether someone must enter a reason code (optional, mandatory, or not required) and specifying which codes are available to choose from when providing a reason (from the reason code group). For example, a production supervisor at an automobile plant can use this app to control how production orders are managed. If a quality issue is found, the **Hold Production** action is triggered, the system then requires a reason code (e.g., defective part identified), and a note must be added to explain the issue in detail. Before resuming production, the system checks if the worker performing the fix is qualified for the task.

> **Tip**
>
> Action handlers defined in the Manage Action Settings app (F2836) are plant- and object type–specific, meaning that they are called every time an action is performed on the specified object, independent of an operation activity.

Manage Buffer Levels (F2706)

Application Type: Transactional

This app helps you to monitor and adjust buffer stock levels for your products. *Buffer levels* are safety stocks that ensure you have enough material to meet customer demand, even if there are delays or unexpected changes. You can choose what to analyze by selecting a product or product group, picking a plant where the product is managed, and set conditions like maximum stock deviation or proposed status (**Adopted, Queued, To Be Processed**). You also can review buffer levels, such as checking if the buffer levels match your needs by using indicators like variability (low, medium, high) and decoupled lead time (DLT). Furthermore, business users can adjust buffer levels by using options like **Adopt** (to accept the proposed level), **Discard** (to reject), or the **Adjust** dropdown to make changes, as well as switching between **Sizing – Internal** and **Sizing – External** to compare different approaches. You also can access logs to check past proposal runs. More importantly, the **Proposal Runs** dropdown

function allows you to initiate or manage proposal runs related to buffer level settings. This function is critical for determining which materials or stock levels require adjustments based on predefined parameters, such as the stock's demand or minimum levels.

> **Tips**
>
> Before the app can be fully utilized for the first time, the business user needs to maintain the area of responsibility.
>
> In the **Maximum Stock Deviation** field at the header selection level, business users can add more than one item.

Manage Buyoff Cycle Templates (F2702)

Application Type: Transactional

This app helps you create, edit, copy, delete, and overall manage buyoff templates used on the shop floor. A *buyoff* is a digital approval process that replaces paper-based approvals in manufacturing, ensuring that tasks are completed correctly before moving to the next step. A buyoff item corresponds to an individual signature required to continue or complete a buyoff cycle. You can create or copy existing buyoff templates, with each template including buyoff items, which are specific steps needing approval. These items are assigned to a sequence and a team with required skills (only qualified team members can approve these steps). During the approval process, if a team member meets all the required skills, they can approve or reject a buyoff item. There is an option to skip a buyoff item if it doesn't need strict approval. The app helps reduce paperwork while maintaining accurate records of completed tasks.

To put this into context, imagine a production engineer is overseeing the assembly of circuit boards. Before a batch is sent to the next stage, a buyoff is needed to confirm that each board meets quality standards. The engineer uses the app to create a template with the required approval steps (with the template specifying that only trained team members can perform the approval). As tasks are completed, team members approve each step digitally, making sure that errors are not missed. The digital process saves time and reduces paperwork.

Tip

To start working with buyoffs, you must activate the relevant function module in Customizing for the **Production Engineering and Operations** process area.

Manage Buyoffs (F2701)

Application Type: Transactional

This app allows you to make sure that products meet quality standards by tracking and approving specific tasks on the shop floor. The app lists buyoffs assigned to operation activities, which are steps that need to be approved before moving forward with line items, including details such as operation activity, wait time, and so on. The hit list can be filtered by status (**Open, In Process, Approved, Rejected**) or other details, like material, order, or serial number. You can make use of the different options available on the initial screen: **Claim** (to take responsibility for a task), **Transfer** (if someone else needs to handle it), or **Skip** (if the task can be skipped). Production engineers can assign buyoffs to tasks using a predefined buyoff cycle template to ensure that each step meets the set quality standards—for instance, when assigning a buyoff to the operation activity in the routing version. In short, the app is a reliable tool businesses can use to confirm that each step in the production process is completed safely and correctly.

Tip

This role is assigned to the standard SAP_BR_ PRODN_SUPRVSR_DISC_CAM (Production Supervisor – Discrete Manufacturing (CAM)) business role in the PP-PEO-BUF (Manufacturing Buyoff) application component.

Manage Change Requests – MRP (F0670)

Application Type: Transactional

This app allows you to create and manage change requests for existing purchase order schedule lines, particularly based on the rescheduling proposals generated by MRP rescheduling checks. The app allows you to send these change requests to suppliers via an external system, requesting adjustments to the quantity or delivery date of a purchase order schedule line to ensure material availability and meet customer demands. Once responses are received from suppliers, you can either apply or discard the changes proposed. Once the purchase order schedule lines are displayed, you can create change requests based on rescheduling proposals. You can modify the proposed delivery dates and quantities or cancel purchase order schedule lines that are no longer required. You also can adjust the reason codes for each change request; these codes are automatically determined by the system or can be manually set. You can add notes to the change requests to communicate additional information with suppliers. The app allows you to preview, edit, or delete notes and provides an easy way to send change requests to suppliers. Once the suppliers respond, you can review the details of any new proposals, including reasons for rejections and costs of change (if provided), though the costs of change are informational and not processed by the system. You can apply or discard these responses depending on the situation. More importantly, the app also supports the discarding of outdated or unnecessary change requests, with a history of discarded requests available for reference.

Manage External Requirements (F0250)

Application Type: Transactional

This app allows you to manage uncovered and delayed items required for sales orders and stock transport orders. It works in conjunction with the Monitor External Requirements app (FO246), where you first make the necessary settings to monitor specific items. Once the desired items are selected, the system proposes solutions to cover the requirements and fulfill the external orders. These solutions can be simulated to preview their effects before selecting the most appropriate one to resolve the issues. You can select different shortage definitions to evaluate how the stock/requirements situation changes for the selected items. The app also enables the creation of purchase requisitions, stock transport requisitions, planned orders, or schedule lines for scheduling agreements. You also can perform rescheduling checks, execute actions based on execution messages, and view an aggregated table in which MRP elements for the same day are combined into one line for better management. You can view additional material and order information; display, edit, and enter notes for materials; and access quick views for more details on MRP elements. The app allows you to edit, convert, or delete MRP elements if necessary and perform MRP runs at the material level. You can preview the effects of proposed solutions and apply the changes, contacting vendors to request modifications to purchase orders and stock transport orders.

The app also provides a list of open change requests for materials, allowing you to navigate to the Manage Change Requests app (FO670) for further management.

be simulated to preview their effects before selecting the best solution. To use this app, you must first access the Monitor Internal Requirements app (FO263) and set up the necessary parameters to display only the items you wish to manage. After selecting the relevant items from the list, you can navigate to this app (Manage Internal Requirements) to apply the proposed solutions.

You can choose different shortage definitions to see how they impact the situation. The app also enables the creation of purchase requisitions, stock transport requisitions, planned orders, or scheduling agreements. You can perform rescheduling checks and execute actions based on the system's messages, and the app aggregates MRP elements for the same day into a single line for easier management. You can view, edit, and enter notes for materials; navigate to quick views for further material details; and manage MRP elements by editing, converting, or deleting them as necessary. You also can execute MRP runs at the material level and preview the effects of proposed solutions in both tabular and chart formats. Once a solution is selected, you can apply the changes, contact vendors to request modifications to purchase orders, and view open change requests in a popup window. Further navigation options are available to manage change requests or to access other related apps for additional information or material data changes.

Manage Internal Requirements (F0270)

Application Type: Transactional

Business users can use this app to monitor and address component shortages for internal orders, such as in-house production orders, process orders, maintenance orders, and network orders. The app provides a detailed view of the stock and requirements situation based on the settings configured in the Monitor Internal Requirements app (FO263). The system offers various solutions for covering these requirements and fulfilling internal orders, which can

Manage Major Assembly Projects (F4027)

Application Type: Transactional

This app is used to plan and track large, complex assembly projects, such as building a new car model or custom machine. Using this app, you can view the list of major assembly projects and get associated details like material, start date, processing status, and so on, as well as create new projects for new models. The **Go To Project Builder** option helps you access more detailed planning tools. You can use the **Assign** or

Unassign options to manage team members and tasks. The **Check** button helps verify information and avoid mistakes. More importantly, the app helps connect different departments and supports, for example, top-down planning, ensuring that tasks are organized from start to finish. Using this app, the process of creating a complex assembly project for a new model helps you automate certain steps to make the process faster and more accurate; each step can be recorded for review, troubleshooting, and improvement. To put this into context, a company designing a new airplane model can use this app to start a project and engineers, planners, and production teams can all access the project to coordinate tasks. They can use the **Check** function to make sure all steps are accurate. For more detailed planning, the project builder can be utilized.

Selecting the **Check** function will redirect you to the Check Major Assembly Projects app (F3769).

Manage Manufacturing Reference Structures (F5647)

Application Type: Transactional

The app helps you view and manage manufacturing reference structures. These are organized collections of parts or components used in manufacturing. You can view a list of manufacturing reference structures that includes details like the reference object, who created it, and if it is marked as obsolete. Filters also can be applied to narrow the list based on criteria like reference object, authorization group, or node type (all nodes, root nodes). You also can click on any reference structure to see more details about its items and drill down into lower-level items to understand the full structure. If a reference object is marked as obsolete, the app will show it clearly. The entire hierarchy of a manufacturing reference structure, including items with their own reference structures, can be visualized.

Tip

The app shows both the absolute position of items and any local occurrences in the hierarchy.

Manage Master Recipes (F5426)

Application Type: Transactional

This app helps you find, view, and manage master recipes for manufacturing products or services in process industries. Simply put, a *master recipe* is a detailed template used in process industries to describe the production process of a product. For instance, for a specific material, it outlines the sequence of operations, resources required, inspection characteristics (for quality checks), and so on. Using this app, you can search for existing master recipes using various filters such as group, plant, product, dates (valid from/to), key date (as a key field), and so on. Master recipes are listed with their overall status (created, released for order, etc.), and filters can be adjusted to narrow the search and make it more specific. You can select specific master recipes to see more details and check which products are linked to a specific master recipe and view their descriptions, plants, and validity periods.

Using the **Create** and **Change** buttons, you can create and change master recipes; these buttons will navigate you to the Create Master Recipe (C201) and Change Master Recipe (C202) apps. Overall, this app helps you quickly find and review master recipes needed for production, manage and track master recipes that are active or outdated, and easily check which products are assigned to each master recipe.

Manage Material Coverage (Version 2) (F0251A)

Application Type: Transactional

You can monitor and manage material shortages within your area of responsibility during the production process. This app displays a list of materials with key fields such as shortage definition, material, and plant, letting you assess whether supply is sufficient to meet demand. You can filter results based on shortage definitions such as MRP standard, stock days' supply, and ordered receipts to check material availability. You also can view and analyze stock and requirements details for each material, identify potential issues, start an MRP run, edit MRP elements, create purchase requisitions, or reschedule orders to resolve shortages. In addition, you can create orders directly within the app by specifying the receipt quantity, availability date, and source of supply in a popup screen.

You also can navigate to the following apps for further analysis and processing:

- Cancel REM Confirmation (MF41)
- Confirm Repetitive Manufacturing (MFBF)
- Display Production Quantities (MF51)
- Maintain Time-Dependent Stock Levels (F5726)
- Manage Repetitive Manufacturing (MF50)
- Monitor Material Coverage (F0247A)
- Monitor Stock/Requirements List (MD04)

Manage Object Qualification Assignments (F2398A)

Application Type: Transactional

This app allows you to assign, modify, and view existing qualifications for work centers, material masters, and users. By managing qualifications, you can ensure that only skilled and authorized personnel operate specific machines, handle materials, or perform designated tasks on the shop floor. You can add a selected qualification to a work center, material, or user. In the **Qualification** screen, you will see the three tabs:

- **Work Centers**
  Here you can use the **Add** button to attach work centers to qualifications. A popup screen enables the entry of the work center, plant, required skill level, and end date as key fields.

- **Material Master**
  Using the **Add** button, you can link a material master to a qualification. Similar to the work center popup screen, the popup screen here requires the material, plant, required skill level, and end date as key fields.

- **Users**
  Here you can view existing business partners. It is important to note that you can qualify users using Transaction PPPM in SAP GUI, where you can select the business partner for the user, add the qualification, and assign the proficiency level.

This is the successor app to the Manage User Certification Assignments app (F2374) and the Manage Certification Categories app (F2373).

Tip

Based on your requirements, you can link numerous work centers and material masters for one qualification; this allows you to easily add qualifications to relevant objects in different plants in one go. This app is in the **Production Execution Process Setup** group in the SAP Fiori launchpad.

Transaction Code

In the SAP GUI system, qualifications for use in this app can be created by using Transaction OOQA or using Customizing menu path **SAP Reference IMG • Personnel Management • Personnel Development • Master Data • Edit Qualifications Catalog.**

Manage PIRs (F1079)

Application Type: Transactional

The app allows you to create and monitor planned independent requirements (PIRs). PIRs serve as forecasts to procure materials or components needed for production before actual sales orders are received, making them critical when customer lead times are shorter than supplier lead times. You also can identify materials with insufficient coverage, locating materials with open PIRs in the past (allowing users to accept or delete them), and spotting material status issues such as deletion indicators at the plant or cross-plant level, unsuitable material statuses for forecasting, or missing forecast period indicators in the material master. You can assess the accuracy of PIRs by comparing forecasted quantities against actual sales demand for each period. On the material details screen, you can create new PIRs with support from available information like the current released forecast, last year's goods issue quantities per period, actual forecast consumption for the current period, and total received sales order quantities.

Manage Planned Orders (F4170)

Application Type: Transactional

This app helps you create, view, firm, unfirm, and delete planned orders. A *planned order* is an internal proposal for meeting material requirements. You can create planned orders manually or use existing ones as references. You also can perform a collective conversion of planned orders into production orders, process orders, or purchase requisitions using the **Collective Conversion to** option. More importantly, fashion-specific information also can be viewed if the relevant settings are activated.

Some selections redirect you to other apps, as follows:

- **Create Order** and **Create via Reference** redirect to the Create Planned Order app (MD11).

- **Production Orders** redirects to the Convert Planned Orders – to Production Orders app (CO41).
- **Process Orders** redirects to the Convert Planned Orders – to Process Orders app (COR8).
- **Purchase Requisitions** redirects to the Convert Planned Orders – to Purchase Requisitions app (MD15).

Manage Process Order Operations (F5323)

Application Type: Transactional

This app allows production supervisors to efficiently monitor and manage process orders by selecting an area of responsibility (AOR), which filters orders shown at the operation level. You assign AORs via the My Area of Responsibility – Production Supervisor app (F7044) or directly from the hit list. Orders are listed by operation, resource, progress, quantity, product, and schedule. Issues like time delays, quality problems, or missing components are color-coded for easy tracking and detailed in the order's long text tab. The app has four KPIs at the top of the initial screen, as follows:

- **Operations by Processing Status**
  Displays the number of process orders categorized by their processing status. You can quickly assess workload distribution and identify if there are any bottlenecks.

- **Operations by Scheduled Start Date**
  Provides a timeline view of operations. This helps you plan and prioritize operations based on dates to make sure that tasks are not overlooked or delayed.

- **Operations by Component Availability Status**
  Displays the status of component availability for each operation. This helps to identify potential disruptions early on by flagging operations that cannot start due to missing components.

- **Operations by Quality Status**
  Monitors the quality performance of process orders. This helps you maintain production

quality standards by quickly identifying operations that need immediate quality interventions.

In the visual filter mode on the initial screen, selecting the **Adapt Filters** option lets you change the chart type for the KPIs (bar, donut, or line chart).

In the **Components** tab, you can toggle between **All** or **Missing** components. For a specific component, you can view the status, check whether backflush has occurred, and see the batch ID. You can select the **Stock Overview** and **Material Documents Overview** buttons to navigate to the Stock Overview and Material Documents Overview (F1077) apps, respectively. In the **Phases** tab, you can select the **Manage Work Center Capacity** button to navigate to the Manage Work Center Capacity app (F3289). The app can indicate when an operation or one or more phases of the operation have been delayed by displaying a message in the **Long Text** tab. You also can access the Manage Product Master Data app (F1602).

> **Tip**
>
> The AOR is set in the My Area of Responsibility – Production Supervisor app (F7044). If the service group (PP_MPE_AOR_SRV) for this application is not published, you can publish it using Transaction /IWFND/V4_ADMIN (SAP Gateway Service Administration) in SAP GUI.

Manage Process Orders (F4587)

Application Type: Transactional

This app helps you monitor, control, and maintain process orders effectively. It provides a clear view of production progress and allows you to address issues quickly to ensure smooth operations. You can toggle between different views for the filter criteria at the top of the initial screen by using the **Visual Filter** and **Compact Filter** icons respectively. On the initial screen,

you can view the KPIs that provide quick insights into production performance, including **Orders by Processing Status**, **Orders by Quality Status**, **Orders by Component Availability Status**, and **Orders by Quantity Status**. You can export the process order table and view process orders to get valuable information such as the product, quantity, scheduled production start and end, and status issues if any arise. For example, business users can see issue details if there is a time delay, quality issues, missing components, quantity issues, and failed material movements. By selecting a specific order, you can edit the order, check components, release the order, and confirm and technically complete the process order.

Within each process order, you can see details such as quantities involved (confirmed yield, confirmed scrap, goods receipt quantity, and open quantity). You can choose to view coproducts, by-products, or all manufactured products linked to a process order. You also can get information related to the work center used for specific operations and get a view of the progress in steps by navigating to the Manage Work Center Capacity app (F3289). For each component for a process order, you can navigate to the Stock – Single Material (F1076) and Check Material Coverage (F0251) apps. For goods movements analysis, you can navigate to the Materials Documents Overview app (F1077).

Another important aspect is that for every process order relevant for in-process inspections, you can see important quality management information such as the associated inspection lot, inspection lot type, number of characteristics, progress of characteristics, usage decision valuation, number of defects (if any), and number of rejected characteristics.

> **Tip**
>
> For specific process orders within the app, a pie chart will show three parameters in different colors: **Open Quantity**, **Confirmed Scrap**, and **Delivered Quantity**.

For every failed material movement for process orders, you can see the associated error message, which helps facilitate faster resolution.

Manage Production Campaign (F7260)

Application Type: Transactional

This app is useful for production supervisors to view, create, and modify production campaigns easily. You can filter and view a production campaign based on production campaign ID, validity start and end dates, fixed lot size, quantity, and so on. For instance, when creating a new production campaign, you can select the **Create** button to open the Create Production Campaign app (PCA1). Here, you can enter details such as the description, validity period start date, material, and plant. Furthermore, you can create production campaigns using the **Create** button, which redirects you to Transaction PCA1 (Create Production Campaign), where you can enter key details such as the description and validity period start date, material, and plant. For production campaigns that require modification, you can use the **Change** icon to facilitate editing (select the **Show More per Row** icon to display this icon).

Furthermore, selecting the production campaign hyperlink will display a popup screen via which you can navigate to the following SAP GUI apps: Create Production Campaign (PCA1), Change Production Campaign (PCA2), and Display Production Campaign (PCA3).

Manage Production Holds (F3101)

Application Type: Transactional

This app helps production supervisors track, review, and manage production holds placed on orders, materials, or serialized products. This app provides a clear view of all active and released holds, helping to resolve production issues caused by missing parts, defects, and so on. You can filter production holds based on hold ID, hold type (e.g., material hold, order hold, or operation activity hold), material, status (active, released), plant, and so on. This allows you to view production holds, showing details such as what object is on hold, who placed the hold and when, reason codes for the hold, defect codes for the hold, and so on.

You can take the following actions:

- **Release Holds**: Removes the hold, with an option to specify a reason and add notes if required.
- **Change Hold Details**: Modify the hold reason and notes for active holds. You also can update the release reason and notes for released holds if permitted.
- **View Detailed Information**: Check the reason codes, defect code (if applicable), and notes entered by the person who placed the hold.

Tip

For shop floor orders, you can split serialized and nonserialized orders to manage affected quantities separately.

Manage Production Operations (F2335)

Application Type: Transactional

This app allows you to monitor the progress of individual production operations within your area of responsibility. You can filter operations based on criteria such as status (created, partially confirmed, closed, deleted), issue type, sequence category (standard, parallel, or alternative sequences), order, and so on. The app lists production operations for orders in the system and provides an overview of operational status, allowing you to quickly identify and address issues like missing components or unfinished previous operations. A chevron icon at the end of each line provides navigation to detailed operation activities, showing information such as available serial numbers, assigned components, inspection characteristics, documents, PRTs, and work instructions.

The **Related Apps** button allows you to navigate to the Manage Production Orders app (F2336), as well as Transactions CO01 (Create Production Order), COHV (Mass Processing Production Orders), COOIS (Monitor Production/Planned Orders), CO08 (Production Order with Sales Order), and CO07 (Production Order Without Material).

Manage Production Operators (F4046)

Application Type: Transactional

This app allows you to manage allocation and monitor production operators. It uses filter criteria such as execution start date, end date, order, plant, work center, person responsible, and more. The app lists all production operators in a tabular format, showing their default work centers, labor status, assigned labor time, and the number of operation activities they are handling. You can drill down into each production operator's profile to view more detailed information, including assigned operation activities (whether directly or through a team), work centers, qualifications, and associated teams. Supervisors also can view basic contact information, such as the operator's mobile number and email address, and their location details, such as default plant and work center. More importantly, you can identify whether an operator is currently idle or actively working; this insight helps supervisors analyze workloads and qualifications and make informed decisions about assigning tasks and setting labor durations.

Manage Production Orders (F2336)

Application Type: Transactional

This app allows you to monitor production orders within a defined area of responsibility (AOR). You can filter production orders based on criteria such as status (delivered, created, locked, closed, technically complete, etc.), issue type (delay, missing components, quality issues, quantity deviations), delay duration (ranging to

more than two days), and many more. Production supervisors can monitor and manage all orders under their responsibility, track production progress, and receive real-time alerts for issues like missing components or quality holds. They can investigate root causes, put orders or materials on hold with notes, and release them once resolved. The app allows you to edit order dates and quantities, release orders, check component availability with priority, and view detailed info on issues, components, schedules, and inspections. It also supports configurable materials, serial numbers, and managing shop floor routings, including splitting orders when needed.

Similar to the Manage Production Operations app (F2335), you also can navigate to other apps via the **Related Apps** button on the initial screen.

> **Tip**
>
> To fully utilize the Manage Production Orders app, you must define an AOR, either as a production supervisor or a work center. If you choose the production supervisor route, it is essential that the production supervisor field be populated for each production order. This can be done manually for each order in the **Assignments** view of the Create/Change Production Orders app or maintained at the material level in the **Work Scheduling** view of Transactions MM01/MM02 (Create and Change Material).

Manage Production Orders or Process Orders (F0273)

Application Type: Transactional

This app helps you manage production or process orders that encounter issues, based on settings made in the Monitor Production Orders or Process Orders app (F0266). You can view delays for materials, view missing components, and monitor the status of operations and milestones. The app provides comprehensive order-management capabilities, including viewing orders with uncovered items and analyzing stock/requirements situations via either a table or a chart view. The app also supports rescheduling checks, aggregated MRP element views, and quick views for deeper inspection of components, operations, milestones, materials, and order details. You can perform MRP runs at the material level, manage MRP elements (such as editing, converting, or deleting), and navigate to related apps like Manage Internal Requirements (FO270) or Transaction COOIS (Monitor Production/Planned Orders) in SAP GUI to resolve shortages.

> **Tip**
>
> You can navigate to this app through the Monitor Process Orders app (F0266A) or Manage Production Orders app (F0266A). For instance, in the Monitor Production Orders app, you can select the **Manage Orders** button to navigate to the Manage Production Orders app.

Manage Production Supply Areas (F6935)

Application Type: Transactional

This app allows you to create, edit, view, and delete production supply areas within a plant. Production supply areas can be filtered based on criteria such as plant, production supply area, storage location, person responsible, last change date, warehouse staging areas, Kanban calculation defaults, and so on. You can create production supply areas by selecting the **Create** button; in the resulting popup, you enter the necessary details such as the production supply area ID, plant, and storage location. When creating a new production supply area, you can view where this production supply area will be used, such as in control cycles, production versions, materials, work centers, and bill of materials (BOM) components. For example, the app can list all, released, or locked control cycles associated with the production supply area. You can differentiate production supply areas in different

countries or regions by specifying addresses using the **Create Address** button. The app supports draft handling, allowing preliminary changes to be saved or discarded. Deletion of a production supply area is only permitted if it is not referenced elsewhere, which can be verified through the where-used lists.

Manage Production Versions (F2568)

Application Type: Transactional

This app displays production versions at the plant level, allowing you to view important information such as production version status, the validity of bill of materials (BOM) and routing versions, and whether the production version has been checked for consistency. You can perform key actions such as creating new production versions, copying existing ones, performing consistency checks, and deleting versions. It also offers the functionality to maintain production version details, particularly for version-controlled BOMs and shop floor routings. You can identify and resolve inconsistencies between BOMs and routing versions.

Manage Reason Code Groups (F2865)

Application Type: Transactional

This app is used to create or delete reason code groups. A *reason code group* is a collection of reason codes that correspond to specific actions or events that occur during production. For example, when a production operator pauses or holds an operation, the system prompts them to select a reason code from a predefined group. These groups are associated with various objects like operation activities, orders, or serialized materials. You can filter the reason code groups based on criteria like code group and catalog, which may include categories such as hold codes or action reasons. Reason code groups are crucial for traceability within SAP S/4HANA as they appear in monitoring apps like Product Genealogy (F2892) and Production Action Log

(F2869), helping you analyze and trace the root causes of actions.

> **Tip**
>
> The code group field is limited to eight characters.

Manage Reason Codes (F2868)

Application Type: Transactional

This app is used to manage reason codes within a manufacturing environment. Reason codes are used to explain the cause or justification for certain actions or situations during production. These codes are defined by the production process specialist using the app and later selected by manufacturing users when executing production activities. You can create and delete reason codes, which are linked to reason code groups, each representing a specific purpose and action. Reason codes are filtered and listed with key details such as code, code group, catalog, and information about who modified the code and when.

> **Tip**
>
> Reason codes are stored in quality management and can be integrated into quality management processes. Reason codes are defined with a limit of four characters for the code field.

Manage Resources (F6176)

Application Type: Transactional

This app allows you to view, create, change, and maintain work centers and their capacities from a centralized screen. The app lists resources with their plant, description, person responsible, work center category, usage, standard value, control key, and more depending on table

settings. Selecting a specific resource for processing will redirect you to Transaction CRC3 (Display Resource) in SAP GUI, where, for example, you can see the resource category, person responsible, usage, and standard value key under the **Basic Data** tab.

> **Tip**
>
> Using the **Show More per Row** icon, you can see immediately important resource information such as cost center hierarchy, standard value key, control key, assigned cost center, and so on.

Manage Routings (F5425)

Application Type: Transactional

This app allows you to efficiently search, view, create, and modify routings within the SAP system. On the initial screen, you can filter existing routings based on criteria such as group, group counter, plant, overall status (e.g., **Created**, **Released for Order**, or **Released (General)**), key date, validity dates, and more. The key date field is mandatory and displays today's date by default, although you can adjust it to refine your search further. You also can view the number of assigned products per routing directly from the initial screen.

Furthermore, selecting a specific routing redirects you to Transaction CA03 (Display Routing). Alternatively, highlighting a routing and selecting either the **Create** or **Change** option will redirect you to Transactions CA01 or CA02 in SAP GUI.

Manage Standard Texts (F2554)

Application Type: Transactional

This app is used to maintain a library of reusable text elements for efficient work instruction authoring. The app lists top-level items with their standard text/category, plant, usage type, version, status, creation date, and more based on settings. There are also options available such as an **Add** dropdown (with options to create a category or standard text), as well as **Edit** and **Delete** buttons. When creating a standard text, you must specify a standard text ID, description, usage type (options include **Copy**, **Reference**, or **Reference or Copy**), and reference type (for instance, **Always Latest Released Version** or **Version as in Routing**). In the **Content** tab, you have access to a rich text editor that allows you to write, upload images, underline, and adjust text color, with more options available. The version history functionality captures and displays who created or modified the standard text, the date of the change, and its status. When creating a category, a popup window appears in which you must enter a category name (up to 40 characters) and a description. We encourage you to first create categories and then add standard texts within them for better organization. If you are navigating within a category, you can use the **Top Level** hyperlink at the top of the screen to quickly return to the main screen.

Besides creating standard texts with embedded rich formatting, you can add hyperlinks, insert images, and even manage text content in multiple languages. The ability to create new versions of standard texts ensures that any updates or versions are properly tracked. Changes can be made to texts that still have the **In Progress** status; however, if a text has already been released, a new version must be created to implement changes either to the header data or the content.

Tip

A notable requirement when creating the standard text ID is that it must be a single, continuous entry without any spaces; otherwise, the system will display an error.

Manage Unassigned EBOMs (F2394)

Application Type: Transactional

This app is used to manage new or changed engineering bills of materials (EBOMs) that are separately imported or created in production engineering and operations (PEO) and not contained in an engineering snapshot. You can view the status of all EBOMs, assign them to a planning scope and change record, and initiate the handover process to manufacturing so that corresponding manufacturing BOMs (MBOMs), routings, and related elements can be updated, created, or deleted. You must select the BOM type and production plant before proceeding; these initial settings can later be adjusted under the shop floor control (SFC) settings. You can enter selection criteria such as material, plant (a key field), EBOM type, assignment type, handover status, and last changed date to filter EBOMs. The app lists EBOMs along with details such as material, plant, production plant, BOM usage, and more, and the associated change records. It allows setting the assignment status filter to **Unassigned** to display only unassigned EBOMs or to **All** to show both assigned and unassigned EBOMs. For each new version or change state of an EBOM, you can determine the planning scope and assign the EBOM to a manufacturing change record to initiate the handover process. If relevant planning scopes or change records already exist, they are displayed in the **Recommended Planning Scope** and **Open Change Records** sections. EBOMs can be assigned manually to an existing planning scope or change record, or you can create new ones directly from the app.

You also can compare an EBOM version with its previous version to view a net change summary, helping you quickly identify and understand modifications between BOM versions. More importantly, you also can view the impacted effectivity for each unitized BOM to showcase an overview of changes affecting production.

Manage Unassigned MBOMs (F2870)

Application Type: Transactional

This app allows you to manage manufacturing bills of materials (MBOMs) by assigning them to change records when changes are made directly to the MBOM, without a corresponding change to the engineering BOM (EBOM). In such case, no EBOM-to-MBOM handover is required, and the MBOM itself triggers the change. You can filter MBOMs based on criteria such as material, plant, BOM status, assignment status, last change date, and more. The app displays MBOM details such as material, plant, BOM usage, the associated change records, and so on. The **Assign to Change Record** and **Create Change Record** options let you either link an MBOM to an existing manufacturing change record or create a new one to include the MBOM in the change process.

The functionality mirrors the Manage Unassigned EBOMs app (F2394), but without the need for a planning scope. The app also allows you to compare an MBOM version to its previous version to identify net changes.

Tip

Although it is possible to use the app without manufacturing snapshots, we strongly recommend implementing snapshots via menu path **SAP Reference IMG • Production • Manufacturing for Production Engineering and Operations • Production Engineering • Engineering and Manufacturing Snapshots**. Here, you can define and assign number ranges for snapshots to fully leverage the SAP S/4HANA production engineering and operations (PEO) capabilities.

Manage User Settings (F2829)

Application Type: Transactional

This app allows you to add, edit, or delete manufacturing users by specifying key details such as the business partner ID, default plant, and default work center. Process owners can assign multiple work centers; however, the system will always prioritize and display operations from the work center marked as default. Overall, this app enables organizations to effectively manage production execution and ensure that operational activities are well aligned with the assigned resources.

Manage Work Center Capacity (F3289)

Application Type: Transactional, Analytical

This app helps you manage and evaluate work center capacities. You can select specific work centers and filter based on criteria such as overdue duration, load type (all, cumulative, overload, underload), bucket type (day, week), and evaluation horizon (date ranges). The app lists work center capacities, displaying information such as capacity category, cumulative load, maximum, first overload, total capacity requirement, and more. The app uses color-coded legends to easily differentiate load amounts: normal load (0% – 80%), critical load (80% – 100%), and overload (above 100%). Settings can be adjusted to show additional details relevant to you, such as utilization, overall percentage load, and average load. You can navigate to detailed views of individual work center capacities, which include a utilization chart and pegged requirements (operations and orders). Work center capacities can be modified, and you can visualize work center loads, detect overload, and manage capacities effectively over different time buckets. You also can personalize load thresholds, defining what constitutes an overload based on your specific requirements. Furthermore, in the detailed views, you can edit shift start and end times, copy shifts across multiple work centers and manage shifts for better utilization across multiple work centers.

More importantly, integration with related apps is provided for managing orders, operations, and variant configurations, particularly for production orders using advanced variant configuration.

Manage Work Center Groups (F4044)

Application Type: Transactional

This app helps you create, organize, and manage work center groups within a manufacturing environment. The app is quite useful for production engineers, planners, and operation managers who work in complex shop floor environments and need to group work centers based on function, type, or even plant location for better control of and visibility into capacity and planning data. You can apply various selection criteria to filter results, and these can be adapted to suit your preferences. The app lists work center groups along with key details such as group name, description, object type, plant, number of work centers, where used, and more. You can easily navigate through the app, making use of the **Create**, **Expand All**, and **Collapse All** options. To create a new work center group, you need to enter the group, group type, and plant. Once a group is created, associated details are linked to individual work centers or even other groups. This feature supports the building of hierarchical group structures, which allows you to assign multiple work centers and work center groups. This is particularly useful for viewing capacity demands and remaining capacities across a logical group of work centers.

The app provides advanced features such as viewing the full work center group hierarchy, filtering results, and assigning or removing work centers. Searching for a work center shows only its immediate parent group. You can nest groups within others and categorize them by group type as defined in the Customizing table. Mass removal of work centers or groups simplifies maintenance. You can update group information or delete groups that are no longer required. Administrative details specific to each group can be viewed at the bottom of the group's detail page. In addition, the **Where Used** column helps you identify the immediate parent group for any selected work center or group. The **Number of Work Centers** column displays the total number of work centers assigned to a group, including those assigned indirectly via subgroups.

Manage Work Center Queue (F2821)

Application Type: Transactional

This app is primarily intended for production supervisors and operators who need to view and manage production operations assigned to specific work centers. The list of operations shown is filtered based on this default work center. If you are assigned to multiple work centers, the app initially displays operations from the default one, but you can switch between work centers to view other assigned operations. This setup can be managed through the Manage User Settings app (F2829). Operations within the app are organized into three categories based on their status: **Not Started**, **In Progress**, and **Finished**. You are supported by graphical progress bars that show how many units are ready, not ready, or already processed for each operation. Issues like missing components, holds, or delays are also flagged with icons, allowing you to identify bottlenecks quickly and decide which tasks to prioritize.

The app displays a configurable, sortable table of operations with key details like production order, material, quantities, and dates. You can access detailed popovers and quickly navigate to related apps, including Perform Operation (F2822). This app supports both shop floor orders and standard production orders. When users begin executing operations from a shop floor order, they are directed to the Perform Operation app or the Perform Operation Activity app (F2898). For standard production orders, you are redirected to Transaction CO11N (Confirm Production Order Operation). The app also allows you to view operation progress and current or previous operation statuses via interactive popovers. The operations are grouped by their status and organized into tabs and subtabs, making it easier to navigate and manage workloads. This setup ensures that you always have a clear overview of your tasks and can respond quickly to changes or issues on the shop floor.

Tip

To use this application, process owners need to make sure that they are assigned to a default work center. Without this assignment, the app cannot be launched.

Manage Work Centers (F6175)

Application Type: Transactional

This app allows you to manage all work center-related operations efficiently. You can create, replace, modify, and perform mass maintenance of work centers and their capacities. The filter criteria allow you to filter and search for work centers using both basic and advanced criteria such as work center ID, plant, work center category, usage, person responsible, and so on. Once the list of work centers is displayed, it includes important details such as standard value key, control key, and more. A set of options within the app, **Create**, **Change**, **Replace**, **Check Usage**, and **Mass Maintenance of Capacities**, allows you to perform various actions.

Selecting the **Create** or **Change** button navigates you to Transaction CR01 (Create Work Center) or CR02 (Change Work Center), respectively. You also can choose a specific work center to view its details in the Display Work Center app (CR03). The **Check Usage** function provides insights into how a specific work center is being used. When multiple work centers are selected, the **Mass Maintenance of Capacities** button opens the Mass Maintenance of Work Center Capacities app (F5381), and you can manage capacity-related attributes across several work centers in bulk.

Mass Adjust Buffers (F3653)

Application Type: Transactional

This app helps to manage stock levels efficiently in a demand-driven replenishment setup. It allows you to review and adjust safety stock, reorder points, and maximum stock levels through buffer proposals. These proposals are essential to ensure that products are available when needed, while avoiding overstocking, excess storage costs, or product obsolescence. You can filter and view products based on criteria such as proposal status, value indicators, variability indicators, and so on. For instance, the app offers visual insights into buffer levels and average daily usage (ADU) for each product over a 90-day period using microcharts. You can choose to adopt or discard buffer proposals, adjust buffer zones and demand figures through absolute or relative values, or even copy these adjustments from other products. When selecting a product, you can access detailed contextual information and perform various actions through navigation to related apps. The app also supports multiple display modes, such as the **Planning View**, which highlights maximum stock, safety stock, and reorder points; the **Execution View**, which shows replenishment activities; and the **Comparison View**, in which you can compare current and proposed values of stock parameters.

In addition, the app supports demand adjustments to modify the ADU and allows you to toggle between a chart view and table view for both buffer and usage information. Selecting specific data points in the charts provides further detail to aid decision-making. If predictive lead times have been calculated for a product, the app ultimately displays both current and proposed lead times in the **Longest Path** view within the product's replenishment network. You also can visualize the complete network flow for a product using the **Product Flow** option.

Tip

The Schedule Buffer Proposal Calculation app (F2837) generates buffer proposals, which are eventually used by this app.

Mass Maintenance of Products (DD) (F2825)

Application Type: Transactional

This app supports demand-driven replenishment by allowing you to display and change product master data in bulk. The app plays an important role in classifying and maintaining product information that influences planning decisions based on decoupled lead times and variability, which are essential aspects of demand-driven replenishment. At its core, this app allows you to manage large volumes of product master data across various dimensions. You can filter and select products using detailed selection criteria such as product, product group, plant, MRP area, procurement type (e.g., buy, make, or transfer), variability indicator (ranging from unclassified to high), decoupled lead time (DLT) indicator, bill of materials (BOM) usage indicator, MRP type, and demand-driven scope, which includes planning and execution contexts like internal or external. Once the selection criteria are applied, the app displays a list of products with key information, including the product name, plant, and value indicator. From here, you can initiate changes using the **Edit** button or adjust the planning scope with the **Change DD Scope** button. This flexibility allows planners to adapt to changes in supply chain strategy without needing to maintain each product individually.

The app also provides positioning features through the **Positioning – Internal** and **Positioning – External** tabs. These views allow you to monitor and analyze products according to where they fall in the supply chain and planning scope. You can assess the readiness of materials for demand-driven replenishment and identify which products have been marked appropriately.

Tip

When using this app for the first time, you might notice that products don't have a lead time (EFG) classification yet. Only products marked as relevant for demand-driven replenishment are classified using their DLT. Once you have selected and marked the relevant products, you can use the Schedule Lead Time Classification of Products (DD) app (F2871) to perform the classification.

Mass Maintenance of Work Center Capacities (F5381)

Application Type: Transactional

To use this app, you need to assign an area of responsibility (AOR) first in the **Area of Responsibility** popup. The different options in the **Area of Responsibility** popup navigate you to certain apps, as follows:

- Clicking **Plant and MRP Controller** navigates to the My Area of Responsibility – Production Supervisor app (F7044)
- Clicking **Work Center Group** navigates to the My Area of Responsibility – Work Center Groups app (F5266)
- Clicking **Work Center Person Responsible** navigates to the My Area of Responsibility for Work Center Person Responsible app (F6716)

You can filter capacities using parameters like work center, plant, capacity ID, and more—including whether a predictive material and resource planning (pMRP) change proposal exists—for better planning integration. The results show key details such as utilization, planner, capacity count, and shift group. You can edit multiple records at once via the **Edit Capacity Header** button, adjusting utilization or capacity numbers across similar entries. Selecting a work center reveals its detailed capacity data. You can view factory calendar information, overload percentage, number of capacities, utilization rates, break duration, and other critical scheduling elements. This helps you analyze if the available capacity aligns with production needs and whether any adjustments are necessary.

You also can navigate to the Work Center Application Logs app (F5931) by selecting the **Logs** button. This provides a list of logs that are categorized by severity, such as canceled, error, warning, information, and success, giving clear visibility into system feedback and the impact of recent changes.

Within the app, you can navigate to Transaction CR12 (Change Capacity) under component PP-BD-WKC (Work Center).

> **Tip**
>
> In the **Intervals and Shifts** tab, you can display extra information such as the standard available capacity, shift sequence, length of cycle, and so on by selecting the **Show Details** button. Furthermore, selecting **Edit Capacity** will allow you to create or delete an interval or add and delete a shift for the work center. For example, when creating an interval, process owners can create an interval for nonworking or working days, create shifts for an entire week, create custom shifts, and even choose to override the factory calendar.
>
> The **Create Intervals with Shifts** button on the initial app screen is similar in function to the **Edit Capacity** option on the **Capacity Details** screen of a particular work center.

Material Scrap (F2035)

Application Type: Analytical

This app allows you to analyze and compare actual scrap figures recorded during production confirmations with expected scrap percentages defined for materials. This app is useful for identifying materials for which there is a significant discrepancy between expected and actual scrap figures. This allows you to implement a feedback loop to update and refine the expected

scrap percentage, which ultimately helps in reducing waste. The selection criteria include the start and end of the performance period, which are key filters for focusing on the analysis. Additional filters for the analysis scope are available, such as MRP area, MRP controller, material, and production plant. At the top of the screen, you can choose to display mini tiles, which offer quick insights into operation scrap and scrap reasons. The **Operation Scrap** mini tile displays information about scrap related to specific production operations, while the **Scrap Reason** mini tile offers a stacked chart that shows the percentage of scrap and rework by work center, using different color coding for clarity.

The app's interactive features let you explore data in depth. With the **Jump To** option and drill up/down, you can navigate between an overview and detailed views. You can toggle data labels, view legends, zoom in/out, and switch between chart types (e.g., bar, line) or between chart and table views for flexible analysis. Several KPIs are central to the app's functionality, including **Yield, Scrap, and Rework Qty by Entry Time** (line chart), which tracks quantities of yield, scrap, and rework over time, and **Planned vs Confirmed Scrap % by Material** (bubble chart), which compares planned scrap percentage to confirmed scrap percentage. Other KPIs include but are not limited to **Scrap Deviation % by Material** and **Scrap Deviation Qty by Material**. The app also offers an **All Dimensions and Measures** tabular view, providing a detailed breakdown of all data related to material scrap. The app uses CDS view C_MFGORDITEMMATERIALSCRAPQ.

One of the most valuable aspects of the app is its ability to identify materials with the greatest deviation between expected and actual scrap percentages. This is particularly important for production planners who want to adjust the expected scrap values in the material master data to improve future forecasting accuracy. The app also allows you to create visual alerts on KPI tiles, making it easier to monitor scrap issues and act when needed.

> **Tip**
>
> For the fashion industry, the app provides specific data related to master production orders, seasonal collections, and stock segments. To display fashion-related data, you must activate the ISR_RETAILSYSTEM business function in the standard SAP GUI system.

Monitor External Requirements (F0246)

Application Type: Transactional

This app helps you manage uncovered requirements originating, for instance, from sales orders and stock transport orders. This app allows you to monitor materials and articles with a focus on timely fulfillment of customer demands—specifically in industries such as fashion, where segmented material and article management is essential. The app presents key information such as material, requirement date, quantity overview, open quantity, missing quantity, and the affected order. For example, using the default filter, time till requirement date, you can narrow down the list to focus on uncovered external requirements within a specific time period. The app also includes various filters and sorting capabilities to help production planners view and manage relevant data efficiently. One of the core features of the app is the ability to select an area of responsibility, which helps refine the list of requirements. You also can personalize the table by choosing which columns to display and how the entries should be sorted. This personalization capability enhances the flexibility and usability of the app, making it easier for planners to view the data that matters most to them. The combination of filters and shortage definitions can be saved as a personalized view and further added as a tile on the home page for quick access.

The app provides quick navigation to the Manage External Requirements app (F0250), where you can view more detailed information about delayed items and explore potential solutions to address any shortages. However, it is important to note that for segmented materials and articles, navigation to the Manage External Requirements app is not supported. In such cases, planners will need to use the app with the segmented material rules in place for the shortage calculations.

Monitor Internal Requirements (F0263)

Application Type: Transactional

This app allows you to monitor internal requirements such as component needs for production orders, process orders, maintenance orders, and network orders. Similar to the Monitor External Requirements app (F0246), this app helps make sure that all required components are available on time and in the right quantities. You can use filter criteria such as shortage definition and time till requirement date to define which internal requirements you want to monitor. The app then displays a detailed components list, showing fields such as component ID, requirement date, quantity overview, open and missing quantity, coverage status, and the affected order. For each line item, you can select the process order hyperlink in the **Affected Order** column to open a popup screen showing critical process order details such as MRP controller, order quantity, goods receipt processing time, and production time. This makes it easy to react quickly to material shortages, prevent production delays, and take necessary actions.

From this popup, you can navigate to Transaction COR3 (Display Process Order) in SAP GUI or access related document information. Navigation to the Manage Internal Requirements app (F0270) is not supported for segmented materials.

> **Tip**
>
> The app also supports segmented materials and articles, incorporating requirement and stock segments based on segmentation rules during shortage calculations.

Monitor Kanban Containers (F5723)

Application Type: Transactional

This app helps production planners and shop floor users monitor and evaluate the status of Kanban containers. You can search for containers using filters such as plant, production supply area, material, and container status (e.g., wait, empty, in process, transit, full, in use, or error). Once filters are applied, the app displays a detailed list showing key information such as container ID, material replenishment, status, quantity, and so on. The real-time visibility capability helps you detect delays, errors, and supply issues early and take prompt corrective action. You also can personalize table columns and sorting criteria, allowing you to create customized worklists tailored to your specific use case—for instance, only monitoring containers in error status or those experiencing delays. These views can be saved using the **Manage Views** feature. Selecting a listed container in the hit list reveals further details of the Kanban container and its control cycle.

Monitor Material Coverage – Net Segments (Version 2) (F0247A)

Application Type: Transactional

This app is quite useful for production and demand planners (e.g., demand planners for retail). It helps you track material availability and identify potential shortages across your area of responsibility. The app allows you to monitor net requirements segments by specifying selection criteria such as shortage definitions, specific materials, and shortage evaluation periods. You can apply these filters to narrow down results and focus on the most critical materials. The app displays a list of materials with their key details, such as material number, description, supplier number, shortage status, quantity, and stock availability. This information allows planners to make quick decisions on replenishment or follow-up actions. Several

options are available within the app, such as **Manage Materials, Start MRP Run, Revoke Acceptance of Shortages**, and a **Configure Chart** icon for visual insights. The app also handles segmented materials and articles, which means that the app considers segmentation rules applied to requirement and stock segments when calculating shortages; this is particularly useful in environments where materials are managed in segments (e.g., batch, quality, or distribution channel).

You can navigate to Transaction MD04 (Monitor Stock/Requirement List) in SAP GUI, which presents pegged requirements in a visual SAP Fiori layout.

This app is the successor to the Monitor Material Coverage – Net Segments app (F0247), which was depreciated as of SAP S/4HANA 2020 FPS02.

Tip

You can use the **Configure Chart** icon to make necessary adjustments to the stock availability chart. For instance, you can choose to display the chart with the horizon in days for the next 28 days.

Monitor Material Coverage (Version 2) (F2101A)

Application Type: Transactional

This app allows production planners to monitor net and individual material requirements for their area of responsibility. You can apply filter criteria that include shortage definitions (e.g., MRP standard, stock days' supply, ordered requirements, ordered receipts), individual segment, shortage evaluation period (e.g., total replenishment lead time, manual planning horizon), and more. The app displays a comprehensive table showing key information such as material number, segment, first shortage date, shortage quantity, and shortage duration. You can assess multiple planning segments such as net segment, individual customer segment, individual project segments, and so on. The system calculates material requirements separately for each segment using the selected shortage definition, which defines what supply and demand elements to consider and under what conditions a shortage is flagged. You can use filters to refine the material list and tailor the app to your operational context.

The app provides visual capabilities such as a stock availability chart per line item and allows you to customize table views and save your personalized views as tiles on the SAP Fiori launchpad. You can utilize the **Show More per Row** icon to display a stock availability chart for every line item listed on the initial screen of the app.

Selecting the Manage Material button will navigate you to the Manage Material Coverage app (FO251). This app is the successor to the Monitor Material Coverage app (F2101), which was deprecated as of SAP S/4HANA 2023 FPS02.

Tip

Note that the stock availability chart shows the availability of stock for the next 21 days by default. However, you have the flexibility to change this by selecting the **Configure Chart** icon and adjusting the horizon to either days or weeks. For days, your potential range is from 7 days minimum to 42 days maximum. For weeks, you can choose either 13 or 26. The preview box will change depending on the settings made before changes are saved.

Monitor Production Orders or Process Orders (Version 2) (F0266A)

Application Type: Transactional

This app was designed for production planners to keep track of the progress of production and process orders within their area of responsibility. The app provides two separate tiles—one for production orders and one for process orders—which helps you monitor each type independently based on your operational needs. It allows planners to assess whether materials are being completed on time for their required

deadlines, whether the necessary components will be available when needed, and whether any delays exist in the scheduled operations or milestones. The app uses a concept called *shortage definition* to determine which orders are flagged as delayed or at risk. You can select a relevant shortage definition and apply additional filters such as material, plant, or production order type to narrow down the list to only the most relevant orders.

Information is displayed in a comprehensive table in which colors indicate the status of each order, material, and component (e.g., green for on track, yellow for warnings, red for issues). You can sort the table, select which columns to display, and even open a quick view of specific materials for more insight. One of the key advantages of this app is personalization. You can save a preferred combination of filters, shortage definitions, and table layouts as a custom view. These saved views also can be pinned to the SAP Fiori launchpad as tiles, allowing for fast and convenient access to frequently used configurations.

You can navigate to the Manage Production Orders app (FO266) and Manage Process Orders app (F4587). This app is the successor to the Monitor Production or Process Orders app (FO266), which was deprecated as of SAP S/4HANA 2022 FPS02.

Monitor Shop Floor Master Data (F4201)

Application Type: Transactional

This app allows you to monitor planned orders that are missing valid shop floor master data in the production version. This is critical: Without the master data, planned orders cannot be converted into production orders or executed. You can filter based on material, plant, MRP controller, production supervisor, and an opening date range to identify planned orders with issues. The app also displays the earliest planned order date for the material. You can navigate to a detailed page where you can view and resolve production version issues, such as creating a shop floor bill of materials (BOM) version or routing, launching the Visual Enterprise Manufacturing Planner (VEMP) to reconcile or create the shop floor BOM version, and editing the planning BOM and routing. You also can assign BOMs to open change records, create new change records, and save search criteria as tiles to prioritize materials for processing.

Monitor Snapshots (F4908)

Application Type: Transactional

This app allows users in engineering and manufacturing to track and analyze snapshots related to product development and production. You can filter by product group, material, snapshot type (external engineering, internal engineering, or manufacturing), snapshot revision (latest, last three, last five, all), creation date, and issue status. The app lists snapshots along with the snapshot type, revision, header material, and plant information. Key actions include preparing Production Integration Portal (PiP) data and change records, viewing key snapshot data, and drilling down into details such as engineering bill of materials (EBOM) or manufacturing BOM (MBOM) versions, change records, documents, and 3D visuals. You also can identify and trigger PiP data or change record preparation manually, create new manufacturing snapshot revisions, filter snapshots with issues, and access the application log for troubleshooting.

You can access related apps like Prepare MBOM Creation (F4909) and Manage Change Records (F2097).

My Area of Responsibility – MRP Controller (F4796)

Application Type: Transactional

This app allows you to manage plant/MRP controller assignments, ensuring that only relevant MRP-relevant app data is displayed based on your defined responsibility. You can filter by area of responsibility (AOR) status (**Assigned** or **Unassigned**), plant, and MRP controller, with the option to refine selections using adaptable filters. The app displays plant/MRP controller combinations with status indicators—red for unassigned and green for assigned—and allows you to assign or unassign responsibilities using dedicated buttons. Assignments can be modified at any time, and default filter values can be set via the user actions menu, applying automatically across all related apps. When using an MRP-relevant app for the first time, you must define your area of responsibility by selecting a plant and MRP controller combination. The app ensures that MRP controllers can efficiently manage material planning by restricting visibility to relevant items only.

My Area of Responsibility – Production Supervisor (F7044)

Application Type: Transactional

The app allows you to define your area of responsibility by managing plant/production supervisor and plant/work center combinations. It features selection criteria for area of responsibility (AOR) status, plant, production supervisor, and work center, with filters to refine the displayed list. The app includes two tabs: **Production Supervisors**, which lists plant/production supervisor combinations, and **Work Centers/Resources**, which lists plant/work center combinations. You can assign or unassign responsibilities using a toggle button per row or select multiple rows and use the **Assign** or **Unassign** button for batch processing. Assigned and unassigned statuses are visually indicated in green and red, respectively.

This app ensures that only relevant orders or operations appear in the Manage Production Orders (F2336) and Manage Production Operations (F2335) apps, based on your selected responsibilities. The app is accessible via the SAP Fiori launchpad or within the Manage Production Orders and Manage Production Operations apps. If an AOR has already been defined within these apps, the system will migrate or merge the data accordingly.

My Area of Responsibility – Work Center Groups (F5266)

Application Type: Transactional

This app provides an overview of plant/work center group combinations, allowing you to manage assignments efficiently. You can filter by area of responsibility (AOR) status, plant, work center group, and work center group type to refine the displayed list. The app includes **Assign, Unassign,** and **Manage Work Center Group** buttons, with the **Manage Work Center Group** button redirecting to the Manage Work Center Groups app (F4044) for deeper control. Key features include filtering the list, viewing assigned and unassigned work center groups by plant, and assigning or unassigning multiple work center groups and plants in a single step, ensuring flexibility in managing responsibilities.

My Area of Responsibility – Work Center Person Responsible (F6716)

Application Type: Transactional

This app provides an overview of plant and work center person responsible combinations and allows you to manage assignments with **Assign** and **Unassign** buttons. The app helps in comparing actual scrap figures recorded in production confirmations against the expected scrap percentage defined in the order operation, enabling a feedback loop to update scrap expectations. It only considers finally confirmed orders and highlights work centers where actual scrap deviates most from expected values. Key functionalities include creating visual alerts on KPI tiles, viewing real-time data across multiple perspectives, and analyzing scrap reasons. You can interact with charts and other representations, navigate with context transfers, and utilize

automatically calculated fields such as expected scrap percentage, actual scrap percentage, actual yield percentage, and actual rework percentage for better decision making.

My Work Queue (F3499)

Application Type: Transactional

This app provides a centralized workspace for production operators in discrete manufacturing to efficiently manage and execute their assigned operation activities. Designed for nonserialized production orders, the app allows users with a relevant role to access and process tasks efficiently. Operators can use the search bar and the filter criteria to filter work queue based on status categories such as **Completed, In Process, Initial, Paused,** or **Skipped**. The app features three tabs—**All Operations Activities, Assigned to Me,** and **Labored On By Me**—to ensure operators can quickly locate relevant tasks. The work queue displays critical details, including operation activity, quality assurance status, operation, and order, allowing operators to track and prioritize their tasks effectively. The app provides several functionalities, such as viewing assigned team members, grouping, sorting, and filtering activities, and searching for specific orders. Operators also can adjust the number of displayed columns via the **Settings** button and execute a selected operation activity directly from the queue. By clicking the chevron icon, you can access detailed operation activity information.

For quality assurance and process control, the app allows operators to filter for quality assurance issues, ensuring that only operation activities without holds or missing components are displayed. Before starting an operation, an operator can check their work queue for trackable tools required for the next activity. If needed, you also can navigate to the Track Tool Usage app (F4762) to access the necessary tool.

Operation Scrap (F2034)

Application Type: Analytical

This analytical app provides detailed insights into scrap and rework trends at the work center level. It helps production managers and quality engineers track, analyze, and optimize production efficiency by evaluating scrap, yield, and rework data over a specified performance period. You can define start and end dates for the performance period, allowing customized filtering of production data. The **Scrap Reason** and **Material Scrap** mini charts provide quick visual insights into scrap reasons and material scrap trends. The following KPIs can be used to visualize data:

- Yield, Scrap, and Rework Qty by Entry Time (Line Chart)
- Yield, Scrap, and Rework % by Entry Time (Line Chart)
- Scrap % by Operation: Planned vs Actual (Bubble Chart)
- Scrap % by Work Center: Planned vs Actual (Column Chart)
- Scrap Qty by Work Center: Planned vs Actual (Column Chart)
- Scrap and Rework Qty by Work Center (Stacked Chart)
- Scrap and Rework % by Work Center (Stacked Chart)
- All Measures and Dimensions (Tabular View)

The app includes various navigation and interaction tools to enhance usability and data analysis. You can quickly jump to other related apps, drill up or drill down for detailed insights at different hierarchy levels, and toggle the legend to customize chart views. The app uses CDS view C_ MFGORDOPERATIONSCRAPQRY to retrieve data. You also can zoom in and out to focus on specific data points and seamlessly switch between chart and table views for a comprehensive analysis of scrap and rework data.

Order Genealogy (F5865)

Application Type: Transactional

This app provides centralized visibility into the entire manufacturing process at the production order level. Designed for production supervisors and quality engineers, this app allows you to analyze, track, and monitor production orders by displaying comprehensive order details including material, plant, confirmed scrap and so on. You can apply filters based on order number, material, and plant to refine your searches. The app offers detailed insights into each production order, summarizing critical data collected throughout the production process. You can view the order status, assembled or disassembled components, performed operation activities, and recorded defects. The app also allows for navigation to inspection lots to review recorded inspection characteristics, helping to maintain quality control. A key function of the app is its ability to log and display production issues, such as holds placed on production orders. It also provides a list of serialized numbers processed, allowing users to track serialized products and navigate to the Product Genealogy app (F2892) for further analysis. Moreover, the app displays assigned production resources and tools, equipment details, and uploaded documents related to a production order.

You can drill down into operation activities, viewing detailed execution records and tracking whether the primary or alternative component was used. If defects were recorded, the app allows you to access the defect history and take corrective actions. You also can print order genealogy reports.

Perform Operation (F2822)

Application Type: Transactional

This app provides a comprehensive overview of operation activities within a shop floor order.

This app enhances transparency in production processes by allowing operators to view key operation details, monitor progress, and take necessary actions to maintain efficiency. From the app, you can begin executing production tasks, track operation activity status, and drill down into detailed information. Operators can access a list of operation activities related to a shop floor order and view crucial data such as segment type, group, status, and the quantity ready or completed. The app enables direct execution of operation activities, but only if all predecessor activities have been completed, thus ensuring a structured and sequential workflow. If execution is allowed, operators are redirected to the Perform Operation Activity app (F2898), where they can carry out specific tasks according to the assigned status and action schema. The app also allows you to analyze existing assignments for each operation activity, including components, work instructions, inspection characteristics, production resources and tools, buyoffs, and qualifications. It also supports retroactive processing, which helps maintain accurate production records. Operators can set or release holds at multiple levels—order, material, operation, or work center—to manage disruptions effectively.

For deeper insights, the app offers a detailed action log, allowing operators to track changes and updates related to an operation. You also can drill down further to examine operation activity assignments such as components to be assembled or disassembled, inspection characteristics, and required documents. Furthermore, you can view assigned production resources and tools and navigate to the Track Tool Usage app (F4762) to manage and claim necessary equipment for the operation.

Tip

The app is assigned to the SAP_BR_PRODN_OPTR_DISC_EPO role (Production Operator – Discrete Manufacturing (Extended Production Operations)).

Perform Operation Activity (F2898)

Application Type: Transactional

This app helps production operators execute tasks efficiently in discrete manufacturing. It offers interactive, step-by-step work instructions with 3D visuals and real-time data. The system checks preconditions, qualifications, and active holds before allowing processing. Operators can track progress, confirm activities, record consumption, report quality results, and manage attachments. Production engineers define shop floor routings through Transaction MSFR4 (Manage Shop Floor Routings) in SAP GUI, and all associated details (components, inspection characteristics, work instructions, buyoffs, and qualifications). Operators can manage the consumption of serialized, batch-managed, and backflushed components, ensuring traceability and compliance with manufacturing requirements.

The app also offers several key features, including the ability to view 3D images of assembled products, select individual components to see their assembly location, view and acknowledge engineering change alerts, and collect inspection characteristics. Operators also can enter manufacturer serial numbers, supplier batch numbers, and place holds or release holds on materials, orders, operations, or work centers as needed. The system supports recording defects, triggering automatic quality notifications, and maintaining an action log for each operation activity.

Another critical functionality of the app is its support for disassembly processes. Operators can view a **Components to Disassemble** list, distinguishing between permanent and temporary disassembly. If a disassembly occurs within the same order, the app allows you to select components previously assembled. Operators can record disassembly and reassembly actions, specifying stock types and storage locations. When a defect is recorded, it is possible to associate it with a specific component, further enhancing traceability and quality control.

This app also acts as an entry point for execution tasks, letting operators scan order and operation details from traveler documents or barcodes. It reflects routing changes accurately and provides quick access to inspections and defect logging via the **Record Defect** button for real-time quality monitoring.

Planner Overview (F2832)

Application Type: Analytical

This app provides an overview of buffer management and replenishment by displaying key metrics through interactive cards. The app allows you to filter and analyze inventory and replenishment data based on key parameters including product, product group, plant, MRP area, and so on. The app has three cards shown by default:

- **Buffer Level Management**
  The KPI for this card is **Breakdown by Value (ABC)**. It displays the number of deviations and categorizes data using breakdown by value. Selecting this app will navigate you to the Manage Buffer Levels app (F2706).

- **Replenishment Planning**
  The KPI for this card is **Breakdown by Lead Time (EFG)**. It highlights the number of buffers that have fallen below the reorder point and utilizes the breakdown by lead time to help planners evaluate replenishment timing. Selecting this app navigates you to the Replenishment Planning app (F6241).

- **Replenishment Execution**
  The KPI for this card is **Breakdown by Variability (XYZ)**. It analyzes how replenishment is impacted by demand fluctuations and component dependencies. Selecting this card navigates you to the Replenishment Execution app (F6242).

pMRP Component View (F4450)

Application Type: Transactional

This app provides a detailed view of all components within a predictive materials and resource planning (pMRP) simulation, which allows business users to analyze material components, assess supply sources, and verify production constraints for efficient supply chain and production planning. You can view material components included in the simulation, check supplier details for externally procured components, verify production lines assigned to in-house manufactured components, and analyze constraint quantities set for specific materials.

Prepare MBOM Creation (F4909)

Application Type: Transactional

The app helps facilitate the handover from engineering bill of materials (EBOM) to manufacturing BOM (MBOM) using engineering snapshots. It guides users through a structured process to ensure a smooth transition of engineering data for manufacturing a material in a specific plant for the first time. It provides a step-by-step process for MBOM preparation as well as 3D model visualization to review the product structure. Let's review the key steps:

1. **Choose preparation basis**
   Select either an **External Engineering Snapshot** or an **Engineering Change Notice** as the basis for MBOM preparation.

2. **Enter EBOM/production details**
   Enter EBOM and production information, such as material plant, alternative, usage, production plant, and production usage.

3. **Select engineering snapshot**
   Review and confirm the engineering snapshot containing EBOM versions and related data.

4. **Select change record**
   Choose whether to create a new change record or update an existing one with EBOM versions, engineering snapshots, and planning scope.

Process Order Action Logs (F5187)

Application Type: Transactional, Reuse Component

This app allows you to view action log details for released or technically completed process orders. This app can be accessed through the Manage Process Orders app (F4587), and you can track and review logged actions related to process orders.

Process Order Confirmation Object Page (F2266)

Application Type: Fact Sheet

This app allows you to display contextual information for process order confirmations. Because this object page is designed to show contextual data, no specific tile is provided for it. Instead, you can access this object page through other apps. In essence, you can view general information about process order confirmations, such as material, plant, dates, times, work center, and so on. You also can access more detailed information, including quantities and confirmation text.

Process Order Object Page (F2263)

Application Type: Fact Sheet

This app shows contextual information about process orders. You can view additional relevant information specific to your business context, including details about operations, items, components, status, and so on.

Process pMRP Capacity Simulations (F3952)

Application Type: Transactional

This app allows you to display and manage the capacity demands of work centers. It provides several key options to effectively perform capacity planning within production processes. You can view the current capacity situation of work centers to identify potential bottlenecks and display work center capacities in a chart for better tracking of utilization. The app allows you to change capacity demands and shift top-level material demands across time periods to effectively handle resource allocation. It also provides the ability to display aggregated work center groups, helping to manage multiple work centers efficiently. Overall, the app supports flexible production scheduling and improves resource utilization across manufacturing operations.

Process pMRP Simulations (F3934)

Application Type: Transactional

Using this app, you can create and manage multiple simulations to assess the impact of simulated changes to capacity or demand on KPIs. The app offers various filtering criteria, including simulation name, reference data, created by, and so on, to manage simulations effectively. It lists active simulations, showing details such as capacity issues, delivery performance, simulation status, and actions that can be taken. The app includes several key features such as demand plan simulation, where you can process multiple simulations, view capacity issues, adjust demand quantities, and release simulations before implementing changes to the operative MRP. The capacity plan simulation provides an overview of the capacity situation at a work center, allowing you to change available capacity and manage capacity issues. The multi-level material simulation allows you to check material components for preproduction feasibility in case of a capacity overload and change the source of supply to resolve issues.

You also can view a simulation summary, which compares all changes made to the simulation plan against the reference plan. The app offers a standard download functionality to export results to Microsoft Excel for further analysis.

Once a simulation is released, planned independent requirements (PIRs) are created for top-level materials and demand-driven materials. You can then check and adjust buffer levels before starting a new operative MRP run. The app provides functionality to view and change material demand quantities, display aggregated work center groups, and change capacity demands. It also allows for shifting top-level material demands, viewing material components, and supplier information, and conducting preproduction or early procurement of materials. You can release one or multiple simulations and manage the global change history for a simulation, including undoing or redoing changes. In addition, the app includes features to refresh reference data and view a prioritized issue list.

Process Production Versions (F6400)

Application Type: Transactional

This app helps manage production version proposals, rejected proposals, and existing production versions. You can filter production versions based on filter criteria such as material, plant, MRP controller, validity start date, lock status, and so on. The app has three main tabs. In the **Production Version Proposals** tab, you can view proposed production versions, including details such as plant, proposed production version, task list group, and reason for proposal. You can choose to reject or accept proposals. The **Rejected Proposals** tab lists rejected proposals and the relevant details, and you can retrieve proposals for further review. The **Existing Production Versions** tab displays all production versions in the system, and you can perform actions such as create, lock, unlock, check, and delete for production versions.

For each selected production version, you can view related information, including the associated tasks list and bill of materials (BOM). The

app also allows you to check whether a production version is allowed for repetitive manufacturing and view any linked production versions.

> **Tip**
>
> In the **Existing Production Versions** tab on the initial screen of the app, you can use the **Overall Check Status** dropdown menu to view all production versions, production versions with successful checks, those with warnings, those with errors, and those that haven't been checked yet.

Process Serial Number Groups (F6773)

Application Type: Transactional

This app enables the collective processing of multiple serial numbers as a group at one or more operation activities within the produce segment type. This app is especially useful for standardized serialized products that undergo similar fabrication processes. You can perform actions once for the entire serial number group, and the system automatically records the data for each serial number individually. The app can be accessed by executing a serial number group from the My Work Queue app (F3499).

Key features of the app include the ability to process serial number groups at one or more operation activities; display the status, size, and details of each serial number in the group; and manage the workflow by starting, completing, pausing, or resuming the group. You also can skip or scrap the group, set and release holds, and register production resources/tools for the group. The app also allows you to record inspection characteristics, record defects, and approve buyoffs for the serial number group. You also can view any related documents, ensuring efficient and streamlined processing of serialized products.

Product Genealogy (F2892)

Application Type: Transactional

This app provides comprehensive traceability of serialized materials in a manufacturing environment. It allows you to track the complete manufacturing history of a serialized product, offering forward and backward traceability from the main material to subassemblies and vice versa. The app supports quality and production processes by displaying a detailed history of operation activities such as components assembled, data collected, and defects recorded during each operation. This visibility aids production supervisors and quality engineers in identifying and addressing critical quality issues. The app includes key features such as searching and displaying serialized materials using filters like serial number, material, production plant, production status, and so on. You can drill down into detailed information, including production holds, operation activities, assembled and disassembled components, defects, and inspection characteristics. The app also provides insights into documents and attachments linked to production orders, as well as variant configuration details for configurable materials. You can even print the product genealogy for a serialized product, which allows you to better analyze and troubleshoot the manufacturing process.

Production Action Log (F2869)

Application Type: Transactional

This app provides detailed monitoring of every action performed during a production order. It allows you to filter using various criteria such as production order, operation, material, serial number, work center, and action to view specific actions taken during production. These actions can include tasks such as setting holds, recording defects, uploading or removing attachments, completing operation activities, and so on. Each action is logged with additional information, including reason codes or notes entered by the production operator, ensuring transparency over the actions carried out. The app allows you to display detailed information for each action, including the contact data of the operator who performed the action.

Production Execution Duration (F2172)

Application Type: Analytical

This app is used to compare planned versus actual operation durations for manufacturing execution. It provides KPIs such as **Processing Duration: Planned vs. Actual**, displayed in both column and bubble charts, allowing you to identify deviations in execution times. You can filter data based on material, planning plant, production plant, work center, and so on while also making use of features such as **Jump To**, drilling up/down, toggling label visibility, showing the legend, and switching between chart and tabular views. The app uses CDS view C_MFGORDEROPEX-ECDURNQRY to retrieve data.

Production Order Confirmation Object Page (F2265)

Application Type: Fact Sheet

This app provides contextual information for production order confirmations at the order level. Because it serves as an information hub, it does not have a dedicated tile and can only be accessed through another app. The app allows you to view detailed confirmation data, including quantities and confirmation texts, while also offering navigation to related business objects and applications. More importantly, the app is useful for tracking and analyzing production order confirmations.

Production Order Object Page (F2261)

Application Type: Fact Sheet

This app provides detailed contextual information about production in SAP S/4HANA. This app does not have a dedicated tile but can be accessed through another app. The app allows you to view and navigate related objects. It displays a list of production orders with detailed order information, including items, serial numbers, operations, components, order status, contacts, production resources and tools (PRTs), and so on. The app is essential for tracking, analyzing, and managing production orders across different industries.

Projected On-Hand Alerts (F6243)

Application Type: Transactional

This app is used for monitoring inventory levels and identifying potential stock issues in a demand-driven replenishment environment. It functions similarly to the Monitor Demand-Driven Replenishment app (F2831), allowing you to track projected stock levels and receive alerts when inventory is at risk of falling below critical thresholds. The app helps you proactively manage stock shortages or excesses by analyzing demand trends and projected supply. It enables decision-makers to take preventive actions before stockouts or overstock situations occur.

When used in conjunction with the Replenishment Planning (F6241) and Replenishment Execution (F6242) apps, it provides a holistic view of demand-driven inventory management and improves real-time stock visibility.

Replenishment Execution (F6242)

Application Type: Transactional

This app helps facilitate the execution of demand-driven replenishment by converting replenishment plans into actionable supply elements. It allows you to filter materials based on the on-hand stock status (such as stock levels above reorder point, below on-hand alert threshold, below, safety stock, below or at reorder point, or out of stock with demand) as well as the procurement type and other filter criteria. The app provides an overview of buffers, displaying key details such as product, description, on-hand buffer status, open supply, and execution status. You can review replenishment proposals and take necessary actions, such as triggering purchase orders, stock transfers, or production orders to ensure inventory availability.

This app works in conjunction with the Replenishment Planning (F6241) and Projected On-Hand Alerts (F6243) apps to maintain optimal stock levels in demand-driven replenishment.

Replenishment Planning (F6241)

Application Type: Transactional

This app is used for demand-driven replenishment based on planning priorities. It allows you to monitor stock levels and take proactive actions to avoid shortages or overstocking. The app allows you to filter and analyze materials using filter criteria such as planning priority status, including conditions such as stock levels above maximum stock, at zero or below with demand, below reorder point, below safety stock, or at maximum stock. The app provides an overview of buffer stock levels, displaying key details such as product, planning priority, net flow position, proposed replenishment quantity, and recommended planning actions. You can take corrective actions via the available functions such as viewing logs for analysis or triggering replenishment through the **Create Supply** button.

It is important to note that the Replenishment Planning (F6241), Replenishment Execution (F6242), and Projected On-Hand Alerts (F6243) apps are related: They are all demand-driven replenishment apps.

Reprocess Failed Material Movements (F3100)

Application Type: Transactional

This app helps production supervisors efficiently manage and correct failed material movements. It allows you to display a list of unsuccessful material movements along with detailed error messages, allowing you to identify the root cause of a failure. You can filter and search for failed material movements based on criteria such as material, plant, application area, message number, and so on. Upon selecting a failed material movement, an exception item view provides key details, including the material, plant, movement type (e.g., goods issue for order [261]), posting date, quantity, batch, and associated error message. The app offers navigation options to investigate and resolve errors by directing you to relevant master data, revision levels, or availability monitoring. You also can take corrective actions by replacing storage locations, batches, or posting dates before reprocessing the material movement.

Resource Object Page (F2264)

Application Type: Fact Sheet

This app provides detailed and contextual information about a resource. Unlike traditional apps, it does not have a dedicated tile on the SAP Fiori launchpad; instead, it is accessed through related applications. The app allows you to view resource capacity and track assigned operations in various stages, including today's scheduled operations, operations in progress, queuing operations, and planned orders. A colored bar highlights delayed tasks, helping you identify bottlenecks. The app also allows you to define and save custom views.

Routing Signoff Task (F4076)

Application Type: Transactional, Reuse Component

Similar to the Routing Version Signoff app (F3768), this app is used to create and manage version-controlled routings for production processes, ensuring that the sequence of operations needed to produce a product is clearly defined. It includes details such as where the work takes place, the resources used at each work center, and key technical specifications like standard times, capacity requirements, and work instructions. You also can use operation activities to further describe the production process. Once a routing version is released, it becomes part of the production execution process and cannot be changed. This app helps maintain accurate and consistent routing information throughout the production process, but it is not always available, and its use depends on the specific system configuration.

Routing Version Signoff (F3768)

Application Type: Transactional, Reuse Component

This app is used to create and manage version-controlled routings for production processes. A *routing version* defines the sequence of operations needed to produce a product, including details like where the work happens, what resources are used, and technical specifications (such as standard times and capacity requirements). It also allows for describing the production process with more detail through operation activities. Once a routing version is released, it can no longer be changed and is used in production execution. The app helps ensure that production follows the correct steps and specifications. However, its availability depends on the system setup, and it might be accessed through other related apps or components.

Schedule Buffer Proposal Calculation (F2837)

Application Type: Transactional

This app is used to manage and monitor background jobs related to demand-driven replenishment. It allows you to generate and update buffer (stock) level proposals for products based on key factors such as average daily usage, decoupled lead time, and buffer profiles, which help to maintain optimal inventory levels. You can create, schedule, restart, and track jobs for recalculating buffer proposals. Jobs can be filtered by status—such as failed, canceled, finished, ready, scheduled, or in process—as well as by date range. One of the critical features of this application is the ability to automate the periodic recalculation of buffer proposals. Because buffer levels fluctuate based on changing demand patterns, recalculating them regularly ensures that inventory remains aligned with actual consumption needs. SAP recommends performing these updates frequently to maintain accurate replenishment planning.

Schedule Copying Total Forecast Runs (F2579)

Application Type: Transactional

This app is designed to manage and monitor background jobs for copying total forecast data using a predefined job template. It provides the ability to create, schedule, and oversee forecast-copying processes efficiently. You can create and schedule jobs for copying the total forecast, modify job parameters according to their business needs, and take actions such as restarting, canceling, or copying existing jobs. You can view detailed logs and results for each scheduled job, check execution steps, and track job history. The ability to monitor job statuses—such as scheduled, in process, failed, or completed—enables proactive issue resolution. You can duplicate a previously scheduled job, making necessary

adjustments while retaining all job content. More importantly, the app provides a clear view of job details, including status, log, results, execution steps, creator information, and so on.

Schedule Kanban JIT Call Output (F3950)

Application Type: Transactional

This app allows you to schedule background jobs for triggering the output of summarized just-in-time (JIT) calls for which the dispatch time is set to scheduled. This dispatch time is determined based on the output control settings. The app is designed to work in combination with output management and enables the automation of Kanban JIT call processing across output channels. You can create and schedule jobs from the **Template Selection** screen, where you select a job template and configure scheduling options to start the job immediately or plan a later execution. You also can set up recurrence schedules for jobs, ensuring that output processing is continuously automated as per business needs. To group summarized JIT calls, the scheduled dispatch time must be utilized.

You can define multiple screen parameters before scheduling a job. Under **Output Data**, you can specify the output type, output channel, and sorting preferences when multiple Kanban-summarized JIT calls are scheduled for output. Under **Processing Mode**, you can select between **First Processing** (initial output), **Repeat Processing** (duplicate and send copies of successful outputs), or **Error Processing** (retry failed outputs). The **Summarized JIT Call** section includes a mandatory plant field, with additional filtering options available. Once scheduled, the jobs list displays planned and executed jobs along with their status and logs. You can filter jobs by various criteria such as date, job ID, run ID, time, description, or creator ID. Filtering options also can be saved as variants for efficient job management. If you need to create a similar job, you can copy an existing job and edit it, retaining all content, including recurrence settings.

Schedule Lead Time Classification of Products (DD) (F2871)

Application Type: Transactional

This app helps you classify demand-driven products by evaluating them based on their decoupled lead item (DLT) using EFG classification across a specified evaluation interval. This classification helps define inputs for buffer settings in demand-driven replenishment. You can schedule periodic reclassification runs to ensure accurate and up-to-date classifications, accommodating seasonal changes, logistics variations, or demand fluctuations. Custom thresholds also can be set for DLT classification based on procurement types such as make, buy, and transfer. The app provides filters to select specific products, define an evaluation interval in days, and ensure the right buffer settings are considered for demand-driven replenishment.

Lead time classification results serve as inputs for the Schedule Buffer Proposal Calculation app (F2837).

Schedule MRP Rescheduling Checks (F5417)

Application Type: Transactional

This app allows you to create and schedule jobs for collecting MRP rescheduling checks related to purchase orders, which ensure timely adjustments based on material planning needs. You can automate the periodic replication of rescheduling checks, with results appearing as proposals in the Create MRP Change Requests app (F5416). The app provides options to create, copy, and cancel jobs, which helps in procurement and MRP processes.

Schedule MRP Runs (F1339)

Application Type: Transactional

This app allows you to create and schedule jobs for executing MRP runs, automating the planning process and reducing annual workload. The app enables regular MRP runs, checking if current material demand is covered by inventory and expected receipts. In regenerative planning, all materials are planned, while net change planning only plans materials that have changed since the last run. If demand is unmet, the system generates planned orders for in-house production or purchase requisitions for external requisitions for external procurement. You can define parameters such as job description, start date, and recurrence, as well as job monitor statuses, including logs and duration.

Schedule Order Conversion Runs (F1718)

Application Type: Transactional

This app allows you to create and schedule jobs for converting planned orders into production or process orders. Planned orders, generated through MRP, represent an internal demand for production and include details such as basic dates, material components, and required quantities, which are copied directly upon conversion. The app also ensures that dependent requirements for components are transformed into reservations. You can monitor job statuses, view logs and results, and cancel jobs if necessary. This app is designed for production planners to effectively perform order conversion to improve production planning efficiency.

Schedule Order Release Runs (F1719)

Application Type: Transactional

This app allows you to create and schedule jobs for releasing production or process orders at the header level, to ensure that an order and all its operations are ready for processing. You can monitor job statuses, view logs and results, and cancel jobs if needed. In essence, the app supports production supervisors in both discrete manufacturing and process manufacturing industries.

Schedule PIR Reorganizing Runs (F2580)

Application Type: Transactional

The app allows you to schedule, monitor, and track the reorganization of planned independent requirements (PIRs) as background jobs. Using predefined job templates, you can adjust requirements, delete outdated PIR records, or clear historical data. The app provides tools for scheduling, monitoring, and filtering job runs to streamline PIR management efficiently.

Schedule pMRP Simulation Creation (F3968)

Application Type: Transactional

This app allows you to schedule the creation of predictive material and resource planning (pMRP) simulations. The app simplifies MRP data, which is then used as reference data for creating multiple pMRP simulations. You define reference data for simulations, which can be copied for further comparisons and adjustments to analyze the impact of potential changes on delivery performance or capacity. The app allows you to simulate capacity issues, make changes, and compare simulations before implementing changes in the actual system. In essence, you define and schedule the creation of simulations based on work center, top level, and so on; easily duplicate simulations based on reference data to compare different scenarios; and simulate changes to capacity, work centers, or material requirements to assess their impact on delivery or capacity.

Schedule Processing of MRP Change Requests (F5725)

Application Type: Transactional

This app allows you to create and schedule jobs to automatically apply or discard MRP change requests that are in **Requested** status, for which no response has been received from suppliers. You can schedule this automatic process based on change request priorities or can manually specify the desired status for matching requests, all according to their configuration settings.

For instance, you can use the SAP_PP_MRP_AUTO_ HANDLE_CHANGE_REQ job (Schedule Processing of MRP Change Requests).

Schedule Product Classification (DD) (F2823)

Application Type: Transactional

This app helps classify products for demand-driven replenishment by evaluating them via ABC, PQR, and XYZ classifications based on various factors, such as goods issue value, bill of materials (BOM) usage, and demand variation. You can schedule periodic reclassification job runs to ensure product classifications remain accurate and up to date. This classification is essential for determining the right buffer settings and deciding whether products are relevant for demand-driven replenishment. Furthermore, you can set thresholds for each product classification. To put this in context, consider a manufacturing company that produces electronic components. The ABC classification is used to categorize products based on their goods issue value, identifying high-value items (A), medium-value items (B), and low-value items (C). These classifications help the business to prioritize inventory to ensure that high-value items are always in stock and available for production, while low-value items have lower stock levels to reduce carrying costs. Let's see how ABC, PQR, and XYZ classifications are used in this context:

- ABC: Evaluate products based on their goods issue value
- PQR: Evaluate products based on their usage across BOMs
- XYZ: Evaluate products based on the variation in their actual demand

---

Tip

For instance, you can use the SAP_SCM_DD_PROD_CLASFY_DEFAULT job template (Product Classification for Demand-Driven Replenishment).

Schedule Sending of MRP Change Requests (F5418)

Application Type: Transactional

This app allows you to schedule the sending of MRP change requests to external systems using the MRPChangeRequest_Out SOAP API. The app allows you to create, copy, and manage jobs by sending these requests to suppliers, while providing detailed information about job statuses, logs, results, and planned start times. You can use the SAP_PP_MRP_CHANGE_REQUEST_TRANSFER job (Sending of MRP Change Request).

Schedule Unassigning of MRP Areas (F5447)

Application Type: Transactional

This app allows you to schedule and manage the deletion of material assignments to MRP areas as a background job. By using a job template, you can select materials marked for deletion (with the deletion indicator set in Transaction MM02) and remove their assignments along with any dependent data. You can use the SAP_PP_MRP_UNASSIGN_MRPAREAS job template (Schedule Unassign of MRP Areas). The app provides options to create, schedule, copy, and manage jobs; view job statutes; and access detailed logs

and results. It also offers a simulation mode to preview changes before deletion.

Scrap Reason (F2216)

Application Type: Analytical

This app helps production engineers track and analyze scrap and rework percentages across work centers. It provides real-time KPIs and visual reports to identify root causes of production losses and improve efficiency. It features a **Smart Business KPI** tile that highlights work centers with the highest scrap and rework—for instance, in the last 24 hours—and allows you to analyze scrap trends based on time, material, plant, and variance reasons. The app provides dynamic charts and automated calculations of scrap, yield, and rework percentages, with filters for selecting specific performance periods. The app has a **Jump To** option and **Show Mini Tiles** (**Material Scrap**, **Operation Scrap**) options to navigate to related apps and dive deeper into scrap and rework data, which ultimately helps production teams to reduce waste and improve efficiency. The app uses CDS view C_MFGORDOP-CONFSCRAPREASONQ to retrieve data.

Set Kanban Container Status (F3717)

Application Type: Transactional

This app allows you to update Kanban container statutes or withdraw quantities, triggering automatic status changes. You also can scan or manually enter containers, which first appear in the entered list, allowing adjustments before saving. In fact, you can toggle between the **Entered** and **Saved** lists (both lists display a number in brackets). Status changes (e.g., **Full**, **In Use**, or **Empty**) and quantity withdrawals can be modified, including batch numbers and target quantities, if needed. Invalid containers are flagged with error messages for correction. Once confirmed, all changes are saved, and containers move to the **Saved** list, helping you effectively perform Kanban inventory management.

Track Tool Usage (F4762)

Application Type: Transactional

This app allows you to track the availability of equipment. It lists equipment with information such as its serial number and claim type (user, work center, operation activity). For specific listed equipment, you can see information about who has claimed the equipment and where it has been claimed. For instance, you can claim, unclaim, hold, or release equipment for use. You have the option to view the equipment list based on the selection criteria, which include material, serial number, system status, unique item identifier, claim type, and so on. The app has numerous key features, which include but are not limited to:

- Track equipment availability and location: See which tools are in use, who has claimed them, and where they are located.

- Claim and unclaim equipment: Assign tools to a user, work center, or operation activity, or release them when no longer needed.

- Search and filter equipment: Find tools using filter criteria such as material, serial, number, claim type, and so on.

Work Center Application Logs (F5931)

Application Type: Transactional

This app tracks and lists logs, detailing the logs for scheduled jobs related to the Mass Maintenance of Work Center Capacities app (F5381). The app logs all activities from the Mass Maintenance of Work Center Capacities app and can be accessed via the **Logs** option that app. You can filter logs based on severity (**Canceled, Error, Warning, Success**), date range (from/to), category, subcategory, and so on. This ultimately helps you get a clear record of job execution, tracking errors, warnings, and successful changes in work center capacities.

Work Center Capacity Application Jobs (F5939)

Application Type: Transactional

This app is used to track and manage jobs scheduled from the Mass Maintenance of Work Center Capacities app (F5381). This transactional application helps monitor and manage work center capacity planning jobs, allowing you to view and track the status of the scheduled jobs related to work center capacity maintenance.

Work Center Object Page (F2262)

Application Type: Fact Sheet

This app provides detailed insights into a work center and its related operations. It helps you manage production by displaying key information and linking it to relevant apps such as Manage Production Operations (F2335) and Manage Production Orders (F2336). For instance, for a specific work center, you can view work center details, accessing key data such as capacity and assigned operations, as well as tracking the operations status by monitoring operations based on their progress.

Chapter 6
Plant Maintenance

In this chapter, we'll cover plant maintenance (also referred to as enterprise asset management [EAM]) apps, which support preventive, corrective, and predictive maintenance processes. These apps cover key features such as maintenance requests, maintenance planning, maintenance notification handling, cost analysis, work permit management, job scheduling, reporting, and so on. Where applicable, we've listed a comparable SAP GUI transaction. Apps are listed in alphabetical order.

Actual Cost Analysis (F3567)

Application Type: Transactional

This app helps you track, evaluate, and understand actual costs arising from ongoing maintenance orders. It provides detailed insights into material and labor expenses; you can assess corrective and preventive maintenance costs across a selected period. You can filter and analyze cost data using various criteria, focusing on key dimensions such as cost element hierarchy, ledger, company code, maintenance plant, catalog profile, construction type, and many more. In fact, you can filter by important criteria such as work center, plant section, general ledger account, controlling area, fiscal year, order type (e.g., maintenance orders), planning indicator (immediate, planned, or unplanned orders), and so on, enabling you to drill into the most relevant cost drivers and trends.

The app provides multiple options for working with the data. You can choose to view details, switch between **View By** modes, select **Show All Items**, select **Display Items Based On Chart Selection** (icon), or make use of visual and compact filters. The main content area offers flexible viewing options where you can toggle between a chart view and table view, or opt for a hybrid view combining both. In the chart view, you can easily explore aggregated cost data, adjust chart types, customize settings, and use drilldown functionality to view costs from different grouping levels. In the table view, you have direct insight into maintenance orders and can tailor the table layout to display dimensions such as company code, maintenance order, ledger, planning indicator, and more depending on the table settings.

> **Tip**
>
> The `EAM_ORDER_ACTUALCOST_MONITOR` OData service should be maintained and activated for the app to function properly.

Create Maintenance Request (F1511A)

Application Type: Transactional

This app allows you to create maintenance requests for technical objects like equipment or functional locations by entering key issue details. Drafts are autosaved if not submitted and can be managed later in the My Maintenance Requests app (F4513). You can explore object hierarchies, scan barcodes for quick identification, and enter failure-related info. Assigned task lists copy operational data into the maintenance order if authorized. You can add long texts using templates and record details like malfunction dates, priority (manual or risk-based), and reporter info. The app requires inputs for both technical and general data, supports document attachments, and organizes them by type. All requests appear in related maintenance request apps. The initial screen includes the following main sections:

- **Technical Data**
  Enter the technical object, location, and malfunction details, including failure mode group, detection method, and effect on production. Options range from no effect to full breakdown. Also indicate if a breakdown occurred and record the malfunction start date and time.

- **General Data**
  Enter the notification context (e.g., emergency or minor work) and a brief description, and optionally select a standard template. A long text field captures detailed issue info, including start time and actions taken. Priority (very high to low) must also be assigned based on urgency.

- **Responsibilities**
  Enter the creator, reporter, and date and time of reporting to ensure traceability and accountability.

- **Attachments**
  Upload, link, or assign documents like design drawings or compliance files here. Attached files can be sorted by name, uploader, size, status, or type for easy management and retrieval.

Finally, after entering all details, you can choose to create a draft or submit the request. The **My Drafts** option displays all the request drafts that have been created but not yet submitted.

> **Tip**
>
> The **My Drafts** option navigates you to the My Maintenance Requests app (F4513), where you can further work on the requests.

Create Mass Time Confirmations (F3925)

Application Type: Transactional

This app allows you to view and post the status of uploaded time confirmations into the system. It becomes accessible only after confirmations have been uploaded through the Find Maintenance Orders and Operations app (F2173). Following the upload, an informational popup appears with a hyperlink, formatted as *Mass-Conf_USERNAME_YYYYMMDD*, which directs users to this app for further processing. You can also access the app using the **Add Time Confirmation** option. Key features include the ability to track the status of time confirmations and efficiently post them into the system, ensuring accurate and timely recording of maintenance activities.

> **Tip**
>
> This app is closely linked to the Find Technical Object (F2072) and Find Maintenance Order (F2175) apps. These apps can also be used together when addressing an unexpected equipment failure. A technician might first use the Find Technical Object app to locate the specific equipment experiencing the issue. After identifying the object, they would then create a maintenance request. A maintenance order can be created from a request, and once it's successfully created, the Find Maintenance Order app allows you to quickly locate and monitor the maintenance order, ensuring timely resolution and visibility into ongoing work.

Create Work Permit Request (F4691)

Application Type: Transactional

This app is designed to help maintenance planners and permit requestors efficiently create and manage safety work permits. You can initiate permits from scratch or generate them by referencing existing permits, templates, or maintenance orders. To begin, you are prompted to enter essential information such as the planning plant, work permit type, and any reference documents, which can be done from the initial screen or in a popover window. The permit creation process is structured across multiple tabs that cover general information, technical objects, orders, partners, and more.

A key strength of the app is its ability to capture comprehensive safety-related details. You can specify the nature of the work, link technical objects like equipment and functional locations, and include additional orders or operations. To reinforce safety measures, the app supports the assignment of personal protective equipment, safety precautions, and relevant safety certificates.

> **Tip**
>
> The UI_WORKPERMIT (Work Permit) service should be activated and maintained for this app.

Display Serial Numbers (F5147)

Application Type: Analytical

This app provides a clear and organized view of serial numbers assigned to materials, based on the most recent goods movements. You can filter, sort, and search the serial number list using criteria such as material, company code, date of last goods movement, work breakdown structure (WBS) element, plant, storage location, stock type, special stock, batch, and more, making it easy to locate specific entries. You also can export the data for external analysis or reporting purposes.

The app is accessible through the Stock – Single Material app (F1076).

Find Maintenance Items (F3621)

Application Type: Transactional

This app allows you to efficiently search for and review maintenance items by applying a variety of filters: ABC indicator, accounting indicator, activity type, actual revenues, allocation methods, asset class, CAD indicator, and more. You also can perform free-text searches to quickly locate specific maintenance items. Once an item is selected, detailed information about it is displayed, including relevant technical and planning data. The app supports navigation to associated maintenance plans or technical objects, making it easy to access related details. You can export the list of filtered maintenance items into a spreadsheet for reporting or further analysis and access linked apps directly from each item for additional processing.

Find Maintenance Notification (F2071)

Application Type: Transactional

This app provides a comprehensive way to search, view, and manage maintenance notifications in the SAP S/4HANA system. You can apply filters such as notification type, maintenance order, technical object, main work center, person responsible, and notification status (**Completed, In Process, Postponed**, etc.) to narrow down results. The list displays key details for each notification, including the notification ID, technical object, priority, breakdown indicator, and creation date. From the results list, you can quickly navigate to the **Maintenance Notification** screen, which displays in-depth information such as planner group, planning plant, work center, notification codes, malfunction details, tasks, activities, attachments, and more.

The app offers several options for taking action, such as assigning notifications to orders, creating new orders, changing responsibility or scheduling, mass editing notifications, setting or unsetting statuses, and exporting data. You can perform actions on one or multiple notifications simultaneously, and mass-editing features allow for bulk updates across selected notifications, with a simulation step to identify errors before changes are applied.

Tip

For mass-editing notifications, you can edit the following sections: **General Data**, **Reference Object**, **Deadline Data**, **Breakdown**, and **Notifications**.

The app keeps a history of changes to the maintenance notification, detailing the edit type (e.g., attribute value change or system status change), short description, details (e.g., the field name that has changed and its old value and new value), and the individual who made the change.

Find Maintenance Order Confirmation (F2174)

Application Type: Transactional

This app allows you to efficiently view and manage maintenance order confirmations through a searchable and filterable interface. You can refine the view using filters such as order number, confirmation status, final confirmation (yes or no), work center, technical object, order status (e.g., released, completed for business), suboperation, posting date, and personnel number. The app displays a comprehensive list of confirmations with key details like order, status, operation, suboperation, and more, all adjustable via table settings. You can copy entries, export the data to spreadsheets, and drill into individual confirmation records to review specifics such as activity type, confirmation text, wage type, personnel involved, and execution timelines. The app also highlights operation status codes like **REL (Released)**, **PCNF (Partially Confirmed)**, and **MCNF (Manually Confirmed)**.

For further actions or deeper insights, you can use the provided hyperlinks and the **Related**

Apps feature to navigate to associated maintenance orders or technical objects.

Find Maintenance Orders (F2175)

Application Type: Transactional

This app allows you to efficiently manage and perform mass changes on maintenance orders by selecting numerous orders and then selecting the **Edit Orders** option. You can search for and filter maintenance orders using filters such as technical object, main work center, order status (e.g., released, marked for deletion, or completed for business), order type, processing context (e.g., **Standard Order** or **Emergency Order**), maintenance item, and more. The app displays a detailed list of maintenance orders with key fields like order ID, order type, notification, and priority, and with the table settings tailored to your preferences. You can execute a range of actions, such as changing responsibilities, updating order statuses, adjusting scheduling, dispatching operations, creating work packs, and copying, printing, and editing orders. The app enables bulk edits for several orders, allowing for transitions for statuses like release, technical completion, business completion, or deletion, with advanced options available for phase-based orders. Work packs can be generated from selected orders by highlighting the work pack name, work pack type, and grouping criteria in the **Create Work Pack** popup screen. Moreover, the app supports reviewing cost data (e.g., estimated and actual), tracking final due dates with historical changes, managing attachments and change logs, exporting order data to Excel, and identifying orders that include task lists.

Specific maintenance order view information is organized into the following tabs:

- **General Information**
  Captures key maintenance order details, including the notification ID, task list, maintenance plan, long text, work center, planner group, responsible person, and critical dates. It also includes equipment, location, and object hierarchy info for full order context.

- **Operations**
  Lists all operations and suboperations with details like ID, description, and control key. You can export the data to Excel for planning or analysis.

- **Organizational Data**
  Shows how the order fits into the organizational and project structure, including company code, business area, cost center, profit center, and project details like work breakdown structure (WBS) element and project definition.

- **Account Assignment**
  Details the maintenance order's financial allocation, including company code, cost center, settlement order, asset, and WBS element if applicable.

- **Costs**
  Lets you track maintenance order finances, comparing value category, estimated, planned, and actual costs for improved budgeting and control.

- **Status**
  Shows the maintenance order's system and user statuses, including IDs and descriptions, to indicate its current progress.

- **Original Files**
  Access maintenance order attachments like documents, images, diagrams, or external references.

- **History**
  Tracks maintenance order changes, including edit type, details, user, and old/new values, ensuring accountability.

Tip

For multiple order selection, you can change the following details simultaneously:

- General data—description, maintenance activity type, priority, system condition
- Reference object—equipment, functional location, material, serial number
- Location data—location, plant section, ABC indicator
- Additional data—responsible cost center

- Responsibilities—main work center, planner group, main work center plant, person responsible

You can change the responsibility using the **Change Responsibility** option. You will need to provide the main work center and main work center plant as key details in the popup screen, with options to add the planner group and person responsible.

Find Maintenance Orders and Operations (F2173)

Application Type: Transactional

This app allows you to search and manage maintenance orders and their operations using various selection criteria such as order type (e.g., refurbishment or maintenance order), specific order, execution, priority, scheduled start date, main work center, final due date, technical object, operation subphase, and processing context (standard order, emergency order). You also can view and utilize logs using the **Application Logs** option. You can perform several actions through options, including changing assignments by specifying the work center, plant, and responsible person; adding time confirmations with details like actual work duration, personnel number, posting date, and final confirmation status; creating work packs by naming and grouping them; printing; changing the order status (releasing, scheduling, completing, canceling technical completion, setting or resetting deletion flags); editing operations via a popup that allows modifications to general, internal, and external data with options to save or simulate changes; and exporting data to a spreadsheet (from the initial screen).

Time confirmations can be added manually or via upload: You can download a template Microsoft Excel file; fill it out with the order details, employee info, activity type, posting date, and confirmation; and then upload it back with the option to post immediately when adding time confirmations.

For each maintenance order operation, you can drill down to detailed information including long texts, order type, work center, control keys, purchasing details, planned schedules, materials, production resources/tools, system and user statuses, and attachments.

Tip

When creating a work pack, you can choose to include operations from other work packs.

Find Maintenance Plans (F3622)

Application Type: Transactional

This app allows you to efficiently search for and manage maintenance plans using a search bar or by applying various filters such as account group, validity dates, maintenance plan ID, plan type, maintenance strategy, and more—customizable based on your preferences, set through the **Adapt Filters** option. Upon selecting a specific maintenance plan, you can view key details such as the plan's category, maintenance type, call object, and status (e.g., **Active**). The **Basic Data** section includes long text descriptions, sort fields, authorization groups, and administrative information like the creator and last change date. The **Items** section lists associated maintenance items, detailing the item ID, technical object, planning plant, order type, and main work center, with the option to export to Microsoft Excel. In the **Planning Data** section, you can view scheduling parameters, including start and end dates, date determination rules, completion details (like shift factors), call control, and cycles—which displays the occurrence, cycle text, offset, counter statuses, and more. The **Maintenance Calls** tab lists all generated calls, indicating call ID, status, scheduling type, due packages, and manual entries. The app supports navigation to related apps for extended review and processing, enabling a comprehensive and flexible approach to maintenance plan management.

Tip

You also can access a customizable **History** tab that displays audit trails such as edit types, details, who made changes, and when using the standard dropdown functionality.

Find Maintenance Task List (F2660)

Application Type: Transactional, Fact Sheet

This app allows you to search for and view various types of maintenance task lists—for instance, general maintenance task lists. You can locate task lists by using the search bar or applying filters such as key date (used as a key field), task list group, task list group counter, work center, maintenance strategy, planning plant, task list type (e.g., general maintenance task list), and status (e.g., created, released, or released for order). Once a task list is selected, you can view essential details such as status, planning and work center plant, planner group, quality management data, administrative data, and validity dates. The app also allows you to explore related maintenance operations, displaying information like operation ID, suboperation, control key, and assigned work center. You also can view assigned maintenance packages and access any attached files.

For extended maintenance planning and review, you can navigate directly to related applications through the **Related Apps** option, making the task list management process wholly integrated.

Tip

For attachments in the **Maintenance Task List** screen, you can sort documents in ascending or descending order by file name, uploaded by, file size, status, and document type.

Note that hierarchical task lists cannot be accessed using this app.

Find Maintenance Task List and Operation (F2661)

Application Type: Transactional, Fact Sheet

This app allows you to search for and display different types of maintenance task lists, including equipment task lists, functional location task lists, and general maintenance task lists. You can find the task lists by filtering based on criteria such as planning plant, work center, and key date, task list type, status, technical object or execution stage (**MAIN [Maintaining/Repairing]**, **POST [Post Execution]**, **PRE [Preliminary]**), and more. Similar to the Find Maintenance Task List app (F2660), hierarchical task lists are not accessible in this app. You can view individual task lists to see their properties, assigned operations, and attachments.

Tip

For attachments in the **Maintenance Task List** screen, you can sort documents in ascending or descending order by file name, uploaded by, file size, status, and document type.

Find Technical Object (F2072)

Application Type: Transactional

This app allows you to search for, view, filter, and manage a comprehensive list of technical objects in the system. Through the search bar and adaptable filters such as location, main work center, technical object type (e.g., equipment or functional location), plant section, system status, company code, manufacturer, object type, and planner group, you can easily identify specific technical objects. The technical objects are displayed in a table view, showing key fields like object type and ID, with options to customize settings, show more or fewer rows, export to Microsoft Excel, or copy and change object statuses. You also can perform mass edits and manage the hierarchical structure of technical objects, including installing or dismantling

subordinate equipment with historical accuracy (past or current dates). Once you select a technical object, you will see important details of that object, including its category, type, current calls, system status, linked maintenance notifications and orders, service orders, maintenance items, historical changes, and attachments. The app also provides visibility into planning-related data such as the planning plant, planner group, and work center, along with user and system statuses.

More importantly, after installing subordinate equipment, you can view the results in logs using the **Application Logs** option, where you can view, for instance, the log details—including severity, description, time stamp, level of detail, and more.

Tips

Using the **Manage Structure** option, you can display, install, or dismantle subordinate equipment. Using the **Change Status** option, you can activate and deactivate objects, as well as set a deletion or reset deletion flag.

Maintenance Backlog Overview (F5105)

Application Type: Analytical

The app is used to provide a comprehensive snapshot of an organization's maintenance workload across various planning stages. You can apply filters such as the reference planning bucket (as a key field), past or future buckets, and specific phases (e.g., initiation, screening, planning, approval, scheduling) to refine the view. The app features multiple cards, each offering distinct insights. The **Maintenance Backlog** card displays KPIs segmented by phase, maintenance activity type, order and notification type, and priority, enabling you to identify bottlenecks or overdue tasks. The **Maintenance Orders** card summarizes open orders requiring attention. The **Quick Links** card provides direct access to related applications, including Find Maintenance Notifications (F2071), Manage

6

Maintenance Notifications and Orders (F4604), Find Maintenance Orders and Operations (F2173), and Find Maintenance Orders (F2175). Additional cards like **Rework Orders** and **External Procurement** offer further insights into specific order categories, helping maintenance planners monitor and act on items requiring rework or external sourcing.

> **Tip**
>
> The app supports drag-and-drop functionality for the cards on the initial screen. You can drag and drop the cards from and to the left, center, or right section, and they will rearrange automatically.

Maintenance Order Costs (F4603)

Application Type: Analytical

This app allows you to monitor and evaluate maintenance-related costs such as estimated, baseline, planned, and actual costs across ongoing maintenance orders. It is especially useful for comparing planned or baseline costs against actual expenditures within a specific period and for particular order types. You can apply a wide range of filters to refine the analysis—including relative date functions, general ledger account hierarchy, ledger, company code, cost center, controlling area, object type, serial number, equipment, assembly, product, and priority. The app also provides the flexibility to view cost data by various dimensions, such as ABC indicator, task list, cost center, equipment, functional location, and more, using the **View By** option.

You can choose from a chart, table, or combined chart-and-table view, with options to switch chart types, open the view in full screen, zoom in and out, or display a legend for clarity. In the table view, you can perform actions such as exporting data to Excel, copying, showing all line items, and maximizing the table for detailed analysis. The app supports analysis of phase-based maintenance orders by enabling cost tracking across subphases and phases.

Maintenance Planning Overview (F2828)

Application Type: Analytical

This app helps you plan, execute, and monitor critical maintenance by highlighting time-sensitive tasks and delays. You can customize views with filters like reference period, work center, planner group, and notification type. It identifies key issues such as unassigned notifications, missing parts, and overdue orders, presenting results as visual cards and charts. You can drill down into lists or contact responsible parties. The app flags unreleased orders, approved requisitions without purchase orders, and released orders with expired end dates that remain open, ensuring timely maintenance management. The following cards are available:

- **Missing Components**
  Highlights nonstock items that are required for maintenance activities but are currently unavailable. Navigates to the Procurement for Maintenance Planner – Purchase Order app (F2827). The component missing filter will be autopopulated with the entry **Yes**.

- **Orders for Planning**
  Lists maintenance orders that have not yet been released. This allows you to monitor planning progress and navigate to the Find Maintenance Orders app (F2175) to further process the orders.

- **Purchase Requisitions Not Approved**
  Focuses on non-stock items and services that are still pending approval. It provides a link to the Procurement for Maintenance Planner – Purchase Order (F2827) app, where you can take appropriate approval actions.

- **Notifications for Screening**
  Displays outstanding maintenance notifications categorized, for instance, by user status. KPIs also can be viewed by priority, activity type, or notification type. You can access more details through the Find Maintenance Notifications app (F2071).

- **Overdue Orders**
  Identifies maintenance orders and operations that have exceeded their planned end

dates but have not been finalized/confirmed. You can follow up on these overdue items by navigating to the Find Maintenance Orders and Operations app (F2173).

- **Purchase Requisitions Not Converted to Purchase Orders**
Highlights nonstock items and services that have approved requisitions but are still awaiting conversion into purchase orders. You can address these by navigating to the Procurement for Maintenance Planner – Purchase Requisition app (F3065). When you navigate to this app from the card, the purchase order filter should be displayed as <empty>.

- **Orders for Completion**
Lists maintenance orders that are ready for final business confirmation or technical closure. These can be processed through the Find Maintenance Orders app (F2175) to ensure proper closure.

- **Purchase Orders Not Approved**
Calls attention to purchase orders related to nonstock items and services that are pending approval. It also provides access to the Procurement for Maintenance Planner – Purchase Order app (F2827) for further action.

- **Quick List**
Offers instant navigation to essential apps like Find Maintenance Notifications (F2071), Find Maintenance Orders (F2175), and more, allowing you to quickly locate and manage notifications and orders across the maintenance process.

Manage Inspection Checklists (F6607)

Application Type: Transactional

This app allows you to view and manage inspection checklists linked to maintenance order operations and technical objects. You can access and process results either at the overall inspection lot level or for individual inspection characteristics. The app allows you to search inspection checklists using free text or filters based on

maintenance orders, technical objects, and checklist details. It provides detailed views of inspection lots, including usage decisions and overall status, as well as specific technical and result data for each inspection characteristic. You can set usage decisions, activate or deactivate inspection lots, and even copy or replace them to perform additional inspections or make corrections without altering the original records. The app also supports submitting inspection lots for approval by locking them and triggering a customizable approval process, provided the relevant system features and authorizations are enabled.

Manage Maintenance Backlog (F4073)

Application Type: Transactional

This app allows you to view and manage all maintenance orders associated with a specific maintenance planning bucket. It presents orders across two main tabs: **All Orders** and **Orders in Preparation**. This allows you, for instance, to change scheduling parameters and submit maintenance orders for scheduling. You can view key order details such as the planning plant, technical object, start and end dates, order status, priority, and final due date. The app also provides insights into stock components, nonstock components, services, and resources. It allows you to monitor material availability and consumption. You can evaluate procurement progress for nonstock items and external services, track internal work center activities, and access application logs for historical insights. Additional features include navigation between current, previous, and next buckets; filtering for orders in preparation; visual indicators of order readiness; and the ability to reschedule or reassign orders across buckets.

This app is accessible through the Manage Maintenance Planning Buckets app (F3888).

Manage Maintenance Items (F5356)

Application Type: Transactional

This app allows you to efficiently create, view, and manage maintenance items through an intuitive interface with a search bar and customizable filters such as ABC indicator, functional location, priority, task list, equipment, assembly, and plan category. Maintenance items are organized into three views: unassigned to a plan, assigned to a plan, and all items. For assigned items, you can perform actions like creating new items, making mass changes, removing items from a plan, copying, exporting to a spreadsheet, and adjusting settings. When creating a maintenance item, you are prompted to enter key details such as the item description, long text, and plant category, with access to activity logs for tracking. For any specific item, the app displays comprehensive information, including the assigned maintenance plan, plan category, task list, reference object type, order type, strategy, serial number, execution factor, location and administrative details, account assignment, compliance calculation method, and related call objects. You also can navigate to related apps for further analysis and follow-up. The app supports mass editing of multiple items and allows you to audit inactive or obsolete items, such as those linked to deactivated equipment or functional locations, and detach them from plans to avoid redundant maintenance. More importantly, call objects generated from the item can be viewed, showing scheduled and completed dates.

Manage Maintenance Notifications and Orders (F4604)

Application Type: Transactional

The app allows you to efficiently monitor, create, and manage maintenance notifications and orders through a flexible interface. A search bar and a wide range of filters such as planner group, notification type, company code, maintenance plant, execution object, priority, plant

section, planning plant, subphase, and more enable you to quickly locate relevant maintenance notifications and orders. Maintenance notifications and orders are displayed in a customizable table view, where you can adjust the layout and fields according to your preferences. Key actions within the app include creating new orders, changing responsibility, submitting for approval, editing orders, assigning or unassigning orders, and updating order or notification statuses. You also have access to application logs, which display critical information such as severity, category descriptions, creation date, and the user who created the log. The app supports exporting data to spreadsheets, making it a central hub for managing maintenance-related activities in an organized and efficient manner.

> **Tip**
>
> You can create and manage the custom table view for the maintenance notifications and orders list using the **Standard** dropdown option on the initial screen of the app.

Manage Maintenance Orders (F5241)

Application Type: Transactional

This app provides a comprehensive way to view and manage maintenance orders. It offers a detailed list of orders, and you can easily narrow down results using contextual filters or a free-text search, allowing you to find specific orders to view or edit. You can create new maintenance orders whenever maintenance is needed on a technical object, from scratch, based on a notification, or by copying details from an existing order. Within each order, you input all the necessary information to plan and execute maintenance, including scheduling, location, materials, services, resources, and costs. The system helps you find the right technical object, select the required parts, check their availability, and organize the work steps with task lists and operations. You also can track material stock, assign production resources, attach documents, and add detailed notes. The app shows the progress of the order through various phases, supports status updates, and allows classification of work stages. Financially, you can settle the order to allocate costs and monitor planned versus actual expenses. For maintenance orders linked to service orders, you also can view billing details.

Manage Maintenance Planning Buckets (F3888A)

Application Type: Transactional

This app allows you to efficiently create, view, and manage maintenance planning buckets. Using a search bar and adaptable filters such as planning plant, reference planning bucket, and type (e.g., event-based maintenance or operational maintenance), you can quickly locate specific planning buckets. The app displays a list of planning buckets with key details such as the planning bucket ID, with start and end dates; the columns can be adjusted via the **Settings** option. You have access to functions like creating new buckets, copying existing ones, exporting to Microsoft Excel, and viewing logs within the app. When creating a planning bucket, you are prompted to provide essential information, including a label, description, general data (such as type, planning plant, and person responsible), time period details, and scope parameters like planner group, plant section, maintenance plant, main work center, and technical object. The app supports two types of buckets: operational buckets for recurring maintenance tasks with recurrence patterns and autoassignment of orders, and event-based buckets for one-time maintenance activities in which multiple orders are grouped and managed together You also can navigate to related apps to manage assigned orders, notifications, or backlog items associated with each bucket.

You can navigate to the Manage Backlog app (F4073) in the **Maintenance Planning Bucket** screen for a specific planning bucket using the **Manage Backlog** option.

This app is a new version of the Manage Maintenance Planning Buckets app (F3888), which is available with the SAP S/4HANA 2023 FPS03 version.

Manage Maintenance Plans (F5325)

Application Type: Transactional

This app allows you to effectively manage maintenance plans by offering a comprehensive suite of functionalities. You can search for maintenance plans using free text or filters and view them based on their status, such as scheduled, created, deactivated, or marked for deletion. The app provides a quick view to access detailed plan information and navigate to related apps for additional actions. It supports creating various types of maintenance plans, including single-cycle, strategy-based, and multiple-counter plans, and allows you to decide whether these plans generate maintenance orders or requests.

Maintenance plans can be activated, deactivated, or marked for deletion, with the option to reverse deletions. Visual indicators, such as microcharts, show scheduling progress, while the change log offers a timeline of plan modifications. You can view maintenance calls generated by the plan and navigate to call object details. Maintenance items, which define the recurring maintenance tasks for specific technical objects, can be created or edited within the app. These items can be linked to task lists outlining the job steps, resources, and time requirements. Maintenance cycles, whether time- or performance-based, are configurable along with scheduling details like shift factors and call controls. The app also supports mass editing of plans and the attachment of documents relevant to various process areas. The app uses CDS view C_MAINTPLANACTVSYSTSTATUSQ.

Manage Malfunction Reports (F2023)

Application Type: Transactional

This app allows you to efficiently report, manage, plan, and confirm repairs for technical objects. It features a search bar and filtering options, including work center, technical object, processed by, notification type (e.g., **IHS Notification**, **Activity Report**, **Work Request**, **Breakdown Work**), effect, and status (created, in process, report closed, requires review, and more). You can adapt filters, create new reports, adjust settings, or export data to Microsoft Excel for further analysis. The app lists malfunction reports with key details such as the issue, assigned processor, status, and report date. The app is organized around three main tiles: **Manage Malfunction Reports**, **Report Malfunction**, and **Repair Malfunctions – My Job List**. The app supports teamwork by allowing multiple technicians to collaborate on the same report, with one designated as the lead to finalize and close the report. When creating a malfunction report, you enter key details like the technical object, brief and detailed descriptions, malfunction effect, and location. You can attach files or images, assign tasks, and use the barcode scanner for easy input. Through the **Process Malfunction Report** screen, you can view and interact with reports through the following tabs:

- **Malfunction**
  Captures key details about the issue being reported. You can view the effect of the malfunction, the failure mode, and historical data for the technical object. This tab provides a quick navigation link to the Find Maintenance Notifications app (F2071) for further review or action.

- **Job Details**
  Focuses on planning and assigning the repair tasks. You can add work items by specifying the work center or plant and assigning the task to a specific person or team. You also can add required parts, either manually or by scanning.

- **Attachments**
  Allows you to enrich the malfunction report by uploading supporting documents or images. You can upload files directly or include relevant web links as attachments.

Tip

Using the **Edit Header** option, you can modify key details of a malfunction report, including the name of the person who reported the issue, the description of the malfunction, the date and time it was reported, and the individual responsible for processing the report.

Manage Safety Certificate Templates (F6772)

Application Type: Transactional

This app allows you to search for, view, and manage safety certificate templates with ease. You can filter templates based on processing status (such as **Closed**, **Marked for Deletion**, or **Approval in Progress**), main work center, technical object, planning plant, template type, and more. The app lists safety certificate templates showing key details like technical object, template ID and type, plant section, and validity dates, all interchangeable via table settings. You have the option to create new templates, copy existing ones, adjust settings, and view application logs. When creating a new template, you must provide essential information such as the planning plant and safety certificate template type, with the option to reference an existing template. For each template, you can access detailed information including work center, technical object type, safety precautions, comments, attachments, status, and more. These templates serve as standardized references for generating safety certificates through the Manage Safety Certificates app (F6678) and can be linked to work permit templates, partners, and multiple pieces of equipment or functional locations. The app also allows you to configure safety measures directly within the templates and supports reuse of both partially and fully approved templates.

Manage Safety Certificates (F6678)

Application Type: Transactional

This app allows you to efficiently create and manage safety certificates across their entire lifecycle in different stages, such as **Planning and Preparation**, **Prepared for Approval**, **Fully Approved**, **Suspended**, or **Closed**. Using the search function and filters such as processing status, main work center, safety certificate number, technical object, object type (e.g., equipment or functional location), authorization group, and location, you can quickly locate specific safety certificates. The app provides options to create new certificates, copy existing ones, adjust settings, export data to a spreadsheet, and access application logs for tracking purposes. When creating a new safety certificate via the popup screen, you must enter key details such as the planning plant and safety certificate type, and optionally select a reference template or work permit. Each certificate can then be populated with important data with safety precautions, linked work permits, assigned partners, and associated technical objects. You also can manage attachments and track changes through the logs. Using the **Application Logs** option, you can view logs detailing the severity (**Error**, **Canceled**, **Warning**, **Information**, **Success**), log number, subcategory, validity dates, and so on.

Tip

You can save a custom table view for safety certificates by using the **Standard** dropdown menu in the initial screen.

Manage Work Packs (F6065)

Application Type: Transactional

This app allows you to efficiently search, manage, and process work packs for output. A *work pack* is a structured collection of job packs that group related work order operations, allowing them to be processed and sent as a single output

request. You can search for existing work packs using filters such as work pack ID, output status (e.g., **Completed**, **Error**, **In Preparation**), or creator. The app lists key details like work pack name, type, and output status, with columns that can be customized via table settings. Available actions include sending output, retrying failed output, marking output as completed, copying, and accessing application logs. You can output work packs via print or email and combine multiple PDF attachments into a single document, while other file types remain separate. Attachments may originate from notification and order headers, operations, or linked technical objects such as equipment and functional locations.

Manage Work Permit Templates (F4692)

Application Type: Transactional

This app allows you to create and manage safety work permit templates across various processing statuses. You can search templates using filters such as processing status, work permit number, technical object, planning plant, technical object type, and work permit template type. The app lists work permit templates with detailed information including template number, location, maintenance plant, room, validity dates, planner group, and authorization group, with customizable table settings to view more options. You can perform actions such as creating new templates, copying existing ones, exporting to spreadsheets, adjusting settings, and accessing application logs. Templates can be created independently or by referencing previously approved ones, and can be populated with important details such as the scope of work, required personal protective equipment, and specific safety precautions. Templates also can be linked to multiple functional locations, equipment, maintenance orders, and relevant safety certificate templates. This functionality ensures standardization and efficiency in permit creation, and the templates can be reused in related apps such as Manage Work Permits (F6579) and Create Work Permit Request (F4691).

Manage Work Permits (F6579)

Application Type: Transactional

This app allows you to create and manage work permits across their entire lifecycle. You can search and filter permits using criteria such as processing status, work permit number, main work center, technical object, work permit type, technical object type, order, planning plant, planner group, and plant section. Work permits can be created independently, by referencing a maintenance order, from a predefined template, or by copying an existing permit. During creation, you provide essential details such as the planning plant and work permit type, and optionally reference an existing permit or template. Once created, the permit can be enriched with specific information about the nature of the work, required safety precautions, personal protective equipment, and other safety measures. You can associate multiple functional locations, pieces of equipment, and maintenance orders with a single permit and link relevant safety certificates. The app supports a structured approval workflow based on the four-eyes principle and allows permits to be printed using customizable output formats. Additional features include application logs for tracking, attachment handling for documentation, filter adaptation, settings customization, and pinning the header for easier navigation.

Mass Schedule Maintenance Plans (F2774)

Application Type: Transactional

This app allows you to efficiently schedule all due maintenance plans within a specified time frame, effectively managing the generation of maintenance call objects such as maintenance orders especially when managing large volumes of data. You can search and filter jobs by status, date range, creator, job ID, and job name. The app lists job records with key details such as job status, log results, job name, planned start time,

creator, and more, all adjustable through table settings. It offers options to create new jobs, copy existing ones, view job details, cancel or restart jobs, and export data to Microsoft Excel. During each scheduling run, the system evaluates all relevant maintenance plans, factoring in strategies, counters, and plan categories to calculate due dates and generate call objects accordingly. It supports various plan types applying their respective cycles and strategies. You can create scheduling jobs by selecting a job template and defining scheduling parameters such as recurrence, calendar, working days, and specific attributes like maintenance strategy, functional location, equipment, and main work center. The app also allows you to review job execution results and logs for better tracking and operational transparency.

The app also supports parallel processing, allowing you to distribute the workload across multiple application servers for more efficient execution. To do this, you need to specify the server group for parallel processing as well as the number of parallel processes.

> **Tip**
>
> You can use job template SAP_PM_EAM_MASS_SCHED_MPLAN_T (Mass Schedule – Maintenance Plans), corresponding to job catalog entry SAP_PM_EAM_MASS_SCHED_MPLAN.

Monitor Maintenance Requests (F1511)

Application Type: Transactional

This app allows you to create and monitor maintenance or repair requests for technical objects with ease and efficiency. A search bar and various filters such as reported by, date, plant section, room, and system status (e.g., **Deletion Flag**, **Completed**, **In Process**) help you quickly locate specific notifications. Additional filters can be added as needed. From the main screen, you have the option to create new notifications, access settings, or export the list to Microsoft Excel for further analysis. The app supports both the creation of new maintenance requests and the tracking of existing ones. Each listed notification includes details like the notification ID, system status, priority text, technical object, description, date and time, and the user who reported the issue. When submitting a request, you provide details about the technical object and the location of the problem, and you also can upload images or damage descriptions, which are saved as document management system (DMS) documents. You can opt to receive updates on repair progress. Contact details for the reporter and creator of each request are also available.

> **Tip**
>
> You also can personalize the tabular view of notifications by using the **Standard** dropdown to create and manage custom views, tailoring the display to your specific preferences.

My Inbox – Maintenance Management (F2953)

Application Type: Transactional

This app allows you to efficiently manage and process maintenance-related tasks from a centralized inbox. You can navigate directly to object lists to work on individual maintenance notifications or maintenance orders, as well as perform mass actions where applicable. The app allows you to view and add comments to tasks for better communication and collaboration. You also can manage attachments by uploading, viewing, or deleting relevant files, ensuring all necessary documentation is readily accessible during task execution.

My Maintenance Requests (F4513)

Application Type: Transactional

This app allows you to view and manage your own maintenance requests. You can see all

requests you've submitted, categorized by status in tabs: **Draft, Submitted, Action Required, Accepted, Rejected,** or **Completed**. Draft requests can be edited or deleted, and submitted or accepted requests can be copied for reuse. If a supervisor needs more information during the screening process, a request may be sent back as **Action Required** for revision. Accepted requests proceed to planning, and rejected ones cannot be resubmitted. The app provides details such as notification type, order number, due dates, and current phase and subphase. For jobs in the execution phase, you can quickly navigate to related apps like Perform Maintenance Jobs (F5104A). The app also supports visualizing technical object hierarchies based on the functional location's identifying level. You can assess and assign priority to maintenance requests, which determines critical scheduling dates like start and due dates. If the technical object or notification type is changed, then the assigned priority is removed. The app also allows you to upload or link relevant documents using predefined or custom document types, which can aid in quicker issue resolution.

> **Tip**
>
> Although the app is optimized for the Reactive (4HH) and Proactive Maintenance (4HI) processes, not all features may function identically for orders outside these scope items, especially if the phase model is not activated as recommended by SAP.

Perform Maintenance Jobs (F5104A)

Application Type: Transactional

This app is designed to help you efficiently manage and execute maintenance jobs that have been dispatched for action. It provides a search bar and a variety of filters such as activity type, assembly, business area, breakdown status, code group, and functional location to narrow down job lists. You can perform key actions using options like **Create Maintenance Request,**

459

Assign to Me, **Quick Confirm**, **Copy**, and various settings to customize the interface. When creating a new maintenance request, you must input essential details including the technical object, current location, effect, breakdown indicator, malfunction start date and time, processing context, description, long text, priority, and reported by/on fields, along with the option to upload attachments, add links, or assign documents.

Once a line item from the job list is selected, you can view detailed information organized into sections. The header shows the technical object, maintenance order, order type, processing context, notification, subphase, system status, breakdown flag, priority, and current work progress, with an option to display the order barcode. The **Job Details** section includes the execution stage, work center, control key, job descriptions, effort details, and job confirmations, along with links to related jobs. The **Malfunction Data** section provides details on the effect, long text, failure mode, and breakdown duration, with the ability to add or copy malfunction data. Through the **Add Malfunction Details** popup, you can specify damage codes, object parts, causes (including root-cause flags), and activity types, with options to add or delete items.

The app allows you to add measurements, plan and record component usage, and attach documents to operations, orders, equipment, locations, or notifications. You can take over jobs, access materials, start, pause, complete tasks, log time, and record failure and measurement data.

Permit to Work Overview (F7304)

Application Type: Analytical

This app offers you a consolidated view of open work permits and safety certificates associated with a specific planning plant. You can filter the data by fields such as planning plant (as a key field), maintenance plant, planner group, and work center and more. The app displays critical metrics, including permit and certificate types, processing statuses, and their distribution across various work centers. You can refine the view by selecting additional location units within the planning plant and applying custom filters. From the overview, you can navigate to related applications for deeper analysis and follow-up actions.

Procurement for Maintenance Planner (Purchase Order) (F2827)

Application Type: Transactional

This app provides you with a detailed view of all purchase order items created for maintenance orders, along with key procurement information. It features a search bar and a range of filters including reference date, release status (e.g., active, in release, release completed, or refused), equipment, whether components are missing, maintenance order numbers, and more. The app lists purchase order items with details like material, release status, item ID, order quantity, reference date, net order value, and other fields based on the selected table settings. You can expand the view using the **Show More per Row** option to display additional information such as maintenance order operation, priority, and purchasing group. Selecting a specific purchase order item opens the Purchase Order app (FO348A), where you can view comprehensive details including supplier name, net value, creation details, purchasing organization, detailed item information, purchase requisition and contract links, goods receipts, supplier invoices, notes, attachments, and approval status.

You can access this app via the **Purchase Orders Not Approved** or **Missing Components** card in the Maintenance Planning Overview app (F2828), or through the **External Procurement** card in the Maintenance Backlog Overview app (5105). The system presents a prefiltered list based on the navigation path to help you quickly identify issues such as unapproved purchase orders or missing components.

Procurement for Maintenance Planner (Purchase Requisition) (F3065)

Application Type: Transactional

This app allows you to view a comprehensive list of all purchase requisition items created for maintenance orders, along with key procurement details. It includes a search bar and various filters such as reference date, release status, fixed vendor, location plant, order type, material, purchasing group, and more to refine search results. The app displays requisition details including material, quantity requested, reference date, maintenance order operation, and other fields based on user-defined table settings. You can customize your view further using options like **Show More per Row**, **Adapt Filters**, **Export to Excel**, **Save View as Tile**, and direct navigation to related applications.

You can access this app through cards in the Maintenance Planning Overview app (F2828), such **as Purchase Requisitions Not Converted to Purchase Orders**. It offers a prefiltered list based on how you accessed the app, making it easier to identify requisitions that need attention, like those pending approval or conversion. For each item, you can view detailed procurement data including release status, supplier, requested materials, requirement dates, and any linked purchase orders, ensuring effective procurement tracking and follow-up.

Report Malfunction (F2023)

Application Type: Transactional

This app allows you to create malfunction reports through the **Create Malfunction Report** screen, where you can enter essential details such as the technical object, a brief description, a more detailed explanation in the long text field, and the effect of the malfunction—whether it has no impact, restricts production, or causes a complete production breakdown. You also can specify the current location, date and time the issue was reported, and the name of the person reporting it, and assign the task to a responsible individual. The app also allows you to upload attachments, add relevant links, and take advantage of built-in scanning capabilities.

Request Maintenance (F1511)

Application Type: Transactional

This app facilitates the process of submitting and monitoring maintenance or repair requests for technical objects. You begin by selecting a technical object—either equipment or a functional location—and entering key details such as the notification type (e.g., work request, activity report, install/dismantle), a brief description, long text for detailed information, current location, the effect on operations (e.g., no effect, production restricted, or breakdown), date and time, and the person reporting the issue. Attachments or links also can be added to support the request. You have the option to submit the notification, view the notification list, or use a default template when entering details. Once submitted, the notification appears in a detailed list of notifications with the notification ID, system status, technical object, description, date, and time based on the table settings. You can then view more information in the dedicated **Notification** screen.

The app features two main options: one for submitting new notifications and another for tracking existing ones, using the **Submit** and **Notification List** options. You can personalize the view, apply filters, and save layouts as variants, which can even be pinned as home page tiles for easy access. More importantly, you also can be notified of the progress.

Schedule Material Availability Check (F2465)

Application Type: Transactional

This app allows you to perform material availability checks for multiple maintenance orders either instantly or by scheduling them as recurring batch jobs. A search bar and filters such as job status (failed, canceled, scheduled, ready, etc.), date range, and job ID help you locate specific jobs. The app displays a detailed list of jobs, including their status, logs, results, steps, job name, creator, planned start time, and other relevant information. Follow a step-by-step process to set up a job:

1. First you select a job template and enter a description.

2. Next, you define the scheduling options, including recurrence patterns.

3. Finally, you configure parameters such as error and exception display settings, log-saving preferences, and parallel processing options like thread count and activation.

You also can define order statuses (e.g., in process, completed) and specific order parameters. The app also includes options to schedule, check, cancel, and manage templates via the **Template** feature. When run as a batch job, the system generates a report that includes any errors or warnings encountered, along with the updated system and user statuses of the relevant orders. This app also can be accessed through the Find Maintenance Orders app (F2175).

Tip

You can use the SAP_PM_EAM_ATP_RIAUFK20_T (PM Orders: Material Availability Check) job template.

Schedule Output for Maintenance Job Packs (F6076)

Application Type: Transactional

This app allows you to create and schedule application jobs for generating output related to maintenance job packs. Using the app, you can select job packs based on various parameters, define specific times and dates for output generation, set recurring schedules, or trigger outputs

immediately. Before scheduling, the app validates the input data. It also provides access to logs for each job, allowing you to track success or failure and filter job entries by criteria such as date, job name, status, or creator. You also can duplicate existing jobs to fast-track the creation of similar ones.

Schedule Output for Maintenance Notifications (F6074)

Application Type: Transactional

This app allows you to create application jobs that schedule the output for maintenance notifications. With this app, you can define when and how maintenance notifications are output, helping streamline communication and documentation across your maintenance processes.

Schedule Output for Maintenance Orders (F6075)

Application Type: Transactional

This app allows you to create and manage application jobs specifically for scheduling the output of maintenance orders. You can define new output jobs by selecting the relevant maintenance orders and configuring output parameters, including timing options such as immediate, scheduled for a specific date and time, or set to recur at defined intervals. Before confirming the schedule, you can review and validate the job setup. Once a job is created, it appears in the **Application Jobs** list, where you can monitor its execution status, view detailed logs to check for success or failure, and apply filters based on job name, creation date, or status. The app also allows you to copy existing jobs, making it easier to reuse configurations for similar tasks. Supported output channels include email and print, ensuring timely and accurate distribution of maintenance orders.

Screen Maintenance Requests (F4072)

Application Type: Transactional

This app provides a centralized platform for managing maintenance requests, allowing you to search and filter by various criteria such as notification ID, planning plant, planner group, work center, priority level, notification type, equipment, and technical object type. Maintenance requests are organized into tabs—**Open, Action Required, Accepted, Rejected**, and **Completed**—with details like notification ID, priority, technical object, and ABC indicator displayed based on the table settings. You can take actions such as accepting, rejecting, changing responsibility, copying, adjusting settings, or exporting data to Excel. The app allows you to view comprehensive request details including status, malfunction data, and equipment hierarchy. Accepted requests move forward for planning and execution, while rejected ones are logged with reasons and locked from further edits. You also can manage output documents, customize them through output control settings, and upload related files.

The maintenance request app includes several key tabs. The **General Information** tab displays core technical and logistical details such as the technical object, location data, and responsible area. The **Hierarchy** tab shows a structured view of related technical objects, open requests, and orders, with links to additional system data. **Malfunction Information** captures details about the issue, including descriptions, causes, and failure data. **Prioritization & Criticality** outlines priority levels and scheduling dates. The **Attachments** tab allows you to view relevant documents or links. **Maintenance Plan Details** shows preventive maintenance links like task lists and plan items. Finally, **Output Items** tracks communication records and allows you to manage output-related actions.

Tip

The app supports logging changes to the maintenance as history for later review. Changes can be viewed in the **Technical Object** screen, detailing the edit type (e.g., **System Status Change**), the field name that was changed, the old value and new value, and the person who changed it.

Technical Object Breakdowns (F2812)

Application Type: Analytical

This app allows you to thoroughly analyze the frequency, causes, and duration of equipment breakdowns, offering key insights into repair trends and downtime patterns. You can apply a wide range of filters such as maintenance plant, planning plant, object type, catalog profile, technical object type, main work center, and ABC indicator to refine the data and focus on specific breakdown scenarios. The app supports flexible viewing options, allowing you to toggle between chart view, table view, and a combined view (table and chart), and it includes tools like zoom, chart screen maximization, and legends to improve data interpretation. In the table settings, you can copy data, configure the table view using the **Settings** option, export to Microsoft Excel, and show all items with detailed columns such as maintenance plant, object type, number of reported breakdowns, total repair time in hours, mean time to repair, and time between repairs. The app calculates key metrics like mean total time to repair and mean time between repairs, and it effectively breaks down data by model number, planning plant, breakdown quarter, activity type, assembly, and more.

You can use the **Visual Filter** option to view the following charts:

- No of Reported Breakdowns by Type of Technical Object
- No of Reported Breakdowns by ABC Indicator for Technical Object
- No of Reported Breakdowns by Maintenance Planning Plant
- No of Reported Breakdowns by Main Work Center for Maintenance Tasks

Tip

The C_MAINTOBJBREAKDOWNQUERY_CDS (Maintenance Object Breakdown Query) service must activated and maintained. The app uses CDS view C_MAINTOBJBREAKDOWNQUERY.

Technical Object Damages (F3075)

Application Type: Analytical

This app is designed to help you analyze recurring damages and identify the technical objects most frequently affected. It provides a centralized view of related maintenance notifications linked to specific maintenance items, enabling better reliability tracking and root-cause analysis. You can visualize damage data either in chart or table format, or both simultaneously, and apply various filters such as maintenance plant, object type, construction type, catalog profile, notification year (e.g., 2024 and 2025, with an option to add more), and so on. The app highlights important details, showing metrics like the number of causes, number of activities, and number of object parts. You can drill down into specific maintenance notifications, technical object labels, and functional locations to gain deeper insight into the nature and frequency of damage. Each listed item includes data such as damage type, technical object, and associated corrective actions. The app offers multiple display options and allows you to switch between table view, chart view, and a combined layout (table and chart). SAP notes that rendering complex data may cause slight delays, but the app remains an invaluable asset for maintenance teams focused on improving operational reliability and minimizing equipment-related issues. You can utilize the following charts using the **Visual Filter** option:

- Damage by Effect
- Damage by Object Type
- Damage by Construction Type
- Damage by Catalog Profile
- Damage by Notification Year
- Damage by Object Part Code Group
- Damage by Damage Code Group
- Damage by Cause Code Group
- Damage by Activity Code Group

You also can select the following chart measures when displaying results in chart view: activities, causes of damage, damage, and object parts.

Tip

The C_DAMAGEANALYSISQUERY_CDS (Damage Analysis Query) service should be activated and maintained for this app. The app uses CDS view C_DAMAGEANALYSISQUERY.

Chapter 7
Materials Management

Materials management in SAP S/4HANA encompasses procurement, inventory management, and logistics, ensuring efficient material flow within an organization. In this chapter, we'll cover the SAP Fiori apps for vital processes such as central procurement, supplier evaluations, sourcing projects, purchasing, good receipts and issues, and more. Where applicable, we've noted a comparable SAP GUI transaction. The apps are listed in alphabetical order.

Adjust Operational Supplier Evaluation Score (F2312)

Application Type: Transactional

This app allows you to adjust the operational evaluation score of a supplier. The app showcases the option to display **All**, **Only Updated**, or **Not Updated** supplier evaluation scores. You can view supplier evaluation scores based on several sorting and filtering parameters, such as quantity score, time score, price score, and inspection lot score, as well as several other factors relevant for process owners. The operational score can be updated by modifying the individual scores that contribute to the operational score for a specific supplier using the **Edit** icon. As a business progresses, the app becomes invaluable for process owners; there will be a need to adjust these scores as more data is collected. This helps you to have the most recent evaluation data. Furthermore, you can remove recently updated scores using the **Discard Updated Scores** button.

Analyze Missing Exchange Rates (F4360)

Application Type: Transactional

This app helps you find and fix missing exchange rates for various purchasing documents such as purchase orders, central purchase orders, purchase contracts, quotations, and so on. Business users can narrow the search using parameters such as validity dates, target and document currencies, purchasing document type, and more. Besides viewing purchase documents with missing exchange rates, you also can maintain rates by using the **Maintain Exchange Rates** option to enter details such as the exchange rate type, currencies, validity dates, quotation method, and exchange rate. Furthermore, you can export the list of purchasing documents with missing exchange rates using the **Export** or **Export As** functionality. Business users will need to make sure that service group UI_CMIS_MYFILES_04 (CMIS My Files UI Service) has been published for the export functionality to work properly. For instance, you can export the purchasing document list as either an XLSX file or a PDF.

This app can be accessed through the **Monitor Missing Exchange Rates** card from the Monitor Purchasing Analytics Operation app (F4339). The app uses CDS view C_MONITORMISSINGEXCHRATE to retrieve data.

> **Tip**
>
> The app applies maintained exchange rates to all purchasing documents with the same or later document dates, whereas previous dates are not affected.

In addition, missing exchange rates entries will disappear from the list once updated.

Analyze Stock in Date Range (F6185)

Application Type: Transactional

This app can be used to track and monitor stock movements for one or more materials across multiple plants and storage locations with a selected date range, which helps in identifying trends to ensure proper inventory management and addressing stock discrepancies. An inventory manager can view stock levels and movements during specific time frames and analyze inventory trends to support forecasting and replenishment decisions. You can narrow the search by using filter criteria such as display currency, analytical date range, plant, material, storage location, stock type (e.g., unrestricted-use stock), special stock type (e.g., project stock, supplier consignment), and so on. Business users can create and schedule export jobs with predefined or custom templates for material stock and postings or material stock with the following options: exclude reversal postings or exclude postings without stock. The **Schedule Export** option navigates you to the Schedule Export of Inventory Analytics app (F7493). You can use the **View By** option to customize chart views using dimensions like material, plant, sales order, supplier, batch, and more. The **Details** option allows you to view in-depth stock data like plant, currency, and stock value, with further options to toggle between chart and table views for flexibility.

For detailed analysis for a selected material, you can track material postings, including stock quantities and values for the selected period, filter data by movement types, and exclude reversal movements. In the **Material Postings** tab for selected material, you can navigate to the Material Documents Overview app (F1077). More importantly, you also can navigate to the Compliance Information (F3226) and Manage Product Master Data (F1602) apps for further processing

and analysis. The app flexibly integrates with the Stock-Single Material app (F1076) or Material Documents Overview app (F1077).

> **Tip**
>
> Access to financial data may be restricted based on user authorization.

Central Procurement Interface Monitor (F3913)

Application Type: Transactional

The app helps monitor and address issues in central procurement processes by displaying failed transactions—for instance, central contracts and requisitions. It allows business to investigate issues or dismiss irrelevant ones. Filter options include purchasing object ID, connected system, purchasing object, and so on. The main function of the app is to investigate or dismiss failures. Using the **Investigate** option allows you to identify the reasons for failed transactions by navigating the relevant application or transaction in the hub system or connected systems. The **Dismiss** option allows you to remove resolved or irrelevant failed transactions from the list.

Central Procurement Job Monitor (F3914)

Application Type: Transactional

This app helps monitor and manage central procurement jobs by listing failed extraction jobs. It allows you to investigate issues, restart jobs or dismiss irrelevant ones using filters such as on connected system, creation date, and so on. Using the **View Job Details** option, you can open the job list to understand the reason for failure and, if needed, create a new job. Using the **Restart Job** option, you can retry failed jobs to resume the process, and the **Dismiss** option allows for the removal of irrelevant or resolved failed jobs from the list.

Central Procurement Operations Monitor (F3912)

Application Type: Analytic

The app helps monitor central procurement operations through various overview cards, highlighting issues that need attention. It allows you to investigate errors, track job failures, and check the compatibility of business operations across connected systems. You can add or remove cards on the initial app screen by making use of the **Manage Cards** option. The following cards can be found on the **Central Procurement Operations Monitor** screen:

- **Central Requisitioning**
  Monitor failed purchase requisitions and source of supply extraction jobs. View and address failed transactions using the Central Procurement Interface Monitor app (F3913) and the Central Procurement Job Monitor app (F3914).

- **Central Purchasing**
  Track failed extraction jobs for purchase requisitions, purchase orders, and text data. Investigate and restart failed jobs, and dismiss resolved or irrelevant failures from the hit list using the Central Procurement Job Monitor app (F3914).

- **Central Contract Management**
  Check failed contract distributions and release order update jobs. Navigate to the Central Procurement Interface Monitor app (F3913) and see central procurement interface failures so that you can investigate or dismiss failures.

- **Features and Systems Compatibility**
  Ensure compatibility of business operations with system requirements/versions.

- **Connected Systems**
  Monitor the online/offline status of connected systems.

Central Purchase Contract Consumption (F3444)

Application Type: Analytical

This analytical app helps you to track the consumption of central purchase contracts, comparing the released quantity against the target quantity from the previous year to date. It supports quick insights through flexible filtering and data visualization options. You can narrow the search results by using filter criteria such as display currency, relative date, calendar month, week, year, company code, and so on, with visualization options such as table and chart views, changing dimensions and measures, and drilling up and down to view information. In essence, you can monitor contract consumption effectively, identifying underutilized or overutilized contracts. Furthermore, you can use the refresh option to update and view the latest data (the time since the last update will be displayed—for example, showing that the available data is 15 minutes old). The app uses CDS view C_CNTRLCONTRCNSMPNQRY to retrieve data.

Using the **Jump To** option, you can navigate to the following apps: Manage Purchasing Categories (F0337), Manage Central Purchase Contracts (F3144), Monitor Central Purchase Contract Items (F3492), Supplier (F0354), and Manage Purchasing Categories (F0337).

Central Purchase Requisition Item Types (F3981)

Application Type: Analytical

This analytical app helps you view and analyze different types of purchase requisition items created in the SAP S/4HANA system (hub) and connected systems like SAP ERP or SAP S/4HANA Cloud. You can view the total of purchase requisition items at a glance; analyze data by supplier, purchasing organization, material group, and more as KPIs; and filter results based on parameters such as relative dates, calendar year, month, week, quarter, vendor, company code, item category, and so on. In terms of data visualization, you can use chart and table view options; drill into specific details for purchase requisitions; adjust, for instance, the chart axes; and even export the data to a spreadsheet for further analysis. Overall, the app helps business users compare purchase requisition item counts and types across purchasing groups, suppliers, and systems; analyze and optimize purchasing processes based on detailed insights; and quickly navigate to related apps for monitoring and processing purchase requisition items. The app uses CDS view C_CNTRLPURREQNITEMTYPES.

You can navigate to the Monitor Purchase Requisition Items Centrally app (F3976) and the Process Purchase Requisitions Centrally app (F3290).

Classification-Based Product Hierarchy (F2351)

Application Type: Transactional

This app can be used to create and maintain product structures in hierarchies based on classifications. This classification system supports better product management, reporting, and integration with processes like sales, procurement, and analytics. The app is used to manage product hierarchical structures for products, assigning products to nodes within the hierarchy using predefined attributes.

This app is useful as it can be used to visualize product hierarchies in a graphical or tabular view and to search and filter products based on their attributes or hierarchy levels. For example, in sales and distribution, product categorization can be used to segment product hierarchies for sales performance tracking, and in operational

planning, products can be grouped based on attributes for procurement, manufacturing, or even distribution planning. The app uses CDS view C_MDPRODUCTHIERARCHYQUERY.

Compare Supplier Quotations (F2324)

Application Type: Transactional

This app allows you to evaluate and compare supplier quotations comprehensively. The app simplifies the supplier evaluation process by providing a side-by-side comparison of multiple supplier quotations for specific requests for quotations (RFQs). You can display all supplier quotations received for a specific RFQ. You can drill down to display the RFQs based on filter criteria such as RFQ, RFQ type, purchase document category, or RFQ description, or even search directly in the search bar. For a selected RFQ, you can view details such as who created the RFQ, its status (e.g., in process), the quotation deadline, the number of invited bidders, and a comparison of the target value and best quotation. Within the app, you can assign scores to suppliers and highlight the best quotation based on overall score or specific attributes based on available data. You also can compare quotations from multiple suppliers for a specific purchase request and view detailed information, such as pricing, delivery dates, payment terms, and conditions. You can only compare and award quotations in the **Submitted** status.

Selecting the **Supplier** hyperlink shows a popup where you can navigate to various apps such as Supplier (F0354), Clear Outgoing Payments (F1367), Display Supplier Balances (F0701A), Manage Supplier Line Items (F0712), Post Outgoing Payments (1612), and many more using the **More Links** option.

Note that appropriate authorization is required for you to access the various related apps.

Compare Supplier Quotations (Sourcing Project) (F4863)

Application Type: Transactional

This app is designed for strategic procurement. It allows you to compare supplier quotations within sourcing projects and helps procurement teams evaluate suppliers based on cost, terms, and other key factors. You will need to enter a sourcing project ID on the initial screen or alternatively use filters to narrow down sourcing projects by version, type, status (published, revised, canceled, etc.), name, and other criteria. You can customize the table view by selecting which columns to display, such as company code, bidders, material group, and document currency. You can compare multiple supplier quotations within a sourcing project by viewing different versions of the same quotation and identifying the best bid for each item across, for instance, submitted and revised quotations. Selected quotations can be opened in the Manage Supplier Quotations app (F4862) to check statuses, proposed delivery dates, notes, attachments, and version history. The app also calculates the total cost of quote items in a sourcing project; based on the comparison, you can create awarding scenarios to select the most suitable supplier and initiate negotiations. Quotations that have not yet been awarded can be revised and edited as needed.

Tip

The UI_SRCGPROJ_COMPARE service (Sourcing Project Quotation Comparison) should be activated and maintained for full functionality.

Confirm Receipt of Goods – New (F4489)

Application Type: Transactional

This app simplifies the goods receipt confirmation process in procurement. This app is targeted at business users responsible for confirming that goods or services ordered have been received or rendered in accordance. You can display a list of purchase orders awaiting confirmation of goods receipt. The app is quite useful for business users who are responsible for confirming that goods and services requested have been delivered or rendered. The app lists all the purchase requisitions even from other users in other departments of an organization with the relevant authorizations. You can easily confirm the receipt of goods or services linked to purchase orders or service entry sheets and indicate full or partial deliveries. The app lists all purchase requisitions with material items relevant for confirmation. All purchase requisitions from numerous process areas are listed and can be confirmed as required. The app supports partial goods receipt confirmations by entering the quantity received and not selecting the **Final Delivery** icon. If the full quantity has been received, you can select **Final Delivery** to indicate completion. The app allows you to confirm goods receipts for one or multiple items consecutively. More importantly, users with the appropriate authorization can view purchase requisitions from different departments, and the system centralizes all procurement requests, making it easier to track outstanding goods receipts.

This app is the successor to Confirm Goods Receipt (F1654), which was deprecated as of SAP S/4HANA 2020 FPS02.

Contract Expiry (F0574)

Application Type: Analytical

This app helps you track and manage expiring contracts effectively. It provides real-time insights into contracts approaching their end date within the next 90 days, which enables businesses to renew, renegotiate, or close contracts in time. This app supports proactive contract lifecycle management, which in turn helps businesses maintain supplier relationships and prevent supply disruptions. The app displays the number of contracts approaching their expiration at the top of the screen (**Contracts Approaching End Date**). You can view details such as contract number, supplier, expiration date, target amount (total contract value), released amount (amount already utilized), and so on. Furthermore, the app visualizes expiring contracts using KPIs: **Supplier**, **Purchasing Group**, **Purchasing Organization**, **Purchasing Document Category**, and more. The following mini tiles are available:

- **Contract Expiry**
  Displays the number of contracts that are nearing their expiration date.

- **Off-Contract Spend**
  Highlights purchase orders that have been created without referencing a contract. Selecting this tile navigates you to the SAP Smart Business Runtime app (F5240).

- **Contract Leakage**
  Identifies purchase orders that should have referenced existing contracts but did not.

- **Quantity Contract Consumption**
  Provides insights into how much of the agreed quantity in a contract has been released versus the total target quantity.

- **Value Contract Consumption**
  Displays the monetary release amount in relation to the target contract value.

- **PO and Scheduling Agreement Value**
  Presents a comparison between actual and planned purchasing spend. Like some other tiles, it also links to the SAP Smart Business Runtime app for further analysis.

You also can navigate to various apps using the **Jump To** option, such as Manage Purchase Contracts (F3144), Supplier (F0354), and My Purchasing Document Items (F0574B). The app uses CDS view C_PURCHASECONTRACTEXPIRY to retrieve data as one of its views.

Contract Leakage (F0681)

Application Type: Analytical

This app provides insights to help track and manage purchases made outside of agreed-upon contracts, a scenario known as *contract leakage*. This app allows you to identify purchase orders that lack a contract reference even when an existing contract could have been used. By viewing this data, you can determine the net value of noncontract purchases, assess the percentage of spend outside contracts (leakage %), and pinpoint suppliers or categories with high leakage rates, which in effect allows organizations to improve contract compliance and improve supplier negotiations and purchasing efficiency. You can apply various filters, such as supplier, purchasing category, company code, plant, material, work breakdown structure (WBS) element, and many more, to refine your analyses.

The app provides KPIs to monitor contract adherence and track procurement trends. It also integrates with other apps, such as Manage Purchasing Categories (F0337) and Purchase Order (F038A), for further processing. The **Refresh** option and mini tiles enable comprehensive usability by keeping data updated and accessible. The app uses CDS view C_PURCHASECONTRACTLEAKAGE to retrieve data.

Contribute to Sourcing Projects (F7757)

Application Type: Transactional

This app helps you manage and work together with others on sourcing projects by providing a central, easy-to-use platform for both strategic and demand-based sourcing activities. It enables process owners to actively contribute to supplier selection, tendering, and contract negotiation activities. Key functionalities include inviting preferred suppliers to participate in sourcing events, adding legal or relevant supporting documents to sourcing projects, and evaluating, comparing, and negotiating supplier quotations. You also can create awarding scenarios to select the most suitable quotations and generate follow-on documents such as purchase orders or central purchase contracts based on the awarded bids. The app provides you with a comprehensive view of assigned sourcing projects, including details such as timelines, tasks, and status updates, ensuring transparency and enhanced collaboration throughout the sourcing lifecycle.

Create Physical Inventory Documents (F3197)

Application Type: Transactional

This app can be used to display lists of materials in specific plants and the corresponding storage locations that fall under three stock types: unrestricted-use stock, blocked stock, and quality inspection stock. Filters can be adapted to view different parameters for these three categories. For example, you can choose to view items counted, special stock, goods issue since last count, and goods receipt since last count, among many others. You can then choose to create physical inventory documents for selected materials or groups of materials. During physical inventory document creation, you can split or not split documents. If document splitting is a necessity, you can choose **Storage Bin**, **Material**, or **Material Group**. This is a useful app for performing physical inventory and for highlighting any identified differences. Also, you can use the **Show More per Row** icon to display more detailed information for the material line items on display, and export functionality to a spreadsheet is available via the **Export to Spreadsheet** icon. Furthermore, you can upload files or add a link using the **Upload** or **Add a Link** option during document creation.

> **Tip**
>
> This app focuses on a limited number of documents to be created at once (currently up to 20), and the inventory status is updated in real time.

> **Transaction Code**
>
> The SAP GUI equivalent is Transaction MI01 (Create Single PI Document).

Create Purchase Order Centrally (F3486)

Application Type: Transactional

This app allows you to efficiently create purchase orders by grouping selected purchase requisition items based on shared attributes such as supplier, company code, purchasing organization, purchasing group, and connected system. Depending on these groupings, one or more purchase orders can be generated. You have several options when creating purchase orders. You can generate purchase orders directly without making additional changes to the details carried over from the requisitions. At the header level, you can input key information like document type, payment terms, Incoterms, and Incoterms location, while item-level fields allow for specifying the material description, quantity, price, tax code, delivery date, and goods receipt options. For connected non-SAP systems, value help for these fields is dynamically provided from the external system.

You can access this app using the Process Purchase Requisitions Centrally app (F3290).

> **Tip**
>
> You need to activate central procurement (scope item SAP S/4HANA Procurement Hub and Scenarios) to use central procurement functions in Customizing using IMG menu path **Materials Management • Purchasing • Central Procurement – Settings in Hub System • Activate SAP S/4HANA Procurement Hub and Scenarios.**

Create Purchase Orders from Central Quotations (F4104)

Application Type: Transactional

This app allows central purchasers to generate purchase orders from centrally awarded supplier quotations. The pricing terms from the central supplier quotation are automatically transferred into the resulting purchase order. These orders are created within the specific connected systems where the initial purchase requisitions originated. You also can view and verify purchase orders within the connected systems, checking essential information such as the purchasing organization, company code, supplier, plant, quantity, price, and more.

Create Purchase Requisition – New (F1643A)

Application Type: Transactional

This procurement app allows you to manage purchasing needs efficiently. Accessible through the My Purchase Requisitions app (F1639A), it allows you to create, view, edit, copy, and delete self-service purchase requisitions. When creating a requisition, you can input detailed information across several tabs. In the **General Information** tab, you can enter material details, quantity, price, plant, storage location, and purchasing group. The **Delivery Address** tab captures address-related fields like address type, city, region, and postal code. In the **Sources of Supply** tab, you can add preferred suppliers, including fixed suppliers. The **Notes** section allows for input of item texts, delivery instructions, and purchase order-related notes. You also can view approval details and attach relevant documents or links in the **Attachments** tab. Beyond creation, the app allows you to track the status and approval flow of your requisitions, confirm receipt of goods, and process returns for defective items. By consolidating key procurement actions in one place, the app enhances operational efficiency and reduces redundancies in the self-service purchasing process.

This app is the successor to the Create Purchase Requisition app (F1643), which was deprecated as of SAP S/4HANA 2023 FPS03. SAP recommends switching to this successor app as soon as possible.

Create Supplier Invoice (F0859)

Application Type: Transactional

This app allows you to create and manage supplier invoices based on received documents, either with or without reference to purchase orders. When referencing purchase orders, you can specify related items, even if they have multiple account assignments. Depending on the document type, you can assign items flexibly—for example, by searching service entry sheets. When invoices are simulated or posted, the system validates quantities and amounts against purchase order or goods receipt data, and if deviations exceed tolerance, the invoice is automatically blocked for payment and may trigger an approval workflow. You can create invoices, credit memos, and debit memos by entering essential data in the **General Information** tab such as company code, invoice date, and posting date. You can adjust exchange rates, input account assignments, define retentions, and include unplanned delivery costs. In the **Purchase Document References** tab, invoices can reference various documents like delivery notes, service entry sheets, and freight orders, including for unplanned items that do not require goods receipts. Invoices also can be created without a purchase order by manually entering all necessary data. The app supports adding general ledger account items in the **G/L Account Items** tab, verifying tax data in the **Tax** tab, editing payment terms in the **Payment** tab, and attaching documents in the **Attachments** tab. You can simulate, hold, or post invoices, with one user able to park a document and another completing the posting. For discrepancies in quantity or price, you can reduce invoices, which triggers both an invoice and credit memo posting, along with a complaint letter.

The app also allows you to view purchase order histories and use advanced editing or display features with proper configuration. Invoices are categorized by status—such as **Posted**, **Held**, **Draft**, or **With Errors**—allowing for different actions like editing, simulating, or discarding. Drafts can be created by you or received from external networks, and incomplete items can be reviewed and corrected before posting.

This is the successor app to the Supplier Invoice (S/4HANA) app (F0346A), which was deprecated as of SAP S/4HANA 1511 FPS02.

> **Tip**
>
> In the table view in the **G/L Account Items** tab, you have options to **Add**, **Copy**, **Delete**, and **Change Tax Codes** for the listed line items.

Cycle Counting – Classification (F4486)

Application Type: Transactional

This app can be used to manage cycle counting classifications for materials at the plant level. The app categorizes materials into three categories: **Unclassified**, **Classified**, and **All** for a specific plant. It is also possible to add or remove an existing cycle counting classification by selecting the **Classify** button and making the required

changes. In general, you can define these settings once or maintain changes whenever necessary. Furthermore, the app allows you to classify numerous materials simultaneously. You can use the app to manage and classify materials for cycle counting based on predefined criteria. It supports efficient inventory management by focusing on high-priority or frequently moved materials and ensuring accurate stock levels.

This app is closely linked to the Analyze Stock in Date Range (F6185) and Create Physical Inventory Documents (F3197) apps.

Dead Stock Analysis (F2899)

Application Type: Analytical

This app provides inventory managers and warehouse specialists with powerful tools to monitor and analyze dead stock over a specified period. You can apply a range of filters, including analytics start and end dates, display currency, material, plant, storage location, material type, and material group, to customize your analysis. The app offers both chart and table views, along with options to zoom in or out, open full screen, and adapt filters. KPIs such as batch, currency, customer, material, sales order, stock type, work breakdown structure (WBS) element, and supplier provide detailed insights into inventory performance. For example, you can view visual charts like **Stock Value in Display Currency (End Period) by Material Number** to easily identify trends. *Dead stock* is defined as stock whose value increases during the analysis period despite consumption postings within the period, and it poses a risk to cash flow and profitability. The app helps you assess whether such materials should be scrapped, discounted, transferred, or discontinued. With features like detailed analytics, filter customization, and export options, the app supports effective inventory optimization across various plants and locations.

This app is closely linked to the Inventory Analysis Overview app (F3366).

Default Settings for Users (F1995)

Application Type: Transactional

This app is important for enabling central procurement scenarios, where user defaults can be managed efficiently across connected systems. It is quite useful for users who want to reduce manual entry, minimize errors, and save time, but it also ensures consistency across processes by ensuring that frequently used data is pre-filled in procurement transactions. On the initial screen, you are required to enter a user ID, with the option to drill down and search by user name. Once the user ID is entered, related details such as basic parameters, account assignments, and assigned catalogs are displayed. You can conveniently manage settings through the **Excel-Based Data Exchange** option, which allows downloading and uploading of configuration data using, for instance, the *Default_Settings_for_User.xlsx* template. Existing settings can be modified or restored using the **Edit** and **Reset** functions.

In the **User Details** section, you must specify a requisition group and shop on behalf type, which can be one of three options: **Company Code Based**, **No Shopping on Behalf**, or **Group Based**. You also can enter procurement settings in the **Basic Parameters** section to define the purchasing context, including company code, delivery date, supplying plant, material group, and more. In the **Account Assignment** section, you must enter an account assignment category and relevant details, and in the **Assigned Catalogs** section, you can manage catalog access for guided buying scenarios. You can use the **Add** or **Remove** options to control which catalogs are available. For each catalog item added, you can choose to select or deselect it as the default catalog, ensuring that preferred sources are highlighted during guided purchasing.

> **Tip**
>
> In **Edit** mode, you can choose to save, cancel, or copy default settings.

Display Inventory Analytics Job Results (F7504)

Application Type: Analytical

This app allows you to view and manage the results of inventory analytics jobs, offering detailed insights through features such as job status tracking and access to generated attachments. On the initial screen, you can filter jobs based on various criteria including search terms, job status (**Failed, Finished, In Process**), planned start date, creation date, and the source app—either Analyze Stock in Date Range (F1685) or Schedule Export of Inventory Analytics (F7493). You can toggle between viewing **My Jobs** or **All Jobs**, and the job list displays key details such as the job name, status, and template used. The app also features an **Application Logs** section, which displays the status and essential information of each job, and an **Attachments** section, where you can download Microsoft Excel files created during job execution. These attachments may include material postings with or without stock and material stock data, depending on the export settings used. Job deletion is managed automatically by the job scheduling framework and cannot be performed manually. Typically, jobs and their related Excel files are retained for 14 days for single-run executions.

> **Tip**
>
> The **Created via App** filter allows you to select jobs from either the Analyze Stock in Date Range app (F1685) or the Schedule Export of Inventory Analytics app (F7493).

Display Purchase Order (F7117)

Application Type: Transactional

This app offers detailed contextual information for purchase orders, allowing you to gain a comprehensive view of its data and status. For enhanced interaction and functionality, you can refer to the related Purchase Order (Version 2) app (F0348A), which serves as a variant of the main application.

Transaction Code

The SAP GUI equivalent for this app is Transaction ME23N (Display Purchase Order).

Display Scheduling Agreements (F7349)

Application Type: Fact Sheet

The app allows you to view detailed information about scheduling agreements within the procurement process. It provides a clear and structured overview of agreement terms, delivery schedules, and associated data. For more advanced functionality, such as creating, editing, or managing scheduling agreements, you can utilize the related Manage Scheduling Agreements app (F2179).

Transaction Code

The SAP GUI equivalent for this app is Transaction ME33L (Display Scheduling Agreement).

Display Supplier Confirmations (F7276)

Application Type: Transactional

This app allows you to display supplier confirmations linked to purchase orders. You can view key details such as the reference purchase order, supplier reference, approver, and creator, along with the processing status. It also captures supplier contact data and provides a detailed view of the supplier confirmation itself, including confirmation number, confirmed price, and other relevant information. This helps ensure alignment between the purchase order and the supplier's acknowledgment.

The app is also closely linked to the Manage Purchase Orders (Version 2) app (F0842A) and Manage Supplier Confirmations app (F5039).

Edit and Approve Purchase Requisition (F4007)

Application Type: Transactional

This app allows you to efficiently review, manage, and edit purchase requisitions during the approval process. It provides an intuitive interface that allows approvers to view comprehensive header and item-level details, such as general information, delivery dates, account assignments, source of supply, and attachments. One key feature of the app is the ability to edit contact information—specifically, the purchasing group and requisitioner—during both header and item-level approvals. For nonlimit and limit items, approvers also can update various fields. The app integrates seamlessly with the Manage Workflow Instances app (F6011), enabling you to add or modify approval steps, view workflow history, and define custom logic for workflow restarts when changes are made. Approvers also can approve, reject, or send requisitions back for rework; add decision comments; and attach supporting notes or documents.

Edit Supplier Invoice Settings (F3813)

Application Type: Transactional

This app enables users, particularly those in finance and accounts payable, to manage and customize supplier invoice processing settings according to specific organizational needs. It allows you to define user-specific parameters for handling supplier invoices, such as displaying journal entries upon posting or parking an invoice, setting default company codes or suppliers in the invoice creation view, and choosing whether to apply default currencies based on financial configurations. The app also includes options to show advanced editing and display features for enhanced navigation within the invoice apps. Beyond these settings, it empowers you to configure deeper invoice management functionalities, such as setting tolerance levels for discrepancies between supplier invoices and purchase orders or goods receipts, automating invoice verification, and adjusting payment terms and discounts. It supports workflow customization by enabling multilevel approval paths based on roles, financial limits, or supplier types. You also can define rules for automatic invoice matching, manage allowed document types like credit or debit memos, and enforce compliance through validation rules.

Global Purchasing Spend (F3679)

Application Type: Analytical

This app allows you to perform detailed purchasing spend analysis using a range of filters such as display currency, relative dates, and customizable criteria. You can explore KPIs across multiple dimensions including supplier, purchasing organization, plant, and material group. The interface offers several interactive options such as refreshing data, using a mini tiles option, drilling down into details, using chart or table views, using zoom functions, and adapting settings for personalized insights. The app consolidates purchasing data from both SAP S/4HANA and connected external systems, giving procurement and finance teams a holistic view of global purchasing performance. It allows you to track spending trends from the start of the previous year, examine planned versus actual spend, detect failed transactions, and evaluate supplier performance and geographical distribution. With powerful visualization and navigation tools, you can dive into spend patterns by supplier, organization, company code, material group, and more. The app uses CDS view C_CNTRLPURCHASINGSPEND to retrieve data.

You can navigate to the following apps: Manage Central Purchase Orders Centrally (F3292) and Monitor Purchase Order Items Centrally (F3676).

Inventory Analysis Overview (F3366)

Application Type: Analytical

This app provides inventory analysts with a comprehensive, real-time overview of key inventory-related metrics and tasks through a customizable dashboard made up of actionable cards, as follows:

- **Stock Value Increase Despite Consumption**
  Displays the top 20 materials that have experienced the highest increase in stock value over the past 365 days, even though there were consumption postings recorded during that time. Navigates to the Dead Stock Analysis app (F2899).

- **More than 100 Days Without Consumption**
  Highlights materials that have a stock level greater than zero but have not had any consumption postings for at least 100 days. This helps identify slow-moving or stagnant inventory, allowing you to take timely action to optimize stock levels and reduce excess inventory. Navigates to the Slow or Non-Moving Materials app (F2137).

- **Monitor Batches by Longest Time in Storage**
  Displays batches organized by the length of their storage period and the corresponding quantities, helping you identify items that have been in storage the longest and may require priority action. Navigates to the Stock – Multiple Materials app (F1595).

- **Monitor Batches by Expiration Date**
  Displays batches arranged by their nearest expiration dates along with their quantities, allowing you to quickly identify items that are approaching expiry and prioritize their usage or removal accordingly. Navigates to the Stock – Multiple Materials app (F1595).

Each card highlights critical insights, such as materials with increased stock value despite consumption, items with no movement for over

100 days, or batches with the longest storage times and earliest expiration dates. These cards are prioritized by relevance to help you quickly identify and act on potential inventory issues. The app supports flexible filtering by parameters such as plant, material group, and currency, allowing you to tailor the dashboard to your specific area of responsibility.

> **Tip**
>
> Cards can be resized, rearranged, shown, or hidden using the **Manage Cards** feature, and you can create and save personalized filter variants for continual analysis. The data powering the cards is refreshed regularly but cached for up to an hour to enhance performance.

Inventory Turnover Analysis (F1956)

Application Type: Analytical

This app enables inventory users to analyze material turnover within a specific plant over a selected period, helping identify materials that may have turnover issues based on user-defined criteria. It supports efficient inventory management by providing detailed insights into how frequently materials are sold or consumed, allowing businesses to optimize stock levels and reduce excess inventory. You can apply dynamic filters such as date range, plant, ABC indicator, and material type to tailor your analysis and visualize key metrics like average inventory value. By offering a clear view of inventory performance, the app aids in making strategic decisions related to procurement, stocking, and pricing at the plant level.

Invoice Price Variance (F0682)

Application Type: Analytical

This app helps you analyze and manage price variances between the invoiced price and the expected or agreed-upon price. It provides real-time insights into any discrepancies between the prices specified in purchase orders and those found on supplier invoices, allowing you to identify and address pricing issues effectively. The app provides a KPI that tracks monthly trends in invoice price variations over the current year to date, based on the weighted material price. Designed for strategic purchasers, it helps analyze and manage discrepancies. You can view KPIs by various dimensions such as **By Plant**, **By Supplier**, **By Purchasing Group**, and numerous other parameters. The app allows filtering of views, visualization through charts or tables, and saving personalized views as tiles. It also supports exporting data to Excel and sharing app links via email. Within the app, you can display the following mini tiles:

- **Invoice Price Variance**: For comparing material price changes (with three thresholds—critical, warning and target)
- **Operational Supplier Evaluation**: Evaluations based on time, price, quantity, and quality
- **Overall Supplier Evaluation**: Across all evaluations
- **PO and Scheduling Agreement Value**: Actual/planned purchasing spend

Furthermore, you can adjust chart filters as either dimensions or measures for display. For example, within the **Measures** section under the **Settings** icon, three options are available: **Average Material Price**, **Invoice Amount**, and **PO Net Amount**. More importantly, you can subscribe to alerts using the **Subscribe** button in the app. The app uses CDS view `C_PURGMATLPRICECHANGE` to retrieve data.

Within the app, you can navigate straight to the following apps: Manage Purchasing Categories (F0337), Purchase Order (F0348A), Supplier (F0354), and My Purchasing Document Items (F0547B).

Job Scheduling and Mass Processing – Physical Inventory (F4550)

Application Type: Transactional

This app allows you to automate the creation and printing of physical inventory documents by allowing you to schedule and manage background jobs. It supports inventory managers in streamlining the inventory process by regularly generating the necessary documents based on predefined or custom job templates. The system determines which materials need to be counted—either annually or based on cycle counting rules—and compiles them into inventory documents. You can define parameters such as start time, recurrence, and control options, and monitor job execution details like status, duration, and logs. Printing also can be automated or done immediately based on needs. The app includes three standard templates (annual inventory, cycle counting, and mass printing) and allows you to create or manage your own templates. It provides status messages throughout the job lifecycle and supports mass processing for large volumes of inventory items. You are guided by system checks and alerts, ensuring that physical inventory documents are created and printed accurately and efficiently in the background.

Tip

For instance, you can use the following job templates:

- SAP_MM_IM_PI_DOCUMENT_CREATEAN (Annual Inventory: Create Physical Inventory Documents)

- SAP_MM_IM_PI_DOCUMENT_CREATECC (Cycle Counting: Create Physical Inventory Documents)

- SAP_MM_IM_PI_DOCUMENT_PRINT (Print Physical Inventory Documents)

Job Scheduling – Manual Reservations (F7554)

Application Type: Transactional

This app allows you to automate the mass deletion of completed manual reservations, eliminating the need to remove each reservation individually. The app defaults to a test run mode to ensure that you always perform a test run before scheduling the actual deletion of manual reservations, so you must switch off the **Test Run** setting to perform an actual deletion. Once the relevant condition is met, the system deletes the reservations. You can track the outcome of each job through a detailed log that shows the status and lists all deleted reservations. The app also supports job monitoring, cancellation, and reviewing results to ensure smooth operation and transparency throughout the process.

Tip

You can use the SAP_MM_IM_RS_MASS_PROCESS template (Deletion of Completed Manual Reservations).

Manage Awarding Scenarios (F5550)

Application Type: Transactional

This app allows you to evaluate and manage different awarding scenarios within a sourcing project by comparing supplier quotations. Awarding scenarios themselves must be created in other apps, such as Manage Sourcing Projects (F4861) or Compare Supplier Quotations (F4863), but this app allows you to search, view, and copy existing scenarios; add or remove quotations; and analyze the total value of submissions. You can mark scenarios as preferred, attach relevant notes or documents, and determine whether a scenario is eligible for awarding. Once finalized, scenarios can be submitted for approval or withdrawn if needed.

Manage Catalog Items (F3149)

Application Type: Transactional

This app supports efficient catalog item management by enabling you to create, modify, activate, or deactivate catalog items using filters like supplier, item, editing status, and material group. It ensures that catalog content is well-maintained, searchable, and ready for procurement use. When creating a catalog item, you are required to input several key details across different sections. Under **General Information**, you must provide the material, item description, material group, an image of the item, associated catalog, base unit of measure, lead time in days, internal catalog name, manufacturer material number, and long text. In the **Supplier Information** section, you enter the supplier's name, validity dates, price per unit, order unit, supplier part number, and any linked purchase contract and contract item. The **Organization Information** tab requires entries such as the plant, company code, and purchasing organization. In the **Text Management** section, you can define the language, item description, search keywords, and long text, with the option to delete any existing entries. You also can filter catalog items by criteria such as price or supplier, view contextual details, and automatically exclude deactivated items from search results.

Manage Central Purchase Contracts (F3144)

Application Type: Transactional

This app allows you to create, modify, and view central purchase contracts that are long-term agreements between an organization and a supplier for the supply of materials or services over a specified period under agreed-upon terms and conditions. These contracts are managed centrally in the SAP S/4HANA hub system and enable you across various company locations to benefit from unified, negotiated pricing and conditions. By leveraging central contracts, organizations can streamline procurement, ensure consistency in supplier terms, and manage purchasing more efficiently across multiple business units.

Manage Central RFQs (F3974)

Application Type: Transactional

This app allows you to display and manage central requests for quotations (RFQs) within the SAP S/4HANA Cloud hub system. You can create central RFQs either from purchase requisitions generated directly in the hub system or from items imported from connected systems. These purchase requisitions created in the hub are then replicated back to the connected systems. The app lets you search for RFQs using various filters such as editing status, RFQ number, company code, and creation date, and you also can edit central RFQs created from the related Process Purchase Requisitions Centrally app (F3290). Central RFQs can be created using different document types: internal sourcing requests (**RQ**), where awarding happens in the SAP S/4HANA system and supplier quotations must be created manually; or external sourcing requests (**RE**), which integrate with SAP Ariba Sourcing to allow suppliers to submit bids and automate the awarding and document creation processes. You can add attachments, notes, and bidders to RFQs, publish them, and create central supplier quotations on behalf of suppliers once the RFQ is published.

The app supports managing supplier data via the Unified Key Mapping Service (UKMS), which can simplify supplier replication between connected and hub systems.

Manage Central Supplier Confirmations (F6634)

Application Type: Transactional

This app allows you to create central supplier confirmations quickly using either a **Quick Create** option or a more detailed **Create** process.

You also can edit, delete, or withdraw existing confirmations as needed. It provides access to important confirmation details such as the related purchase order, connected system ID, creator, purchase order date, and supplier contact information. At the item level, you can review confirmation lines and schedule lines for more detailed insights. The app also offers powerful search capabilities, with filters including editing status, purchasing document, supplier reference, processing status, creator, and supplier, making it easy to find and manage central supplier confirmations efficiently.

Manage Central Supplier Quotations (F3975)

Application Type: Transactional

This app allows you to visualize and manage central supplier quotations that originate from the Manage Central RFQs app (F3974). You can modify quotation details, update pricing information, and submit quotations on behalf of suppliers. The app allows you to award quotations and then create follow-on documents like purchase orders and central contracts based on the awarded quotations. It provides a comprehensive list of all received central supplier quotations, with filtering criteria such as editing status, quotation type, or RFQ number. You can view detailed information, edit or delete quotations in preparation, and navigate to related RFQs and supplier details. When creating purchase orders from awarded quotations, the system groups orders based on purchasing organization, group, company code, and connected system, handling errors by placing orders on hold if necessary. Central purchase contracts also can be created with prefilled contract type, quantity, and distribution line details from the preceding RFQ while allowing you to add further contract specifics before releasing and distributing the contracts to connected systems. This app helps facilitate the management of supplier quotations and efficient procurement processes at a central level.

Manage Cost Breakdown Templates (F6666)

Application Type: Transactional

This app allows you to create and manage cost breakdown templates that suppliers use to provide detailed cost information for sourced items. These templates help organizations compare and analyze costs more effectively before making purchasing decisions. You can view, edit, activate, or deactivate templates, with only one active version allowed at a time; activating a new version automatically revises the previous one. Templates support multiple entries for fields like material number and company code, and they can be integrated into sourcing projects, where standardized cost data ensures transparency and control. The app is designed to standardize cost breakdown management across procurement and project systems, helping organizations maintain consistency and improve cost comparisons. It requires specific business functions and authorizations to work properly and supports detailed configurations such as plant-independent cost conditions and single validity periods for cost breakdown entries.

Manage Manual Reservations (F4839)

Application Type: Transactional

This app enables users—specifically, warehouse clerks—to create and manage manual reservations for goods movements within a warehouse. These reservations can cover goods issues, goods receipts, or transfer postings tied to specific purposes and time frames, helping to plan and organize stock movements efficiently. It simplifies the reservation process by anticipating necessary tasks, allowing you to create, search, view, and delete reservations. It supports advanced available-to-promise (ATP) availability checks to ensure stock availability during reservation and requires appropriate authorization for all actions. You also can copy reservations and integrate with related apps for broader warehouse and profitability management. You can adapt the filters to display line items of manual reservations. More importantly, you can create manual reservations by selecting the **Create Reservation** button in the app. Key default fields include the movement type, plant, and base date.

> **Tip**
>
> Within the **Reservation** creation screen, you must click the line item in the **Reservation Items** section to enter edit mode in order to be able to populate the **Quantity** field.

Manage Model Product Specifications (F5079)

Application Type: Transactional

This app allows you to create and organize structured groups of related service and material items that are often ordered together to complete specific tasks. By using these grouped sets, called *model product specifications*, you can simplify the purchasing process by using them as templates to create purchasing documents like purchase contracts. The app lets you create, view, and edit these model product specifications, import them from spreadsheets, and easily generate purchase contracts based on all or part of these specifications. More importantly, when creating model product specification items, you can choose an item set, service, or material.

> **Tip**
>
> To create model product specifications using this app, item hierarchies need to be activated; if they are not, contact your SAP Fiori administrator.

Manage Negotiations (F5551)

Application Type: Transactional

This app lets you select multiple supplier quotations and create negotiations to secure the best possible terms for sourcing materials and services. You can base negotiations on either the quantity or the price of the items. This feature is accessed through the **Negotiations** section within the Manage Sourcing Projects app (F4861) and works only with submitted supplier quotations. Designed to support procurement teams, the app helps simplify the negotiation process, helping you manage each stage effectively to engage with suppliers, discuss terms, and finalize agreements that deliver the best outcomes for the business.

Manage Physical Inventory Count (F5430)

Application Type: Transactional

This app allows you to enter or update count results for items in physical inventory documents, and you also can create a new physical inventory document and enter counts in one step through ad hoc counting. You start by searching for inventory documents that have materials ready to be counted, then navigate to the specific document you need. On the item level, items are grouped as either uncounted or counted, and for uncounted items, you can directly enter count results and save them to complete the counting step. The app supports barcode scanning on mobile devices with cameras to speed up counting, as well as exporting inventory documents to spreadsheets for further use. It provides an easy-to-use, top-down interface from list level to item level, helping inventory managers efficiently manage the counting process and improve stocktaking. The app lists all physical inventory count documents at plant and storage location levels. More importantly, you can scan a barcode using the

Scan function or perform a random count using the **Ad Hoc Count** function.

Manage Preferred Supplier Lists (F4333)

Application Type: Transactional

This app helps you create and manage lists of suppliers recommended for use during the sourcing process. It supports both standard supplier lists, which are accessible to all users and linked to at least one material group and company code, and user-specific lists that you can customize based on the business scenario(s) and assign to a single material group. The app has two tabs: **Standard Supplier List** and **User-Specific Supplier List**. You can create new supplier lists by selecting relevant material groups and company codes or choosing all company codes if needed. The app allows you to search, view, and edit supplier lists. You also can set user-specific lists as public to share with others, convert them into standard lists by adding more details, and delete lists under certain conditions, such as being the owner or if the list isn't currently in use. You can activate lists so that they become usable in sourcing projects or deactivate them to prevent further use. The app also supports proposing additions or removals of suppliers and company codes within lists, and you can track whether the proposals have been accepted or rejected. Creating a supplier list involves filling in details like description, material group, and responsible purchaser, then adding suppliers and reviewing proposed changes in dedicated tabs.

Tip

Process owners can assign a supplier list to all company codes or specific company codes in the **Supplier List** creation screen, by either selecting **All Company Codes** or specifying a preferred company code(s) in the **Selected Company Code** section under the **General Information** tab.

Manage Procurement Projects (F4860)

Application Type: Transactional

This app allows you to effectively manage procurement projects, which are used to plan product demand at the plant level. Each procurement project can include multiple plants, and for each plant, you can define specific production start and end dates. Using the app, you can create new procurement projects, assign plants to them, and set key production timelines. You also can search for and manage existing projects, apply filters to refine the search, and view detailed change logs for transparency. Once a project is ready, you can activate it to make it available for use in a sourcing project.

Manage Product Master Data (F1602)

Application Type: Transactional

This is a cross-functional app that allows you to create, maintain, and manage product master data across various business processes. It serves as a centralized platform for product information, which ensures that data is accurate and consistent across various process areas such as procurement, sales, production, compliance functions, and many more. The app is used by multiple business roles, most notably procurement teams, sales and distribution teams, warehouse managers, production planners, and compliance officers, among others. You can create, edit, copy, and delete product master records as well as support mass processing for bulk updates. The filter criteria include but are not limited to product, product, GTIN, product group, product category, ANP code, assortment grade, authorization group, purchasing group, product for kit-to-order, and ABC indicator.

The app has various tabs in which you can define product attributes. These tabs collectively support detailed product setup and management. **Basic Data** captures foundational information like division, legacy product numbers, multilingual descriptions, product groups, and categories, serving as the starting point for product identification and classification. **Product Compliance** determines whether a product is subject to regulatory requirements, which is essential for risk management and compliance. **Components** allows you to manage product parts, which is particularly useful for kits or items with bills of materials (BOMs), ensuring all subproducts are correctly recorded. **Classification** enables assignment of attributes such as color, style, or weight to aid in product categorization and reporting. Finally, **Configuration** supports product customization using the **Configure Variant** option, ideal for configurable products with variable features based on customer needs and more.

> **Tip**
>
> Before saving, you also can use the **Check** option to check that there are no errors before proceeding further. The **Edit Header** option at the top of the screen allows for the editing of the product header, which includes the product type, product group, base unit of measure, reference product, change number, and revision level.

Manage Purchase Contracts (Version 2) (F1600A)

Application Type: Transactional

This app allows you to efficiently create, manage, and maintain purchase contracts. It supports the entire lifecycle of a purchase contract—from creation and editing to tracking, renewal, and deletion—improving visibility and control across procurement activities. You can initiate new contracts using predefined templates, select contract types (value or quantity), define conditions, and manage key data such as delivery terms, payment terms, reference quotations, and validity at the purchasing organization level. Items can be added using a flat or hierarchical structure, with support for various

item categories, including **Standard**, **Consignment (K)**, **Material Group (W)**, and **Subcontracting (L)**. Each item can be enriched with details like price, quantity, delivery information, supplier confirmation, and account assignments. You can renew contracts, manage templates, withdraw from approval, and copy, delete, and even create purchase contracts. To create new purchase contracts, you can select the **Create** function. More importantly, the app also allows you to copy, block, delete, or restore contract items, manage partner functions, handle notes and attachments, and schedule automated updates to ensure accurate pricing.

This is the successor app to the Manage Purchase Contracts (F1600) and Purchase Contract (FO350A) apps, which were deprecated as of SAP S/4HANA 1511 FPS02.

> **Tip**
>
> You can create purchase contract templates by using the **Manage Templates** button with a validity period. These templates can be edited or removed as required by process owners.

Manage Purchase Orders Centrally (F3292)

Application Type: Transactional

This app provides a centralized platform for managing purchase orders across systems. You can visualize the entire purchase order process flow, including follow-on documents and service entry sheets. Notifications alert you to overdue items, which you can review or dismiss through the **Situations** tab. The app displays all purchase orders from the hub or connected systems and indicates where outputs are created. With responsibility management enabled, purchase order filters are autofilled based on team settings; otherwise, you can manually filter using criteria like supplier, status, or plant. Unified key mapping is enabled by default, allowing searches based on supplier and material mappings, but can be disabled for regular searches using the **Adapt Filters** option. Purchase orders can be created or edited directly in the hub or connected systems. The app supports print previews (PDF) of purchase order outputs, with business add-in (BAdI) enhancements for form template selection. Output management is extensive: you can manually send, resend, add, reset, or delete outputs. Output status, logs, recipient info, and channels (email, print, Electronic Data Interchange [EDI]) are all accessible. If errors occur, outputs can be retried or marked as completed. You also can refresh purchase data, view notes and attachments, monitor rework status, and trace sourcing project details for replicated purchase orders.

Manage Purchase Orders (Version 2) (F0842A)

Application Type: Transactional

This app allows you to efficiently create, search, view, and manage purchase orders. It allows you to manually order both materials and services, view purchase orders based on their header-level information, and easily identify overdue items. It supports various item categories such as standard, consignment, subcontracting, third-party, and limit items for unplanned materials and services. Purchasers can create new purchase orders from scratch, search and filter existing ones, and save personalized views using variants. Each purchase order's details can be reviewed and edited based on its current status. The app also flags advanced purchase orders that contain unsupported features, prompting you to switch to classic apps for editing. You can delete unsaved purchase orders, copy items or entire purchase orders, and manage item-level blocks or deletions. Statuses such as **Draft**, **In Approval**, **Rejected**, **Sent**, and **Follow-On Documents** help track purchase order progress and approval.

The app supports referencing existing purchasing documents and adding supplementary information for suppliers via texts. It accommodates lean services and unplanned materials with configurable limits and expected values, supports multiple delivery schedule lines, and handles both standard and customer-defined account assignment categories. You can preview outputs, view change logs, and monitor purchase order commitments. Attachments can be added at both header and item levels, including those carried from reference documents. Additional functionalities include specifying acceptance at origin, customizing delivery addresses, setting tax dates, and managing tax codes.

This is the successor to the Manage Purchase Orders app (F0842), which was deprecated as of SAP S/4HANA 1909.

Manage Purchase Requisitions (F2229)

Application Type: Transactional

This app helps you create and manage purchase requisitions by enhancing visibility, control, and sourcing accuracy. At the list level, the app displays all purchase requisitions, and you can copy or even delete purchase requisitions based on status. You can create purchase requisitions by selecting the **Create** button to open the **Purchase Requisition** creation screen. Business users often lack timely updates on new or revised contracts, which can lead to assignments of open purchase requisitions to outdated contracts. To resolve this, the app notifies you about newly negotiated or updated contracts, allowing you to directly assign appropriate sources of supply to purchase requisition items. You can select document types predefined in configuration and manage both flat and hierarchical item lists. Limit items, used for unplanned materials or services, can be created with defined expected values, time periods, and overall limits.

The app fast-tracks requisition creation with features like catalog ordering with automatic price scales, item copying, and autofilled delivery addresses based on plant or storage data. You can link contracts to service items, factor in sourcing criteria like carbon footprint, and edit rejected requisitions for resubmission. The app supports audit trails, workflow previews, and configurable approvals. Commitments are tracked per item-account assignment, and the app limits updates with purchase order creation. Prices carry over as gross or net, tax codes validate against plant data, and valid sources of supply can be assigned unless restricted. Purchasing groups are autovalidated, attachments can be added via the document management system (DMS), and validation messages help ensure accuracy and compliance throughout.

> **Tip**
>
> Purchase requisition document types are maintained in Customizing. By default, the system shows three types: framework requisition (**FO**), purchase requisition (**NB**), and outline agreement requisition (**RV**).

Manage Purchase Scheduling Agreements (F2179)

Application Type: Transactional

This app allows purchasers to create, manage, and view purchase scheduling agreements to ensure timely delivery of goods at the right place and in the required quantities. Within the app, you can maintain key details such as the supplier, materials, schedule lines, and release documentation. It offers a clear overview of all scheduling agreements, highlighting those nearing expiration to prompt timely renewal. You can search for agreements using various filters like document ID, status, supplier, agreement type, and validity period—and even save your preferred filter settings as variants. You can renew an agreement, release it, withdraw it from approval, copy it, and delete it. You can create new purchase scheduling agreements and enter downstream information such as agreement type, company code, purchasing organization, supplier, and purchasing group, among others, in the **Purchase Scheduling Agreement** creation screen.

The app is closely tied to the Display Scheduling Agreements (F7349) and Manage Sources of Supply (Version 2) apps.

> **Tip**
>
> When creating or editing purchase scheduling agreements, you can save custom views by adapting filters for purchasing scheduling agreements items and partners, selecting the **Standard** dropdown icon, selecting the **Save As** button, renaming the view, and saving the view either as public or default.

Manage Purchasing Info Records (F1982)

Application Type: Transactional

This app allows you to efficiently view, create, edit, and manage purchasing info records, which are essential for maintaining the relationship between a company and its suppliers. These records specify pricing, terms, and conditions for the procurement of materials or services from specific suppliers over defined validity periods. You can search for info records based on parameters such as info record number, material, material group, supplier, and plant. Info records can be created for various procurement types, including standard, consignment, pipeline, and subcontracting. To create an info record, you must specify a material or material group along with supplier details at the plant or purchasing organization level. When creating an info record without a material, the sort term becomes mandatory to further classify the material group. During creation, you enter key data such as general information, purchasing data, delivery and quantity, conditions, and references. You also can maintain shipping instructions and select predefined instructions via the value help. The app also allows for enabling info records for automatic sourcing and extending existing records to additional plants or purchasing organizations. This is done by entering the material and supplier combination and then assigning the new organizational data.

Manage Quota Arrangements (F1877)

Application Type: Transactional

This app allows you to view, manage, and create quota arrangements for materials based on a defined validity period. Quota arrangements are purchasing documents used to allocate material requirements across multiple sources of supply by assigning specific quotas or percentages to each source.

At the list level, you can display all existing quota arrangements and filter them by criteria such as material, plant, editing status, and validity. During the creation process, you can define whether the quota arrangement is for internal or external procurement using the **Create** button at the item level. To ensure data integrity, the system prevents creating overlapping arrangements for the same material and plant, and it validates that the sources of the supply—suppliers or plants—are correctly linked to materials and relevant records. The app also supports copying items from existing arrangements using the **Copy from Another Quota Arrangement** function, which helps speed up setup and maintain consistency across similar scenarios. You can define minimum split quantities and maintain base quantities to regulate how demand is distributed, even when new sources are added. Editing is possible at the header level (validity and minimum split) and at the item level (base and max quantities, among other details), but the validity period cannot be changed once the arrangement is in use. Drafts can be deleted, and you can navigate to linked material or supplier object pages. For additional sourcing configuration, the related Manage

Sources of Supply (Version 2) app (F0840A) proves invaluable.

Manage Renegotiations (F5714)

Application Type: Transactional

This app allows you to initiate price renegotiations for items in existing central purchase contracts, helping organizations capitalize on shifting market conditions. You can define key parameters for each renegotiation, such as start and end dates, response deadlines, expected savings percentages, and the specific pricing condition type to be adjusted. Before a renegotiation can begin, the system performs a price check to ensure that the selected items have valid pricing conditions for the specified time frame. Contracts containing hierarchical items are currently excluded from renegotiation. You can propose new validity periods for updated pricing, set criteria for future renegotiations, edit existing renegotiation entries by adjusting conditions or item lists, delete unneeded entries, and finally submit the renegotiation to proceed with the process.

Manage Reservation Items (F5601)

Application Type: Transactional

This app allows you to manage reservation items efficiently by creating, viewing, editing, copying, or marking manual reservations as deleted. The app supports both manual goods movement reservations and stock transfer reservations generated by MRP, though automatically generated dependent reservations cannot be modified. You can search and filter reservation records, perform advanced available-to-promise (ATP) availability checks, and tailor your display using filter criteria on the initial screen. When creating a new reservation, a popup prompts you to specify the movement type, plant, and base date before proceeding to the **Reservation** creation screen. Attachments, including uploaded files or links, also can be added for documentation purposes. All actions are governed by the system's authorization logic to ensure secure processing.

> **Tip**
>
> You cannot change dependent reservations that are automatically generated by the system. Furthermore, similar to the Manage Manual Reservations app (F4839), you can upload or add a link as an attachment to the reservations for future reference.

Manage RFQs (F2049)

Application Type: Transactional

This app allows you to efficiently create, manage, and process requests for quotations (RFQs) within SAP S/4HANA, playing a crucial role in procurement operations. You can initiate RFQ processes and send them either to SAP Ariba Sourcing or directly to suppliers via email or mail. When integrated with SAP Ariba, suppliers can submit bids online, and once a bid is awarded, the system automatically generates a follow-on document—either a purchase order or contract—which can be accessed through corresponding SAP apps. You also can send RFQs to SAP Business Network for Procurement to request price, quantity, or shipping details and receive supplier responses in SAP S/4HANA. The app allows for full customization of RFQs, including maintaining item details (materials or services), delivery addresses, payment terms, and bidder information. Features like automatic bidder proposal suggestions, attachment handling, process flow tracking, and change history provide enhanced transparency and control. You also can edit RFQs, assign legal transactions, and cancel or copy documents as needed. The app provides a user-friendly interface for managing quotations and viewing historical

changes, and it is integrated with the Manage Supplier Quotations app (F1991).

> **Tip**
>
> In the **Request for Quotation** creation screen, the **Get Bidder Proposals** button allows you to get information regarding existing suppliers for material specified in the **Items** tab. To retrieve existing suppliers in the system, the prerequisite is the purchasing organization.

Manage Rules for Automation of Business Processes (F4042)

Application Type: Transactional

This app supports business process automation by enabling you to define, create, and manage automation rules for purchasing documents in the SAP S/4HANA system. You can search for automation rules and use filter criteria such as changed by, created by, draft ID, rule ID, and more based on your preferences. When creating a rule, you input data across several structured tabs. In the **General Information** tab, you must specify a mandatory rule ID, optional description, rule status, validity dates, and starting business object (e.g., `PrmtHbRpldPurchaseRequisition` or `PurchaseRequisition`). Under **Connected System**, you assign relevant automation systems and can copy or delete entries. The **Selection Criteria** tab allows for definition of automation conditions using parameters like system ID, node type (e.g., `PurchaseRequisitionItem`), field names (e.g., `SUPPLIER`), and logical operators (e.g., equal to, not equal to, between). In the **Activity** tab, you define the automation activity with a unique ID and description. The app supports custom fields and grouping logic for more advanced rule configurations, and SAP offers integration with the `MMPUR_BUSPROC_AUTMN_ACTN_CMPLX` business add-in (BAdI) to define complex actions.

Tip

The MM_PUR_MANAGE_BUS_AUTOMN_RULES_SRV OData service (Manage Business Rules for Process Automation) should be maintained and activated in the frontend.

Manage Service Entry Sheets (F2027)

Application Type: Transactional

This app allows you to create, edit, delete, and manage service entry sheets to confirm the completion of ordered services. Each entry sheet is tied to a specific purchase order, allowing you to select planned or unplanned lean service items, add multiple items at once, and even record consumable materials used during the service execution—provided at least one service item is included. You can maintain account assignments directly in the service entry sheet, modify them if needed, and add internal or external notes at both the header and item levels. The app also supports attachments and links for better documentation.

Once completed, service entry sheets can be submitted for approval. If a flexible workflow is active, the approval process is automatically routed; otherwise, a contact person must be entered. Approved sheets generate a goods receipt document, which can then trigger invoice creation. If needed, you can revoke approval (**Withdraw Approval**) and make changes, especially if only the goods receipt exists. Additional capabilities include viewing and editing taxes and pricing, setting the **Final Entry** indicator for completed services, and reviewing the process flow diagram to track related documents. The app supports a lean procurement process like material procurement, improving efficiency and visibility across the service management lifecycle.

Manage Source Lists (F1859)

Application Type: Transactional

This app allows you to efficiently create, edit, delete, and manage source lists used in procurement processes. You can search for and view source lists by criteria such as plant, material, or supplier and more based on the filter criteria. It allows you to use purchasing info records or contracts as valid sources of supply for purchase requisitions and purchase orders. You can create new source lists manually or by copying from existing ones, with available sources filtered based on the selected material and plant. Suppliers can be assigned to source lists, and the status of a source (such as **Fixed** or **Blocked**) can be adjusted as needed. The app also supports generating source lists directly from existing sources of supply, such as contracts or info records, using the **Generate** option. Filters can be applied to tailor the view at the list level, and at the item level, you have full flexibility to copy, generate, delete, or create entries, making the source list management process efficient and comprehensive.

Tip

In the **Source List** creation screen, you can choose to assign a status to source list: **Empty Status, Blocked Status,** or **Fixed Status.**

Manage Sources of Supply (Version 2) (F0840A)

Application Type: Transactional

This app allows you to efficiently search, view, create, and manage sources of supply for specific materials over a defined validity period. Acting as a central tool for supply management, it supports the creation and maintenance of source lists, contracts, and purchasing info records. You can search for sources of supply using filters such as material, supplier, material group, plant, purchasing organization, quota arrangement, and validity period. The app also helps determine and simulate default source assignments for purchase requisitions based on material, plant, and date, displaying the applicable source or multiple options if no single default can be determined. These capabilities enhance procurement efficiency and organization by ensuring that materials are sourced from reliable suppliers under appropriate conditions. You can adapt filters to display existing lists of sources of supply.

Within the app, you can select the **Create** button to navigate to the following screens:

- **Source List** creation screen within the Manage Source Lists app (F1859)
- **Purchase Contract** creation screen within the Manage Purchase Contracts app (F1600A)
- **Purchasing Info Record** creation screen within the Manage Purchasing Info Record app (F1982)
- **Quota Arrangement** creation screen within the Manage Quota Arrangements app (F1877)
- **Purchase Scheduling Agreement** creation screen within the Manage Purchase Scheduling Agreements app (F2179)

This is the successor app to Manage Sources of Supply (F0840), which was deprecated as of the SAP S/4HANA 1511 FPS02 version.

Tip

It's important to note that the list of sources of supply in the app can be filtered based on **Relevance** and **Weighted Relevance**. Relevance is shown using a star ranking system from 1 to 5 stars, while weighted relevance provides a more detailed assessment displayed as a decimal value. This weighted average takes multiple factors into account to reflect the overall suitability of a source more accurately.

Manage Sourcing Projects (F4861)

Application Type: Transactional

This app allows you to efficiently manage sourcing projects by allowing you to display, create, edit, and publish them. It is designed to help identify suitable sources of supply for complex bundles of materials or services, streamlining planning and reducing costs. You can view a list of all existing sourcing projects, search for specific ones, and navigate to detailed information. The app allows you to create new projects, copy or delete existing ones, and update multiple fields across several projects at once. Within a sourcing project, you can add or copy items— including those originating from purchase requisitions— link related items, and indicate the demand status (**Confirmed** or **Uncertain**) and settlement status.

You can define commercial terms such as payment terms, Incoterms, Incoterm locations, and currency, and carry out pricing using conditions and commodity pricing methods. The app supports attaching documents and notes at various levels, from the entire project to individual items, and allows suppliers to contribute notes and attachments to their quotations. You can work with lists of preferred suppliers, monitor the status of supplier quotations, and control access by locking or unlocking suppliers after a deadline or declining participation on their behalf. The app also supports simulations, negotiations, awarding scenarios, and cost breakdown spreadsheets. It allows you to manage sustainability data requests, export and reimport item data through spreadsheets, and track all changes via the version history feature.

Manage Stock (F1062)

Application Type: Transactional

This app allows you to perform a variety of stock changes within the SAP S/4HANA system at the plant level. It supports critical functions such as the initial entry of stock balances during system setup. This includes stock types like unrestricted-use, blocked, and quality inspection stock. The app facilitates stock adjustments during live operations, such as posting goods issues for cost centers when materials are used internally, and scrapping damaged or expired materials with or without serial numbers. You can select multiple items and post them in bulk, generating a single material document for all postings. In addition, the app supports barcode scanning with internal or external devices to speed up and simplify material identification using simple or composite barcodes like global trade item numbers (GTINs). It includes features for checking stock coverage based on consumption data over the past weeks and supports reason codes to explain the cause of stock movements. Stock data can be filtered by storage location and batch numbers using advanced search and drilldown functionality. Materials also can be searched using the **Add Storage Location** option to view materials belonging to specific storage locations.

The app allows navigation to related apps like Compliance Information – For Products (F3226), Stock – Single Material (F1076), Material Documents Overview (F1077), and Manage Inspection Lots (F2343).

> **Tip**
>
> You can search for specific batches using the search functionality using the drilldown functionality in the **Materials** field.

Manage Stock Reporting Procedures (F7068)

Application Type: Transactional

This app allows you to define procedures for sending stock reports to suppliers by specifying how frequently the reports should be generated: daily, weekly, or monthly. After selecting the desired reporting frequency, you can define the exact day of the week or month when stock data should be collected. SAP recommends that you choose the frequency that aligns with the background job schedule set by the system administrator to ensure the timely generation and delivery of reports. You can create, edit, delete, or copy procedures, and quickly identify who created a procedure by using the **Show More per Row** option.

> **Tip**
>
> Stock reporting procedures that are created using the **Create Procedure** option in the Manage Stock Reporting Subscriptions app (F7069) are listed in the hit list of this app.

Manage Stock Reporting Subscriptions (F7069)

Application Type: Transactional

This app allows you to manage stock reporting subscriptions by linking suppliers to defined stock reporting procedures, ensuring they receive regular notifications about stock levels and goods movements. You can filter subscription records using criteria such as editing status, supplier, plant, and company code. The main screen displays a list of stock reporting subscriptions, including details like the material stock subscription ID, plant, reporting procedure, and onboarding date. Available actions include creating or editing a procedure, creating or copying a subscription, deleting subscriptions, adjusting settings, and exporting data to Microsoft Excel.

When using the **Create Procedure** option, you define the procedure name and select a reporting frequency—for instance, daily, weekly, or monthly. This navigates you to the Manage Stock Reporting Procedures app (F7068). Once a frequency is chosen, a **Collection Data** section appears, prompting you to define the reporting period details. After setting up the procedure, you can create subscriptions by navigating to the **Subscription for Procedure** screen, where you enter key subscription item details such as the supplier, plant, and onboarding date. Onboarding and offboarding dates can be used to control when suppliers begin or stop receiving notifications; it is advisable to use offboarding dates instead of deleting subscriptions to preserve historical report references.

Manage Supplier Confirmations (F5039)

Application Type: Transactional

This app allows you to efficiently create, manage, and track supplier confirmations (order acknowledgements) to ensure that goods are delivered in the agreed quantities and on time. You can filter supplier confirmations using various criteria such as editing status, creator, purchase order, supplier reference, and processing status (e.g., new, accepted, in approval, rejected, or withdrawn), with the option to adapt filters as needed. The main screen lists supplier confirmations by confirmation number, purchase order, processing status, and creator. You can perform actions such as withdrawing a confirmation, using the **Quick Create** or **Create** options, deleting, copying settings, and exporting data to Microsoft Excel. The **Quick Create** function requires you to input a reference purchase order to quickly generate a confirmation, while the **Create** option takes you directly to the **Supplier Confirmation** screen. Here, you can enter header-level details, including the reference purchase order and supplier reference, as well as create individual confirmation items by providing relevant information in the **Supplier Confirmation Item** screen. Once all required data is entered, you can confirm the ordered items.

Manage Supplier Quotations (F1991)

Application Type: Transactional

This app allows you to manage and review supplier quotations related to various requests for quotation (RFQs) with detailed visibility and control. You can filter quotations by criteria such as supplier, purchasing group, quotation ID, RFQ type (e.g., external sourcing request, request for quote), status (e.g., **Awarded**,

Cancelled, In Approval), quotation submission date, and creation date. The main view lists supplier quotations along with key attributes such as quotation ID, associated RFQ, supplier name, and status. Selecting a specific line item redirects you to the **Supplier Quotation** screen, which is organized into several informative tabs. The **General Information** tab includes fields like supplier, quotation deadline, RFQ ID, and follow-on document type. **Delivery and Payment Terms** outlines payment terms and Incoterms. The **Items** tab details the quotation items including quantity, awarded quantity, net order price, and material information, with options to adjust table settings or export to Excel. The **Notes** and **Attachments** tabs allow for multilingual note entries and file uploads. The **Approval Details** tab shows the workflow status, processors, recipients, and scenario information.

From this **Supplier Quotation** screen, you can take various actions depending on the quotation's status, such as marking it complete, creating a purchase order or contract, viewing the quotation history, or navigating to related apps like Manage Workflows for Supplier Quotations (F2704). The **History** option allows you to track changes made to the quotation, including who made changes, when, and what data was modified. However, it excludes changes to notes, attachments, approvals, or process flows and only reflects backend database changes.

Manage Supplier Quotations – Sourcing Project (F4862)

Application Type: Transactional

This app lets you view all the supplier quotations that have been received for various sourcing projects, along with detailed information about each quotation. You can easily search for specific quotations and see key details such as the quotation ID, creation date, currency, status, and information about the last changes made, including who made them and when. The system automatically assigns the quotation ID and creation date and tracks any subsequent changes, which cannot be altered manually. For each quotation, you can review item details like quantity, material number, material group, and pricing. You also can add proposed delivery dates. If quotations are still in the **In Preparation** or **Under Revision** status, you can submit, delete, or edit them on behalf of a supplier. For quotations already submitted, you can revise them, which creates a new version while keeping previous versions read-only for reference. The app calculates total amounts at the item, item set, and overall quotation levels based on pricing and quantities. The currency for the quotation is initially copied from the sourcing project but can be changed if needed. You also can manage pricing using conditions and commodity pricing. In addition, the app displays a version history of quotations, allowing you to track revisions with either a table or graphic view.

Tip

Keep in mind that supplier quotations also can be made on behalf of suppliers through the Manage Sourcing Projects app (F4861). Revisions can be made multiple times, but only before the submission deadline.

Manage Suppliers for Sourcing (F4536)

Application Type: Transactional

This app allows you to assign suppliers to a sourcing project by selecting a specific combination of material group and company code. You do this by updating the supplier list directly within the Manage Sourcing Projects app, which is the only way to access this functionality. When you begin this process, the system automatically makes available the preferred supplier list that corresponds to the selected material group and company code for the sourcing project. You can choose to view the supplier list in more detail by selecting the **Display Supplier List** option. There, you can review the list of suppli-

ers linked to the sourcing project. If no standard supplier list exists for the chosen combination, the system generates a blank list, giving you the option to either manually add new suppliers or copy them from a user-defined supplier list.

Manage User-Defined Criteria for Supplier Evaluation (F3812)

Application Type: Transactional

This app allows you to define and manage custom (user-defined) evaluation criteria for rating suppliers. You can assign scores to individual suppliers based on specific, user-defined criteria, which are visible in a comprehensive list within the app. By selecting an entry from the list, you can view detailed information related to that criterion. A key function of the app is the ability to add, edit, or delete counts for criteria associated with combinations of suppliers, purchasing categories, and material groups, all within specified date ranges. You also can leave comments when modifying existing criteria and can activate or deactivate a particular count. The app supports several filter options, including editing status, supplier, material group, purchasing category, validity dates, and criteria, to streamline the management and analysis of supplier evaluations.

Manage Workflow for Purchase Requisitions (F2705)

Application Type: Transactional

This app allows you to configure approval processes at both the item and header levels for purchase requisitions. You can define preconditions and approval steps that determine when and how workflows are triggered. In central procurement, workflows can be configured for connected systems like SAP S/4HANA and SAP ERP, with start conditions based on factors like amount, document type, or company code. Steps can include reviews or approvals, assigned manually or via business add-ins (BadIs), with support for automatic approvals, rework requests, and deadline settings. Reviewers are notified but can only monitor. HR integration is needed for role-based recipients, and requestors can be excluded from approval roles. Workflow errors can be managed in the Flexible Workflow Administration app (F5343). Active workflows can't be edited but can be deactivated or copied. You can prioritize workflows to avoid conflicts, and SAP provides a default fallback approval. Retriggering rules for changes can be customized, and active scenarios appear as defaults in the app.

> **Tip**
>
> SAP recommends using these new workflow scenarios, which should be activated in Customizing, as follows:
>
> - **Overall Release of Purchase Requisition** (WS00800157), object abbreviation PRHeader
>
> - **Release of Purchase Requisition Item** (WS00800173), object abbreviation PRItem
>
> - **Overall Release of Central PR** (WS02000434), object abbreviation CntrPRHeader
>
> - **Release of Central PR Item** (WS02000438), object abbreviation CentralPRItm
>
> To use these, the corresponding old workflows first must be deactivated.

Manage Workflow for Supplier Lists (F4760)

Application Type: Transactional

This app allows you to configure and manage approval workflows for supplier lists used in sourcing processes. When a supplier list is submitted, the app ensures it follows a structured approval or rejection process based on defined conditions. Approvers can be assigned in various ways, including roles such as the manager of the person initiating the workflow. Once an active workflow's start conditions are met—such as when a mandatory supplier is deleted from the sourcing supplier list—the approval process begins. The app allows you to view all workflow definitions and their details, prioritize the order in which workflows are evaluated using the **Define Order** option, and guarantee that only one workflow is triggered per submission. SAP recommends setting a fallback workflow as the last in the order, ensuring there's always a response if no other workflows apply. You can create new workflows or modify copies of existing ones by configuring attributes like descriptions, validity periods, preconditions, approval steps, recipients, and exception handling. Activated workflows cannot be edited, but they can be deactivated or duplicated for changes. Only inactive workflows can be deleted, so active ones must be deactivated first if they are no longer needed.

> **Tip**
>
> You can activate the WS02000090 scenario (**Workflow for Sourcing Supplier List**) in Customizing. By default, the default workflow, **Automatic Release of Sourcing Supplier List**, will be displayed on the initial screen of the app.

Manage Workflows for Awarding Scenarios (F5716)

Application Type: Transactional

This app allows you to configure and manage workflows for approving awarding scenarios in a streamlined and controlled manner. When an awarding scenario is submitted, it must go through an approval process, which is triggered if a workflow is active and its defined start conditions—such as a scenario's total amount exceeding a certain threshold—are met. The app provides flexibility in assigning approvers, including roles like the workflow initiator's manager. You can view all existing workflows and their details, set the order in which workflows are evaluated, and ensure that only one appropriate workflow is triggered per scenario. SAP recommends defining a fallback workflow as the last option to catch any scenarios that don't match earlier conditions. Workflows can be newly created or copied and customized, with configurable properties such as validity period, start conditions, approval steps, recipients, and exception handling actions. Once a workflow is activated, it becomes locked from editing, though it can be deactivated or copied for changes. Inactive workflows can be deleted if no longer needed.

> **Tip**
>
> You can activate the WS01800213 scenario (**Workflow for Awarding Scenarios**) in Customizing. By default, the default workflow, **Automatic Release of Sourcing Project Awarding Scenarios**, will be displayed on the initial screen of the app.

Manage Workflows for Central Purchase Contracts (F3401)

Application Type: Transactional

This app allows you to configure and manage flexible workflows for the approval and review of central purchase contracts and their hierarchies in SAP. It allows you to define specific approvers or reviewers, including individuals or roles such as managers or custom-defined agents, and set up workflows that are triggered when predefined start conditions—like document type, purchasing group, or grouping ID—are met. Each workflow consists of one or more steps, where you can define whether a step is for approval or review, assign recipients, and determine whether steps are mandatory or optional. You can prioritize workflows by defining the order in which their conditions are evaluated, ensuring only one relevant workflow is triggered per document. The app also supports

email notifications to inform workflow initiators and recipients about for instance, approval status, with templates configurable through a dedicated app. You can activate, deactivate, or copy workflows.

> **Tip**
>
> You can activate the WS00800346 (**Workflow for Central Purchase Contract**) scenario in Customizing. By default, the default workflow, **Automatic Release of Central Purchase Contract**, will be displayed on the initial screen of the app.

Manage Workflows for Central Supplier Confirmations (F6635)

Application Type: Transactional

This app allows you to configure and manage approval workflows for central supplier confirmations by defining key workflow parameters such as validity dates, workflow names, descriptions, start conditions, and review steps. You can create detailed workflows that include step-specific conditions, deadlines, and rules about whether to exclude restricted agents as determined by a business add-in (BAdI). The app also supports the creation of alternative preconditions to handle varying approval scenarios. It accommodates automatic, single-step, or multi-step approval processes triggered by specific business events. Once a workflow is activated and its conditions are met, the system initiates the approval flow accordingly. Integrated with responsibility management, the app can automatically assign approvers based on predefined team attributes, enhancing efficiency (if enabled). You also have access to various management options such as creating new workflows, copying existing ones, deactivating outdated workflows, defining the evaluation order, and sorting workflow entries for easier navigation and control.

Tip

You can activate the WS01800277 scenario (**Workflow for Cntrl Supplier Confirmation**) in Customizing by using IMG menu path **ABAP Platform • Application Server • Business Management • SAP Business Workflow • Flexible Workflow • Scenario Activation**. By default, the default workflow, **Automatic Release of Central Supplier Confirmation**, will be displayed on the initial screen of the app.

Manage Workflows for Centrally Managed Purchase Orders (F3316)

Application Type: Transactional

This app allows you to design and manage approval workflows for centrally managed purchase orders. It allows you to define specific conditions—such as company code, purchasing group, or net order value—that trigger a workflow, ensuring only one applicable workflow runs at a time. You can set up both approval and review steps, assigning recipients either directly or through predefined business roles using responsibility management. The app supports various step types, including automatic approvals and reworkable steps, allowing approvers to send purchase orders back for corrections. Each step can include deadlines, preconditions, and optional settings, with email notifications sent if deadlines are missed. Exception handling is built in, letting businesses decide how to proceed when steps are rejected or rework is requested. You also can reuse or customize workflows by copying existing ones and define team roles for approvers and reviewers through the Manage Teams and Responsibilities – Procurement app (F2412).

Tip

You can activate the WS00800333 scenario (**Workflow for Centrally Managed PO**) in Customizing. By default, the default workflow, **Automatic Release of Centrally Managed Purchase Order**, will be displayed on the initial screen of the app.

Manage Workflows for Centrally Managed Purchase Requisitions (F3794)

Application Type: Transactional

The app for managing centrally controlled purchase requisitions allows administrators to create, configure, and manage workflows for approval and review processes. These workflows can be set at either the item or header level and can follow an automatic, one-step, or multistep approval path, depending on the conditions defined. You can specify start conditions such as item category, material group, purchasing group, purchasing organization, plant, and more. Workflows only trigger when all preconditions are met, and the system checks workflows in the predefined order to ensure only one is activated. Within each workflow, you can define step sequences involving approvers and reviewers, where approvers make decisions and reviewers only monitor progress. Workflow steps can be configured with specific properties such as marking them optional or excluding restricted agents based on business add-in (BAdI) logic. Approvers and reviewers can be assigned based on specific users, roles, or custom logic defined via a BAdI. Deadlines can be applied to steps using workday-based time frames like days, hours, or minutes, with escalation options like overdue marking. Exception handling is customizable; for instance, when a

7

step is rejected or rework is requested, the system can be configured to cancel, repeat, or restart the workflow. Recipients of rework tasks also can be determined based on user roles or custom logic. The app also allows you to copy existing workflows, activate or deactivate them, and navigate directly to workflow details. You can select the following step types:

- Automatic iRelease of Centrally Managed Purchase Requisition Item
- Release of Centrally Managed Purchase Requisition

Tip

If there are currently no scenarios available (with the corresponding information that modification is not allowed and workflows cannot be modified) when using the app for the first time, we recommend following these steps to resolve the issue: You need to activate the sourcing project scenario in Customizing using menu path **SAP Reference IMG • ABAP Platform • Application Server • Business Management • SAP Business Workflow • Flexible Workflow • Scenario Activation**, select **New Entries**, proceed to input scenario WS02000494 (**Workflow for Centrally Managed PR Item**), select the **Active** checkbox, and save the entry.

Manage Workflows for Purchase Contracts (F3043)

Application Type: Transactional

This app allows you to configure flexible workflows that streamline the approval and review processes for purchase contracts. With it, you can create new workflows, define their validity periods, and set start conditions based on criteria such as company code, purchasing group, or contract document type. The system checks workflows in the order specified, ensuring only one is triggered when multiple workflows have matching conditions. You can configure step

sequences with both approval and review steps. Approval steps allow for automatic release or for reworkable contract handling—where approvers can send contracts back to creators for adjustments. Each step can be made optional, exclude certain agents like the contract creator, and have user- or role-based recipients. Advanced logic, such as agent determination via business add-ins (BadIs), is also supported. Review steps notify reviewers who monitor the workflow without taking approval actions. Both step types allow for deadline settings with overdue notifications and email alerts, provided the necessary email templates are configured. Exception handling includes actions for rejected contracts or rework requests, with configurable outcomes like repeating steps or restarting workflows. Administrators can activate, deactivate, copy, or modify workflows, but standard SAP workflows should remain active and positioned last in the order. Workflow restarts can be triggered by critical changes to contracts unless configured otherwise. Notifications can be sent on approval or rejection, and approvers or reviewers can be managed through the Manage Teams and Responsibilities – Procurement app (F2412). If no active workflow preconditions are met, the system defaults to an automatic contract release. You can choose the following steps:

- Automatic Release of Purchase Contract (default)
- Release of Purchase Contract
- Release of Reworkable Purchase Contract

Manage Workflows for Purchase Orders (F2872)

Application Type: Transactional

This app allows you to design and manage workflows that streamline the approval process for purchase orders. You can define workflows based on specific start conditions and configure the sequence in which these conditions are evaluated. Only one matching workflow is triggered at a time, ensuring efficient and conflict-free processing. Workflows can be newly created or copied and modified, with settings for properties like validity periods and detailed preconditions—for instance, a threshold for the total net amount of the purchase order, automatically converted to the relevant currency. Each workflow can include multiple steps with defined recipients and preconditions. You can order these steps, remove unnecessary ones, and set deadlines, including specifying actions if deadlines are missed. Steps also can be marked optional; if no approver is found or an error occurs, the step is skipped, and the workflow moves on. The workflow also can exclude specific agents, like the purchase order creator, from receiving approvals. Recipients can be determined using various roles, such as cost center or project responsible persons, or defined using the Manage Teams and Responsibilities – Procurement app (F2412), where appropriate roles must be assigned to team members.

Several step types are supported, including **Automatic Release of Purchase Order** and **Release of Purchase Order**. Email notifications can be configured for different workflow outcomes, and email templates can be managed separately. Exception handling allows for responses like canceling, repeating, or restarting the workflow depending on rejection or rework events. In addition, workflows can be set to restart based on critical changes to the purchase order, using custom logic defined through extensibility tools.

Manage Workflows for RFQs (F3232)

Application Type: Transactional

This application enables business process specialists to configure and manage workflows to fast-track the approval process of requests for quotations (RFQs). It offers flexibility in defining how and when workflows are triggered, including the ability to assign approvers based on roles like the manager of the workflow initiator. Once a workflow with one or more steps is activated and its defined start conditions are met—such

as a specific net value for an RFQ—the approval process is automatically initiated.

Using the app, you can view a list of existing workflows and examine detailed configurations for each. You can create new workflows or copy and modify existing ones to suit your business requirements. Each workflow allows you to define a series of steps, assign recipients, and set preconditions for each step. You also can determine the order in which steps should be executed and remove any unnecessary ones. Workflows are prioritized based on the order of their start conditions; when multiple workflows could be triggered, only the first one with matching conditions is activated. Example step conditions that can be selected include **Initiator of Workflow**, **RFQ Creator**, **RFQ Document Type**, and more. In addition, you can define workflow properties like descriptions and validity periods, activate or deactivate workflows as needed, and easily navigate into the configuration details of each workflow for further adjustments.

> **Tip**
>
> The default workflow, **Automatic Release of RFQ**, is activated automatically and displayed on the initial screen.

Manage Workflows for Scheduling Agreements (F3044)

Application Type: Transactional

This app allows you to configure workflows to streamline the approval process for purchase scheduling agreements. You can define specific preconditions—such as document type, purchasing group, company code, plant, or material group—to trigger the overall release workflow. Once a workflow with one or more steps is activated and the start conditions are met, the approval process is automatically initiated. You can manage the order in which workflows are evaluated to ensure only the first applicable one is triggered, even if multiple workflows share similar conditions. For workflow steps, you can select the following step types:

- **Automatic Release of Scheduling Agreement**
- **Release of Scheduling Agreement**

The app allows you to build workflows from scratch or copy and adapt to existing ones. Each workflow can contain approval and review steps, for which you define the step name, type (approval, review, or automatic release), recipients, step-specific conditions, deadlines, and exception-handling actions. Recipients can be assigned using direct user selection or role-based determination, such as managers or persons responsible for cost centers or projects. Roles must be properly configured using the Manage Teams and Responsibilities – Procurement app (F2412), which also allows you to assign team functions that the workflow uses to find the correct approver or reviewer.

> **Tip**
>
> If there are currently no scenarios available (with the corresponding information that modification is not allowed and workflows cannot be modified) when using the app for the first time, we recommend following these steps to resolve the issue: You need to activate the sourcing project scenario in Customizing using menu path **SAP Reference IMG • ABAP Platform • Application Server • Business Management • SAP Business Workflow • Flexible Workflow • Scenario Activation**. Select **New Entries** and proceed to input the scenario WS00800305 (Workflow for Scheduling Agreement), select the **Active** checkbox, and save the entry.
>
> Also, SAP strongly recommends keeping this standard workflow (**Automatic Release of Scheduling Agreement**) active and placing it last in the order, ensuring a backup is always available if no custom workflows apply.

Manage Workflows for Service Entry Sheets (F3173)

Application Type: Transactional

This app allows you to set up flexible workflows for service entry sheets, allowing them to define customized approval processes. The workflow can follow an automatic, one-step, or multistep approval structure, and approvers can be assigned through roles, specific users, or even teams. When a workflow is activated and its defined start conditions are met, the approval process begins automatically. You can select the following step types:

- **Automatic Release of Service Entry Sheet**
- **Release Service Entry Sheet**

> **Tip**
>
> SAP recommends using the new flexible workflow functionality offered by the SAP S/4HANA system. Furthermore, SAP recommends using the default workflow, **Automatic Release of Service Entry Sheet**, as a fallback option.

Manage Workflows for Sourcing Projects (F5289)

Application Type: Transactional

This app allows you to configure workflows that streamline the approval process for sourcing projects. The app provides flexible options for determining the approver, such as assigning specific users or the manager of the workflow initiator. When a workflow with one or more steps is activated and its defined start conditions are met, the approval process is automatically triggered. You can view a list of existing workflows, examine detailed configurations, and define the sequence in which start conditions are checked. Only the first matching workflow in the sequence is triggered, even if multiple workflows share similar conditions. New workflows can be created or copied from existing ones and customized with properties like descriptions and validity periods. Start conditions are defined through preconditions, such as setting a threshold for the sourcing project's target value. Workflow steps can be configured by assigning recipients, setting step conditions, and defining exception handling actions if needed.

Once a workflow is activated, it can no longer be edited, though it can be copied or deactivated. If changes are needed, you must duplicate the workflow, apply the updates, deactivate the original, and then activate the new version. Workflows that are no longer needed can be deleted, but only if they are inactive; active workflows must be deactivated first. You can choose the following step types:

- Automatic Release of Sourcing Project
- Release of Sourcing Project

Manage Workflows for Supplier Invoices (F2873)

Application Type: Transactional

This app allows you to configure workflows that facilitate the approval process for supplier invoices, particularly those that are blocked or parked. For example, a blocked invoice may require approval from one or more responsible individuals before it can be released for payment. The workflow can be tailored to the organization's needs, offering flexible options for selecting approvers—either by assigning specific users or using roles such as the cost center responsible. Once a workflow is activated and its start conditions are met, the approval process is triggered automatically.

You can choose between different workflow scenarios. Workflows can be newly created or copied from existing ones, and configured with descriptions, validity periods, start conditions (e.g., specific company codes, account assignment categories like cost center or work breakdown structure [WBS] element, and blocking reasons such as quantity or price variances). The workflow setup also includes defining step sequences, assigning recipients, setting preconditions for each step, and optionally specifying actions for exception handling. If integrated email templates are maintained, the app allows you to send automatic email notifications based on workflow results.

Manage Workflows for Supplier Quotation (F2704)

Application Type: Transactional

This app allows you to create and manage flexible workflows specifically for supplier quotations, allowing for the customization of approval processes. Using the search bar and available options like **Create**, you can access the **New Workflow** screen, where you can input a workflow name, add a description, set validity dates, and define start conditions. Workflow steps can then be configured with optional step names, selected step types, specific conditions, deadlines, and exception handling parameters.

The app supports a variety of approval models, including automatic, single-step, or multistep workflows. Approvers can be assigned directly or determined dynamically—for example, based on cost center responsibilities. Once a workflow is activated and its preconditions or start conditions are met, it triggers the appropriate approval process. The app ensures only the first workflow that matches the start conditions is executed, avoiding any conflicts or overlaps. Within each workflow, you can reorder, add, or remove steps, assign recipients, and define step-specific conditions. Start conditions include **Quotation Follow-On Document Category**, **Initiator of Workflow**, **Quotation Document Type**, **Supplier Quotation Creator**, and more. The app also offers functionality to view, activate, deactivate, or duplicate existing workflows, making it easier to maintain or adapt processes as business needs evolve.

Tip

If you do not specify the step name when creating workflow steps, the system will automatically assign the step type instead.

Manually Retriggering Business Process Automation (F5741)

Application Type: Transactional

This app allows you to view errors that occurred during the execution of business process rules and provides the ability to manually retrigger those rules for reprocessing. The app is closely tied to the Manage Rules for Automation of Business Processes app (F4042). For example, if an automated rule to create a purchase order from a purchase requisition fails due to missing supplier data, the error will appear in this app. After correcting the data in the source document, you can manually retrigger the rule to execute the process again without needing to restart the entire workflow.

Mass Changes to Central Purchase Contracts (F3792)

Application Type: Transactional

This app allows you to manage multiple central purchase contracts and their hierarchies efficiently in both online and offline modes. Through filter options like supplier, central purchasing organization, central company code, connected company code, and distribution status (such as distributed, in distribution, or error in distribution), you can narrow down the contracts you want to work with. In online mode, you can directly edit several contracts and their hierarchies within the app's interface, using tabs organized into central contract headers, items, hierarchy headers, and hierarchy items. You can choose to clear, keep, or replace values (limited to one new value per field) across selected records. The app also allows mass editing of header and item distribution percentages across multiple contracts, either uniformly or individually. Offline, you can export contract data into Microsoft Excel, make necessary changes or additions, and reupload the file to apply the updates. More importantly, when version-dependent fields are updated, the system automatically generates new versions of the affected contracts. Additional options such as mass edit, mass add, upload, download, and monitor mass changes support seamless data handling, while the interface can be adapted to show only relevant fields based on user needs. The **Monitor Mass Changes** option navigates you to the Monitor Mass Changes – Central Purchase Contracts app (F4400). The following tabs are available:

- **Central Purchase Contract Headers**
  Provides an overview of central contract header details, including ID, type, supplier, validity dates, and target value. You can copy contracts, manage distribution, adjust display settings, and export data to Microsoft Excel.

- **Central Purchase Contract Items**
  Shows line-item details of central contracts, including item ID, product type, material group, and target quantity. You can copy items, manage distribution, and customize the view as needed.

- **Central Purchase Contract Hierarchy Headers**
  Displays parent-level data of hierarchical contracts, including supplier, validity dates, and header info. You can copy headers, manage distribution, adjust settings, and export to Microsoft Excel.

- **Central Purchase Contract Hierarchy Items**
  Shows item-level details in contract hierarchies, such as product type, material group, and target quantity. You can copy items, manage distribution, adjust views, and perform mass edits across contracts.

Mass Changes to Centrally Managed Purchase Requisitions (F7377)

Application Type: Transactional

This app allows you to effectively manage the process of updating multiple centrally managed purchase requisition items at once. You can filter requisition items based on criteria such as requisition number, connected system, company code, plant, purchasing organization, supplier, material, product group, and item category. After applying filters, you can select multiple requisition items, provided they originate from the same connected system and choose to clear, replace, or keep specific field values. During each mass update, only one new value can be assigned per field. The app includes a **Mass Edit** button to initiate updates and a **Monitor Mass Change** function to track the status of scheduled mass change jobs. You also can activate features like responsibility management to focus on items assigned to you. The interface is adaptable, allowing you to customize visible data fields based on organizational requirements. Purchase requisition item details such as the item number, material, item category, plant, and company code are clearly displayed for review and selection.

> **Tip**
>
> The `UI_PHRPRQITMMASSUPDT` OData service should be activated and maintained in the SAP S/4HANA system.

Mass Changes to Purchase Contracts (F2669)

Application Type: Transactional

This app allows you to perform mass updates to multiple purchase contracts, including both header-level and item-level fields. You can filter and view purchase contract items grouped by contracts using criteria such as supplier, purchasing organization, purchasing group, and material group. Once the relevant contracts are selected, the **Mass Edit** function enables changes to fields like terms of payment, purchasing group, exchange rate, fixed exchange rate, target value, and purchasing document name, as well as partner-related fields. Only one new value can be assigned per field across the selected contracts. After specifying the updates, the **Apply Mass Changes** button executes the modifications. To track progress and status, the **Application Jobs** feature displays job details including scheduling, parameters, and execution results.

> **Tip**
>
> You also can download and upload Microsoft Excel files for data exchange within the app.

Mass Changes to Purchase Orders (F2593)

Application Type: Transactional

This app allows you to efficiently apply changes to multiple purchase orders at once, including both header- and item-level fields. By using filter criteria such as purchase order number, supplier, company code, purchasing organization, purchasing group, material group, and plant, you can view and select relevant purchase order items grouped by their respective orders. Through the **Mass Edit** feature, you can update fields like payment terms, discount periods, net payment terms, discount percentages, and purchasing group. However, only one new value per field can be applied across the selected purchase orders. Once updates are defined, the **Apply Mass Changes** button processes the changes. To monitor these changes, the app provides a change log via the **Monitor Mass Changes** button, which shows the status and details of each update job, including scheduling, parameters, and general information, offering full visibility and traceability for mass changes.

Mass Changes to Purchase Requisitions (F2594)

Application Type: Transactional

This app simplifies the process of editing multiple purchase requisition items simultaneously. You can apply filters such as purchase requisition number, fixed vendor, plant, material group, company code, item category, purchasing organization, and desired vendor to narrow down the list of requisition items. The app then displays details like purchasing group, requirement tracking number, storage location, and indicators for goods receipt and invoice receipt. Once the relevant items are selected, you can utilize the **Mass Edit** function to assign a single new value per field such as quantity, material group, short text, or storage location across all selected entries. These changes are finalized using the **Apply Mass Changes** option. In addition, the app offers the Mass Change Log feature, which helps you track the progress and results of their changes by providing information on job parameters, schedules, and execution details. Options like application jobs, copying, and export to Microsoft Excel further enhance usability.

Mass Changes to Purchasing Info Records (F2667)

Application Type: Transactional

This app allows you to manage multiple purchasing info records efficiently, either in online or offline mode. You can apply filters such as purchasing organization, plant, and supplier, with the flexibility to adapt filters based on specific needs. The interface includes two tabs, **Purchasing Info Records** and **Purchasing Organization**, with the number of entries indicated in brackets for each. In online mode, you can directly update general and purchasing organization data for multiple records from the list view. This includes making price adjustments using the **Change Price** option within the **Purchasing Organization** tab and performing bulk updates on fields like deletion indicators, short texts, supplier material numbers, and regular suppliers using the **Mass Edit** feature. Navigation to detailed object pages provides further insight into suppliers, materials, or individual info records. Mass change activities can be monitored via the Monitor Mass Changes – Purchasing Info Records app (F4394), accessible through the **Application Jobs** option.

In offline mode, you can download purchasing info record data as a Microsoft Excel file to create, update, or delete large volumes of records. This approach enhances flexibility and efficiency for bulk operations. After modifications are made offline, the updated file can be uploaded back into the system. Whether changes are made online or offline, all actions are recorded in change logs for transparency and auditing. Additional options include copying data, adjusting settings, exporting to Microsoft Excel, and uploading/download operations.

Mass Changes to Scheduling Agreements (F2668)

Application Type: Transactional

This app allows you to make mass changes to multiple header and partner fields for purchase scheduling agreements efficiently. You can easily update fields like payment terms, fixed exchange rates, delivery addresses, incoterms, purchasing group or partner data across several agreements using the **Mass Edit** option. The app allows you to filter and display scheduling agreements based on criteria such as supplier, purchasing organization, purchasing group, and material. Headers and items are shown on separate tabs, allowing targeted changes. You can choose whether to clear, replace, or keep existing values, with the option to enter only one value per field. In addition, the app lets you view the status of scheduled mass change jobs using the **Application Jobs** option and provides integration with the Monitor Mass Changes – Purchase Scheduling Agreements app (F4392) for real-time tracking. The app is customizable, so business users can adjust the views to show only the data relevant to their tasks.

Material Documents Overview (F1077)

Application Type: Transactional

This app allows you to track and review material documents. You can filter results based on stock changes (such as all changes, no stock change, stock decrease, stock increase, or transfer postings), as well as criteria like plant, storage location, stock type, material document number, document year, material, posting date, and document date. The resulting list displays key details including the material document number, year, material, plant, storage location, and posting date. When you select a specific material document from the list, you are taken to a detailed **Material Document** screen. Here, you can view additional information such as the

number of reserved items associated with the document. If necessary, you also can reverse a goods receipt directly from this screen by using the **Reverse** option; this action requires entering a new posting date and optionally adding reference notes before finalizing the reversal. More importantly, the app supports material documents for batch-managed materials with split valuation. The **Related Apps** option allows you to navigate to the Display Document Flow app (F3665), Display Process Flow – Accounts Payable app (F2691), and more. The following tabs are available:

- **General Information**
  Displays key material document details like document and posting dates, creator, document type, and transaction type. It also shows time stamps, delivery note, bill of lading, and related notes for quick context.
- **Items**
  Lists material document line items, including document number, material, movement type, quantity, and plant. It features search, display customization, and Microsoft Excel export options for easier analysis.
- **Attachment**
  Lets you upload supporting documents or links, such as delivery notes or certificates, to store with the material document.
- **Process Flow**
  Shows a visual flow of related documents, linking the material document to items like accounting entries and their statuses. You can zoom in, zoom out, or reset the zoom view to trace the full process easily.

This is the successor app to the Material Documents app (F1807), which was deprecated as of SAP S/4HANA 1511 FPS02.

> **Tip**
>
> Based on your authorizations, you can navigate to the numerous related accounting apps by selecting **Accounting Document** in the **Process** flow tab for further processing.

Material Price Variance (F3294)

Application Type: Transactional

This app allows you to monitor and analyze material price variances by comparing the actual prices paid for materials against the prices maintained in the system. It provides insights into these variances across different procurement documents such as info records, purchase orders, contracts, and invoices. You can apply filters like display currency, material, creation date range, company code, plant, supplier, purchasing organization, and purchase contract to narrow down the data for analysis. The app supports a variety of object types, including invoices and moving average prices, making it easier to trace the root causes of pricing deviations. With both chart and table views available, you can view aggregated data, zoom in on specific details, and get information into related documents for deeper insight. The app also includes features for customizing filters, navigating to related apps, and exporting data for reporting.

Monitor Central Purchase Contract Items (F3492)

Application Type: Analytical

This app allows you to efficiently monitor central purchase contract items. You can refine your analysis by applying a range of filters, including supplier (a key field), display currency, company code, plant, material, material group, logical system, and more. The **Details** option allows you to explore data in greater detail, such as moving from an overview of target values by material group to specific materials within each group. In addition, you can track the status of change requests related to each contract item. The change request statuses include the following: **Active Document Has Enabled Change Request**, **Active Document Has No Enabled Change Request**, **Change Request Document Is Disabled**, **Change Request Document Is Enabled**,

and **Change Request Not Applicable (Historic Version)**. If an intermediate document related to a change exists, the system marks the relevant change request status and displays information from the last approved version. You also can select individual contract line items in the table view for detailed views and initiate contract renewals directly from the app, setting updated validity periods and target values as needed.

Monitor Compatibility of Central Procurement Operations (F4618)

Application Type: Transactional

This app provides detailed insights into the compatibility of central procurement business operations within the hub system and their alignment with the versions of connected systems necessary for smooth functionality. It presents a comprehensive list of central procurement operations and indicates whether each operation is compatible, incompatible, or not configured, based on attributes such as connected systems, API names and versions, software components, support pack levels, and more. The app supports monitoring for a variety of business operations, including refreshing central purchase requisitions and orders, bulk extraction of procurement data, creating purchase orders from requisitions, managing value help and process flows, handling sources of supply, and actions like reassigning, closing, or changing the status of requisition items.

Monitor Follow-On Document's Log (F5300)

Application Type: Transactional

This app supports you in tracking and analyzing follow-on documents within Central Sourcing, providing essential visibility into central procurement activities. It presents a list of application logs, each capturing critical details such as severity level (e.g., error, warning, success), number of items involved, creation date, category (e.g., MMPUR_CQTN [central supplier quotation]) and subcategory descriptions, and the creator's name. You can filter the logs using a variety of criteria including severity, date range, category, subcategory, and external reference. Filters also can be adapted to suit user preferences, allowing for more tailored log monitoring. This functionality helps ensure that you can quickly identify issues, successes, or relevant changes in sourcing processes by reviewing the logs generated by follow-on documents, such as central supplier quotations or related sourcing actions.

Monitor Job Scheduling for Supplier Evaluation Score History (F5062)

Application Type: Transactional

This app allows you to create and manage custom criteria used to evaluate and rate suppliers. You can assign scores to specific suppliers based on these user-defined criteria. A list of all existing criteria is displayed within the app, and you can click on individual entries to view detailed information. In addition, the app allows you to monitor scheduled jobs that have failed, providing basic details such as the job name, scheduling options, creation date, and start and end times. By selecting a failed job, you are directed to the Schedule Persistency of Supplier Evaluation Scores app (F3814), where you can analyze the issue, troubleshoot it, and attempt to restart the job.

The app uses CDS view C_SUPLREVALSCRHIST-JOBMNTR to retrieve data.

Monitor Jobs for Model Product Specifications (F7440)

Application Type: Transactional

This app allows you to monitor and manage jobs generated during the import of model product specifications through the Manage Model Product Specifications app (F5079). For instance,

when a spreadsheet is uploaded to create or update model product specifications, the system automatically schedules an import job. You can track the status of these jobs, whether scheduled, in process, completed, or failed, using filters such as status and date range. The app also allows you to create new jobs directly by following a step-by-step process: selecting a job template, defining scheduling options (such as recurrence), and entering relevant parameters. Additional features include options to adapt filters, view job details, cancel or restart jobs, export results to Excel, and save customized views for future use. This functionality ensures a transparent and efficient import process by allowing you to review errors, validate outcomes, and manage data updates effectively.

Tip

When using the app for the first time, if you see the error **Usage of item hierarchies is disabled. Contact your administrator.**, we recommend that administrators take these steps: Using Transaction SFW5 (Switch Framework: Change Business Function Status), activate business function MM_PUR_SRVCPROC_ITM_HIER (Item Hierarchy in Service Procurement) by selecting the business function under **CLIENTS_FUNCTIONS** and selecting Ctrl + F3. Then, using Transaction SFW1 (Switch: Initial Screen), activate the MM_PUR_SRVCPROC_SFWS_ITM_HIER switch and select Ctrl + F3.

Note that a warning message will be displayed when activating the Item Hierarchy in Service Procurement business function. Only make activation changes as and when required by the business.

Monitor Mass Changes – Central Purchase Contracts (F4400)

Application Type: Transactional

This app allows you to monitor mass change jobs created and scheduled using the Mass

505

Changes to Central Purchase Contracts app (F3792). It also allows you to track modifications made to multiple fields. You can filter jobs by attributes such as status, creation date, planned start time, document category, error and warning messages and job name. By selecting a job under the **Mass Change Job** or **Simulation Jobs** tab, you can view its general details, system messages, and any documents affected by errors or warnings.

Monitor Mass Changes – Centrally Managed Purchase Requisitions (F7451)

Application Type: Transactional

This app allows you to track and manage mass change jobs created and scheduled using the Mass Changes to Centrally Managed Purchase Requisitions app (F7377). It also allows you to view modifications made across multiple fields through the **Mass Edit** option in the Mass Changes to Centrally Managed Purchase Requisition Items app. You can filter jobs by criteria such as status, creation date, planned start time, and job name. Under the **Mass Change Jobs** tab, you can access general job details, system messages, and documents affected by warnings or errors. The app displays various job statuses, including **Scheduled** (not yet active), **Released** (ready for execution), **Ready** (pre-execution phase), **Active** (currently running), **Finished** (completed successfully), and **Canceled** (terminated due to errors).

Monitor Mass Changes – Purchase Contracts (F4393)

Application Type: Transactional

This app allows you to track and manage jobs that have been created and scheduled through the Mass Changes to Purchase Contracts app (F2669) through the **Mass Edit** option. You can

monitor the progress of these mass change activities, view detailed job results by selecting the **Application Jobs** option, and perform actions such as copying or canceling a job as needed in this app.

Monitor Mass Changes – Purchase Orders (F3332)

Application Type: Transactional

This app allows you to monitor and manage both simulation and mass change jobs related to purchase orders. It offers visibility into job details such as job ID, planned start time, and any system-generated error or warning messages. You can filter jobs by various criteria, including status (e.g., attended, failed, in process), job name, creation date, planned start time, and document category, like purchase orders. The interface is divided into two tabs, **Simulation Jobs** and **Mass Change Jobs**, where you can review job progress and results. Through integration with the Mass Changes to Purchase Orders app (F2593), you can initiate and track changes by selecting the **Monitor Mass Changes** option. Each job logs changes made across multiple fields and provided tools to address issues by skipping documents with errors or warnings. Once you are satisfied with the simulation results, you can apply the changes, which are then reflected and tracked under the **Mass Change Jobs** tab in this app for future reference.

> **Tip**
>
> Simulation jobs are triggered by making mass edits, for instance to purchase order headers, and selecting the **Simulate Job** option in the Mass Changes to Purchase Orders app (F2593). Note that simulation jobs are then listed in the **Simulation Jobs** tab in this app.

Monitor Mass Changes – Purchase Requisitions (F4391)

Application Type: Transactional

This app allows you to oversee jobs that have been created and scheduled using the Mass Changes to Purchase Requisitions app (F2594) through the **Mass Edit** app. It allows you to monitor the status of these mass change jobs, view the outcome by accessing the log through the **Application Jobs** option, and provide options to copy or cancel a job, when necessary.

Tip

For instance, you can use the Mass Update Purchase Requisition template (SAP_MM_PUR_MASSPRBG_T).

Monitor Mass Changes – Purchase Scheduling Agreements (F4392)

Application Type: Transactional

This app lets you track jobs created and scheduled through the Mass Changes to Purchase Scheduling Agreements app (F2668) using the **Mass Edit** option. You can monitor the progress of these mass change jobs, view detailed results by clicking the **Application Logs** icon, and can copy or cancel any job as needed.

Tip

For instance, you can use the Mass Changes to Purchase Scheduling Agreements template (SAP_MM_PUR_MASSSABG_T) to create jobs.

Monitor Mass Changes – Purchasing Info Records (F4394)

Application Type: Transactional

This app allows you to search and filter jobs by criteria such as status (**Canceled, Failed, Skipped, User Error, Ready, In Process**) and date range. It lists jobs with details including status, log, results, steps, job name, planned start, and created by. You can create new jobs, copy existing ones, cancel jobs, restart failed jobs, and adjust settings. The app is designed to monitor jobs created and scheduled through the Mass Changes to Purchasing Info Records app (F2667), enabling you to track job progress, view detailed results by accessing logs, and manage jobs efficiently.

Monitor Purchase Contract Items (F2423)

Application Type: Analytical

This app allows you to effectively monitor and analyze purchase contract items by providing both visual charts and detailed tables that summarize key values, such as target quantities and amounts, grouped by material group. You can filter data using parameters like supplier, purchase contract number, material, company code, plant, and display currency. The **View By** option allows for deeper insights for instance, starting from high-level target amounts by material group and narrowing down to individual materials. In table view, each entry includes specific details such as the contract number, item number, material, material group, and target quantity. You can view more information by selecting a contract item and choosing the **Details** option, which links to related purchase orders and additional contract data. The app also supports exporting data to Microsoft Excel, copying contract information, and renewing existing contracts.

For further context, you can navigate directly to related apps like Manage Purchase Contract (F1600A) to explore contract-level details, such as through the **Details** option.

Monitor Purchase Order Down Payments (F2877)

Application Type: Analytical

This app allows you to track and analyze purchase order down payments by applying a variety of filters, including supplier, company code, plant, order type, category (such as mandatory down payment, voluntary down payment, or no down payment), and status (for instance, overdue or partially posted). A display currency must be specified to view the relevant financial data. You can sort and drill down into the information by key metrics such as due date, down payment percentage, and invoice quantity. For a given supplier, the app displays details such as the total down payment amount, posted down payments, and the net order value. The app includes visual filters and a details function for deeper analysis, such as examining trends by purchasing group or supplier.

You also can navigate to related apps to explore contextual details about associated purchase orders, materials, and suppliers such as Clear Outgoing Payments (F1367), Post Outgoing Payments (F1612), and more based on authorization. More importantly, you can make a down payment by using the **Create Down Payment Request** option, where you can either simulate or make postings accordingly.

Monitor Purchase Order Items (F2358)

Application Type: Analytical

This app allows you to efficiently monitor and analyze purchase order items and their statuses by applying various filters such as display currency, supplier, material, plant, order status, delivery forecasts, and more. You can view purchase orders based on different dimensions like who created them, final invoice status, material, plant, and storage location. The app supports both chart and table views, allowing you to toggle between visual insights and detailed data. In chart view, you can track key metrics such as net order value, value still to be delivered, and value pending invoicing. Example charts include **Item Count by Next Delivery Status**, **Item Count by Upcoming Deliveries**, and **Item Count by Order Status**. It also includes a supplier delivery reliability overview covering the past 180 days. Selecting a purchase order item provides detailed item-level data, and you can further navigate to related apps like Manage Purchase Orders (F0842A) to gain deeper context about suppliers and materials. Additional options like maximizing charts, customizing settings, and copying or exporting data enhance usability and analysis.

Monitor Purchase Order Items Centrally (F3676)

Application Type: Analytical

This app enables centralized monitoring of purchase order items, presenting data by default in both chart and table views, with aggregated values grouped by material group. You can filter purchase order items based on various criteria such as display currency, supplier, material, or purchasing organization. The app supports drill-down functionality, allowing you to explore detailed data across dimensions, such as viewing the total value by material group and then narrowing it down to individual materials. In addition, you can click on a purchase order to access its detailed information and use visual filters to analyze total values by material group, supplier, and more, ensuring comprehensive visibility into purchasing activities.

Monitor Purchase Requisition Items (F2424)

Application Type: Analytical

This app allows you to efficiently search for and monitor purchase requisition items using a range of filter options such as display currency (as a key field), purchase requisition number, purchasing organization, purchasing group, material group, plant, and more. You can select individual requisition items to view important details including the source of supply, account assignment data, and general item information.

You also can navigate to other related apps such as Analyze Stock in Date Range (F6185), Inventory KPI Analysis (F3749), Manage Stock (F1062), Display Supplier Balances (F0701A), and more using the **Material** and **Supplier** hyperlinks.

Monitor Purchase Requisition Items by Account Assignment (F2422)

Application Type: Analytical

This app allows you to view and analyze purchase requisition items based on their account assignments, with the data initially grouped by default according to cost centers. You can apply various filters such as purchase requisition number, account assignment category, cost center, asset, desired vendor, material group, company code, and plant to refine the results. The app displays key details including requisition item number, material, account assignment category, sequence number, distribution percentage, and the distributed net value. By selecting a specific cost center, you can drill down into the corresponding purchase requisition items to view detailed information such as assigned materials, quantities, and account distribution. You also can click individual requisition items to access more comprehensive details, including source of supply and general item attributes. Additional options like adapting filters, copying entries, exporting data to a spreadsheet, and customizing table settings.

Monitor Purchase Requisition Items Centrally (F3976)

Application Type: Transactional

This app allows you to effectively monitor purchase requisition items originating from both the central SAP S/4HANA system and connected external systems. With a range of filters such as display currency, asset, chart of accounts, company code, material, and more, you can tailor their view to focus on specific data sets. The app provides visibility into requisition items that are either awaiting approval or lack follow-on documents, helping you keep track of pending procurement actions. It also summarizes the total value of purchase requisitions and supports drilldown functionality for deeper analysis by material, material group, or other relevant dimensions. You can interact with visual filters to assess requisition value by purchasing category, status, delivery date range, or delivery status. Additional options such as zoom controls, full-screen view, and detailed views enhance the user experience. Furthermore, you can navigate directly from the app to individual purchase requisition details for more in-depth evaluation or action.

Monitor Purchase Scheduling Agreement Items (F3143)

Application Type: Transactional

This app allows you to effectively monitor spending associated with purchase scheduling agreements by providing insights into key financial metrics such as the released amount and the amount still pending invoicing. You can apply various filters including material, supplier, material group, scheduling agreement, validity status, and company code to refine the data. By default, the information is grouped by material group, but you can choose to view it by company code, material, plant, delivery status, or other criteria. The app offers both table and chart views, with options to open in full screen, adjust settings, or display a legend for better visualization. You can switch between compact and visual filters and drill down into detailed data for deeper analysis. Personalized views can be saved as variants for easy access in future sessions, enabling effective tracking and analysis of scheduling agreement expenditures. The app relies on several CDS views:

- C_SCHEDAGRMTDELIVSTATUSVH
- C_SCHEDAGRMTINAPPRVLSTSVH
- C_SCHEDGAGRMTDELIVSCHEDULE
- C_SCHEDGAGRMTITEMMONITOR
- C_SCHEDGAGRMTVALIDITYSTS

Monitor Purchasing Analytics Operations (F4399)

Application Type: Transactional

The app allows you to get a view of important purchasing analytics in the SAP S/4HANA system. The app has numerous cards, as follows:

- **Monitor Missing Exchange Rates**

  Provides a concise overview of purchase orders that are missing exchange rates necessary for currency conversion. It displays details such as the document currency, target currency, the number of affected documents, and the dates of the oldest and most recent entries. The card also links directly to the Analyze Missing Exchange Rates app (F4360).

- **Supplier Evaluation Score History**

  Shows the number of failed jobs related to supplier evaluation score history. It helps you identify issues in the background processes that update or maintain historical evaluation scores for suppliers. When you click this card, you are redirected to the Supplier Evaluation Score History app (F5062).

- **Extraction of PO for Central Procurement Analytics**

  Tracks the number of jobs executed to extract purchase order data for use in central procurement analytics. It serves as a monitoring tool to help you verify whether the extraction jobs are running successfully or if any have

failed. Clicking the card opens the Extraction of Purchase Orders for Central Procurement Analytics app (F5063).

> **Tip**
>
> You can choose to add or remove the cards on the initial screen via the **Manage Cards** option. Furthermore, you can use the **Refresh** option to get the latest results as required.

Monitor Purchasing Info Records Price History (F2988)

Application Type: Analytical

This app allows you to efficiently track and analyze purchasing info records along with their historical pricing details, such as price, date, and units. You can apply filters including material, material group, supplier, plant, and info record category (such as consignment, pipeline, standard, or subcontracting) to refine your search. The interface includes tabs for both info records and related purchase orders, each displaying key data like supplier, plant, and pricing. Selecting a purchase order navigates you to the Manage Purchasing Info Records app (F1982), where you can access detailed information such as validity periods, last purchase price, and ordering units. Features like price analysis allow you to view average prices and identify price trends or variances over time. The app also supports navigation to contextual object pages, making it easy to explore deeper insights related to materials, suppliers, and pricing terms.

Monitor RFQ Items (F2425)

Application Type: Analytical

This app helps purchasers track and manage items within requests for quotation (RFQs) by offering a range of filters such as RFQ number, RFQ type, purchasing category, supplier, status (**In Approval, In Preparation, Published,**

Rejected), quotation deadline, material group and more. You can analyze data in both chart and table views, switch between chart types, and adjust settings to tailor the display. The app provides a search bar for quick access and supports full-screen mode for detailed analysis. By selecting specific RFQ items, you can view related supplier quotations and key details, while also navigating to contextual information about the associated supplier and material. This allows for a more complete and efficient management of the RFQ process.

Monitor Schedule Extraction of Purchase Orders for Central Procurement Analytics (F5063)

Application Type: Transactional

This app allows strategic buyers to define and manage custom criteria for evaluating suppliers. You can assign scores to suppliers based on these personalized criteria, enabling tailored performance assessments. The app displays a list of all existing user-defined criteria, and you can click each entry to view its specific details. In addition, the app helps track any failed scheduled extraction jobs by listing them along with key information such as job name, creation date, and time frames. By selecting a failed job, you can access more detailed error information and navigate to the Schedule Extraction of Central Purchasing Data for Analytics app (F4897) to investigate, resolve, and restart the job as needed. The app uses CDS view C_POCNTRLAN-LYTSFAILEDJOB to retrieve data.

Monitor Subcontracting Documents (F3095)

Application Type: Transactional

This app provides a centralized view to monitor and manage subcontracting purchasing documents, such as purchase orders and scheduling agreements. Through a search bar and a variety

of filters including subcontractor, third-party supplier, assembly, plant, company code, material group, next delivery or shipping status, and document type, you can refine the data displayed to suit specific tracking needs. The interface is divided into two main tabs: **Purchasing Document Items** and **Purchasing Document Items for Third-Party Suppliers**. Each listed item shows key information such as the purchasing document, assembly, order quantity, and quantity still to be delivered. You can customize table views, export data to spreadsheets, and easily navigate to detailed object pages for assemblies, components, and suppliers. The app also supports viewing an assembly flow diagram, particularly for documents generated through MRP, enhancing visibility into delivery timelines and overall subcontracting progress.

Tip

You can use the **Save as Tile** option to create a customized tile with predefined filter criteria. You will need to enter a title, subtitle, and description, and select the page on which the tile will be displayed, and you can view a preview as data is populated in real time.

Monitor Supplier Confirmations (F2359)

Application Type: Analytical

This app allows you to monitor purchasing document items such as purchase orders and scheduling agreements that are still awaiting confirmation from suppliers. It provides a clear overview of the confirmation status and allows you to drill down into specific document items to view details such as delivery schedules, account assignments, and general information. You can apply filters like purchasing document number, organization, group, supplier, and confirmation category to refine the view. The app also offers navigation options to related suppliers and material information. For industries like fashion and vertical business, additional details such as stock segment, generic article, season, collection, theme, and product characteristics can be displayed, provided the business function ISR_RETAILSYSTEM is activated in the SAP S/4HANA system.

Tip

It's important to note that only items pending confirmation are displayed; once a supplier confirms an item, it no longer appears in the app.

**My Inbox –
Approve Awarding Scenario (F5932)**

Application Type: Transactional

The app allows you to review and approve awarding decisions directly from the inbox. The app helps facilitate the approval process by presenting all relevant details and supporting documents for each award scenario within a single interface. Approvers can make informed decisions quickly, take necessary actions such as approve or reject, and ensure timely processing of procurement or sourcing-related approvals—all without navigating away from the inbox.

**My Inbox –
Approve Purchase Contracts (F0400A)**

Application Type: Transactional

The app allows you to efficiently review and approve purchase contracts directly from their inbox. It provides all relevant contract details in one view, allowing approvers to make informed decisions without switching between applications. You can take actions such as to approve, reject, or forward contracts.

**My Inbox –
Approve Purchase Order (F0402A)**

Application Type: Transactional

The app allows you to review and act on purchase order approvals directly from their inbox. It provides a clear overview of each purchase order's details for instance supplier, items, and pricing, enabling approvers to make informed decisions. You can approve, reject, or forward purchase orders, helping fast-track the procurement workflow and ensure timely processing.

**My Inbox –
Approve Purchase Requisitions (F0401A)**

Application Type: Transactional

The app allows you to review and act on purchase requisition approvals directly from their inbox. It provides all relevant details of the requisition, such as requestor, materials, quantities, and cost centers, allowing approvers to make quick and informed decisions. You can approve, reject, or forward requisitions, ensuring efficient handling of procurement requests and maintaining smooth operational workflows.

**My Inbox –
Approve Service Entry Sheets (F2446)**

Application Type: Transactional

This app allows you to review and approve or reject service entry sheets directly from their workflow inbox. It displays essential details such as the vendor, service descriptions, quantities, and associated purchase orders, enabling approvers to make informed decisions quickly. The app facilitates the approval process, helps ensure accurate service confirmations, and supports timely payments by integrating into the overall procurement and service management workflow.

**My Inbox –
Display Central Purchase Contract
(F7345)**

Application Type: Transactional

The app allows you to view the details of central purchase contracts directly from your workflow inbox. It provides key contract information such as supplier details, contract validity, items, pricing, and organizational data. More importantly, the app allows you to access and review contract data efficiently as part of the workflow tasks in central procurement processes.

**My Inbox –
Display Centrally Managed Purchase
Requisition (F6985)**

Application Type: Transactional

This app lets you review and approve centrally managed purchase requisitions at both the overall and item levels. You can utilize different views to visualize key approval details. For overall approvals, the app shows the full requisition information and for item-level approvals, it displays only the relevant item sections. If something needs to be changed, you can send the requisition back to the central purchaser for rework and add a comment or question using the **Decision Note** field. For instance, if a user is reviewing a requisition for office supplies and notices a missing item description at the item level, they can return it to the purchaser with a note reading, "Please clarify
the item details for line 1."

**My Inbox – Display Centrally Managed
Purchase Requisition Items (F6984)**

Application Type: Transactional

The app allows you to view both overall and item-level approval attributes through the multiselect list in both the master-detail view and the expert view. For overall approvals, you can

access all sections of a centrally managed purchase requisition as they appear on the **Process Purchase Requisitions Centrally** object page. Likewise, for item-level approvals, the app displays all relevant item sections, mirroring the item-level object page in the same application. In addition, you can send a centrally managed purchase requisition back to the central purchaser for rework and include comments or queries using the **Decision Note** field in the **Submit Decision** popup.

**My Inbox –
Display Sourcing Project (F5934)**

Application Type: Transactional

This app allows you to view detailed information about sourcing projects directly within your inbox, providing quick access to project status, details, and related documents. The app is designed to give an overview as well as in-depth insights. It supports sourcing professionals, category managers, and procurement teams in managing strategic sourcing activities such as identifying, selecting, and engaging suppliers for the procurement of goods or services.

**My Purchase Requisitions –
New (F1639A)**

Application Type: Transactional

This app offers an all-in-one solution for you to create and manage self-service purchase requisitions with minimal navigation. At the top of the screen, a search bar allows quick access to existing items, while action buttons such as **Create Item**, **Create Limit Item**, **View All**, **Confirm**, **Return**, **Copy**, **Delete**, and **Edit** make interaction efficient and intuitive. When you select **Create Item**, you are directed to the **Purchase Requisition** screen within the My Purchase Requisitions app (F1643A), where you can input detailed data such as material, supplier material number, valuation price, quantity, delivery date, purchasing group, plant, storage location, and account assignment. You also can specify delivery address details, add sources of supply including proposed ones, write notes (e.g., item text, delivery instructions, approval details), and upload attachments. Beyond creation, you can track requisition status, confirm goods receipt, and initiate returns when necessary, making it a complete tool for managing purchase requisitions efficiently.

This is the successor app to the My Purchase Requisition (F1639) and Create Purchase Requisition (F1643) apps, which were deprecated as of SAP S/4HANA 2023 FPS03.

Tip

You can view the shopping cart using the **Cart** icon at the top of the screen to view items in the **Purchase Requisition Overview** screen. Here you can view the total amount of requested items, order, view the cart, or cancel.

My Purchasing Document Items (F0547B)

Application Type: Transactional

The app offers a comprehensive overview of your purchasing document items, categorized by supplier. It allows you to track and manage purchase requisitions, purchase orders, goods receipts, and supplier invoices through dedicated tabs: **Overview**, **Purchase Requisitions**, **Purchase Orders**, **Goods Receipts**, and **Supplier Invoices**. The app provides flexible filtering options, for example, by material group, purchasing group, plant, status, and purchase order, enabling you to focus on specific document types or suppliers. You can search by purchase orders to view related documents or filter by status to find items that are, for example, canceled or completed. The app highlights discrepancies between ordered values and supplier invoice values, and presents detailed data on

each document item, including how quantities and values are distributed across account assignments.

You can adapt the interface to show only the data you need and navigate to relevant SAP Fiori apps, such as Manage Purchase Orders (F0842A), Monitor Purchase Requisition Items (F2424), Purchase Requisition Item Types (F2016), Purchase Requisition (F1640), and more. If authorized, you also can post goods receipts or create purchase orders directly from the app. In addition, SAP highlights that if long material numbers are enabled in the backend system, these are automatically shown in the **Material** field for enhanced visibility.

This is the successor app to the My Purchasing Document Items app (F0547A), which was deprecated as of SAP S/4HANA 1809 FPS02.

Non-Managed Spend (F0571)

Application Type: Analytical

The app allows you to identify the total amount spent, excluding taxes, on suppliers where no purchase order reference exists, compared to the overall supplier spend. This analysis is based on data from accounting documents and helps highlight uncontrolled or ad hoc spending. The KPI is presented across different views, such as by **Supplier**, **G/L Account**, **Material Group**, **Company Code**, and more. You can filter data, visualize it in chart or table formats, and export the results to Excel or share the app link via email.

In addition, the app connects you to related apps like My Purchasing Document Items (F0900) and Supplier (F0354), and it offers trend and document-based insights to support strategic purchasing decisions.

Tip

Within the app, you can customize the chart measures to be displayed in the **Chart** view according to the business preference. For instance, you can use **Non-Managed Spend %**, **Non Managed Spend Ratio**, and **Non Managed Spend**.

Off-Contract Spend (F0572)

Application Type: Analytical

This app enables business users to analyze the organization's total purchasing spend and determine how much of it is attributed to purchase orders that do not reference a contract since the beginning of the previous year. These off-contract purchases indicate procurement activities conducted without formal agreements, offering insights into areas that could benefit from improved contract coverage and strategic sourcing. You can filter results by display currency and date and evaluate KPIs such as supplier, purchasing category, purchasing group, plant, work breakdown structure (WBS) element, company code, and cost center. The app supports both chart and tabular views, along with customization options like chart type selection, zooming, legend display, export to spreadsheet, subscription to updates, and mini tiles for a compact view.

You can drill down into detailed data or quickly access related apps such as Manage Purchasing Categories (F0337), My Purchasing Document Items (F0547B), Supplier (F0354), Manage Purchase Order (F0842A), and Purchase Order (F0348A) to take further action based on the analysis.

Operational Supplier Evaluation (Version 2) (F1662A)

Application Type: Analytical

This app calculates a supplier's performance score within an organization by evaluating key operational metrics, specifically quantity variance, price variance, time variance and more. The score reflects the performance from the beginning of the year up to the current date and is based on a weighted average. You can apply filters to narrow down the data view and set default values using various filter parameters as reference. The app presents data in both chart and table formats, allowing for visual analysis and straightforward interpretation. Results can be shared via email or exported as a Microsoft Excel file. In addition, you can save specific filtered views as custom tiles for easy access on the launchpad. This is done by selecting the arrow icon on the screen, choosing **Save as Tile**, and entering the relevant details. Views within the app are organized across various categories, including supplier, purchasing group, purchasing organization, purchasing category, material group, documents, and trend analysis, making it easier to evaluate supplier performance from multiple angles.

The app also allows you to navigate to related apps such as Manage Purchasing Categories (F0337), My Purchasing Document Items (F0547B), and Supplier (F0354).

This is the successor app to the Operational Supplier Evaluation app (F1662), which was deprecated as of SAP S/4HANA 2021 FPS02.

Overall Supplier Evaluation (F2019)

Application Type: Analytical

This app calculates a supplier's overall score within an organization by combining both operational performance data and questionnaire-based feedback. The score reflects performance from the start of the year to the current date and is derived from a weighted average. The supplier score KPI can be analyzed from multiple perspectives, including by supplier, purchasing category, score vs. spend, and trends over time (such as by supplier or purchasing category). It also highlights the top and bottom 10 performing suppliers. Data can be viewed in both chart and table formats, shared via email, or exported to Microsoft Excel. In addition, you can save filtered results as custom tiles on your launchpad for quick access.

The app also provides navigation to related apps such as Manage Purchasing Categories (F0337), My Purchasing Document Items (FF0547B), Supplier Factsheet (F0354), and more.

**Overdue Materials –
Goods Receipt Blocked Stock (F2347)**

Application Type: Analytical

This app offers a focused overview of overdue materials for which goods receipts have been posted into nonvaluated goods receipt blocked stock, helping you identify and address incomplete goods receipt processes or investigate discrepancies. Using the search bar and filters such as days since posting date, plant, material, and purchase order, you can narrow down the list of overdue items. The results display key details like plant, material, purchase order quantity, and item numbers from purchasing documents. You can adjust the layout by choosing how many items appear per row, modifying table settings, or exporting the data to a spreadsheet for further analysis.

Hyperlinks within the list, such as those for **Material** and **Purchase Order**, enable seamless navigation to related apps, allowing you to quickly perform actions like checking stock levels or posting goods receipts. The app displays all relevant purchase orders with stock held in blocked stock even if the delivery completion indicator hasn't been set, ensuring no overdue item is missed. Display preferences can be tai-

lored by organizing data through grouping options and column personalization.

Tip

It's important to note that financial stock values may appear as zero for some users, depending on their authorization levels, as SAP restricts access to sensitive financial information.

**Overdue Materials –
Stock in Transit (F2139)**

Application Type: Analytical

This app provides a comprehensive view of materials in transit under stock transport orders, helping you monitor overdue items and ensure the stock transfer process is completed efficiently. The interface includes a search bar and a variety of filters such as material, goods issue/goods receipt document, days in transit, issuing and receiving plants, storage location, and purchase order. The app is divided into three main tabs: **Delivery Open**, **Delivery Completed**, and **All Items**, each showing the number of relevant entries. Within each tab, materials are listed with key details like shipping duration, posting date, days in transit, purchase order number, material ID, and quantities customizable via table settings. By selecting a line item, you can navigate to the **Stock Transport Order Item** screen, which displays detailed information such as total goods issued and received, delivery status, quantities, process flow, and various timeline deltas (e.g., between purchase order and goods issue, or goods issue and receipt). The app supports both standard and advanced intercompany stock transfer processes and allows filtering by predefined or custom movement types. You also can choose to display either all related goods movement postings or only the latest, helping reduce clutter and streamline analysis.

A standout feature of the app is its predictive analytics model, which forecasts delivery dates based on historical company data. Integration with related apps and navigation to process-specific screens adds further value, making this a robust tool for managing stock transport logistics.

> **Tip**
>
> You can use the **Show More per Row** option on the initial screen to view extra details such as the forecast delivery date and to check if delivery has been completed or not.

Overdue Purchase Order Items (F0343)

Application Type: Analytical

This app helps you monitor and analyze overdue purchase order items by comparing the actual delivery dates with the scheduled ones at the item level. It displays the total number and value of overdue purchase order items at the top of the screen, allowing you to quickly identify delays. KPIs can be broken down by dimensions such as supplier, plant, purchasing group, purchasing category, company code, and material group. You can apply various filters such as display currency and relative dates (for instance, previous year to date) and customize your view using options like drilldown, chart or table view, zoom, full screen, and export to spreadsheet. The app also indicates the supplier confirmation status of each item, showing whether a confirmation is expected, received, or not required. Additional features include subscribing for updates, showing mini tiles, and saving customized views as launchpad tiles.

You can easily navigate to related apps such as Purchase Order (F0348A), Manage Purchase Orders (F0842A), Supplier (F0354), Manage Purchasing Document Items (F0900), and Monitor Supplier Confirmations (F2359).

Overview Inventory Management (F2769)

Application Type: Analytical

The app provides a snapshot of the most important and timely information and tasks relevant to the business case. Presented through a set of actionable cards, it allows you to quickly identify priorities, make faster decisions, and take immediate action. The following cards are available:

- **Recent Material Documents**
  Displays a list of material document items and provides detailed information for each selected item. It also offers the option to reverse a material document if needed, helping you manage and correct inventory transactions efficiently. It redirects to the Material Documents Overview app (F1077) when the card is selected.

- **Overdue Materials – GR Blocked Stock**
  Displays overdue materials for which a goods receipt (GR) has been posted into the nonvaluated GR blocked stock, helping you identify items that may require further action or investigation. Redirects to the Overdue Materials – GR Blocked Stock app (F2347).

- **Stock Value by Stock Type**
  Provides an overview of material stock values, organized by different stock types. Redirects to the Stock – Multiple Materials app (F1595).

- **Stock Value by Special Stock Type**
  Gives you an overview of your material stock values categorized by special stock types.

- **Warehouse Throughput History**
  Provides a clear view of all goods movements within a company. It allows you to explore various dimensions and key figures and lets you drill down to specific material document items to find detailed information quickly and easily. It redirects to the Goods Movement Analysis app under the inventory management application component.

7

- **Monitor Purchase Order Items**
  By using the view switch, you can easily toggle between viewing overdue items and items that are currently awaiting approval, allowing for better tracking and prioritization of tasks. It redirects to the Monitor Purchase Order Items app (F2358).

- **Overdue Materials – Stock in Transit**
  Presents the top 10 stock transport order items experiencing the longest delays, where the goods receipt process is still pending. It helps identify critical bottlenecks in stock transfers that require immediate attention. Redirects to the Overdue Materials – Stock in Transit app (F2139).

Overview Inventory Processing (F2416)

Application Type: Analytical

The app provides a centralized and real-time snapshot of key inventory-related information and tasks, displayed through a set of actionable cards. These cards highlight the most relevant and urgent matters, allowing you to focus on what needs immediate attention, make faster decisions, and take prompt action. You can filter the entire overview page by parameters such as plant or storage location, which dynamically updates the displayed cards to reflect only relevant data, including overdue materials, outbound deliveries, and warehouse throughput. Selecting a card header typically navigates to the corresponding app, while clicking an item lets you drill down into specific details. The following cards are available:

- **Recent Inventory Counts**
  Provides item-level details for physical inventory strategies like annual or cycle counting and links to related inventory document apps for header and item views.

- **Overdue Materials – Stock in Transit**
  Shows the top 10 stock transport order items with the longest pending goods receipts, highlighting critical delays. Links to the Overdue Materials – Stock in Transit app (F2139).

- **Inbound Delivery List**
  Shows all completed and open inbound deliveries for tracking shipping status and key details.

- **Outbound Delivery List**
  Lists completed and open outbound deliveries for monitoring shipping activities.

- **Overdue Materials – GR Blocked Stock**
  Displays overdue materials for which a goods receipt (GR) has been posted into the nonvaluated GR blocked stock, helps to identify items that may require further action or investigation. Redirects to the Overdue Materials – GR Blocked Stock app (F2347).

- **Monitor Purchase Order Items**
  The view switch lets you toggle between overdue and awaiting-approval items for better task tracking. Links to the Monitor Purchase Order Items app (F2358).

- **Recent Material Documents**
  Lists material document items with details and offers an option to reverse documents. Links to the Material Documents Overview app (F1077).

- **Warehouse Throughput History**
  Shows all company goods movements, you can explore various dimensions and key figures and drill down to specific material document items to find detailed information quickly and easily.

> **Tip**
>
> If the cards are not displayed by default, you need to make sure that the **Plant** key field is populated so as to display the SAP Fiori cards with metrics from the system.

Parts Per Million (F4885)

Application Type: Analytical

This app allows you to evaluate the delivery performance of purchase orders by comparing actual delivery dates with the promised delivery dates at the item level. It helps identify overdue purchase order items that have not yet been delivered, allowing you to assess delivery reliability across key dimensions such as supplier, plant, purchasing organization, material group, region, and purchasing category. KPIs are presented based on these categories to provide actionable insights. You can filter data using relative date ranges and customize filters to refine the analysis. The app also provides visibility into supplier confirmation statuses—whether confirmation is not expected, expected but not yet received, or already received. The app supports various features including chart and table views, data export to Excel, and link sharing via email.

You can further enhance efficiency by creating personalized tiles on the SAP Fiori launchpad and can navigate directly to related applications such as Material Documents Overview (F1077), Manage Purchase Orders (F0842A), and Supplier (F0354) for deeper analysis and follow-up actions.

Physical Inventory Document Overview (F0379A)

Application Type: Transactional, Fact Sheet

This app helps you view and manage physical inventory documents by allowing you to search using criteria like storage location, count status, or planned count date. It presents information from a broad list of documents down to detailed data for each inventory item. At the list level, you see all documents matching the defined filter criteria. Selecting a document reveals detailed inventory and process data, attachments, and change history. On the item level, you can view specific count results, inventory data, and situation handling information. The app is especially useful for inventory managers to track physical counts, compare counted quantities against book values through a micro chart, and see the status of each item (such as counted or adjusted). You can initiate recounts or post counted quantities for individual or multiple items, and after posting, you can navigate to related material documents. When posting differences, you can add reasons if needed.

The app also allows you to logically mark documents or items as deleted (making them inactive but still visible for archiving) or fully delete them from the system if authorized, keeping in mind deletion actions cannot be undone. In addition, you can print entire documents or selected items, either individually or via background jobs.

> **Tip**
> The MM_IM_PHYS_INV_DOC_SRV OData service (Physical Inventory Document) should be maintained and activated for this app.

PO Output Automation Rate (F4378)

Application Type: Analytical

The app allows you to analyze the total spend associated with purchase orders that lack a contract reference over a defined period by output channel, typically from the previous year to date. It focuses on identifying items purchased without, for instance, any negotiated pricing agreements thus classifying these as noncontract purchases. For instance, you can choose to display information for the previous year to date for automated versus manual outputs in a chart view. By using this app, you can measure the proportion of procurement activities that occur outside of established contractual arrangements, helping to monitor and reduce off-contract spend within your organization. You can proceed to navigate to related apps, switch between tabular and chart view, or adjust chart settings to view the different measures and dimensions.

Post Goods Receipt for Inbound Delivery (F2502)

Application Type: Transactional

This app allows you to post goods receipts with reference to an inbound delivery, making it an essential tool for receiving specialists in daily warehouse operations. It provides a list of relevant inbound deliveries from suppliers, supporting stock procurement, direct consumption (with single or multiple account assignments), and vendor consignment stock. The app displays account assignment details but doesn't recalculate quantities in partial receipts. It also enables shelf-life checks when required fields are maintained in the material master and supports the entry of production or expiration dates. You can record reasons for goods movements and manage batch creation and editing if the appropriate roles are assigned. For materials with split valuation, the app can assign or display valuation types, though consistency across documents requires manual input beforehand. In subcontracting scenarios, it provides a component view showing related materials and supports adjustments to component quantities or units. It also handles serial numbers at both the header and component level. Barcode scanning is supported by using internal or external devices, with options for simple or GS1-standard barcodes, and manual entry is available. The app can print slips and labels, display detailed storage-level data, and adjust stock values using the current material price regardless of the reporting date.

Post Goods Receipt for Process Order (F6352)

Application Type: Transactional

This app allows you to post goods receipts specifically tied to process orders. On the initial screen, you can either enter a production order manually or use the scan function to retrieve it. The app includes multiple tabs for easier navigation. The **General Information** tab captures key fields such as printing options, notes, delivery note, document date, and posting date. The **Items** tab displays details like material, open quantity, plant, storage location, and stock type (e.g., unrestricted-use, quality inspection, or blocked). A **Split Item** (+) function within the **Distribution** column allows you to split the same material and enter the necessary details for each split. The **Attachments** tab enables uploading supporting documents or adding links. This app simplifies the goods receipt process by listing all relevant process orders and associated materials ready for posting.

Post Goods Receipt for Production Order (F3110)

Application Type: Transactional

This app allows you to post goods receipts specifically tied to production orders. On the initial screen, you can either enter a production order manually or use the scan function to retrieve it. The app includes multiple tabs for easier navigation. The **General Information** tab captures key fields such as printing options, notes, delivery note, document date, and posting date. The **Items tab** displays details like material, open quantity, plant, storage location, and stock type (e.g., unrestricted-use, quality inspection, or blocked). A **Split Item** (+) function within the **Distribution** column allows you to split the same material and enter the necessary details for each split. The **Attachments** tab enables uploading supporting documents or adding links. This app simplifies the goods receipt process by listing all relevant production orders and associated materials ready for posting.

Post Goods Receipt for Purchasing Document (F0843)

Application Type: Transactional

This app allows you to post goods receipts with reference to various types of purchasing documents, such as purchase orders or scheduling agreements. When goods are delivered for a specific purchasing document, referencing it during the goods receipt process is essential to ensure that all relevant departments are aligned. This linkage automatically updates inventory levels and triggers the necessary financial postings as soon as the receipt is posted. The app includes several functional tabs. The **General Information** tab captures key details such as delivery note, document date, posting date, and print options (no print, individual slip, individual slip with inspection text, or collective slip). The **Items** tab provides a searchable list of materials, displaying information such as open quantity, delivered quantity, plant, storage location, and stock type (e.g., unrestricted-use, quality inspection, blocked, or goods receipt blocked stock). The **Attachments** tab allows you to upload relevant documents or include helpful links to support the receipt process.

> **Tip**
>
> The app supports the scanning functionality through the **Scan** option on the initial screen of the app.

Post Goods Receipt Without Reference (F3244)

Application Type: Transactional

This app allows you to post goods receipts without needing a prior purchase order or document reference, making it especially valuable for warehouse clerks handling day-to-day operations. It streamlines the process of recording incoming goods directly into the system, helping maintain accurate and up-to-date inventory records. The app is organized into several tabs: the **General Information** tab includes fields such as printing options (**No print**, **Individual slip**, **Individual slip with inspection text**, or **Collective slip**), note, delivery note, document date, and posting date. The **Items** tab lets you enter item details like ID, material, quantity, plant, and storage location, with options to create, copy, or delete entries. The **Attachments** tab allows you to upload supporting documents or insert relevant links, ensuring complete and traceable goods receipt entries.

Post Subsequent Adjustment (F5476)

Application Type: Transactional

This app is designed to facilitate correction postings for subcontracting components when a subcontractor consumes a different quantity of materials than originally planned. On the initial screen, you must enter either a purchasing document or a supplier using the purchasing document filter, with the option to scan documents for added convenience. Once in the **Post Subsequent Adjustment** screen, you can view key information such as notes, delivery notes, document and posting dates, and any attachments. The **Items** tab displays all listed materials, and by selecting a material, you are taken to the **Goods Receipt Item** screen where you can specify overconsumption or underconsumption. Here, you can input necessary details like batch number, quantity, reason code, goods recipient, and unloading point. The search functionality helps quickly identify relevant subcontracting purchase orders, while automatic filters prevent the selection of unrelated or invalid orders. Built-in validation ensures data integrity by generating error messages if incorrect documents are entered, making the adjustment process more accurate and efficient.

Price History for Central Purchase Items (F5711)

Application Type: Analytical

This app allows you to view both chart and table representations of the price and condition history for items in central purchase contracts. The information is organized based on key organizational attributes such as supplier, company code, central purchasing organization and group, and plant. To use the app effectively, you must provide required input fields including display currency, condition change valid from/to dates, and the central purchase contract item.

In the chart view, selecting data points lets you drill down into condition change details by date. Key features include the ability to view comprehensive details for each central contract item, such as the contract number, item, plant, material, short text, creation date, company code, base price, net price, effective price, and whether a price change reason is noted. You can click the hyperlinked effective price to access further details such as condition, amount, and price change reason. In addition, selecting the version history provides insights into previous versions of an item, displaying key fields like version number, price values, and condition change periods. These results can be filtered using multiple criteria, including condition change dates and plant.

The app also supports exporting all data to Microsoft Excel and performing advanced operations like generating custom charts to visualize price trends over time. The backend functionality is powered by CDS view C_CntrlPurContrItemPriceHist, ensuring robust data retrieval and analysis capabilities.

Process Central Purchase Contracts (F5715)

Application Type: Transactional

This app allows you to search for items within central purchase contracts and initiate price renegotiations with suppliers to capitalize on shifting market conditions and support year-on-year savings goals. You can select individual or multiple items across one or more central purchase contracts and initiate renegotiation through two primary paths: creating a new sourcing project or starting a direct renegotiation. By applying filters, you can refine the item selection based on the type of renegotiation required. Once selected, these items are processed further into related applications such as Manage Sourcing Projects (F4861) or Manage Renegotiations (F5551), where you can eventually revise the pricing in the original central contracts. The app supports two key actions: **Create Sourcing Project**, which allows you to initiate sourcing for multiple items in a contract (typically used for direct materials sourcing if configured by an administrator), and **Create Renegotiation**, which lets you renegotiate selected items by choosing the central contract renegotiation type.

> **Tip**
>
> It's important to note that renegotiation is currently not supported for contracts with hierarchical item structures.

Process Purchase Requisitions Centrally (F3290)

Application Type: Transactional

This app allows you to efficiently manage and act on purchase requisition items from multiple systems in one central location. You can filter requisitions based on various criteria such as purchasing group, supplier, material, approval status, processor, requisition number, and whether requisitions are open or from connected systems. The app also supports features like responsibility management, recommended material group suggestions, the option to show deferred items, and quick filtering for items assigned to the user. Filters can be adapted as needed to refine the displayed data. The requisition items are presented in a table format that shows details such as purchase requisition ID, purchasing organization, material, and other attributes depending on the user's table settings. Action buttons in the interface include **More Actions, Add to Document, Create Document, Assign Processor**, and **Export**, all available via convenient dropdown menus.

For bulk updates, the **Mass Change** function allows you to seamlessly navigate to the Mass Changes to Centrally Managed Purchase Requisition Items app (F7377), enabling efficient processing of multiple items at once.

Process Purchase Requisitions (Version 2) (F1048A)

Application Type: Transactional

This app allows you to efficiently view and manage purchase requisitions by providing a centralized interface with search and filter capabilities, including criteria such as material group, requisition and delivery dates, plant, purchasing organization, release date, and processing status (e.g., **RFQ Created**, **PO Created**, **Contract Created**, or **Not Edited**). The app displays detailed information for each requisition, including ID, material, quantity, and total value, with customizable table settings. You can take direct actions such as creating purchase orders, contracts, or RFQs from within the app using the **Create Purchase Order**, **Create Contract** and **Create RFQs** options respectively. A key feature is the hierarchical display of requisition items, which shows item sets and subitems in a structured format—especially useful for complex scenarios where supply sources can be inherited across levels. Access is controlled based on user authorization for specific purchasing groups, organizations, and plants. You can assign sources of supply like info records, contracts, scheduling agreements, or suppliers at both individual and hierarchical levels.

The different buttons and hyperlinks on the initial screen come in handy in navigating to other related apps such as Manage Purchase Contracts (F1600A), Manage Purchase Requisitions (F2229), Manage RFQs (2049), Buffer Positioning (F3282), and more.

This is the successor app to the Process Purchase Requisitions app (F1048), which was deprecated as of SAP S/4HANA 2023 FPS03.

Procurement Overview (F1990)

Application Type: Transactional, Analytical

This app provides a centralized dashboard containing cards that highlight the most critical procurement tasks and information relevant to the user, enabling quicker decision-making and immediate action. For list cards, clicking on the header opens the related app, while clicking on an individual item reveals more detailed information. For analytical cards, a single click provides a deeper dive into analytical views. Filters can be customized using the **Adapt Filter** function, and all displayed data will reflect the selected filters. To preserve a personalized setup, you can save your chosen layout and filters as your default view using the **Save As** option under the **Standard** dropdown. The following cards are available:

- **Monitor Purchase Contracts**
  Monitors purchase contracts by consumption, expiry, and target value, offering a quick performance overview. Links to the Manage Purchase Contracts (F1600A) app for details.

- **Monitor Purchase Requisition Items**
  Shows purchase requisition items by delivery date and value for easy tracking of upcoming or high-value items.

- **Purchasing Spend**
  Analyzes purchasing spend by supplier, material group, and trends, providing insights into costs. Links to the Purchase Order Value and Scheduling Agreement Value app (F1378).

- **Non-Managed Spend**
  Shows nonmanaged spend as a percentage of total spend, categorized by material group and supplier. Links to the Non-Managed Spend app (F0571) for details.

- **Purchase Requisition Item Types**
  Displays purchase requisition item types, including manual free-text entries, helping analyze and improve structured purchasing. Links to the Purchase Requisition Item Types app (F2016).

- **Monitor Supplier Confirmations**
  This card tracks supplier confirmations by delivery date and value, highlighting missing confirmations and variances. Links to the Monitor Supplier Confirmations (F2359) app.

- **Supplier Performance Monitoring**
  Shows supplier performance with KPIs like evaluation trends and scores, helping track

reliability. Links to the Operational Supplier Evaluation app (F1662A).

- **Off-Contract Spend**
  Shows off-contract spending as a share of total spend, with KPIs by material group and supplier, highlighting compliance gaps. Links to the Off-Contract Spend app (F0572).
- **Purchase Requisition Touch Rate**
  Shows the purchase requisition touch rate, tracking manual adjustments after submission. Links to the Purchase Requisition Touch Rate app (F2018).
- **Monitor Purchase Order Items**
  Monitors purchase order items by delivery date and value, highlighting overdue or pending approvals for quick action.

Product Sourcing Overview (F6890)

Application Type: Transactional, Analytical

This app provides you with analytical insights into sourcing projects and supplier quotations using interactive chart views in cards. It features a search bar and offers several filtering options such as display currency, relative date functions, central purchasing group, sourcing project type, and more. The app includes dashboard cards like **Sourcing Projects by Status** and **Supplier Quotations by Status**, which help you quickly assess project and quotation statuses. By visualizing data in charts, you can easily identify priorities, make informed decisions faster, and take immediate action to drive procurement processes forward.

Tip

The UI_PRODUCT_SOURCING OData service needs to be maintained and activated.

Purchase Contract Changes (F7439)

Application Type: Transactional

This app offers valuable insights into changes made to purchase contracts over time, helping you track and analyze modifications across various contract fields. The default period shown is from the previous year to date, but you can customize filters to refine your view. Data can be analyzed using different KPIs, such as supplier, company code, material group, and purchasing category. Change information is categorized by these dimensions, enabling detailed analysis and improved visibility into contract updates.

In addition, you can easily navigate to related apps like Manage Purchasing Categories (F0337), Purchase Contract (F0350), and Supplier (F0354) for further context and actions.

Tip

You can view the number of purchase contract changes, including number of items added, price changes, total purchase contract item changes, and quantity changes.

Purchase Contract Item (F1665)

Application Type: Fact Sheet

This app provides you with contextual insights into individual purchase contract items by presenting a clear overview of essential data. It displays key details such as the supplier, associated materials, purchase orders, account assignment category, material group, product type group, target quantity, net order price, price unit, and plant. You also can see whether an item is blocked or deleted, along with other related data. The app highlights important business information like net order price and supplier details, making it easy to grasp critical contract terms at a glance.

Purchase Contract Items By Account Assignment (F2421)

Application Type: Analytical

This app allows you to view and analyze purchase contract items with various customization options, including copy, settings, export to Microsoft Excel, filter adaptation, and personalized view configurations. Purchase contract items are displayed based on their account assignments and are grouped by default according to cost centers. By selecting a specific cost center, you can access detailed information such as the material, supplier, account assignment category, sequence number, and quantity. For each contract item, you also can review distributed values and quantities. Selecting a line item navigates you to the Manage Purchase Contracts app (F1600A), where you can view additional details like approval status, required actions, supplier evaluation scores, validity dates, and the overall status of the purchasing document. The app also allows you to explore contextual information related to the material and supplier, providing a comprehensive view of each contract's execution and relevance.

Purchase Order Average Approval Time (F7441)

Application Type: Transactional

This app allows you to determine the average time taken to create a purchase order, calculated based on data from the previous year to date. You can filter results based on currency, date range (default is year-to-date but this can be changed), company code, material, plant, purchasing organization, supplier and more using the **Adapt Filters** option. You can make use of KPIs such as supplier, material group, company code, and so on. More importantly, you can navigate to other related apps for further processing and detailed insights—for example, Manage Purchasing Categories (F0337), My Purchasing

Document Items (F0900), Purchase Order (F0348A), and more.

Purchase Order Average Delivery Time (F1380)

Application Type: Analytical

The app allows you to monitor the average delivery time of orders to suppliers. Using the app, you can analyze data in both chart and tabular formats, view the amount of goods received along with average and weighted delivery times, and share the app link via email or export data as an Excel file. You also can filter views, personalize the app by saving filtered views as tiles, and have these tiles appear on the launchpad for easy access.

In addition, the app allows navigation to related apps such as Supplier (F0354), Purchase Order (F0348A), Manage Purchase Orders (F0842A), and more.

> **Tip**
> The average delivery time is weighted in days, and the average order-to-delivery time is displayed at the top of the screen.

Purchase Order Changes (F3791)

Application Type: Analytical

This app displays the total number of item changes in purchase orders from the previous year to date, helping you to identify all modifications made to the items. It allows you to track the total changes across multiple attributes in purchase orders, including number of items added, quantity, price, purchasing group, storage location, delivery date and more. You can navigate to related apps and drill down to view detailed information in the chart view. For instance, you choose to view results from the previous year to date in a specific company code to view all the purchase order changes to date. More importantly, the refresh option allows you to get the latest information as changes are made in the system.

Purchase Order Items by Account Assignment (F2420)

Application Type: Analytical

This app allows you to display purchase order items organized by their account assignments, with items grouped by default according to their associated cost centers. For each purchase order item, you can view details such as the material, account assignment category, sequence number, and quantity assigned. The app allows you to filter and view purchase order items by criteria like purchase order number, account assignment category, cost center, and work breakdown structure (WBS) element. By selecting a cost center, you can access detailed information for each assigned purchase order item, including material, plant, and account assignment category descriptions. In addition, you can drill down into individual purchase order items to see account assignment specifics and data, and general data, as well as navigate to contextual information related to the material. More importantly, you can choose to copy purchase order items, adjust table settings using the **Settings** option, and export data to Microsoft Excel.

> **Tip**
>
> At the purchase order items list level, you can save a preferred view and set it as a default or public view.

Purchase Order Value and Scheduling Agreement Value (F1378)

Application Type: Analytical

This app allows you to retrieve the order value for all purchase orders within a specified time frame and calculate KPI values based on selected filter criteria such as material, supplier, and plant. For instance, you can choose to view the scheduling agreement amount and purchase order net amount for a specific supplier over a period of one year. Its key business value lies in allowing you to determine the current total value of all purchase orders in the system. You can analyze data in both chart and tabular formats, share results via email, or export them as Microsoft Excel files. Filtered results can be saved as personalized tiles on the launchpad for future reference. The app lists the following mini tiles:

- **PO and Scheduling Agreement Value:** Actual/ planned purchasing spend
- **Off-Contract Spend:** Purchase orders without contracts
- **Contract Leakage:** Point of sale ignoring existing contracts
- **Non-Managed Spend:** Invoices without purchase orders
- **Operational Supplier Evaluation:** Time, price, quantity, and quality

In addition, the app provides navigation to related apps such as Manage Purchase Scheduling Agreements (F2179), Manage Purchasing Categories (F0337), My Purchasing Document Items (F0900), Supplier (F0354), Manage Purchase Orders (F0842A), and Purchase Order (F0348A).

Purchase Order (Version 2) (F0348A)

Application Type: Transactional

This app allows you to search and view a list of purchase orders using a search bar and filters for quick access. Each purchase order entry displays key details such as the order name, status, company code, document category, Incoterms, supplier information, and more. You can navigate to this app using the Purchase Order Average Delivery Time app (F1380).

You can view important details in several tabs for instance The **General Information** tab displays foundational data like the purchase order number, creator, net value, Incoterms, company code, purchasing group, and organization, providing an overview of the transaction and organizational context. The **Items** tab lists all included items with details such as item ID, product type, short description, and material group, with customizable layouts for user preferences. **Limit Items** shows purchase order items managed by value limits instead of quantity, useful for framework orders or capped service purchases. **Supplier Contact Data** provides detailed supplier information including address, contact numbers, email, salesperson, and internal references. The **Supplier Confirmation** tab tracks supplier commitments with confirmed delivery dates, quantities, and service periods. **Purchase Requisition Items** traces items from the requisition process, showing requisition numbers, descriptions, material details, and delivery dates. **Purchase Contract Items** lists items linked to broader contracts with details like item number, document type, and quantities. The **Goods Receipts** tab captures receipt-related material document info for inventory tracking and verification. **Supplier Invoices** displays invoice data including references, dates, status, and amounts to support financial reconciliation. The **Notes** tab allows you to add or view additional order-related information such as terms, deadlines, and warranties. Finally, the **Attachments** tab enables uploading of documents or links like specifications, contracts, and delivery notes for reference and compliance.

This is the successor app to the Purchase Order app (FO348), which was deprecated as of SAP S/4HANA 1511.

Purchase Orders Created After Invoices (F7442)

Application Type: Transactional

This app allows you to gain insights into instances where invoices were received before the corresponding purchase orders were created, helping to identify and analyze deviations

from standard procurement processes. You can choose to view results using various filter criteria such as currency, date range, supplier, posting date and so on depending on user preference using the **Adapt Filters** option. More importantly, you can choose to view information using various KPIs such as company code, purchasing group, and more. The app highlights the number of purchase orders at the top of the initial screen.

Purchase Requisition (F1640)

Application Type: Transactional, Fact Sheet

This object page app provides contextual information about the purchase requisition business object, offering you a comprehensive overview of relevant procurement data. The app allows you to view details such as the supplier, ordered items, material information, and approval status. Key facts like gross value and number of items are prominently displayed to support decision making. In addition, the app serves as a central point for navigating to related business partners, master data, or documents, helping you access all relevant procurement information efficiently.

Purchase Requisition Average Approval Time – Flexible Workflow (F3980)

Application Type: Analytical

This app allows you to determine the average time taken to approve a purchase requisition from the moment it is created, specifically for requisitions that follow the flexible workflow approval process. The calculation, which defaults to data from the previous year to date, presents the average approval time in days. You can highlight your currency of choice as well the indicate the lowest, medium and highest price as the filter criteria. The app has the following mini charts:

- **Purchase Requisition Item Changes**: Total number of item changes
- **Purchase Requisition Item Types**: Number of items by type
- **Purchase Requisition No Touch Rate**: Number of high-, low-, and no-touch items
- **Purchase Requisition Avg Approval Time**: Approval time based on value
- **Purchase Requisition to Order Cycle Time**: Cycle time based on value

In addition, the app supports both standard and central purchase requisitions. You also can navigate to related applications such as Manage Purchasing Categories (F0337), Manage Purchase Requisitions Centrally (F3290), and Supplier (F0354), for further context and actions.

Purchase Requisition Average Approval Time – Release Strategy (F2014)

Application Type: Analytical

This app is designed to help you monitor the time it takes for a shopping cart to be converted into a purchase order, using year-to-date data. It offers flexible filtering options such as display currency, relative dates, and cost categories (low, medium, high), allowing you to tailor the data view. KPIs, including supplier, purchasing group, and purchasing category, are visualized through interactive chart and table formats. You can drill down into or roll up data, adjust settings, switch chart types, open views in full screen, and export data to spreadsheets. The app also supports trend analysis and various categorization KPIs.

In addition, you can navigate directly to related apps like Manage Purchasing Categories (F0337), Manage Purchase Requisitions Centrally (F3290), and Supplier (F0354) for deeper insights and actions.

Purchase Requisition Item Changes (F2015)

Application Type: Analytical

This app provides insight into the total number of item-level changes such as quantity, price, purchasing group or even supplier changes made to purchasing documents from the previous year up to the current date. The app allows you to view KPIs across various dimensions, including supplier, purchasing group, organization, category, and material group, with both tabular and graphical display options. Trend views by supplier, purchasing group, and purchasing category help identify patterns over time. You can apply filters, export results to Excel, or share the app link via email.

The app also supports navigation to other related apps such as Manage Purchasing Categories (F0337), My Purchasing Document Items (F0900), and Supplier (F0354). More importantly, you can view the total number of changes at the top of the initial screen.

Tip

The app allows you to view different item changes to purchasing documents such as quantity, price, purchasing group, supplier, material group, and purchase requisition item changes.

Purchase Requisition Item Types (F2016)

Application Type: Analytical

This app is used to track the number of free-text purchase requisition items created through the SAP GUI system from the previous year to date (relative date function as a key field). It also provides visibility into the number of material items created in the same manner, as well as catalog, material, and free-text items created using the app. The KPIs can be analyzed through various views, such as by supplier, purchasing group, purchasing organization, purchasing category, material group, and in different trend formats. You can filter data, display it as charts or tables, export it as Excel files, or send the app link via email. The app has the following mini tiles:

- **Purchase Requisition Item Changes**: Total number of changes
- **Purchase Requisition No Touch Rate**: Number of high, low and no touch items
- **Purchase Requisition to Order Cycle Time**: Cycle time based on value
- **Purchase Requisition Avg Approval Time**: Approval time based on value

The app also supports navigation to related applications such as Manage Purchasing Categories (F0337), My Purchasing Document Items (F0900), and Supplier (F0354), providing deeper insight into purchasing behaviors.

Tip

By default, the app uses the **Item Type** KPI but you are able to use other KPIs to view data using different dimensions.

Purchase Requisition Item (Version 2) (F0349A)

Application Type: Transactional, Fact Sheet

The app provides detailed contextual information about the purchase requisition item business object and allows you to navigate to related apps. The app displays an overview of purchase requisition item data, including supplier, material, delivery, quantity requested, validity dates, and more.

This is the successor app to the Purchase Requisition Item app (F0349), which was deprecated as of SAP S/4HANA 1511 FPS02.

Purchase Requisition No Touch Rate (F2018)

Application Type: Analytical

This app allows you to monitor the percentage of purchase requisition items that have been automatically processed from the shopping cart, covering data from the previous year to the current date. The KPIs can be viewed across various dimensions such as supplier, purchasing group, organization, category, material group, and item type, as well as trend and document views. You can apply filters, visualize the data in chart or table format, export it as a Microsoft Excel file, or send the app link via email. The app also supports personalization by allowing you to save filtered views as tiles directly on the SAP Fiori launchpad.

The app lists the following mini charts:

- **Purchase Requisition Item Changes**: Total number of item changes
- **Purchase Requisition Item Types**: Number of items by type
- **Purchase Requisition to Order Cycle Time**: Cycle time based on value
- **Purchase Requisition Average Approval Time**: Approval time based on value

In addition, it provides navigation options to related apps including Manage Purchasing Categories (F0337), My Purchasing Document Items (F0900), and Supplier Fact Sheet (F0354) for deeper insight.

Purchase Requisition to Order Cycle Time (F2017)

Application Type: Analytical

This app helps you analyze the average cycle time, measured in days, between the creation of a shopping cart and the point when a purchase order is issued to the supplier. It provides insights from the previous year to the current date. More importantly, necessary authorization objects must also be maintained in Transaction SU22. The KPI can be viewed across various dimensions such as supplier, purchasing group, purchasing organization, purchasing category, material group, and document trends. You can filter the data, view it in chart or table format, and export the results as a Microsoft Excel file or send a link via email. The app also allows you to personalize your dashboard by saving filtered views as tiles on the SAP Fiori launchpad.

You can choose to view the following measures in the chart view, with the option to move the measures up or down in the **View Settings** screen:

- Days Low-Cost,
- Days High-Cost
- Days Very High-Cost
- Average Cycle Time

In addition, you can navigate directly to related apps, including but not limited to Manage Purchasing Categories (F0337), My Purchasing Document Items (F0900), and Supplier Factsheet (F0354), for further exploration of procurement data.

Purchasing Documents by Requirement Tracking Number (F2905)

Application Type: Transactional

This app allows you to monitor purchasing documents such as purchase requisitions, requests for quotation (RFQs), purchase contracts, scheduling agreements and even purchase orders using their requirement tracking number, with the ability to apply additional filter criteria such as material group, supplier, or purchasing organization. The app provides a comprehensive view of purchasing documents organized by requirement tracking number.

From the list of documents, you can directly navigate to the object pages of related applications such as Manage Purchase Requisitions (Professional) (F2229), Manage RFQs (F2049), Manage Purchase Contracts (F1600A), Manage

Scheduling Agreements (F2179), and more. Each navigation option provides contextual information related to the selected material or supplier, offering deeper insight into the purchasing process.

Tip

The different tabs list the total number of documents individually in brackets on the initial screen of the app.

Purchasing Group Activities (F1660)

Application Type: Analytical

The app provides visibility of all completed purchase orders, contracts, goods receipts, and scheduling agreements linked to a specific purchasing group and category within a selected time frame. It excludes deleted items and only considers documents and items that are completed but not released. The app also allows you to calculate the number of goods receipts associated with a specific purchase order and item, along with the total invoice and net purchase order amounts. KPIs in this app are available in various views such as by **Purchasing Group**, **Trend**, **Trend by Amount**, **Trend by Purchasing Group**, and **Document**. You can apply filters to tailor the displayed views, which can be visualized in either chart or tabular formats. In addition, you can export the data as a CSV file or share the app link via email. Personalized tiles also can be created on the SAP Fiori launchpad by filtering views, selecting the save as tile option, entering relevant details, and saving them. The personalized tile then appears on the launchpad for quick access. The app has the following mini charts:

- **Purchasing Group Activities:** Purchasing docs by purchasing group
- **Purchase Requisition Item Changes:** Total number of item changes
- **Purchase Requisition Item Types:** Number of items by type

- **Purchase Requisition No Touch Rate:** Number for high-, low-, and no-touch items
- **Purchase Requisition Average Time:** Approval time based on time
- **Purchase Requisition to Order Cycle Time**

You can navigate to the following apps using the **Jump To** option: My Purchasing Document Items (F0900), Supplier (F0354), Manage Purchasing Categories (F0337), and Manage Purchase Orders (F0842A).

Purchasing Info Record (Version 2) (F0351A)

Application Type: Fact Sheet

This object page app provides a comprehensive overview of the purchasing info record business object, displaying key contextual data such as the supplier, material, and other relevant business facts like the supplier material. The app serves as a central point to view detailed purchasing info record data and supports navigation to related apps.

This is the successor app to the Purchasing Info Record app (F0351), which was deprecated as of SAP S/4HANA 1511.

Purchasing Spend (F0683)

Application Type: Analytical

The app enables you to analyze and compare purchasing spend across different dimensions by applying specific comparison filters such as supplier, material group, purchasing group, supplier country/region, and purchasing organization. It allows you to assess a subset of data in relation to the total spend and visualize the results both in percentage and net value. The app provides visual and tabular insights, allows exporting results to Microsoft Excel, and lets you share the analysis via a direct link. KPIs are

presented in views like **By Supplier, By Material Group, By Service Performer, By Purchasing Category, By Company Code,** and **Document.** The app displays the following mini charts:

- **Purchasing Spend:** Filter and compare purchasing spend with thresholds (critical high, warning high, critical low, warning low, target)
- **PO and Scheduling Agreement Value:** Actual/planned purchasing spend
- **Off-Contract Spend:** Purchase orders without contracts
- **Non-Managed Spend:** Invoice without purchase orders
- **Operational Supplier Evaluation:** Time, price, quantity and quality

You also can navigate from this app to related apps like Supplier (F0354), Purchase Order (F0348A), Manage Purchase Orders (F0842A), Manage Purchasing Categories (F0337), and My Purchasing Document Items (F0900) for deeper insights.

Quantity Contract Consumption (F2012)

Application Type: Analytical

This app allows you to track the consumption percentage of quantity-based contracts from the beginning of the previous year to the current date. It also provides insights into the target and released quantities for each contract. Specific authorization objects must also be configured in Transaction SU22. The app displays contract consumption KPIs across various dimensions, including by supplier, purchasing group, organization, category, material group, and cost center, as well as document and trend views.

You also can navigate to related apps such as Manage Purchasing Categories (F0337), Supplier (F0354), Manage Purchase Contracts (F3144), and Manage Purchase Orders (F0842A).

Redistribute Workload (F2504)

Application Type: Transactional

This app enables purchasers to reassign purchasing documents from one purchasing group to another using the **Reassign Purchasing Group** option, helping to manage workload distribution based on personnel availability. By using various search filters such as creation date, document number, document type, purchasing organization, purchasing group, and document category, you can locate relevant documents and manually reassign them as needed. Supported document types include contracts, purchase orders, purchase requisitions, and requests for quotation. Once reassigned, the change log can be viewed to track updates made to the purchasing group using the **Change Log** option.

> **Tip**
>
> SAP warns that reassigning the purchasing group overwrites all selected documents of the current group during the reassignment process.

Request for Quotation Types (F4149)

Application Type: Analytical

This app provides a comprehensive interface for viewing and analyzing requests for quotation (RFQs) created in the SAP S/4HANA system. It supports various RFQ types, including external price requests where prices are fetched from systems like SAP Ariba; external sourcing requests where bidding and awarding occur externally and results are returned to SAP S/4HANA; and internal sourcing requests, fully managed within the SAP S/4HANA environment. You can apply flexible filters such as display currency and relative date ranges, and analyze RFQs using KPIs like **Supplier**, **Purchasing Organization**, **Purchasing Group**, **Company Code**, **Status**, **Document**, and **Trend**. The app allows you to drill down or up for deeper insights, toggle between

chart and table views, show legends, adjust settings, expand to full screen, and export data to a spreadsheet. For instance, in the **Chart View**, you can view some of the following measures, which can be selected or deselected in settings (you can choose to display all measures):

- **Internal Sourcing Requests by Suppliers**
- **External Sourcing Requests by Suppliers**
- **External Price Requests by Suppliers**
- **Internal Sourcing Request Target Value by Suppliers**
- **External Sourcing Requests Target Value by Suppliers**
- **External Price Requests Target Value by Suppliers**

You can use the **Jump To** option to navigate to the following apps: Manage RFQs (F2049), Monitor RFQ Items (F2425), and Supplier (F0354).

Reservation (F5690)

Application Type: Analytical

The reservation object page allows you to display the details of a reservation, such as the account assignment, the goods recipient, and the items that have been reserved. The app can be accessed through the Manage Reservation Items app (F5601) by selecting a listed reservation item.

This app includes several tabs that provide essential information and support tracking. The **General Information** tab captures key reservation details like movement type, base and check dates, goods recipient, and verification company code for accurate logistics and financial tracking. In **Account Assignment**, you enter the cost center and order number to ensure proper budgeting and internal accounting. The **Administrative Data** tab logs who created or last changed the reservation and when, ensuring traceability. **Reservation Items** lists each material reserved, including quantities, plant, storage location, and batch, along with whether the item is allowed to be issued. The **Process Flow** tab offers a visual view of the reservation's lifecycle, showing document status and linked material documents. Finally, the **Attachments** tab lets you upload related files like approvals or instructions, keeping all supporting information in one place.

> **Tip**
>
> In the **Process Flow** tab, you can select, for instance, the reservation document or the material document to navigate to the respective related apps for each selection.

Return Delivery (F1996)

Application Type: Transactional

This app, accessible via the **Return** option in the My Purchase Requisition app (F1639A) or through SAP Fiori configuration, allows you to manage the return of items from a purchase requisition that have already been confirmed by the supplier. It is particularly useful when goods are received in damaged or unsatisfactory condition, allowing you to initiate a return process. The app presents a list of purchase orders associated with a given requisition, provided that at least one item is eligible for return. In central procurement scenarios, where the SAP S/4HANA system functions as a hub, it can be integrated with connected systems such as SAP ERP. When purchase requisitions are replicated to these connected systems, the corresponding purchase orders are automatically generated. Return deliveries can then be created in the connected systems to send the goods back to the supplier. Notably, purchase orders referencing multiple requisitions are excluded from this app.

Employees with the appropriate business role can access the **Return Delivery** functionality either through the SAP Fiori launchpad or the My Purchase Requisitions app (F1639A), view confirmed items, and process returns by entering the return quantity, ensuring it does not exceed the delivered quantity, and selecting an appropriate return reason.

Schedule Automation of Business Processes (F3983)

Application Type: Transactional

You can schedule jobs to automate business processes in the hub system based on activated rules. You can create a job by selecting the **Create** button, and then either start it immediately or schedule it for a later time. Scheduled jobs run in the background, and you can define recurrence settings if needed. After specifying all the necessary parameters and verifying the input data, the job can be scheduled, and it will appear in the **Application Jobs** list. This list allows you to view the status of each job—for example, whether it was successfully completed or canceled—and you can filter the results by date, job description, or job creator. You also can copy an existing job, modify its details, and use it to schedule a new one. In addition, you can access job logs directly from the **Application Jobs** list for further insights into job execution.

> **Tip**
>
> You can utilize the **Schedule Automation of Business Processes** job template (SAP_MM_BUS_PROC_AUTOMATION_T), corresponding to the SAP_MM_BUS_PROC_AUTOMATION job catalog. In Transaction SAPJ, you can view the job template names and the associated job catalog entry names.

Schedule Export of Centrally Managed Purchase Orders (F6733)

Application Type: Transactional

This app allows you to manage scheduled jobs efficiently by offering flexible filter options such as a search bar, job status (including in process, ready, canceling, finished, canceled, or skipped), and date range selection, which can all be adjusted to suit individual preferences. You have several action options, including creating a new job, copying existing jobs, viewing job details, canceling or restarting jobs, modifying settings, and exporting job data to Microsoft Excel for further analysis. The job creation process follows three key steps: first, defining the job template ID and providing a relevant description; second, setting up scheduling preferences and a recurrence pattern; and third, optionally entering parameters that further tailor the job to specific requirements.

> **Tip**
>
> You can use the SAP_MM_PR_MATGRP_TRAIN-ING_T job (**Schedule Export of Centrally Managed Purchase Orders**), which corresponds to job catalog entry SAP_MM_PR_MATGRP_TRAIN-ING.

Schedule Export of Inventory Analytics (F7493)

Application Type: Transactional

This app allows you to schedule background jobs to export material stock and posting data for a specific period, which helps manage system load by moving large data queries to less critical times of the day. The system provides status updates when the job is created and executed. In addition, you can use the Analyze Stock in Date Range app (F6185) to initiate a background job using the **Schedule Export** option, with the resulting data automatically forwarded to the Schedule Export of Inventory Analytics app (F7493) for further handling. To review job data and outcomes, the Display Inventory Analytics Job Results (F7504) app is available. When scheduling, you can select a predefined job template—Export of Inventory Analytics: Stock in Date Range (SAP_MM_IM_INVTRY_ANLYTS_EXPRT_SIDR)—and define the job's start time. Detailed parameters also can be specified, including selection criteria and any necessary exclusions.

> **Tip**
>
> The job named **Export of Inventory Analytics: Stock in Date Range** (SAP_MM_IM_INVTRY_ANLYTS_EXPRT_SIDR) corresponds to job catalog entry SAP_MM_IM_INVTRY_ANLYTS_EXPRT.

Schedule Extraction of Central Purchasing Data for Analytics (F4897)

Application Type: Transactional

You can now enhance the performance of certain purchasing analytics apps by migrating purchasing data, such as purchase orders and scheduling agreements. This is achieved using this app that allows you to create and schedule jobs to extract central purchasing data for analytics purposes. The available extraction types include **Process Extraction** (to extract all purchasing records from the hub system into an optimized analytical table), **Reprocess Error Records** (to extract records that previously failed during extraction), and **Reprocess Extraction** (to extract specific records based on replication dates). Key parameters such as the number of tasks, package size per task, and the server group name must be specified. After scheduling, job results are displayed in the **Application Jobs** list, where you can monitor job success or cancellation and filter the list based on attributes like date, job description, or creator. You also can copy and modify existing jobs and access detailed logs for each scheduled job.

Schedule Import of Catalog Data (F2666)

Application Type: Transactional

The app includes a search bar and allows filtering by job status, such as failed, canceled, in process, or scheduled. You can choose from options like scheduling a new job, checking job status, using templates, or canceling jobs. When scheduling a job, the first step involves entering the job template ID and description. In the second step, you can choose to start the job immediately or specify a start time and recurrence pattern. In the parameters section, you define import details under catalog data, such as selecting the web service—for example, `Material_Master` for material master data—and then proceed to select a material group and specific materials for extraction. The app is useful for Central Procurement scenarios as recommended by SAP.

> **Tip**
>
> You can use job template `SAP_MM_OCI_EXTRACTION_T` (**Import Catalog Data**).

Schedule Import of Purchasing Documents (F3268)

Application Type: Transactional

This app allows you to import purchasing documents such as purchase orders and purchase requisitions into the SAP S/4HANA system, which acts as the hub system. These documents can originate from connected SAP systems like SAP ERP, and SAP S/4HANA Cloud as well as non-SAP systems. The app supports importing all purchasing documents created either in the connected systems or in the hub system itself. In addition, it enables the transfer of associated descriptions, notes, and attachment details from the connected systems into the hub. The app allows you to create jobs using various templates, including ones for importing purchase requisitions, purchase orders, associated texts, central purchase requisitions relevant to the hub system, purchase order history, and currency-related fields. For example, the purchase requisition import template enables the extraction of requisitions from connected systems, while the import of central purchase requisitions supports requisitions created within the hub system for processing without replicating them to connected systems. The system

offers three import types depending on the selected job template: **Full Import** (which extracts all relevant data), **Delta Import** (which extracts only changed or new records since the last import), and **Ad-Hoc Import** (which allows immediate extraction based on specific criteria). Note that only delta and ad hoc imports are supported for importing purchase order history, and delta import is only available after a successful full import.

You can schedule jobs to run immediately or set them for later execution in the background, including setting up recurring schedules. During job setup, you can define parameters such as date ranges and import categories like purchase requisitions or purchase orders. The system also provides input help for many fields to ensure valid entries.

> **Tips**
>
> To use this functionality, you must activate the **Central Purchasing** scenario in Customizing under **Central Procurement** settings. For any system compatibility issues, you should contact your system administrator and refer to the documentation on monitoring compatibility of central procurement operations.
>
> You can use job template `SAP_MM_PO_HIST_EXTR_T` (**Import History for Purchase Orders from Connected Systems**).

Schedule Import of Release Orders (F3314)

Application Type: Transactional

This app allows you to import release orders created in connected systems into the SAP S/4HANA system, which serves as a hub. These connected systems can be SAP systems such as SAP ERP, SAP S/4HANA Cloud, or SAP S/4HANA itself, as well as non-SAP systems. Release orders can be generated for each distributed contract item in the connected systems, using either the full or partial target quantity or target value specified in the distributed contract. You can view the details of these imported release orders in the Manage Central Purchase Contracts app (F3144).

When creating a new job, you can choose to start it immediately or schedule it to run later. The job runs in the background and can be set to recur according to a defined schedule. There are three types of imports you can select from: **Full Import**, which imports all release orders; **Delta Import**, which imports only new or changed documents since the last import; and **Ad-Hoc Import**, which allows you to import documents as needed, such as immediately after creation without waiting for a scheduled import. Note that a delta import can only be executed after a full import has been successfully completed. Furthermore, you need to keep in mind that any documents imported via **Ad-Hoc Import** and not included in earlier delta imports will be reimported during the next delta import. Once you have specified all the necessary parameters and verified the data, you can schedule the job, which will then appear in the **Application Jobs** list. Here, you can track job status, including whether it was successful or canceled, and filter the job list by criteria such as date, job description, or creator. You also can copy existing jobs to create new ones, modifying the parameters as required. Logs for every scheduled job are accessible in the **Application Jobs** list.

After the jobs are complete, the Manage Central Purchase Contracts app (F3144) lets you review the release orders for all items within a selected central contract.

> **Tips**
>
> To use this app and schedule import jobs, the **Central Purchase Contracts** function must be activated within the **Central Procurement** setup.
>
> You can use job template SAP_MM_CCM_CALL-OFF_EXTRACT_T (**Import Release Orders from Connected Systems**).

> **Schedule Jobs for Central Purchase Requisitions (F7579)**
>
> Application Type: Transactional

This app is designed to facilitate the import of central purchase requisitions from connected backend systems into the SAP S/4HANA procurement hub, specifically to support local currency conversion. It streamlines the data integration process by allowing you to schedule and manage import jobs efficiently. Within the app, you can create new jobs, copy existing ones, view detailed job information, cancel or restart jobs, and adjust various settings to suit your needs. Each job is displayed with key information such as job name, planned start time, execution status, results, and the user who created it. In addition, you can filter the job list using criteria like status, date range, or keywords in the search bar, while detailed logs provide visibility into job progress and outcomes for better monitoring and control. You are required to populate the following information:

1. **Template Selection**
   Here, you need to enter the job template and description.

2. **Scheduling Options**
   The second step focuses on determining when the job will run. You can choose to start the job immediately or schedule it for a later time. In addition, there is an option to set recurrence.

3. **Parameters**
   You need to enter the source currency, connected system ID, and company code.

> **Tip**
>
> You can use job template SAP_MM_PUR_SSP_LCC_NON_REP_PR_LIST_T (**Display Non-Replicated Purchase Requisitions**).

Schedule Migration of Purchasing Data for Analytics (F3815)

Application Type: Transactional

Business can now improve the performance of certain purchasing analytics apps by migrating purchasing data, including purchase orders and scheduling agreements. This app allows you to create and schedule jobs to handle this data migration using the job template called **Schedule Migration of Purchasing Data for Analytics** (SAP_MM_PUR_ANA_MIGRT_JOB_T). When creating a new job, you can choose to start it immediately or schedule it to run later in the background, with options to set the job to recur on a defined schedule. Migration options include **Initial Load**, which migrates all purchasing records; **Reprocess Records**, which migrates any records that failed during the initial load; and **Delta Load**, which selectively migrates purchasing records filtered by criteria like purchase order date, suppliers, or specific purchase orders. You also need to specify parameters such as log level (**Advanced** for detailed logs or **Basic** for simple logs), tolerance limit (the maximum allowed migration failures before the job is flagged as failed), number of tasks to divide the job into, package size for records per task, and the server group name. The results of scheduled jobs can be viewed in the **Application Jobs** list, where you can track whether jobs were successful or canceled, filter the list by various criteria, copy existing jobs to create new ones, and access detailed logs for every scheduled job.

Tip

SAP recommends that new users enable performance optimization by selecting the **Activate Performance Optimization** option in step 3 in the **Performance Optimization for New Implementation** section; otherwise, the analytics app performance will not improve.

Schedule Output for Central Purchase Contracts (F5429)

Application Type: Transactional

The app allows you to manage the automated scheduling of output for central purchase contracts. You can search and filter jobs based on status, date range, and other parameters. When setting up a job, you can select a predefined job template and provide a job description. Scheduling options allow you to define how frequently the output should be triggered using a recurrence pattern. Key parameters such as output channel, output type, supplier, central company code, contract type, contract number, and document date can be specified, while the application object type is automatically populated based on prior selections. The app also offers useful actions including creating new jobs, copying existing ones, viewing job details, canceling scheduled jobs, and restarting any failed jobs,

---

Tip

For instance, you can use job template SAP_MM_PUR_CCTR_OUTPUT_RUN (**Schedule Output for Central Purchase Contracts**).

Schedule Output for Centrally Managed Purchase Orders (F5471)

Application Type: Transactional

This app allows you to schedule application jobs that send outputs for centrally managed purchase orders directly to the relevant suppliers. You can create and schedule these jobs according to specific needs, either starting them immediately or setting them to run later in the background. In addition, you can configure the jobs to recur based on a defined schedule.

When creating a job, you provide parameters tailored to business requirements. These include details about the output data such as the application object type (e.g., **Procurement Hub Replicated Purchase Orders** (PRMTHB_RPLD_PUR-

7

CHASE_ORDER)), the output channel used to send the report, and the output type that specifies the business document involved. You also specify centrally managed purchase order data, including the connected system ID (a required field with a list of available systems), supplier ID, company code, purchasing organization, purchasing group, document type, purchase order number, and the document's creation date. After entering and verifying all the necessary information, you can schedule the job, which will then be listed in the **Application Jobs** list. This list allows you to monitor the status of your scheduled jobs, showing whether each was completed successfully or canceled. You also can filter the jobs based on criteria like date, job ID, description, or creator. If you need to schedule a similar job, you can easily copy an existing one and adjust the parameters as needed. Logs for every scheduled job are also accessible directly from the **Application Jobs** list for detailed tracking and troubleshooting.

> **Tip**
>
> You can use job template SAP_MM_PUR_CPO_OUTPUT_RUN (**Schedule Output for Centrally Managed Purchase Orders**).

Schedule Persistency of Parts per Million Data (F5301)

Application Type: Transactional

This app allows you to view the history of supplier evaluation scores and manage the persistence of these scores by scheduling jobs. You can create and schedule jobs to either save or delete the supplier evaluation score history. Selecting the **Create** button, you can start a job immediately or schedule it to run later, with the option to have the job recur according to a set schedule. You also can provide optional parameters tailored to your business needs, such as selecting a relative date function or specifying a start and end date for persisting the scores. After defining the necessary criteria and verifying the

data, you can schedule the job, which will then appear in the **Application Jobs** list. This list allows you to monitor the status of scheduled jobs, filter them by various criteria like date, description, or creator, and check whether they were completed successfully or canceled. In addition, you can copy existing jobs to create new ones with similar settings and view detailed logs for each scheduled job directly within the **Application Jobs** list.

> **Tips**
>
> You can use job template SAP_MM_PUR_ANA_MIGRT_PPM_TO_UDC_T (**Persist Parts Per Million to User-Defined Criteria**).
>
> You also can choose to schedule a job in simulation mode by selecting the **Simulation Mode** checkbox in step 3: (**Parameters**) – **Extraction Parameters**.

Schedule Persistency of Supplier Evaluation Scores (F3814)

Application Type: Transactional

The app allows you to view the history of supplier evaluation scores and manage the persistence of these scores by scheduling background jobs. It provides an easy way to ensure that supplier evaluation data is stored or deleted as needed. You can create and schedule jobs directly within the app to either save or remove supplier evaluation score histories. To get started, simply select **Create** to open the **New Job** screen, where you can start the job immediately or schedule it to run later. When setting up a job to persist supplier evaluation scores, you can specify parameters such as a relative date function or a start and end date range, depending on the business case. After defining all necessary criteria, you can verify the data for accuracy before scheduling the job. Once scheduled, the job will appear in the **Application Jobs** list where you can monitor the status. The list allows you to see whether jobs were completed successfully or cancelled.

Schedule Price Updates for Central Purchase Contracts (F5722)

Application Type: Transactional

The app allows you to schedule and manage price updates for central purchase contracts, specifically in response to renegotiations. It helps automate the update process, ensuring that new pricing agreements are consistently and accurately reflected in the system. You can create and schedule jobs that update prices in central purchase contracts based on the outcomes of renegotiation activities. To initiate this, simply select **Create** to launch the **New Job** screen. Here, you can select various scheduling options to tailor the execution of the job according to business needs. For instance, you can choose whether the job should run only once or follow a recurring schedule. The app offers flexibility to either start the job immediately or schedule it for a future date, depending on operational requirements. Before finalizing, you can verify that all the specified data and parameters are correct. Once everything is confirmed, the job can be scheduled.

Schedule Purchasing Jobs (F1702)

Application Type: Transactional

The app allows you to schedule and monitor recurring purchasing-related activities as background activities. The system provides several

job templates that cater to different business needs within the purchasing process. To schedule a job, simply select **Create** which opens the **New Job** screen. Here, you can choose to start the job immediately or set it to run at a later time. You also can define recurrence settings for jobs that need to be executed periodically. After entering all required parameters and validating the information, you can schedule the job. Scheduled jobs will then appear in the **Applications Jobs** list, where you can track and manage jobs.

Schedule Release of Purchase Order Outputs (F7083)

Application Type: Transactional

You can generate output types such as dunning reminders and reminders for purchase order acknowledgements. The app allows you to create and schedule background jobs to send these outputs for purchase orders. To start, you can create a new job using the **Create** button and either run it immediately or schedule it for later execution. Scheduled jobs run in the background, and you can set them to recur according to the business needs. Parameters can be defined based on specific business requirements. Once the job is scheduled, it appears in the **Application Jobs** list, where you can monitor its status – whether it ran successfully or was canceled. The list can be filtered by various criteria, including date, job ID, description, and job creator. You also can copy existing jobs and modify them as needed for new scheduling and view detailed logs for each scheduled job directly within the app. The app has output channels (you can choose multiple options) as follows: Electronic Data Interchange (EDI), email, intermediate document (IDoc), print, and Extensible Markup Language (XML).

> **Tip**
>
> You can use job template SAP_MM_PUR_PO_OUTPUT (**Schedule Purchase Order Output**).

Schedule Release of Supplier Lists for Sourcing with Adaptations (F6130)

Application Type: Transactional

The app is designed to help you automate the release of supplier lists in sourcing projects once their due date has passed. By scheduling background jobs, this app ensures that sourcing supplier list work items are completed in a timely and efficient manner without requiring manual intervention. You can perform various actions such as creating a new job, copying an existing one, viewing job details, canceling, restarting, or adjusting settings. Each scheduled job is displayed along with relevant information including job name, planned start time, status, results (such as failed or successful), and the user who created it. The app also maintains comprehensive application logs that help track the execution and outcomes of each job, providing transparency and control over the supplier list release process.

Schedule Replication of Config Data to Connected System (F6781)

Application Type: Transactional

This app allows you to edit specific fields at both the header and item levels of distributed purchase outline agreements, such as purchase contracts and purchase scheduling agreements, directly within their connected systems. Instead of returning to the hub system to make changes to a central purchase contract or its hierarchy, you can now make these edits locally. To support this functionality, the system includes

tools for creating and scheduling jobs that replicate the relevant configuration data from a central environment to the connected systems. You can initiate the replication process by selecting the **Create** button to set up a new job. From the **New Job** screen, you can choose to start the job immediately or schedule it for later. The system supports background processing and recurring job schedules, enabling automated data synchronization at intervals that suit the business needs. More importantly, before scheduling the job, you can validate the data entered to ensure accuracy. Once scheduled, the job will appear in the **Applications Jobs** list.

Schedule Supplier Invoice Jobs (F1683)

Application Type: Transactional

This app allows you to schedule and monitor recurring supplier invoice-related activities by running them as background jobs. The app offers a variety of job templates tailored to specific supplier invoice scenarios. You can create a new job by selecting **Create**, and you can define job-specific parameters. After configuring the necessary settings, you can validate the input data and proceed to schedule the job. Once scheduled, it will appear in the **Application Job** list for tracking.

Schedule Transfer of PR for Intelligent Approval (F4155)

Application Type: Transactional

The app allows you to manage and automate the transfer of purchase requisition approval data. It provides a streamlined interface where you can view and control all scheduled jobs related to purchase requisition data transfers. At the top of the app, you can filter and search for jobs using various criteria, such as job status (e.g., failed, canceled, finished, in process, ready), job ID, job name, and the user who created the job. The main display lists all jobs along with key details such as their status, log access, job results, job name, planned start time, and the creator's name. You can perform actions like creating a new job, copying an existing one, canceling or restarting jobs, and adjusting settings. When creating a new job, selecting the **Create** button takes the user to the **New Job** screen, which follows a three-step process: selecting the job template and naming the job, configuring the scheduling options, and entering any optional parameters. Once these details are provided, the job can be scheduled, and it will be executed based on the specified timing and recurrence settings. For each step, the following information is required:

1. **Template Selection**
   In the first step, you need to specify the job template.

2. **Scheduling Options**
   The second step focuses on determining when the job will run. You can choose to start the job immediately or schedule it for a later time. In addition, there is an option to set recurrence.

3. **Parameters**
   The third step involves specifying optional parameters that further refine the scope or behavior of the job. For instance, the **Test Connection** checkbox enables transfer of data between two systems. You can use the **Check** option to see if everything is okay.

For created jobs, you can view job details in the **Job Details** screen in three tabs: **Scheduling Options**, **Run Details**, and **Parameters**. You can view whether jobs failed or succeeded.

This app was deprecated as of the SAP S/4HANA 2023 FPS02. SAP recommends using the successor app, Application Jobs (F1240).

> **Tip**
> The **Template** option allows you to save the populated entries as a template and even manage existing templates according to user preferences.

Schedule Update of Phase Switch (F6121)

Application Type: Transactional

The app allows you schedule jobs that update the phases of sourcing projects in line with the phase model defined in the configuration. You can specify a time interval during which the scheduled job will retrieve all active sourcing projects, evaluate whether any project phases qualify to advance to the next stage, and execute the phase transition when the predefined conditions are fulfilled. To get started, select **Create** to open the **New Job** screen. Here you can choose from a variety of scheduling options and define how often the job should run (whether as a one-time event or on a recurring basis). You have the flexibility to launch the job immediately or set it to start later, depending on the business requirements. Once all relevant parameters are entered, you can verify the accuracy of the data before scheduling the job. After scheduling, the job will appear in the **Applications Jobs** list where you can monitor its progress and status.

> **Tip**
> You can use job template SAP_MM_PUR_SRCG-PROJ_PHS_UPDATE (**Sourcing Project Phase Switching**).

Schedule Update of Price Details for Central Purchase Contracts (F5688)

Application Type: Transactional

This app allows you to schedule background jobs that determine and update price trend and price history information for items in central purchase contracts which helps ensure that pricing data remains current and accurate across procurement documents. To begin using the app, you can create and schedule a job tailored to your specific needs. From the initial screen, you can choose to create and choose a template titled **Update Price Details for Central Purchase Contracts**. In the second step, scheduling preferences can be set either to run the job immediately or at a later time. Jobs also can be configured to run on a recurring basis. Once all relevant parameters are defined, you can validate your entries and proceed to schedule the job, and successfully created jobs will then appear in the **Applications** list. In the third step, you select a load mode that best suits your data requirements, as follows:

- **Initial Load**

  Prepares price trend and history for central purchase contracts that are valid on a specific date. This date is mandatory and serves as the basis for selecting relevant contract records.

- **Delta Load**

  Captures any updates to pricing information that occurred after the initial load. This ensures that any recent changes to price trends or history are reflected in the system.

- **Ad-hoc Load**

  Allows you to run a one-time data update by specifying a particular central purchase contract item number. The system then retrieves and processes pricing data for that item alone.

Once the job setup is complete, you can run a validation check to ensure there are no errors. You can choose to schedule the job or save the configuration as a template for future use. All scheduled jobs can be tracked in the **Application Jobs** list, where you can see the execution status—that is, whether the job was successful, cancelled, or failed. This list can be filtered by various criteria such as date, job description, or user who created the job.

> **Tip**
>
> Loads can only be performed on contracts that are in an active or released or approved status. Contracts that are deleted, not released, or not approved are excluded from the update process.

You can copy existing jobs, which is useful for quickly setting up new jobs based on previous configurations. Failed jobs can be restarted if needed or cancelled if no longer relevant.

Scheduling Agreement Consumption (F3192)

Application Type: Transactional

This app allows you to gain a comprehensive view of how scheduling agreements are being utilized. By displaying consumption data in direct comparison to key measures such as the released amount, goods receipt amount, and invoiced amount, the app provides valuable insights into how agreements are progressing over time. One of the core features of the app is the ability to display consumption percentages of scheduling agreement. This allows you to quickly assess whether deliveries and invoices are aligned with those that have been released under each agreement. The app also allows you to analyze consumption based on various dimensions, such as time, validity window, supplier, and material group. This granular visibility helps you identify patterns and trends in supplier performance and material demand. More importantly, you also can perform customized analyses by applying a variety of filters to the data set. This targeted filtering capability allows for specific scenario reviews, such as evaluating consumption for a particular supplier or material category. The app is powered by CDS view C_SchedgAgrmtItmCnsmpn, ensuring that the information is presented in real time, is accurate, and is consistent with underlying transactional data in the system.

> **Tip**
>
> The C_SCHEDGAGRMTITMCNSMPN_CDS OData service (Scheduling Agreement Item Consumption) should be maintained.

Settings for Release Strategy (F2274)

Application Type: Transactional

This app lets you view and define release strategies for purchase requisitions. It opens with a list of existing strategies, which can be reviewed or edited. You also can create or modify release groups, which determine how strategies apply across document types or organizational areas. Within each group, you define release codes (approvers) and build strategies based on criteria like purchasing organization, requisition type, or total value. The app supports assigning multiple values to each strategy for complex approvals and allows up to three approval levels, each with a release indicator reflecting approval status. Available indicators depend on the selected release group to ensure consistent configuration. In the **Release Strategy** screen, you will need to define the following information:

- **General Information**
  Captures basic release strategy data, requiring a unique ID and description to define and distinguish the strategy.

- **Classification**
  Defines the criteria for triggering a release strategy by assigning a release group and setting characteristic values like document type or total value. You can create, delete, or copy entries and adjust settings for detailed control.

- **Releases Prerequisites and Status**
  Defines the approval structure for the release strategy, listing approvers, their release codes, and responsibilities. You set release indicators to show approval status, ensuring a clear, controlled approval flow.

Settings for Release Strategy Classes (F2273)

Application Type: Transactional

This app is used to manage release classes specifically configured with class type O32, which is used in defining document approval processes

such as release classes and their associated characteristics, which form the basis of release conditions in purchasing and other business processes. The app lists all existing release classes created under class type O32. The overview displays important details such as the class ID (for instance, **Release Procedure for Purchase Order (REL_PUR)**), description (for instance, **Release Procedure for Purchase Order**), validity period, and current release status. You can easily search, filter, or sort through the list to quickly locate a specific release class. Selecting a class opens a detailed view where you can see comprehensive information, including general attributes and a list of assigned characteristics. These characteristics define the conditions under which the release strategy is triggered, such as purchase organization, document type, or net order value. This detailed view also includes the current release status and any version, allowing you to track changes over time.

The app also allows you to change the release status of any existing class. This is helpful for managing the lifecycle of a release, setting it to active when it's ready for use, marking it inactive when no longer relevant, or labeling it as under development during configuration and testing.

Tip

In **Edit** mode, you can add, copy, or delete characteristics as well as sort the characteristics based on the filter criteria such as table name, status, class maintenance, and more based on the user's preferences.

Settings for Web Services (F1994)

Application Type: Transactional

This app allows you to configure and manage connections to external web-based product catalogs using the Open Catalog Interface (OCI), facilitating seamless integration between SAP S/4HANA and supplier catalogs for catalog-based procurement. You can view, create, edit, copy, or delete web service configurations and define catalog parameters, such as the call structure and default indicators. The app supports importing catalog data, using HTTP GET, and maintaining partner details. It enables validation of catalog item prices during the creation of purchase requisitions to ensure price accuracy. You can choose to enable or disable data import from OCI 4.0 or 5.0 catalogs, and for secure access to OCI 4.0 catalogs, the app supports OAuth 2.0 authentication.

When a web service is selected, you are navigated to **Settings for Web Services** screen, which displays two tabs: **General Information** and **Call Structure**. The **General Information** tab captures key settings for connecting to an external product catalog. You enter a description and configure options like using the catalog as a supply source, enabling cross-catalog search, supplier details, HTTP GET usage, and external availability. Additional flags manage message display and data import. The **Price Validation** setting controls how catalog prices are handled: disabled, verified, or autoupdated in procurement documents. The **Call Structure** tab defines how SAP connects to external catalogs by setting technical parameters. You manage a list of call structure items, each representing part of the HTTP request. For each item, you specify the parameter name, value, and type (such as fixed value, return URL, SAP field, or full URL) to ensure proper catalog integration during procurement.

Slow or Non-Moving Materials (F2137)

Application Type: Analytical

The app provides you with powerful tools to monitor, investigate, and react to materials that show little or no movement over a defined period. By highlighting combinations of materials and locations where stock levels remain constant without any consumption, the app helps identify materials tying up unnecessary capital. You can track slow-moving materials using a **Slow-Moving Indicator**, which normalizes consumption trends over a year for better

comparison, and define specific consumption groups to tailor the analysis to relevant goods movement types. This enables inventory managers to take immediate follow-up actions, such as scrapping or transferring slow-moving stock, and optimize inventory levels to improve working capital efficiency.

The app also offers advanced analysis through detailed charts and predictive analytics, forecasting the future trend of the **Slow-Moving** indicator based on historical company data. By visualizing monthly consumption patterns and slow movement over the last 12 months, inventory managers gain deeper insight into stock behavior and can proactively manage potential stock risks. You can find results using filters such as number of days with low consumption, reference date, currency, inventory consumption group, product type, material, plant, product group, storage location, and more.

> **Tip**
>
> On the initial screen, you can switch between chart or table view or view both at the same time using the **Chart**, **Table**, and **Chart and Table** view options.

Spend Variance (F1377)

Application Type: Analytical

This app helps you analyze and compare purchase order, goods receipt, and invoice values across different procurement dimensions such as supplier, material, cost center, plant, and company code. Since the start of last year, it has enabled organizations to identify variances between these values, offering valuable insights into potential discrepancies and helping optimize purchasing efficiency. You can filter data using key fields like display currency and relative date and personalize views with adaptable filters. The app presents KPIs related to suppliers, purchasing categories, materials, cost centers, work breakdown structure (WBS) element, purchasing category, with interactive features such as drilldowns, jump-to navigation, chart type customization, and refresh options. Data is viewable in both charts and tables, and you can save customized views as tiles, export results to Excel, or share via email. Visual tools like mini charts display key procurement metrics such as spend variance, off-contract spend, contract leakage, and operational supplier evaluation helping you monitor thresholds and identify areas requiring attention.

The mini charts offer key insights into procurement performance. **Spend Variance** compares purchase orders, goods receipts, and invoices to highlight value mismatches, using visual thresholds to flag discrepancies. **PO and Scheduling Agreement Value** shows actual versus planned spend over time, helping track budget adherence. **Off-Contract Spend** reveals the percentage of purchase orders made without referencing contracts, indicating compliance issues. **Contract Leakage** tracks purchase orders that bypass available contracts, showing missed opportunities to leverage negotiated terms. **Operational Supplier Evaluation** scores suppliers on time, price, quantity, and quality, helping assess performance and improve supplier management.

You can navigate to other apps, using the **Jump To** option, as follows: My Purchasing Document Items (F0900), Manage Purchasing Categories (F0337), Purchase Order (F0348A), Supplier (F0354), and Manage Purchase Orders (F0842A).

Stock – Multiple Materials (F1595)

Application Type: Analytical

This app provides a comprehensive view of material stocks across various plants and storage locations, helping inventory managers track and manage inventory efficiently. You can filter data by material, plant, storage location, base unit, and reporting date, with the option to adapt filters as needed. The app displays key details such as material ID, description, plant, special stock type, and stock levels (unrestricted, quality inspection, and blocked). Selecting a

material ID hyperlink enables navigation to related apps, including the Product app (F2773), which offers in-depth product details like type, group, category, unit of measure, related plants, purchasing data, and sales order information. Stock data can be presented in table format, and additional insights include stock values, storage location data, and carbon footprint metrics, supporting organizations focused on sustainability through integration with SAP Sustainability Footprint Management. The app also features advanced search capabilities to filter for active materials or those without deletion indicators, and it calculates stock values using current or historical prices. You can export results to spreadsheets, send material details via email, and collaborate using Microsoft Teams, making the app a powerful tool for real-time inventory analysis and strategic decision making.

Stock Reporting Overview (F6266)

Application Type: Transactional

This app enables real-time view of stock reports, making it easier to monitor inventory levels across different materials and locations. You can drill down into individual reports for detailed insights into inventory status and history and navigate directly to related material documents for a deeper review of specific stock movements. The app supports customizable filters such as stock report number, report year, supplier, and processing date through the **Adapt Filters** option. Additional functions include deleting, duplicating, triggering outputs, copying stock reports, and adjusting settings for a more personalized experience.

Stock – Single Material (F1076)

Application Type: Analytical

This app provides a detailed and interactive view of material stock across various plants and storage locations, helping inventory managers effectively monitor and analyze stock levels.

You can view stock by plant or storage location and filter by reporting date. It displays various stock types, including unrestricted-use, blocked, quality inspection, restricted-use, returns, stock in transit, tied empties, transfer stock (both plant and storage location), and valuated goods receipt blocked stock. The app supports both tabular and chart formats, offering a clear comparison of stock type quantities. Additional features include barcode scanning for accurate data capture and the ability to pin the header for easier navigation. The app also allows you to jump to related apps for deeper stock analysis and operational tasks. The **Open In** option allows you to navigate to other related apps (50+ apps as listed in the **Define Link List**).

You can navigate to the following apps: Manage Stock (F1062), Transfer Stock – In Plant (F1061), Transfer Stock – Cross-Plant (F1957), Material Document Overview (F1077), Stock – Multiple Materials (F1595), and Display Serial Numbers (F5147).

Subcontracting Cockpit (F2948)

Application Type: Transactional

The subcontracting app offers a comprehensive view of both open and completed purchasing documents related to subcontracting. You can filter data by subcontractor, plant, component, assembly, material group, and shipping status to manage and track processes more efficiently. Key functionalities include posting goods issues, calculating stock balances, viewing stock transfer reservations, and creating component deliveries. For optimal performance, SAP recommends limiting the selection to not more than 10,000 documents when performing actions such as posting or delivery creation. It supports only plant-based shipping point determination. You can customize filters such as shipping date, component, or purchasing group via the **Adapt Filters** option and access various features like expanding/collapsing views, adjusting settings,

and managing stock requirements. The document list displays essential details including component, plant, subcontractor, stock balance, document item, shipping date, and shipping status.

Supplier Evaluation by Price (Version 2) (F1663A)

Application Type: Analytical

The app is used to evaluate suppliers based on the price variance between purchase order and invoice amounts. Both higher and lower price deviations are factored into the score, calculated over the previous year to date. Filters like purchasing organization, material group, plant, and purchasing group can be applied, with default values configurable. Data is displayed in charts and tables, can be exported or emailed, tiles can be personalized, and navigation is available to related apps.

This is the successor app to the analytical app, Supplier Evaluation by Price (F1663), which was deprecated as of SAP S/4HANA 2021 FPS02.

Supplier Evaluation by Quality – for Quality Notification (Version 2) (F3295A)

Application Type: Transactional

This app is used to evaluate suppliers based on quality complaint scores, highlighting suppliers with fewer complaints and higher quality performance. Filter criteria such as purchasing organization, material group, plant, and purchasing group can be applied, with default values configurable. You can view data in charts or tables, export results, email app links, personalize tiles, and navigate to related apps.

This is the successor app to the Supplier Evaluation by Quality app (F3295), which was deprecated as of SAP S/4HANA 2021 FPS02.

Supplier Evaluation by Quality (Version 2) (F2309A)

Application Type: Analytical

This app allows you to assess a supplier's overall quality score based on inspection results, using data from inspection lots linked to purchasing documents. The evaluation considers performance from the previous year to date. Access requires the Materials Management – Purchasing Strategy catalog role, which is part of the Purchaser business role, and specific authorization objects must be set in Transaction SU22. The app provides KPIs across various views—by supplier, purchasing group, organization, material group, and plant, among others. You can apply filters, set default values, analyze data in charts or tables, and export or email results. It also allows tile personalization for the SAP Fiori launchpad and offers navigation to related purchasing apps. The app uses CDS views C_OPERATIONALSUPLREVALQRY_CDS and C_SUPLREVALBYQUALITYQRY.

This is the successor app to the analytical app, Supplier Evaluation by Quality (F2309), which was deprecated as of SAP S/4HANA 2021 FPS02.

Supplier Evaluation by Quantity (Version 2) (F1661A)

Application Type: Analytical

This app calculates supplier scores based on quantity variance between ordered and delivered quantities, considering both excess and shortages, over the previous year to date. More importantly, for better performance, SAP recommends migrating purchasing data using the Schedule Migration of Purchasing Data for Analytics app (F3815). KPIs are displayed by supplier, material group, purchasing group, purchasing organization, purchasing category, plant, document, and trend. You can filter, analyze data as charts or tables, export results, personalize tiles, and navigate to related purchasing and supplier apps. Access requires the Materials Management

– Purchasing Strategy catalog role (SAP_BCR_MM_PUR_STRATEGY) and maintaining relevant authorization objects.

This is the successor app to the analytical app, Supplier Evaluation by Quantity (F1661), which was deprecated as of SAP S/4HANA 2021 FPS02.

Supplier Evaluation by Questionnaire (Version 2) (F2234A)

Application Type: Analytical

The app calculates supplier scores based on completed questionnaires, comparing actual scores with target scores over the previous year to date. Available through the Materials Management – Purchasing Strategy catalog role, the app displays KPIs by supplier, purchasing category, and document. You can apply filters, analyze data in charts or tables, export or email results, personalize tiles on the launchpad, and navigate to related purchasing and supplier apps. Access requires maintaining the authorization objects.

This is the successor app to the analytical app Supplier Evaluation by Questionnaire (F2234), which was deprecated as of SAP S/4HANA 2020 FPS02.

Supplier Evaluation by Time (Version 2) (F1664A)

Application Type: Analytical

The app calculates supplier scores based on delivery time variance – considering both early and late deliveries over the previous year to date. The app provides KPI views by supplier, purchasing group, organization, category, material group, plant, document, and trends. You can apply filters, set default values, analyze data in charts or tables, export or email results, personalize launchpad tiles, and navigate to related purchasing and supplier apps.

This is the successor app to the analytical app, Supplier Evaluation by Time (F1664), which was deprecated as of SAP S/4HANA 2021 FPS02.

Supplier Evaluation by User-Defined Criteria (Version 2) (F3842A)

Application Type: Analytical

The app calculates supplier evaluation scores based on active user-defined criteria, typically over a one-year period (this can be adjusted using the **Relative Date Function** field as a key field). Scores are weighted according to settings from the Supplier Evaluation Weighting and Scoring app (F2551). You can apply filters like region, supplier, material group, plant, and purchasing group, and set default filter values. The app allows you to view KPIs by supplier, purchasing category, region or document in chart or table format, export results, send links by email, and personalize launchpad tiles.

Navigation to related apps like Manage Purchasing Categories (F0337) and Supplier (F0354) is also available using the **Jump To** option.

This is the successor app to the analytical app, Supplier Evaluation by User-Defined Criteria (F3842), which was deprecated as of SAP S/4HANA 2021.

Tip

The mini tile, **Overall Supplier Evaluation (Across All Evaluation)** navigates you to the Central Fiori Component for Smart Business App (SAP Smart Business Runtime) (F5240) app, where you can view extra mini tiles for granular insights.

Supplier Evaluation Score History (F3811)

Application Type: Transactional, Analytical

The app is used to search for and view historical supplier evaluation scores by material group, supplier, calendar year, and calendar month, job run ID and more. It helps track supplier performance over time. The app displays scores in three tabs: **Operational Score** (with options to delete or copy items), **Overall Score** (with

options to delete or copy items), and **Aggregated Score** (with option to copy items). You can view operational scores (quantity variance, inspection lot, and quality notification scores) and overall scores (including questionnaire and user-defined criteria scores) in separate tabs. Data can be exported to Microsoft Excel for further analysis.

Tip

To schedule a job, you can use the Schedule Persistency of Supplier Evaluation Scores app (F3814), to save evaluation data. The app also allows viewing job details such as job ID, execution time, and user, and it supports deleting score history when needed. This app requires the MM_PUR_SE_SCORE_HIS_SRV OData service.

Supplier Evaluation Scores Output (F5061)

Application Type: Transactional, Analytical

This app is designed to help you create, manage, and distribute supplier evaluation scores based on defined criteria. It features two main tabs: **Output Messages** and **Supplier Evaluation Historical Scores**. The **Output Messages** tab allows you to perform actions such as deleting, creating, activating scores, or copying output configurations. It displays details such as supplier name, score output type, and evaluation trigger dates. The **Supplier Evaluation Historical Scores** tab lists information such as job ID, supplier, and overall score, and supports actions like initiating output and copying records. When you create a new score using the **Create Active Scores** option, you must input details such as the supplier, evaluation period, and currency. Upon saving, the system confirms that the score creation has started in the background.

The app offers extensive capabilities for handling output messages, including viewing, sorting, and filtering them by supplier, score creation date, time stamp, and output status (e.g.,

Error, Output Sent, Prepared for Output, Scores Created). Outputs are categorized as either Active (manually created within the app) or Historical (generated via scheduled jobs). You can initiate new outputs, resend scores, print or email them, and view the documents in PDF format, which summarize key scores across material groups. In addition, outputs can be exported to Microsoft Excel, and you can drill down into detailed aggregated scores.

In the Output Message screen, you can select the Supplier hyperlink which will navigate them to numerous apps as displayed in the Define Link List popup screen (choose up to 20 apps).

> **Tip**
>
> You can add multiple suppliers when creating active scores within the app.

Supplier Evaluation Weighting and Scoring (F2551)

Application Type: Transactional

This app is used to set up how supplier evaluations are compiled. It allows you to define weighting factors for different evaluation criteria such as time, price, quantity, quality, and questionnaires, and quality notification. You also can configure how the scoring is calculated for criteria such as time, price, and quantity. Supplier evaluations are based on purchasing categories. Again, you can create new weighting and scoring settings for purchasing categories by defining variance percentages and scores. You edit the scoring for active criteria such as time, price, and quality notification directly by selecting the criterion name. The details page shows how variance percentages map to variance scores. Note that scoring edits are not available for quality and questionnaire criteria. You also can adjust weighting factors directly within a purchasing category. In addition, you also can maintain supplier classifications, for instance, using the Classification option to define categories A (80–100), B (60–79), and C (0–59).

> **Tip**
>
> SAP recommends that if the weighting factor changes, it's important to review and possibly adjust the scoring too. Furthermore, the Adjust Weighting Factor option allows you to define weighting factor, activate/deactivate criterion or set the criterion to an equal percentage.

Supplier Invoice Inbound Automation Rate (F4518)

Application Type: Analytical

This app allows you to track the percentage of supplier invoices generated automatically, manually, or through self-billing. By default, it analyzes data from the previous year up to the current date, though you can adjust the time frame to fit your specific business needs. You can explore the data in greater detail through drill-down options, visualize results in both chart and table formats, export the information to a spreadsheet, and save customized views as tiles on the launchpad. KPIs include Overall, By Invoicing Party, By Company Code, By Invoicing Party and Entry Type, and Document.

In addition, you can navigate directly to related applications such as Supplier Invoice List (F1060A) and Supplier (F0354) via the Jump To option. The app also provides interactive options like toggling data label visibility, showing legends, adjusting settings, switching to full-screen mode, and choosing between chart or tabular views.

Supplier Invoice Items by Account Assignment (F2631)

Application Type: Transactional

This app allows you to display supplier invoice items based on their account assignments during invoice verification. You can view key details such as account type, status, company code, posting date, amount, and quantity for each item. The app allows you filter invoice items by criteria such as invoice number, company code, posting date, invoicing party, status, and general ledger account, cost center, work breakdown structure (WBS) element, and so on. Selecting an invoice from the existing list, you are redirected to the Create Supplier Invoice app (F0859), where you can perform further actions such as reversing or navigating to journal entries.

In the **Supplier Invoice** screen, selecting the **Invoicing Party** hyperlink at the top of the screen will display a popup screen showcases company data (invoicing party name, address), and contact details (email address, phone number). From here, you also can navigate to the Supplier app (F0354).

Supplier Invoice List (Version 2) (F1060A)

Application Type: Transactional

The app allows you to search for and manage supplier invoices. You can filter invoices, customize the filter layout, sort results, hide or show columns such as Approval Status, and delete parked or held invoices—and create a new supplier invoice directly from the app. When you select an invoice, you are redirected to either the Create Supplier Invoice (F0859) or Create Supplier Invoice (MIRO) apps, depending on the complexity. Actions vary by status; for example, you can edit held invoices, complete and post draft invoices, copy existing invoices or reverse posted invoices. Invoices created here automatically generate journal entries in Finance. You can use the **Journal Entries** option to navigates to the Display Process Flow – Accounts Payable (F2691), where you can define the document type as **Incoming Invoice** as well as defining the key fields – **Incoming Invoice** and **Fiscal Year**. You also can export an invoice list to a spreadsheet for reporting or further processing.

This is the successor app to the Supplier Invoices List (F1060) app, which was deprecated as of SAP S/4HANA 1709.

> **Tip**
>
> The MM_SUPPLIER_INVOICE_LIST_ENH_SRV service (Supplier Invoices List Enhanced) should be activated and maintained.

Transfer Stock – Cross-Plant (F1957)

Application Type: Transactional

This app is designed to facilitate the transfer of stock from one plant to the next. It is particularly useful for warehouse clerks and inventory managers who handle day-to-day stock movement across different locations. The app is centered around a simple filter based on material and offers quick navigation options to related functions such as Manage Stock (F1062), Transfer Stock – In Plant (F1061), Material Documents Overview (1077), and Manage Inspection Lots (F2343), helping you maintain efficient logistics operations. This app allows you to post both within-plant and cross-plant stock transfers. It is equipped to handle serialized stock as well as handling stock transfers from one stock type to another. If you want to handle batch-managed materials during the transfer, you must first ensure that the batches are assigned to a plant reference using the Manage Batches app (F2462) or through an initial batch posting.

For materials with shelf-life expiration dates or production dates, the app enforces data validation. If these dates differ between the issuing and receiving batches, the system issues a warning before posting. You can only edit these fields for the receiving batch when they are initially blank. The app does not automatically handle batches that have no plant reference, so preparatory steps are essential. One of the app's key features is the ability to create stock transport orders (STOs) with a single click. This option becomes available when the proper system configuration has been maintained and when the selected stock is neither serialized nor classified as special stock.

Tip

Authorization checks are built into the app to prevent unsupported transfer postings, especially for materials associated with quality management processes. If you attempt to move materials linked to an inspection lot, you need the relevant quality technician authorization. In such cases, you can navigate directly to the Manage Inspection Lots app (F2343) to continue the process.

Transfer Stock (In-Plant) (F1061)

Application Type: Transactional

This app enables warehouse clerks, inventory managers, and other logistics personnel to perform internal stock transfer postings within the same plant. It is especially useful for managing day-to-day warehouse operations involving the movement of inventory between storage locations or stock types within a single plant. On the initial screen of the app, you are presented with a material filter bar where you can search for specific materials and select the relevant plant using a dropdown list. This selection is crucial, as the entire transfer posting process revolves around the plant and its associated storage locations. You also can get an overview of the stock situation for a specific material at the top of the screen where stock is grouped into unrestricted-use stock, blocked stock, and stock in quality inspection with their associated quantities. The app also provides access to supporting app options such as Manage Inspection Lot (F2343), Stock – Single Material (F1076), and Material Documents Overview (F1077), which allow you

to navigate to other apps to review inventory data and inspection status before and after the transfer. One of the core features of this app is the ability to perform stock transfers with or without shelf-life expiration dates. When handling materials that are batch-managed or that have shelf-life constraints, you must enter the production date and shelf-life expiration date before posting the transfer. Although these dates are critical for validation during the transfer process, they are not displayed in the header area of the app once the posting is complete; instead, they are recorded in the material document generated by the posting.

The app supports batch-managed and non-batch stock transfers within the same plant, ensuring shelf-life checks for batches and accurate movement data for all transfers. You can move stock between storage locations, stock types (e.g., blocked to unrestricted), and special stocks like consignment or project stock. An item list feature allows bulk processing of multiple materials in one transfer, generating a single material document for efficiency. Furthermore, the app provides visibility into stock coverage by displaying the range of coverage based on unrestricted-use stock consumption from the last 30 days. This helps inventory managers make informed decisions about transferring stock where it is needed most. This app supports a variety of movement types, especially movement type 311 (transfer between storage locations in one step) and its reversal, 312. For these two movement types, you can print material document slips. For other movement types, the printing option is disabled and set to **No print** by default. The availability of movement types is authorization-dependent, and you may see different options based on your assigned roles.

You can choose to open the material in other related apps using the **Open In** option. You also can define the number of apps (from more than 50 apps) that you can navigate to via the **Define Link List** popup screen.

Unused Contracts (F0575)

Application Type: Analytical

This app allows you to monitor and manage contracts that have remained unused or underutilized over a specified period—typically from the previous year to date (this can be changed). The focus is on contracts that have been created but remain unreleased or unchanged, signaling that these agreements have not led to procurement activity. Key figures such as target amount and released amount for these contracts are shown to help you assess the extent of underutilization. The app includes several interactive options that enhance usability and navigation. You can refresh the data, with a time stamp showing when the data was last updated, drill up or down, zoom in/out on visual charts. You can interact with the KPIs through filters and analyze data at multiple organizational levels, making it easier to pinpoint where and why contracts are being ignored or underused. The app displays the following mini tiles that give access to supporting metrics:

- PO and Scheduling Agreement Value
- Off-Contract Spend
- Contract Leakage
- Quantity Contract Consumption
- Value Contract Consumption

The app uses CDS view C_UNUSEDPURCHASECON-TRACTQ to retrieve data. You can utilize the **Jump To** option to navigate to Manage Purchase Contract (F1600A), My Purchasing Document Items (F0900), and Supplier (F0354).

Upload Supplier Invoices (F2452)

Application Type: Transactional

This app allows you to efficiently upload and manage supplier invoice files stored locally. The primary purpose of this app is to simplify the process of bringing external electronic invoice documents into the SAP S/4HANA system, where each file uploaded is automatically converted into a draft invoice, which is especially useful for organizations that receive invoices in bulk from suppliers in formats such as PDF, XML, or scanned images. On the initial screen of the app, you can use the search bar or filter criteria for refined searches based on key fields such as processing status, creation date, fiscal year, company code, and many more. The app has mainly two options: to either **Open Uploads** or **Create Uploads**. Selecting the **Create Uploads** option, a popup window is triggered where you can enter a company code, add an upload description, and attach one or multiple invoice files. Each file uploaded is automatically linked as an attachment to a newly created invoice draft within the system. This automation reduces manual data entry and speeds up the invoice creation process significantly.

The app supports retries for open uploads, allowing you to handle failed attempts without starting over. For follow-up actions, you can navigate directly to the Manage Supplier Invoices (F0859) and Supplier Invoices List (F1060A) apps.

Value Contract Consumption (F2013)

Application Type: Analytical

The app is designed to help you monitor performance and consumption of value-type contracts over time. It provides deep insights into how much of a contract's target value has been utilized. It also allows you to identify the consumption percentage of value contracts from the previous year to date, along with the target amount and released amount, which helps business to stay on top of contract performance. Again, you can adapt filters such as **Display Currency** and **Relative Date Functions** to refine data views according to business needs which makes it easier to analyze contract data across different periods and currencies. The app has interactive features to support detailed data exploration. You can choose to refresh the data, show or hide mini tiles, and navigate using options such as **Jump To**, **Drill Down** or **Drill Up** for further processing and granular insights. The app also supports dynamic axis changes in the chart view, allowing you to visualize and interpret data flexibly. Furthermore, KPIs are available by several dimensions, including supplier, purchasing group, purchasing organization, trend by supplier, and many more, allowing you to segment performance data and identify areas of high or low contract usage.

The mini tiles provide key insights into procurement performance. **PO and Scheduling Agreement Value** compares actual procurement spend against planned budgets over a specific period, helping monitor cost variances and support informed decision making. **Off-Contract Spend** highlights purchase orders made without referencing existing contracts, flagging potential noncompliance and missed savings from negotiated terms. **Contract Leakage** tracks purchase orders that bypass available contracts, revealing gaps in contract usage and helping enforce procurement policies to reduce maverick spending. The app uses CDS view C_VALUECONTRACTCNSMPN to retrieve data.

Using the **Jump To** option, you can navigate to the following apps: Supplier (FO354), Manage Purchase Orders (FO842A), Purchase Order (FO348A), and Manage Purchase Contracts (F1600A).

Chapter 8
Quality Management

Quality management functions in SAP S/4HANA encompass quality planning, quality inspection, quality improvement, and quality analytics. In this chapter, we'll discuss the SAP Fiori apps used by quality planners, technicians, engineers, and managers across the logistics supply area. Where applicable, we've noted the comparable SAP GUI transactions. The apps are listed in alphabetical order.

8

Characteristic Analytics (F3383)

Application Type: Analytical

The app displays the number of inspections for results recorded in the last 365 days. Recorded results are based on time, work center, inspection characteristics, and other parameters as defined in the **Select View** dropdown functionality on the initial screen. You can view a chart that shows the number of accepted inspections, rejected inspections, and skipped inspection lots, as well as the rejection rate as a percentage. The rejection rate (%) is calculated from the number of rejected inspection lots versus the number of accepted inspections lots. You can also see mini charts within the app, and you can choose a chart suited for the analysis from the **Selected Chart Type** button. Also, you can adjust the chart layout and graphical output by changing the dimensions and measures for the respective axes. For example, you can include the inspection characteristic lower limit, mean value, or any other parameter as specified in the settings area. In addition, you can navigate between chart and table views.

From within the app, you can jump to the Characteristic Detailed Analytics app (F3382) and the Nonconformance Detailed Analytics (F3583) app.

Tip

Business users can use the **Show Mini Charts** option to display granular information related to inspection lots. For example, with the **Quantitative Characteristics** mini chart, you can view the number of inspection recordings made for a characteristic and see the units that are over or above tolerance.

Characteristic Detailed Analytics (F3382)

Application Type: Analytical

This app can be used by business users to create analysis paths in order to comprehensively analyze inspection characteristics for specified categories within the app. For example, when creating a new analysis path using the + button, you can select the **Total No. of Inspections** category and then select **Top 10 Suppliers with Most Inspections** to see the suppliers with the most material inspections done. You also can choose from numerous categories, such as the process capability index, units below tolerance, and nonconforming units. Business users also can save, delete, or print the analysis paths using options found by selecting the **Related Options** button. You can navigate between viewing the information either in a chart view or a list view.

555

Within the app, business users can navigate to the Nonconformance Analytics app (F3584), the Nonconformance Detailed Analytics app (F3583), and the Display Results History app (F2428).

Display Inspection Methods (F0311A)

Application Type: Fact Sheet

The app allows you to display and analyze existing inspections methods as defined in the system. *Inspection methods* are used in the SAP S/4HANA system to specify techniques, procedures, or tools used to inspect materials or processes. These methods are assigned to inspection characteristics in inspection plans, and they provide detailed instructions for how to perform inspections. Here you can find information about inspection methods linked to master inspection characteristics (MICs), documents, and administrative data. The key purpose of the app is to display and review details of the inspection methods and to help clarify how inspections are performed for specific characteristics or materials.

Editing an inspection method using the **Edit** button navigates you to the Change Inspection Method app (QS33).

Display Inspection Points (F0309A)

Application Type: Transactional

An *inspection point* is a distinct record of inspection results linked to a specific inspection operation; as such, this app allows you to list all the existing inspection points and view all the contextual information related to the inspection points. Inspection points are used in quality management to define specific stages or steps in a business process in which inspections or quality checks are required. They can be part of inspection plans, outlining the steps and criteria for inspecting materials or processes to ensure they meet quality standards. Inspection points can be displayed based on inspection selection

criteria such as inspection lots, inspection lot type, work center, and so on.

For example, in production, inspection points can be linked to specific operations in the routing or master recipe, which ensures that inspections are carried out at the correct step in the production process.

Display Master Inspection Characteristics (F2219)

Application Type: Transactional

This app allows you to view master inspection characteristics (MICs) based on the selection criteria. Selection criteria include but are not limited to plant, status, version, and characteristic type. For example, at the MIC list level, you can list all released MICs. You also can filter MICs to either quantitative or qualitive characteristics. This app is important as it lists the MICs used in task list types—for example, inspection plans at the plant level. Business users can proceed to edit MICs, which will redirect them to the Change Master Inspection Characteristic app (QS23). For a specific MIC, a business can view either quantitative or qualitative data, the inspection method, and numerous other parameters relevant for quality management. It is also possible to navigate to the Change Master Inspection Characteristic app (QS23) by selecting the **Display More** button.

Tip

It is also important to note that MICs are created by the quality planner, who distinguishes between qualitative and quantitative characteristics.

Transaction Code

The equivalent transaction in SAP GUI is Transaction QS24 (Display Master Inspection Characteristic).

Display Quality Info Records (F2256)

Application Type: Transactional

This app lists quality info records for suppliers at the plant level. For example, selection criteria include the material, revision level, supplier, and plant. You can view comprehensive details about quality info records, which serve as a control mechanism for purchasing materials from suppliers. For suppliers that require approval before delivering materials, a quality info record becomes invaluable. Quality teams can release, block, or restrict quality info records based on material quality performance. Within the app, you can see released or blocked quality info records, which affect downstream procurement processes. Documented information relating to the combination of a material and a supplier in a quality information record is provided as well. You can create a quality info record whenever you need a quality assurance agreement or a supplier release for a material.

In the quality info record for procurement, you can store the following quality-based control information for the combination of a supplier and a material:

- **Release Date**: Corresponds to the latest date when you can order the material from the supplier.
- **Release Quantity**: Order a maximum of X material from the supplier. If there are multiple purchase orders, then the system calculates the total of all purchased quantities.
- **Blocking Function**: The selected function—for example, the creation of a purchase order or quotation—is blocked for quality reasons. If the purchaser creates a purchase order and this function is blocked, they will receive a message.
- **Inspection Control**: You can set up source inspections.
- **Quality Agreement**: Link documents to this info record.

Tip

This app is closely linked to the Manage Quality Info Records app (F2256A). These apps work together to help maintain quality controls in the procurement process. You can use the Display Quality Info Records app (F2256) to view existing quality info records for a specific material, plant, and vendor combination. This helps you check details like validity periods, whether quality agreements are in place, or if inspections are required. If changes are needed or if no record exists, you can switch to the Manage Quality Info Records app (F2256A) to create, update, or delete the quality info record.

Transaction Code

The transaction in SAP GUI is Transaction QI03 (Display Quality Info Record).

Display Results History (F2428)

Application Type: Transactional

This app allows users who need to track and analyze historical quality inspection results. It provides an overview of past inspection results for materials, batches, and inspection lots. Quality management teams can use the app to analyze past inspection results and improve quality control; in procurement, business can assess supplier trends; and in production, the app is used to monitor internal production quality over time.

You can choose between two aggregation levels: the inspection plan characteristic or master inspection characteristic (MIC) level. For example, for quantitative characteristics, you will see graphical, statistical, and general information. Statistical information and KPIs include process capability information, tolerance limits, variance, and mean value, whereas general information includes a partial sample, total sample size, and so on.

557

Display Sampling Procedures (F2255)

Application Type: Transactional

This app allows you to display sampling procedures at the sampling procedures list level. For example, sampling procedures can be filtered based on sampling types: 100% inspection, fixed sample, percentage sample, and use sampling scheme. Sampling procedures define the method for determining the sample size when inspecting materials or products. Sampling procedures are critical in quality management as they determine how samples are taken for inspection during the quality control process. For example, sampling procedures can be applied in goods receipt inspections or in-process inspections. In the SAP S/4HANA system, there are four sampling types: the fixed sample size, the percentage sample, 100% inspection, and the sampling according to a sampling scheme. A sampling procedure is ultimately assigned to each inspection characteristic.

You can also edit existing sampling procedures using the **Edit** button; this will redirect you to the Change Sampling Procedure app (QDV2).

Display Samples (F2254)

Application Type: Transactional

This app allows you to display a list of samples from various process areas in the SAP S/4HANA system. Sample selection criteria include the material, plant, batch, sample category, and so

on. The app lists samples at the samples list level. For example, you can list goods receipt samples (01) based on the sample type.

Opening a specific sample for detailed information allows you to make edits by redirecting you to the Change Sample app (QPR2).

Inspection Lot Analytics (F3239)

Application Type: Transactional, Analytical

This app allows business users to visualize the number of usage decisions made in the last 365 days. Inspection lot results can be filtered to showcase different parameters based on business requirements such as the number of accepted lots, rejected lots, lots skipped, or even lots that were not skipped on the axes of your choice. Mini charts can also be displayed to showcase inspection lot key figures, the mean quality score, and frequencies. You can filter inspection results to get information about the different set parameters. For example, in the **Mean Quality Score** mini chart, you can choose to view results based on a time series, inspection type, material and plant, supplier, or even customer.

Within the app, you also can navigate to the following apps: Inspection Lot Detailed Analytics (F3273) and Manage Inspection Lots (F2343).

Inspection Lot Detailed Analytics (F3273)

Application Type: Transactional, Analytical

Like the Characteristic Detailed Analytics app (F3382), this app allows you to create analysis paths for inspection lots for which usage decisions have been made in the last 365 days. You can create a new analysis path by indicating the category that the inspection lot should fall under. For example, predefined categories include but are not limited to mean process, time, skip rate, rejection rate, mean quality score, and so on. Once an analysis path has been created, you can view the inspection lot information and filter based on set requirements. You can save your most frequently used analyses and then use them as references as more data is collected.

Within the app, you can navigate to the following apps: Manage Inspection Lots (F2343), Characteristic Analytics (F3383), and Characteristic Detailed Analytics (F3382).

Tip

With the **UD Made On – Date Function** field, you can select a specific date range within which usage decisions were made, allowing for more targeted filtering and analysis of inspection results.

Inspection Lot (S/4HANA) (F2180)

Application Type: Transactional

This app allows you to analyze and monitor inspection lots by displaying critical information based on the selection criteria. You can display inspection lots to view information such as stock postings, usage decisions, and inspection lot types. In terms of monitoring, you can track inspection lots based on their statuses—for example, in process or completed. You also can display lists of serial numbers for inspection lots with serialized materials and print inspection reports and sample drawing instructions.

You can access this app by selecting an inspection lot from the list in the Manage Inspection Lots app (F2343).

Inspection Operation (S/4HANA) (F2181)

Application Type: Fact Sheet

This app allows you to display and list inspection operations. In SAP S/4HANA, *inspection operations* are specific steps or activities defined within an inspection plan that detail how and when quality inspections should be carried out during a business process. Inspection operations ensure that quality checks are performed at the right stages of a process, and they provide a structured approach to managing inspections. You can drill down into specific operations to view all the relevant and related information. It is also important to note that an inspection operation can have more than one inspection characteristic.

You also can navigate to the following apps: Display Inspection Lot (QA03), Display Inspection Method (QS34), and Display Master Inspection Characteristic (F2219).

Manage Control Charts (F2810)

Application Type: Transactional

Using this app, you can display and manage control charts. A *control chart* displays process data over time, along with upper and lower limits that delineate the expected range of variation for the process. You can use and adapt these charts based on criteria such as material, plant, work center, master inspection characteristic (MIC), violation type, statistical process control (SPC) criterion, status, and so on. You also can switch between visual and compact filters.

If you select a chart, you can also perform various actions. The initial screen shows **Action Limit Violations by Period (Inspection End Date)** and **Control Chart** sections. Here, the app allows control charts to be calculated and activated, recalculated, and completed based on your preferences. Again, by selecting a specific control chart from the list, you can view the related contextual information. In the **Action Limit Violations by Period** section, you can view deviations by comparing the measured values against action limits. Using the **View By** button, you can visualize results based on the chart type, control chart, characteristic, inspection end time, valuation, and so on.

This app is closely tied to the Quality Engineer Overview app (F2360) through the **Action Limit Violations** card.

Manage Defects (F2649)

Application Type: Transactional

This app allows you to see detailed information about defects as they are recorded in the various process areas at plant level. Selection criteria can be based on various inputs: defective material, plant, lot origin, customer, supplier, defect class, work center, and so on. You can monitor and manage defects created during the inspection process, and you can filter defects using a compact and visual filter. For instance, with the visual filter, you can select defects according to different analysis criteria.

In the **Defects** area, you can view defects at the plant level or even based on predefined views, such as notification type, supplier, defect location code, or group. The **Defects** list level displays the defects list based on the filtering criteria. You can also maintain defects in notifications directly using the **Transfer Defects to Notification** button. This way, you can transfer defects to quality notifications and continue further processing within the notification. For example, you can transfer the affected quantity of the material to quality inspection stock and create an inspection lot for in-depth analysis. In this case, you can choose to transfer defects to either an existing quality notification or a new one. The defect data is transferred to a notification item. You can employ this approach if the defect doesn't require formal processing, such as the 8D method. Alternatively, you can initiate a problem-solving process from a defect using the 8D method by choosing the **Start Problem-Solving Process** button within the defect screen, which allows you to access the Resolve Internal

Problems app (F4197) and initiate the 8D process. The results can be displayed as a table and as a chart, showing the occurrence of defects over a calendar year or calendar week. You can add further dimensions as well, such as the defect status or defect code. The visual filter allows you to filter defects according to the following KPIs:

- **Number of Occurrences by Defects Material**
- **Number of Occurrences by Inspection Lot Origin**
- **Number of Occurrences by Defect Code Group and Code**

This app is closely tied to the Quality Engineer Overview (F2360) app through the following SAP Fiori cards: **Inspection Lots with Defects**, **Top Defective Materials**, and **Top Defects**.

Manage FMEAs (F4340)

Application Type: Transactional

The app is similar in functionality to the failure mode and effects analysis (FMEA) cockpit in quality management in the SAP S/4HANA system. The app provides a centralized platform for managing all FMEA activities, ensuring consistency and ease of access to information. The app lists previously created FMEAs in the resulting hit list. The app has numerous capabilities, from creating new FMEAs to changing existing FMEAs to modeling FMEA structures using graphical nets. The app allows for the execution of the FMEA process steps using a graphical framework. It provides a flexible work list from which you can open an FMEA that is displayed as the focus element with its higher-level and lower-level elements (depending on the structural model). On the details page of this focus FMEA, you can access detailed information about the FMEA as well as various graphical views to show the related elements, the functions, and/or the failure modes. The app can be used easily to perform structural, functional, and failure analysis—which falls in line with the automotive industry standard for performing FMEA. For example, you can build an FMEA model by first creating an element and then assigning it to a function, possible failures, and their associated causes and effects—aptly named the *cause-effect chain*. The **Analysis** tab shows the different graphical nets: **Elements**, **Functions**, **Failure Modes**, and **Functions/Failure Modes**.

The risk optimization view in the app supports risk analysis and optimization for an FMEA study. *Risk optimization* is also known as just *optimization* in the SAP S/4HANA system. The app also supports exporting FMEA lists to Microsoft Excel.

Tip

The Manage FMEAs app can help teams in the automobile industry perform structural, functional, and failure analysis—as recommended as part of the new Verband der Automobilindustrie (VDA) regulations—using the elements, functions, failure mode, and functions/failure mode views, together aptly named *graphical nets*.

Transaction Codes

Relevant (although not equivalent) transactions in SAP GUI are Transactions QM_FMEA-MONITOR (FMEA – Monitor) and QM_FMEA (FMEA – Cockpit).

Manage Inspection Lots (F2343)

Application Type: Transactional

This app lists all inspection lots based on selection criteria such as inspection lot origin, inspection lot number, results recording status, serial number, serialized material, and so on. Inspection lots contain all the data for the inspection, such as material, order, inspection specifications, inspection lot size, inspected quantity, and, if required, quality management order.

For example, you can choose to display the inspection lot list based on the inspection lot

origin and status. You also can create, record results for, and make usage decisions for inspection lots. The app is closely linked to the Quality Engineer Overview app (F2360) through the **Inspection Lots Without Inspection Plan** card. It is also closely linked to the Quality Technician Overview app (F2361) through the **Inspection Lots Without Inspection Plan** card.

Tip

You can use the **Show More per Row** and **Show Less per Row** options to display more or fewer inspection lot details on the initial screen.

Manage Inspection Plans (F3788)

Application Type: Transactional

This app lists existing inspection plans using selection criteria such as plant, group, group counter, material, overall status, valid-from date, and so on. At the inspection lists level, you can create, copy, show a preview of, and delete inspection plans. Clicking the **Show Preview** button will showcase the materials and operations associated with the selected inspection plan. In addition, you can assign materials to the inspection plan in edit mode using the **Assign Materials** option in the **Material Assignment** tab.

Tip

In this app, you can use calculated characteristics, which allow the system to derive inspection results from other characteristic values using predefined formulas.

Manage Quality Info Records (F2256A)

Application Type: Transactional

This app allows you to create, display, and manage quality info records based on the selection criteria: editing status, supplier, supplier block, flagged for deletion, and so on. You can also analyze inspection lots by using the **Analyze Inspection Lots** button on the initial screen, and by adapting the filters, the app can list quality info records in the selection criteria. You can remove a block, flag a record for deletion, and even create a quality info record. The **Analyze Inspection Lots** function navigates you to the Inspection Lot Analytics app (F3239) from the initial screen. Of great importance as well is the ability to filter settings at the quality info record list level, including inspection control, production part approval process (PPAP) level, PPAP status, quality management control key, revision level, and so on.

Tip

Colors are used to highlight some quality info records. The ones highlighted in red are flagged for deletion, the ones in yellow have exceptions, and ones in blue are in draft status.

Manage Quality Levels (F2914)

Application Type: Transactional

Using this app, you can display and manage lists of quality levels based on selection criteria such as inspection lot origin, plant, and material. The inspection severity, dynamic modification level, plant, skip, customer, supplier, and so on can be maintained by adjusting the filters at the header level. The app has two tabs on its initial screen, for inspection lots and for characteristics at the plant level. It is important to note that you can reset or tighten quality levels within the app.

One of the key important features of this app is that you can view the last six inspection lots for each quality level and edit the inspection stage for the next inspection. By choosing the **View History** option, you can navigate to the Quality Level History app (F2915). This app is closely linked to the Quality Engineer Overview app (F2360) through the **Quality Level for Characteristics**, **Quality Level for Inspection Lots**, and

Inspection Severity of Next Inspection Stage cards.

Manage Quality Tasks (F3381)

Application Type: Transactional

This app allows you to display and manage existing quality tasks based on selection criteria such as quality task status, task processor, defect, processor assignment, and so on. Listed quality tasks can be displayed in either a chart or table view, or in both at the same time. Business users can set tasks in process, assign a processor, or complete tasks as required. This app is closely related to the Quality Engineer Overview app (F2360) through the **Quality Tasks by Processor Assignment** and **Quality Tasks by Planned Order** cards.

Tip

You can toggle between the **Show All Items** and **Show Items Based on Chart Selection Only** functions on the initial screen at the quality task list level.

Manage Usage Decisions (F2345)

Application Type: Transactional

This app lists existing inspection lots in the system from various process areas—for instance, goods receipt and in-process inspection lots. Inspection lots can be listed based on selection criteria; for example, inspection lots can be based on the results recording status (not started, in process, not required, completed with violations, or completed). You can make a usage decision for an inspection lot by using the **Make Quick Usage Decision** option. By selecting an existing inspection lot, you can add a usage decision for the inspection lot as well as change, display, and manage usage decisions for each inspection lot. For inspection lots for which a usage decision has been made, you can view or change that usage decision. For inspection lots for which a usage decision has not been made, you can make a usage decision. Furthermore, you can post the inspection lot stock, view the results history, view defects (if any), view the quality score history, and view the material documents.

You can navigate to the Stock – Single Material app (F1076) by selecting the **Go to Stock Overview** option in the **Material Documents** tab. More importantly, you can post stock of materials from the inspection lot stock to available stock types, choose single inspection units (serial numbers), and selectively post them to different stock types.

> **Tip**
>
> When a usage decision is reset from the **Usage Decision** screen, the fields for the usage decision code and quality score are initialized. The inspection lot status is changed from **UD** to **UDRE**.

Manage Workflows for Quality Task Processing in QM (F4977)

Application Type: Transactional

This app allows you to configure and manage workflows that facilitate the processing of quality tasks. From the initial screen, you can view existing workflows, create new ones, copy or deactivate workflows, and define their execution order. A predelivered workflow such as **Assign Processor to Quality Task** can be used out of the box or copied and customized to meet specific business requirements. When creating a new workflow, the **New Workflow** screen prompts you for key details like the workflow name, description, and validity dates, and it lets you define one or more workflow steps by entering a step name, a step type, recipients, and additional parameters. Once activated, these workflows drive automated task assignments that appear in the end user's My Inbox app,

ensuring that quality tasks are processed promptly and consistently.

Nonconformance Analytics (F3584)

Application Type: Transactional, Analytical

This app displays the number of defects created for the previous 365 days. Business users have the flexibility to display the defect information using the chart view for numerous KPIs: customer, defect category, material and plant, time series, and so on. You can select the **Show Mini Tiles** option to display mini tiles at the top of the screen for defect analytics and notification analytics. Hovering over the **Defect Analytics** mini tile will show you the number of defects with or without notifications.

Furthermore, from the initial screen, you can jump to the Manage Defects (F2649) and Nonconformance Detailed Analytics (F3583) apps. Quality engineers can analyze defects based on, for example, time, material, plant, and more. The app encompasses data from quality notifications. For example, you can see the monthly defect counts for a material over time to assess potential improvements in product quality.

> **Tip**
>
> The key figures field at the top of the Nonconformance Analytics app shows the total number of defects to date.

Nonconformance Detailed Analytics (F3583)

Application Type: Transactional, Analytical

This app allows you to create analysis paths for nonconformance in certain categories—namely, defects, quality notifications, mean processing time of quality notifications, and overruns quality notifications. For example, defects can be analyzed based on calendar year, inspection characteristics, master inspection characteristics

(MICs), plant, material, work center, defect category, and so on. This app is used for comprehensive and thorough nonconformance analysis, particularly for quality engineers.

Business users can also access the app through the Nonconformance Analytics app (F3584).

Process Quality Tasks (F3250)

Application Type: Transactional

This app lists quality tasks that have been created, have been completed, or are still being processed. Business users can list these tasks based on selection criteria such as task code, task code group, task processor, and so on. You can select a specific quality task for further processing.

Selecting a specific quality task will direct you to the **Quality Task** screen. In edit mode, you can make changes to the existing quality task here. Details such as the task code, task code group, task status, task processor, and time effort can be found in the **Task Info** tab. Necessary changes can be made to the quality task as the assigned task processor works on resolving, say, a generic defect. You can add a planned end date and add relevant notes. Once the quality task is complete, the assigned task processor can change the status to completed, which will automatically populate the **Completion** tab with the completion date and time.

Tip

Team members assigned to act on quality tasks documented in the 8D methodology steps who are using the Resolve Internal Problems app (F4197) can also use this app to process quality tasks and provide feedback.

Quality Engineer Overview (F2360)

Application Type: Analytical

This app shows a comprehensive overview of quality-related tasks for quality engineers on its

initial screen. The information is displayed in a set of cards. On the initial screen, you can display card information based on the selection criteria, which include the inspection lot origin, material, sales organization, supplier, inspection type, plant, posted-to inspection lot, inspection type, purchasing organization, and so on. The following cards are available:

- **Inspection Lots Without Usage Decision**
  View the number of inspection lots without a usage decision in the header, followed by a chart that shows the percentage distribution of inspection lots based on their results recording status. This card navigates you to the Manage Usage Decisions app (F2345).

- **Inspection Lots Without Inspection Plan**
  Displays open inspection lots without an inspection plan. This card navigates you to the Manage Inspection Lots app (F2343).

- **Inspection Lots Ready for Usage Decision**
  Displays the number of inspection lots ready for usage decision processing. This card navigates you to the Manage Usage Decisions app (F2345).

- **Inspection Lots with Defects**
  Displays the number of defects in inspection lots. This card navigates you to the Manage Defects app (F2649).

- **Top Defective Materials**
  Displays the number of defects created for defective materials. It is important to note that you can filter by date, for the current date up to the previous 29 days.

- **Top Defects**
  Displays the number of the top defects for a specific defect code and defect code group. This card navigates you to the Manage Defects app (F2649).

- **Inspection Severity of Next Inspection Stage**
  Displays the number of quality levels created for inspection lots and characteristics. This card navigates you to the Manage Quality Levels app (F2914).

- **Quality Levels for Characteristics**
  Displays the number of quality levels created for characteristics. This card navigates you to the Manage Quality Levels app (F2914).

- **Quality Levels for Inspection Lots**
  Displays the number of quality levels created for inspection lots. This card navigates you to the Manage Quality Levels app (F2914).

- **Quality Tasks by Processor Assignment**
  Displays the number of open quality tasks by processor assignment. This card navigates you to the Manage Quality Tasks app (F3381).

- **Action Limit Violations**
  Displays the number of times action limits have been violated. This card navigates you to the Manage Control Charts app (F2810).

- **Q-Info Records with Exceptions**
  Displays the number of quality info records with exceptions. Quality info records can be filtered by release date or block level. This card navigates you to the Manage Quality Info Records app (F2256A).

- **Quality Tasks by Planned End**
  Displays the number of in-process quality tasks with planned end dates. This card navigates you to the Manage Quality Tasks app (F3381).

> **Tip**
>
> You can hide a card or reset the filters for cards via the **Manage** cards functionality, found by selecting the **More** icon on each card.

Quality Level History (F2915)

Application Type: Transactional

This app provides a historical overview of the quality level of materials, batches, or vendors over time. Quality levels are used to evaluate and monitor the performance of materials, batches, or suppliers based on inspection results, defects, and other quality-related data. The app displays the skip rate of inspection lots for the previous 180 days. The information in the quality level determines which inspection stage will be used for the sample determination of the next inspection lot. The system updates the fields in the quality level differently during

dynamic modification. This depends on whether dynamic modification occurs at lot creation or when the usage decision is made. In the quality level header—showing the number of inspections that have occurred since the last stage change and the number of inspections that were unsuccessful—these counters are updated each time a quality level is updated. The system uses these counters to determine when an inspection stage change should occur. Business users can display inspection lots based on skip history, inspection severities, and inspection stages, and can list inspection lots with the corresponding characteristics. The app calculates the skip rate of inspection lots and inspection characteristics and analyzes the quality levels. You can display quality level information using the following KPIs:

- **Skip History: Inspection Lots and Characteristics**

- **Inspection Severities by Inspection Lots and Characteristics**

- **Inspection Stages by Inspection Lots and Characteristics**

- **List: Inspection Lots and Characteristics**

The app uses CDS view C_QUALITYLEVELANALYSIS. Within the app, you can navigate to the Manage Usage Decisions app (F2345) and the Manage Inspection Lots app (F2343).

Quality Technician Overview (F2361)

Application Type: Analytical

This app provides a centralized overview of tasks and responsibilities for quality technicians or inspectors, enabling them to efficiently manage and execute their quality-related activities. The app displays this quality information on a set of cards shown on its initial screen. The following cards are available:

- **Inspection Lots Without Inspection Plan**
  Displays open inspection lots without an inspection plan, sorted by creation date. This card navigates you to the Manage Inspection Lots app (F2343).

- **Inspection Lots with Open Results Recording. No Inspection Points**

  Displays the total number of inspection lots in which results recording is in process or not started. This card navigates you to the Record Inspection Results app (F1685A).

- **Inspection Lots with Open Results Recording for Inspection Points**

  Displays the total number of inspection lots for which results recording has not started or is in process. This card navigates you to the Record Results for Inspection Points app (F2689).

- **My Tasks**

  Displays the quality tasks assigned to users that are in either the created or in process status, sorted by planned end date. The app navigates you to the Process Quality Tasks app (F3250).

> **Tip**
>
> You can save and manage customized views by using the **Standard** dropdown on the initial screen of the app.

Record Defect (F2929)

Application Type: Transactional

With this app, business users can record defects. In the **Defect Info** tab, you can input the defect description, defect code (from existing proposals), and defect code group, and you can assign a reference number. The initial app screen holds the following tabs: **Defect Info, Tasks, Attachments, Output Overview, Administrative Data**, and **Changes**. After creating, for example, a generic defect, you can proceed to create tasks and assign a task processor to investigate the defect and give feedback. Key functions also include setting defects to complete, in process, and not relevant. Generic defects can be recorded using the Record Defect app (F2929) and the Process Defect app (F2929).

In the **Administrative Data** tab, documented changes in relation to the defect, such as the

business user who has made changes at a specific date and time, are displayed. Once a defect has been created, business users can start the problem-solving process; that is, the **Start Problem-Solving Process** option is immediately displayed at the top of the screen. If printouts are required, you can make use of the print functionality by selecting the **Print** button. For attachments, business users can sort order lists in ascending or descending order. Files can be sorted by file name, file size, status, and document type.

> **Tip**
>
> Selecting the **Start Problem-Solving Process** option will kickstart the 8D methodology to resolve the defect. Refer to the description of the Resolve Internal Problems app (F4197) to understand more about the 8D methodology.

Record Inspection Results in Table Form (F3365)

Application Type: Transactional

This app allows you to record inspection results for several master inspection characteristics (MICs) for different inspection lots in a table on a single screen using the **Record Multiple Results** option. You can use this function if you must inspect one characteristic for several materials.

The app lists inspection lots for which results are to be recorded. Business users can display inspection lots based on selection criteria such as plant, material, and MICs.

> **Tip**
>
> At the inspection lot list level, you can filter by **Progress of Characteristics** to view the accepted, rejected, and open characteristics based on color codes.
>
> You also can record results for various inspection lots for different inspection characteristics all at once.

Record Inspection Results (Version 2) (F1685A)

Application Type: Transactional

This app lists inspection lot sets for results recording. Similar to the Record Inspection Results in Table Form app (F3365), quality technicians also can record multiple results for numerous inspection lots using the **Record Multiple Results** option. Inspection lots are listed based on selection criteria such as plant, material, results recording status, characteristics status, inspection lot origin, and so on.

On the initial screen, the app shows three tabs: **Inspection Lots**, **Operations**, and **Master Inspection Characteristics**. In the **Operations** tab, the app lists the operations based on the selection criteria. For a specific operation in edit mode, you can display inspection characteristics and add more characteristics, as well create defects as required by the business. For ease of use, you can use the characteristics dropdown function to filter characteristics to those that are open, required, accepted, rejected, skipped, or unplanned. The **Master Inspection Characteristics** tab lists the master inspection characteristics (MICs) based on the selection criteria. Similar to the functions in the **Operation** tab, in edit mode, you can add characteristics and create defects. The **Inspection Lots** tab lists the accepted, open, rejected, and skipped inspection lots. You can adjust list columns by adjusting the filter criteria in the settings. To display more important details, such as the lot end date and the progress of characteristics (accepted, open, rejected, and skipped), you can select the **Show More per Row** icon to showcase these parameters for the listed inspection lots.

This app replaces the Record Inspection Results app (F1685), which was deprecated as of SAP S/4HANA 2022 FPS02.

Tip

For attachments, business users can sort order lists in ascending or descending order. Files can be sorted by file name, file size, status, and document type.

Record Results for Inspection Points (F2689)

Application Type: Transactional

This app allows you to record inspection results at the inspection point level during quality inspections. *Inspection points* in this case are specific steps or criteria within an inspection plan for which measurements are taken and tests are performed. Selection criteria can be based on inspection lot, order, work center, plant, material, and so on. The app lists inspection lots at the orders list level on the initial screen. It allows you to enter and save inspection results for each inspection point; these results can include measurements, pass/fail status, or other qualitative or quantitative data. You can also drill down to get detailed information for each inspection point, such as characteristics to be inspected, target values and tolerances, and measurement units. You can also display the progress of characteristics based on color codes and can reject or accept the processing statutes of the inspection characteristics.

Resolve Internal Problems (F4197)

Application Type: Transactional

This app allows business users to manage and resolve internal quality issues within the business via a step-based approach. The app lists defects to be examined via problem-solving processes. It allows you to document, analyze, and track corrective actions for internal problems that arise during production, procurement, or general business processes. You can report on and document issues found during, say, inspections. This provides a way to assign and track corrective actions, which can be assigned to responsible teams for resolution. The app uses the 8D methodology, which is a standardized approach to problem-solving, to determine defect causes and take steps to resolve them. It is only recommended to use this approach if your business employs the 8D methodology. Within the app, business users can create quality tasks that will then be sorted out by the assigned team members. The 8D methodology steps are as follows: Assemble an 8D team (D1), problem description (D2), containment actions (D3), root causes (D4), defined corrective actions (D5), implemented corrective actions (D6), preventive actions (D7), and congratulate your team (D8).

Tip

On the left-hand side of the **Resolve Internal Problems** initial screen, you can scroll down to see a summarized overview, which includes the problem-solving steps, general data, referenced defects (if any), attachments, and administrative data for ease of reference during the problem-solving process.

Chapter 9
Project System

With Project System, you can manage projects from planning to execution in an enterprise. It offers tools for structuring projects using work breakdown structures (WBSs), scheduling tasks, tracking costs and revenues, managing procurement, and monitoring project progress. In this chapter, we'll explore various SAP Fiori applications for this component, presenting them in alphabetical order for ease of navigation. Each app entry describes the app's purpose, a brief overview of its functionality, its key features, and key tips and tricks.

Change Network Activity Status (F0539)

Application Type: Transactional

This app allows project logistics controllers to modify the system and user status of a network activity in Project System. You can have all subobjects inherit a user status and analyze an inheritance summary for lower-level elements, facilitating accurate tracking of project progress. Before applying changes, you can analyze the summary of all lower-level objects linked to the selected network activity to prevent unintended changes.

In addition, you can navigate here from the Network Activity Overview app (F1970). Thus, with this app, you can maintain a network activity status across related subobjects consistently.

Transaction Code

The equivalent transaction in SAP GUI is Transaction CJ20N.

Change WBS Element Status (F0292)

Application Type: Transactional

This app allows project financial controllers to modify the system and user status of a work breakdown structure (WBS) element while displaying contextual status details. You can search for a WBS using search parameters such as project definition or WBS element. The app supports inheriting a user status across all subobjects and provides an inheritance summary for lower-level elements. Before changes are applied, it is recommended to review the status. The app also supports building your own user-defined validations of the user status. In addition, the app integrates with SAP Jam, thus enabling collaborative discussions and streamlined project management.

Transaction Codes

The equivalent transactions in SAP GUI are Transactions CJ20N or CJ02.

Confirm Network Activity (F0296)

Application Type: Transactional

This app lets project logistics controllers confirm the processing status of activities within a network, documenting their progress and completion. You can search for networks using search parameters such as network, activity, element, and split number. Network activity can be confirmed using actual start and actual finish dates. By capturing activity confirmations, you can track execution status and use the recorded data to forecast future progress, ensuring accurate project planning and monitoring. Thus, the app allows you to confirm network activity within a project network.

Tip

Explore the BAPI_NETWORK_CONF_ADD BAPI for confirmation in an integration scenario or for enhancing the logic for additional requirements.

Transaction Code

The equivalent transaction in SAP GUI is Transaction CN25.

Confirm Project Milestone (F0295)

Application Type: Transactional

This app allows project financial controllers and project logistics controllers to confirm milestones within a project, marking transitions between phases in Project System. Once a milestone is completed, you can set the actual milestone date and confirm its achievement. Thus, this activity ensures that you can track project progress in real time.

Transaction Code

Transaction CJ20N allows you to confirm project milestones in SAP GUI.

Manage Project Procurement (F2930)

Application Type: Analytical

This app lets you monitor purchase orders and purchase requisitions associated with a project or work breakdown structure (WBS) element in

the display currency. You can apply filter criteria using fields such as material, supplier, project manager, purchase order, purchase requisition, purchase order delivery status, and delivery date. The app displays results in the **Purchase Orders** and **Purchase Requisition** tabs. The app displays important information such as the value to be invoiced and values to be delivered. In addition, microcharts help visualize KPIs by comparing delivered versus ordered quantities for purchase orders and ordered versus total quantities for purchase requisitions. Furthermore, you can select a project to access a range of related actions and navigate to detailed contextual information for WBS elements, materials, suppliers, purchase order items, and purchase requisition items.

Milestone (F0286A)

Application Type: Fact Sheet

This app provides a structured overview of milestone data, including key events across project phases, responsible personnel, applicants, and control details. Milestones are displayed in three tabs: **General Information**, **Network Activity/ Network**, and **WBS Element/Project** for the selected **Scheduled Date**. The app also displays associated networks, network activities, projects, and WBS elements to provide comprehensive tracking capabilities. In addition, you can navigate to other apps via the **Related Apps** option for further business context and expanded analysis.

Transaction Code

The equivalent transaction in SAP GUI is Transaction CJ20N.

Milestone Overview (F1975)

Application Type: Analytical

This app displays detailed master data for milestones, which serve as key indicators for tracking events across different project phases. You can search for milestone data using parameters such as project definition, work breakdown structure (WBS) element, level, network, and activity. You also can view selected milestone details through object pages and navigate to a confirmation app to record the actual completion date and confirm the milestone. The app leverages CDS view C_MilestoneWithVersion for robust data integration.

Tip

This analytical app provides a high-level perspective of your project progress through summarized data. In contrast, the Milestone app (F0286A) is a fact sheet that provides detailed master data and related object information for individual milestones, enabling precise tracking and management of each milestone event.

Transaction Code

The equivalent transaction in SAP GUI is Transaction CN53N.

Network (S/4HANA) (F0288A)

Application Type: Fact Sheet

This app provides a detailed overview of network master data and related objects, representing the flow of a project or task within a project. The app displays data in the **General Information, Status, Networking Activities,** and **WBS Element/Project** tabs. The **General Information** tab displays fields such as responsible personnel, applicants, organizational data, and key dates, along with system and user status details. In addition, you can view associated network activities, linked projects, and work breakdown structure (WBS) elements while also navigating to relevant apps for additional business context and insights.

Transaction Code

The equivalent transaction in SAP GUI is Transaction CN23.

Network Activity (S/4HANA) (F0289A)

Application Type: Fact Sheet

This app provides a detailed view of network activity master data, representing tasks within a network and serving as the foundation for logistics planning and execution. It includes a **General Information** tab that shows information such as responsible personnel, applicant, key dates, and activity details, along with status and user status. In addition, you can track associated networks, linked milestones, material and service components, purchasing document items, and related projects with work breakdown structure (WBS) elements. Furthermore, the app enables navigation to other applications for expanded business context using the **Related Apps** function.

Tip

This app is designed for granular network activity information, whereas the Network (S/4HANA) app (F0288A) is designed to handle the entire network structure within a project.

Transaction Code

The equivalent transaction in SAP GUI is Transaction CN23.

Network Activity Overview (F1970)

Application Type: Analytical

This app provides a detailed view of master data for network activities, which represent tasks within a network and serve as the foundation for logistics planning and execution. You can display network activity using search fields such as project definition, work breakdown structure (WBS) element, level, network, and activity. In addition, you can access object pages to display key details, navigate to related apps for an overview of associated objects, and manage activity status or confirmations. The app also leverages CDS view C_NETWORKACTIVITYWITHVERSION for data for network activity.

Transaction Code

The equivalent transaction in SAP GUI is Transaction CN47.

Network Overview (F1973)

Application Type: Analytical

This app provides a detailed view of master data for networks, which define the flow of projects or tasks within a project. You can apply filters using parameters such as project definition, work breakdown structure (WBS) element, level, network, and activity. In addition, you can display selected network details through the object page and navigate to related apps for insights into associated objects such as network activities and milestones. The app leverages CDS view C_PROJNETWORKWITHVERSION for data processing and integration.

Tip

This app offers a broader perspective by presenting high-level master data for entire networks, allowing you to view the overall structure and relationships between activities, milestones, and associated objects. In contrast, the Network Activity Overview app (F1970) provides details of individual network activities, displaying their master data, statuses, and execution details for precise tracking and logistics planning.

Transaction Code

The equivalent transaction in SAP GUI is Transaction CN46.

Product Component (F0540A)

Application Type: Fact Sheet

This app provides a detailed overview of the master data and related object details for material or service components assigned to a network activity. This app supports project-based procurement. In addition, the app displays a **General Information** tab, which includes information such as responsible personnel, control data, and bill of materials (BOM) details, along with assigned purchase document items. Furthermore, you can track associated networks, network activities, projects, and work breakdown structure (WBS) elements while navigating to related apps for extended business insights.

Transaction Codes

The related transactions in SAP GUI are Transactions CS03, CS01, CS02, and CS15.

Product Overview (F1971)

Application Type: Analytical

This app provides a detailed view of the master data for material or service components assigned to a network activity. You can filter data by project definition, work breakdown structure (WBS) element, network, activity, product, or product type group to differentiate between a material and a service. When you click the **Go** button, the app displays selected component details through the object page and supports navigation to related apps for further insights. The app utilizes CDS view C_ProjMatComponentWithVersion for data processing.

Project Claim Overview (F6497)

Application Type: Transactional

This app allows you to view a list of project claims in a structured table format and access individual claim details through an object page. With this app, you can search for a project claim using parameters such as the claim ID, claim type, project, work breakdown structure (WBS) element, and claim purpose. You can navigate to relevant object pages for specific claims by clicking hyperlinks and can explore related apps to view associated project elements such as WBS elements, sales orders, deliveries, customers, and suppliers. In addition, you can navigate to SAP GUI Transaction CLM3 to view the claim details in the system.

Project Cost Line Items (F6992)

Application Type: Analytical

This app lets you analyze actual cost line items and planned cost details across various project components, including work breakdown structure (WBS) elements, internal orders, production orders, service order items, service contract items, assigned networks, and plant maintenance orders with header account assignment. You can search for line-item details using fields such as plan category, additional plan category, project definition, appended plan only, controlling area, general ledger account hierarchy, WBS element, cost component, service order, and sales order item. The app displays amounts in both global and user-specific currencies. The app reads plan costs from table ACDOCP (for financial planning), while actual costs are derived from table ACDOCA (the Universal Journal). Thus, the app facilitates comprehensive financial tracking and comparison of WBS elements.

This app replaces the deprecated Project Cost Report – Line Items app (F2538).

Transaction Code

The corresponding transaction in SAP GUI is Transaction CJI3.

Project Cost Overview (F6991)

Application Type: Analytical

This app lets you analyze actual and planned costs across various project components. You can search for costs using fields such as plan category, additional plan category, project definition, appended plan only, controlling area, general ledger account hierarchy, work breakdown structure (WBS) element, network, and sales order. The app's scope includes cost objects such as WBS elements, internal orders, production orders, service order items, service contract items, assigned networks, network activities and elements, and plant maintenance orders with header account assignment. The app reads plan costs from table ACDOCP (for financial planning), while actual costs are derived from table ACDOCA (the Universal Journal). Thus, the app facilitates comprehensive financial tracking and comparison.

This app replaces the deprecated Project Cost Report – Overview app (F2513).

Tip

As a prerequisite to use the general ledger account hierarchy, you must ensure that general ledger accounts are assigned for the selected hierarchy via the Manage Global Hierarchies app (F2918).

Transaction Codes

The corresponding transactions in SAP GUI are Transactions CJI3, CJI4, and S_ALR_87013542.

Project Definition Overview (F1976)

Application Type: Analytical

This app provides comprehensive master data for project definitions, which form the basis for all planning, execution, and monitoring tasks in Project System. You can search for project definitions using search parameters such as the project definition and person responsible. The app displays detailed information for selected project definitions through an object page and allows you to navigate to related apps to explore associated objects such as work breakdown structure (WBS) elements, networks, and network activities. The app leverages CDS view C_PROJECTWITHVERSION for robust data integration.

Transaction Codes

The equivalent transactions in SAP GUI are Transactions CN42 or CN42N.

Project Definition (S/4HANA) (F0290A)

Application Type: Fact Sheet

This app provides a comprehensive view of project definition master data, forming the foundation for planning, execution, and monitoring tasks in Project System. The app presents data in four tabs: **General Information**, **Status**, **WBS Elements/First Hierarchy Level**, and **Networks**. The **General Information** tab displays key information such as responsible personnel, applicants, project key dates, and organizational data. The **Status** tab displays system and user status details. With the **WBS Elements First Hierarchy Level** tab, you can track associated work breakdown structure (WBS) elements at the first hierarchy level, linked networks, and relevant user fields. In addition, you can navigate via **Related Apps** to other apps for expanded business insights.

Transaction Codes

The corresponding transactions in SAP GUI are Transactions CJ20N, CJ06, CJ07, and CJ08.

Project Network Graph (F5130)

Application Type: Transactional

This app provides an interactive graph view of network activities and their relationships. You can search for project network graphs using fields such as project definition, work breakdown structure (WBS) element, and network. With this app, you can identify critical activities, dependencies, and the confirmation status. The app also lets you group by network or WBS element, detect loops, and navigate via successors and predecessors. In addition, you can highlight activities based on selected criteria, search for specific objects, and view detailed information. Furthermore, the app supports navigation to related apps for further analysis into networks, network activities, and WBS elements, while offering an overview of assigned objects such as milestones, subnetworks, and material components. The apps' customization options include adding or editing attributes displayed within graph nodes, with the flexibility to switch between different view types.

Transaction Code

The equivalent transaction in SAP GUI is Transaction CJ20N.

Project Schedule (F5611)

Application Type: Transactional

This app lets you visualize project object hierarchies and monitor their timelines through a Gantt chart representation. You can search for project schedules using parameters such as project definition, work breakdown structure (WBS) element, and network. The table view displays a configurable project tree, while the chart area presents key date sets, including basic, actual, and forecast dates. In addition, you can specify the default hierarchy level on project opening, explore assigned network activity objects, view relationships within networks or plant maintenance orders, and access detailed side-panel insights. Furthermore, the app supports custom shape coloring in the Gantt chart via UI Theme Designer, navigation to relevant object pages and applications, and saving preferred configurations for future use.

Transaction Code

The equivalent transaction in SAP GUI is Transaction CJ20N.

PS-Texts (F5612)

Application Type: Transactional

This app lets you create and update Project System texts, allowing for long text entries in free-text format and the addition of attachments. The app presents data in the **Long Text**, **Attachments and Links**, **Assignments**, and **Administrative Date** tabs. With this app, you can assign Project System texts to work breakdown structure (WBS) elements, standard WBS elements, network activities, and standard network activities. The app supports searching, viewing lists of Project System texts and attachments, accessing text details, copying metadata to create new texts, and deleting existing entries for efficient project text management. Thus, the app facilitates adding structured documents and texts to Project System objects.

Schedule Project Cost Forecast Jobs (F7455)

Application Type: Transactional

This app lets you schedule jobs to estimate the remaining costs and completion costs based on the plan, forecast, commitment, and actual values during the execution phase. The app uses the Project Cost Forecasting predelivered job template to estimate the cost. In addition, you can set scheduling options for immediate or recurring processing. Furthermore, you can define job parameters, including the project scope, forecast version, and cost distribution timing. You can also analyze the status of scheduled jobs and the results of completed jobs for project cost planning.

Schedule Project Version Creation (F2143)

Application Type: Transactional

This app lets you schedule the creation of project versions triggered by system or user status changes. You can schedule a new job either immediately or on a specified schedule via the scheduling options. The app offers users various parameters such as project, work breakdown structure (WBS) element, network/order, activity, materials in network, level, version key, and version group. Once executed, the app automatically generates snapshots based on object status and predefined time intervals, providing a report that captures the project's state at a specific moment. These snapshots serve as valuable benchmarks for analyzing project progress over time.

Transaction Code

The corresponding transaction in SAP GUI is Transaction CN71.

WBS Element Overview (F1974)

Application Type: Analytical

This app provides a detailed view of master data for work breakdown structure (WBS) elements, which organize project tasks in a hierarchical structure. You can search WBS elements with filter fields such as project, WBS element, and level. You can access selected WBS element details through the object page; navigate to related apps for additional information related to associated networks, network activities, and milestones; and modify the system or user status using a dedicated app. The app leverages CDS view C_WBSElementWithVersion for data processing and integration.

> **Tip**
>
> This analytical app aggregates data across multiple WBS elements, presenting key performance indicators and summary metrics. In contrast, the WBS Elements (S/4HANA) app (F0291A) is a fact sheet app that provides a comprehensive, detailed view of individual WBS element master data, including all related object details for in-depth analysis and navigation.

> **Transaction Code**
>
> The equivalent transaction in SAP GUI is Transaction CN43.

WBS Elements (S/4HANA) (F0291A)

Application Type: Fact Sheet

This app provides a structured overview of work breakdown structure (WBS) element master data, supporting hierarchical organization of project tasks. The app displays key details such as responsible personnel, applicant, organizational data, dates, and control information, along with system and user status updates. In addition, you can track associated project definitions, linked milestones, networks, orders, and assigned purchase document items, as well as view relevant network activities and user fields. Furthermore, the app facilitates navigation via **Related Apps** to other apps for further analysis and contextual information.

> **Transaction Codes**
>
> The corresponding transactions in SAP GUI are Transactions CJ20N, CJ01, CJ02, and CJ13, and CN43N.

Chapter 10
Flexible Real Estate Management

Flexible real estate management (often abbreviated RE-FX) enables enterprises to manage their real estate portfolios throughout the entire lifecycle of an asset. This component supports processes such as property acquisition, lease administration, contract management, space planning, and financial analysis while providing a dual view of master data—with both architectural and usage views. This allows you to perform detailed tracking of diverse property types. In this chapter, we'll explore various SAP Fiori applications from flexible real estate management, presenting them in alphabetical order for ease of organization.

Cancel Valuation for Contracts (F6276)

Application Type: Transactional

This app can reverse valuation entries for contracts when adjustments or corrections are needed in flexible real estate management. This app is particularly useful in scenarios in which a valuation run has produced results that are either incorrect or no longer applicable due to changes in contract conditions. The app allows you to cancel the last unposted valuation step based on predefined valuation rules. As a result, the app ensures erroneous or outdated valuation data is removed before final financial postings. In addition, you can also examine detailed valuation logs and enforce strict eligibility criteria for cancellation.

Transaction Code
The corresponding transaction in SAP GUI is Transaction RECEEPRV.

Check Real Estate Object (F5967)

Application Type: Transactional

This app helps ensure the integrity and consistency of real estate master data across the real estate management lifecycle. The app systematically validates key attributes and relationships of real estate objects such as properties, buildings, and spaces. In additioan, the app examines essential details and configurations. It identifies discrepancies, missing links, and data anomalies that could disrupt subsequent processes like contract management, period end posting, or occupancy tracking. Furthermore, the app provides detailed processing logs and error reports, which provide actionable insights for corrections, thereby supporting data quality, regulatory compliance, and streamlined integration with relevant modules.

Transaction Code
The corresponding transaction in SAP GUI is Transaction RE80.

Contract Cash Flow (F5491)

Application Type: Transactional

This app provides a detailed forecast of the expected cash flow for contracts. The app facilitates analysis of projected payments and receipts based on contract terms. The app is delivered with an analytical list with two tables to display relevant cash flow information. You can refine the cash flow forecast by contract number and contract dates. Thus, the app lets you perform liquidity planning and perform ad hoc reporting for visibility into trends.

Tip

Consult SAP Note 3014244 to learn about additional app features and restrictions.

Transaction Code

The corresponding transaction in SAP GUI is Transaction REISCDCF.

Contract Management (F5273)

Application Type: Transactional

This app enables comprehensive management of contract portfolios by allowing users to search, filter, create, and edit contracts. You can maintain key details such as general information, terms, partners, objects, conditions, posting parameters, and reminders. The app also supports contract activation using the **Activate Contract**, **Check Contract**, and **Refresh Contract** buttons, access to other apps via **Related Apps**, and a timeline view showing the current contract period, upcoming renewals, and notification dates for possible notices. In addition, you can display current term information, manage the system status, and configure user status settings for enhanced tracking and compliance. All these tasks can be performed via several sections of the app, such as **Display of the Contract**

Header, General Information, Term, Partner Assignment, Object Assignment, Condition, Posting Parameters, Reminders, and Registration Entry.

Tip

Consult SAP Note 2982936 to learn about additional app features and restrictions.

Transaction Code

The equivalent transaction in SAP GUI is Transaction RECN.

Contract Mass Check (F5968)

Application Type: Transactional

This app performs comprehensive integrity checks on contract data across multiple contracts. By automating the validation process, it identifies discrepancies and anomalies in contract records that could affect subsequent mass updates or changes. The app provides advanced filtering options, allowing you to focus on specific contract groups or criteria, and it produces detailed reports that highlight deviations from expected standards.

Transaction Code

The corresponding transaction in SAP GUI is Transaction RECNCHECK.

Group Real Estate Object for Occupancy (F5592)

Application Type: Transactional

This app enables the grouping of real estate objects such as buildings, floors, spaces, and land for occupancy planning. The app provides a table to filter synchronized objects by key date, company code, object type, and location, offering an overview of available assets. The app's predefined variants include the following:

- The **All Objects** function includes all objects matching the filters, regardless of status.
- **Vacant Objects** displays objects that are not grouped.
- **Only Grouped Objects** displays objects that are grouped.
- **Grouped on Lower-Level Objects** shows objects whose subobjects contain groupings.
- **Nongrouped Objects** filters for objects that are not part of any grouping.
- **Occupied Objects** lists only grouped objects that have a contract assignment within the defined filters.

In addition, you can create new groups or assign selected objects to existing ones directly from the list, while an embedded chart visualizes measurements. Furthermore, the app supports intent-based navigation for quick access to related applications and includes variant management for customized table views.

Tip

Consult SAP Note 3084357 to learn about additional app features and restrictions.

Group Real Estate Object for Use (F5397)

Application Type: Transactional

This app enables the grouping of real estate objects such as buildings, floors, spaces, and land for usage enablement. The app provides a table to filter synchronized objects by key date, company code, object type, and location, offering an overview based on the selected date. The app's predefined variants include the following:

- The **All Objects** function includes all objects matching the filters, regardless of status.
- **Grouped Objects** displays only grouped objects that meet the filter criteria.
- **Grouped on Lower-Level Objects** shows objects whose subobjects contain groupings.

- **Nongrouped Objects** filters for objects that are not part of any grouping.
- **Grouped Objects with Contract** lists only grouped objects that have a contract assignment within the defined filters.

In addition, you can create new groups or assign to existing ones directly from the list, while an embedded chart visualizes measurements. The app supports intent-based navigation for quick access to related apps and includes variant management for customized table views.

Tip

Consult SAP Note 2998630 to learn about additional app features and restrictions.

Transaction Codes

The related transactions in SAP GUI are Transactions RE80, REISRO, REISBE, REISBU, and REISPR.

Manage Contract Term (F6580)

Application Type: Transactional

This app lets you perform contract renewals or issue notices in accordance with the renewal and notice rules maintained in contract management. You can search for contracts using various dimensions such as editing status, company code, contract type, real estate contract, person responsible, contract end date, notification by, contract renewal status, and contract notice status. Upon searching, the app presents a worklist of contracts subject to the renewal rules and offers a timeline view that displays the current contract term, upcoming renewals (both optional and automatic), and potential notice dates. The app presents this data in the **With Renewal Rule** and **With Notice Rule** tabs. In addition, the app shows simulated renewals and notices, supports the approval or rejection of renewal options, allows you to create contract notices, and facilitates memo management for renewal options and created notices. The app has four major sections: **List of Contracts, Display of the Contract Header, Notices,** and **Renewals**.

Tip

Consult SAP Note 3206996 to learn about additional app features and restrictions.

Transaction Code

The related transaction in SAP GUI is Transaction RECN.

Manage Occupancy Groups (F5593)

Application Type: Transactional

This app enables management and maintenance of occupancy groups by displaying associated real estate objects and supporting contract and cost object assignments based on group type. The app is delivered with several sections, such as **General Information, Measurement Chart, Object Relation, Contract Assignment,** and **Cost Object Assignment**. In addition, you can assign real estate contracts, which can be processed using the Contract Management (F5273) and Valuation Management (F6170) apps. The app features include a time-dependent data approach that identifies past, current, and future activity; intent-based navigation for quick access to related apps; and customizable UI settings for field adaptations and layout adjustments. In addition, you can create app variants, define table views, leverage embedded charts for data visualization, and utilize the draft feature for maintaining incomplete entries.

Tip

Consult SAP Note 3084320 to learn about additional app features and restrictions.

Transaction Code

The related transaction in SAP GUI is Transaction REISROOC.

Manage Real Estate Output Requests (F7482)

Application Type: Transactional

This app provides a centralized view of outputs sent via various channels (such as print or mail) and displays their current statuses. Once such requests are sent, you can search for them using the contract, company code, process ID, and app object type. The app also lets you drill down into each output request to review related contract details, process details, and associated output items. The app presents information in four tabs: **Contract, Process, Invoice,** and **Output Item**. In addition, for each output item, the app presents comprehensive information that includes the output item status, output type, output item action and its settings, dispatch time, sender information, and form templates used.

Transaction Code

The corresponding transaction in SAP GUI is Transaction REBDRO.

Manage Rentable Objects (F7694)

Application Type: Transactional

This app lets you display worklists for rentable object master data and manage rentable object records in flexible real estate management. You can search for rentable objects using filter criteria such as usage category, rentable object usage type, and company code. The app provides a structured view of relevant data in **Architecture Object Usage** and **Measurement** views, allowing you to track, update, and maintain rental object details for streamlined real estate management.

Tip

After a rentable object is created, you must assign it to a usage category, which determines its utilization and relationship with other real estate objects.

Transaction Code

The corresponding transaction in SAP GUI is Transaction REBDRO.

Manage Usable Objects (F7693)

Application Type: Transactional

In flexible real estate management, *usable object master data* refers to real estate objects that can be assigned to contracts or other business processes. This app lets you display worklists for usable object master data and manage usable object records. You can view the status of all objects in the worklist, apply filters to refine lists, and perform actions such as creating, modifying, or deleting usable objects. For example, you can search with fields such as object type, status, assignment, and company code. Usable objects can only be deleted if they are not linked to other dependencies, such as rentable objects or contracts.

Manage Usage Enablement Groups (F5465)

Application Type: Transactional

This app lets you manage and maintain usage enablement groups, offering a structured approach to handling time-dependent data. You can search for enablement groups using various fields such as editing status, enable use group, company code, controlling object type, group type, usage type, and person responsible. The app features indicators that distinguish between past, current, and future activity statuses. In

addition, you can customize the interface by adapting fields, labels, and layouts, and you can create and share app variants. The app supports intent-based navigation for seamless access to related applications, variant management for tailored table views, embedded charts for data visualization, and a draft feature for maintaining unfinished work. All these are delivered with the app's **Display of the Group Header**, **General Information**, **Measurement Chart**, **Object Relation**, **Financial Integration**, **Contract Assignment**, and **Internal Cost Allocation** sections.

Tip

Consult SAP Note 3011768 to learn about additional app features and restrictions.

Perform Contract Valuation (F6277)

Application Type: Transactional

This app lets you execute comprehensive valuation runs for real estate contracts based on preconfigured valuation rules and parameters in flexible real estate management. This process calculates the current value of each contract by analyzing historical data, cost accruals, depreciation, and market conditions. In addition, the app supports valuation to align with accounting standards such as IFRS 16 or US GAAP ACS842. By generating detailed valuation cash flows, the app not only provides insights into a contract's financial performance but also facilitates adjustments when contract changes occur. In addition, the app supports mass execution of valuation runs and subsequent posting to the Universal Journal (table ACDOCA). Thus, this app allows you to perform continuous reassessments of the value of a contract.

Transaction Codes

The related transactions in SAP GUI are Transactions RECEPR and RECEEP.

581

Periodic Posting for Internal Cost Allocation (F6114)

Application Type: Transactional

This app streamlines the recurring posting of internal costs associated with real estate, automating the allocation of expenses such as maintenance, utilities, and other service costs to designated cost centers or settlement units. By facilitating regular, time-dependent postings, the app ensures that internal cost flows are accurately tracked and aligned with overall financial planning and reporting processes. Customizable parameters allow organizations to define posting periods, apply specific cost allocation rules, and integrate seamlessly with broader SAP Fiori functionalities, thus minimizing manual intervention and reducing errors. This capability is especially beneficial for companies managing extensive real estate portfolios, where systematic and timely internal cost allocation is critical for budgeting, performance analysis, and operational efficiency.

Real Estate Accounting Objects Plan and Actuals Report (F5362)

Application Type: Web Dynpro

This app provides a structured comparison of planned and actual financial values for real estate portfolio objects. You can search for real estate objects' plan and actual data using filters such as company code, fiscal year, currency type, and object type. In addition, the app supports portfolio-level financial reporting; offers detailed breakdowns by company code, fiscal year, currency type, and object type; and calculates absolute and percentage variances. Thus, the app lets you perform accurate cost and variance analysis across various locations and asset types. The app also integrates with general ledger and controlling components.

Transaction Code

The corresponding transaction in SAP GUI is Transaction REISPLCT.

Real Estate Cash Flow Report (F5363)

Application Type: Web Dynpro

This app provides an enhanced overview of the cash flow generated from contracts, allowing you to analyze financial trends and portfolio performance. The app focuses on contract counts across various dimensions, offers analytical details into the contract portfolio, and enables filtering of charts by company code and contract type for targeted analysis. The app utilizes CDS view C_RECASHFLOWQ to support data retrieval and reporting.

Real Estate Contract Object Assignment Report (W0201)

Application Type: Web Dynpro

This app offers a structured view of objects assigned to real estate contracts in flexible real estate management. The app lets you track and analyze various assigned objects, including usage objects, cost centers, work breakdown structure (WBS) elements, orders, equipment, functional locations, and object groups. The app supports lease-out contracts, object grouping based on shared conditions, and flexible validity period definitions. In addition, you can customize permitted object types per contract and navigate detailed object screens to modify assignments and analyze contract-related data.

Transaction Code

The related transaction in SAP GUI is Transaction RECN.

Real Estate Contract Validity Report (W0200)

Application Type: Web Dynpro

This app provides a comprehensive overview of contract validity periods in flexible real estate management. You can search for contracts using search parameters such as company code, validity status, and contract type. In addition, the app supports fixed and indefinite contract terms for lease-in and lease-out agreements, and it allows renewal and notice simulations across multiple contracts. The app also presents contract validity in a graphical form to visualize a contract's duration. You also can navigate to other contract processing apps for further analysis. Thus, the app lets you analyze contract terms, renewal options, and notice periods for effective real estate lifecycle management.

Real Estate Objects Plan and Actuals Report (F6760)

Application Type: Web Dynpro

This app provides a comprehensive overview of the planned budget versus actual performance for individual real estate objects within flexible real estate management. The app aggregates key financial metrics such as forecasted revenues and costs, actual expenditures, and the resulting variances. With its drill-down capability, you can examine discrepancies between plan and actual figures, identify trends, and uncover areas requiring managerial intervention. In addition, the report supports multidimensional analysis, allowing filtering by various criteria such as location, object type, and dates. Thus, the app allows you to perform detailed analysis of real estate objects for better financial control.

Transaction Code

The corresponding transaction in SAP GUI is Transaction REISPLCT.

Reconciliate Real Estate Object (F5960)

Application Type: Transactional

This app facilitates the reconciliation of real estate objects by automatically aligning master data with corresponding financial postings. The app validates records to identify discrepancies; uses filtering options based on the company code, object type, and key date to refine the process; and integrates with the Universal Journal (table ACDOCA) for financial postings.

Transaction Codes

A similar check can be performed using Transactions RE80 and RECNCHECK in SAP GUI.

Reverse Contract Postings (F6275)

Application Type: Transactional

This app lets you reverse previously posted contract entries in flexible real estate management. The app helps in situations in which posting errors or changes in contract conditions necessitate adjustments. In such circumstances, the app first validates that the original postings are eligible for reversal by checking that no dependent processes are impacted. If the app determines the posting is eligible for reversal, it then generates reversal postings with inverted amounts to offset the original entries. In addition, you can view detailed logs and audit trails for troubleshooting errors or incorrect postings.

Transaction Code

The corresponding transaction in SAP GUI is Transaction RERAPPRV.

10

Update Objects in Worklist (F5969)

Application Type: Transactional

This app lets you perform mass updates on real estate objects within flexible real estate management. The app lets you filter and select objects based on key criteria and then update critical attributes such as object statuses or organizational assignments. Thus, the app facilitates bulk updates by reducing manual efforts.

Transaction Codes

The corresponding transactions in SAP GUI are Transactions RE80 and MASS.

Valuation Management (F6170)

Application Type: Transactional

This app lets you manage and maintain valuations for a contract portfolio in accordance with IFRS 16 or US GAAP ACS 842 accounting standards. The app features a comprehensive timeline displaying the current contract term, upcoming renewals (both optional and automatic), and potential notice dates, along with current term details. This app includes **Display of the Contract Header, Valuation, Condition,** and **Valuation Condition Assignment** sections. In addition, you can access valuation logs by process step, delete the last unposted valuation step per the governing rule, maintain memos and change log entries for valuation rules, and search and filter contracts. Furthermore, the app allows for the creation and editing of valuation rules, performing valuations to generate cash flows, and managing supplementary details such as conditions and contract term information, with direct navigation to related apps and an enabled draft feature to streamline the workflow.

Tip

Consult SAP Note 3118697 to learn about additional app features and restrictions.

Transaction Codes

A similar check can be performed using Transactions RECEPR, RERAPPRV, and RECEEPRV in SAP GUI.

Chapter 11

Cross-Functional Apps

In this chapter, we'll detail all the most useful and relevant SAP Fiori apps available that don't fit naturally into one specific functional area. There are literally thousands of these apps, and this chapter would take up most of the book if we detailed all of them. As a result, we will feature on the most useful and relevant ones, delving into topics such as system administration and SAP Application Interface Framework, cross-functional areas such as the material master and situation handling, and mass-change and data loading apps. As with the other chapters, this chapter will not cover SAP GUI transactions that have been given the SAP Fiori look and feel, instead focusing exclusively on SAP Fiori apps. Also, in keeping with the other chapters, the apps will be ordered alphabetically for ease of reference.

Analyze Consents Import Logs (F3283)

Application Type: Transactional

As part of the consent administration suite of apps, it is necessary to link any personal data that has been collected to a predefined purpose, like a contract. When importing consent records as part of data privacy management, application import logs are created. This app allows those logs to be displayed according to the filter criteria specified, including subcategory, date to and from, created by, severity, external reference ID, log handle, log number, and time to and from. Detailed information for the import log can be viewed by clicking individual lines in the detailed list.

Tip

The details included when using this app are entirely dependent upon the settings made in the Import Consents from File app (F3306).

Analyze Query Log – Enterprise Search (F2571)

Application Type: Transactional

Business process specialists can use this app to analyze logs created as a result of enterprise searches carried out in the SAP Fiori launchpad. The results can be shown in graphical or tabular format. Metrics such as the number of executed searches per day are available. You can also drill down into the data to achieve a greater degree of granularity for the selected data details.

Analyze Your Selection Process (F7829)

Application Type: Transactional

As part of the data migration suite of SAP Fiori apps, this app can be used to execute a simulation process for a specific migration object instance. As part of the process, the results of the simulation can be viewed, including any error messages collected. It is also possible to copy existing analysis runs so as to create simulation processes for other data objects. This is normally one of the first steps carried out in the data migration process in SAP Fiori, using the Migrate Your Data app (F3473). The selection steps in the migration process (e.g., checking the existence of an instance in source system, identifying data using selection logic, extracting data from source system) are displayed in the app in a process flow diagram.

App Support (F4914)

Application Type: Transactional

This app allows system administrators to troubleshoot any issues related to configuration and authorization errors for SAP Fiori apps. All SAP Fiori app types are supported by the app, and all users can use it to see their own logs. System users can view logs from all users. The app is launched from the user actions menu when you're already in an SAP Fiori app by selecting the **App Support** option. You will find app information (e.g., app title, type, ID, and version) and info about the embedded server (depending on your SAP Fiori deployment, this could be embedded or hub, and you will see the configuration checks and statuses). You can also view different types of errors: authorization, gateway, and runtime.

Tip

You need the S_FLP_AS role assigned to be able to view and download logs.

Transaction Code

In terms of authorization checks, the equivalent transaction in SAP GUI is Transaction SU53.

Application Logs (F1487)

Application Type: Transactional

Administrators can view application logs to show errors in the navigation and performance of the SAP Fiori application by using this app. The logs can be filtered by severity, date, category, subcategory, and external reference. The severity is displayed via a color-coded exclamation mark or checkmark next to each row of the list of application logs. By default, the details shown include the severity, the number of items in the log, the category description, subcategory description, creation date, and created by, but adding other display columns is possible.

Assign Recipients to Users (F2949)

Application Type: Transactional

Communication scenarios in SAP often require recipient IDs to be set up. This app can be used by administrators to assign user IDs to the recipient IDs for the purpose of monitoring interfaces and data messages. For example, a recipient ID name such as "EDI_SD_ORDER_RECIPIENT" may be created. This app can be used to assign user IDs to that recipient ID. Available features in the app include the option to add and remove user IDs from recipient IDs, define which types of messages a user can see in the message dashboard (e.g., warnings or errors), and define the type of user (technical or standard; a technical user can see all messages in process or those that had a technical error).

Business Event Queue (F2392)

Application Type: Transactional

This app lets administration users view the number of events raised for a given business object. Available filters for the selection of the data include the business object type, business object task, subscriber, date, and time. The data is retrieved using CDS view `C_BEHQUEUEANALYT-ICS`.

Tip

A technical job automatically runs in the background to clear the queue every day and select only events that were created in the last three days.

Business Event Subscription (F2569)

Application Type: Transactional

In the SAP system, business object events exist, such as `BusinessPartner.Created`, used to show whenever a business partner is created. This app can be used to manage events using SAP Event Mesh. Administrators can use this app to activate and deactivate a subscription to specific business object events. In addition, the priority of the subscriptions can be adjusted. For example, administrators could open the app and subscribe to the `BusinessPartner.Created` event, then specify an endpoint, such as a customer relationship management system, so that SAP Event Mesh can pick up the change and send it to the target platform. Useful events include, for example, `SalesOrder.Changed`, `BusinessPartner.Deleted`, `PurchaseOrder.Created`, `GoodsMovement.Posted`, and `Product.MasterUpdated`. The data is retrieved using CDS view `C_BusEvtSubscrpnMgr`.

Tip

Using SAP Event Mesh requires an SAP Business Technology Platform (SAP BTP) account on Cloud Foundry.

Change Activities Dashboard (F5761)

Application Type: Analytical

Change management is a central component within SAP S/4HANA via which you can manage

and coordinate activities related to organizational change management. This app shows chart visualizations of change activities, with options for filtering by change activity type, change activity status, overdue status (yes or no), responsible assigned (yes or no), and due date. The data is retrieved using CDS views `C_ChgMgmtActyOverviewQuery` and `I_ChgMgmtActyOverviewCube`.

Change Requests and Activities – Detailed Analysis (F5469)

Application Type: Analytical

Change management is a central component within SAP S/4HANA via which you can manage and coordinate activities related to organizational change management. This app shows a detailed analysis via graphical representations of data related to change requests. For example, a bar chart can be used to display the number of change activities by person responsible, and a line chart can be used to display the number of change activities by due date. Filter criteria for selection are available, including status, plant, leading object, person involved, due date, and change type.

Navigation to the Manage Change Requests (F3406) and Manage Change Activities (F3405) apps is supported.

Change Requests Dashboard (F5760)

Application Type: Analytical

Change management is a central component within SAP S/4HANA via which you can manage and coordinate activities related to organizational change management. This app shows an analytical view of change requests, represented as graphical visualizations. Selection filters are available, including change request type, status group, leading object type, entered date, and impact. The resulting data is shown in tabular and graphical format.

Check Electronic Document Consistency (F6680)

Application Type: Transactional

Automatic creation of electronic documents in SAP, especially billing documents, is central to many integrated processes. With this app, the consistency of electronic documents in an organization can be checked against those stored in the relevant tax authority's system. It is possible to navigate from the list generated directly to the Resolve Electronic Document Inconsistencies app (F7043) and to the Manage Electronic Documents app (F4306).

Classification-Based Product Hierarchy (F2351)

Application Type: Transactional

This app allows master data specialists to create, view, and edit product hierarchies based upon classification data. This is closely related in functionality to the Manage Product Master Data app (F1602), in which you can change parent nodes by changing classification data. The difference here is that it is possible to create product hierarchies based upon the classification data via this app, whereas the Manage Product Master Data app allows you to change the classification data to drive allocation to a specific product hierarchy. The process here is to use this app to create a link between the classification data and the product hierarchy, then navigate to the Manage Product Master Data app to amend the classification data, thus allocating the product to the new hierarchy. Available features include the ability to display products in a specific product hierarchy, create a subnode in a hierarchy, copy an existing product in a hierarchy as a subnode, and navigate to the Manage Product Master Data app to change the hierarchy assignment.

Configure Personalized Search (F2800)

Application Type: Transactional

Search results carried out in the enterprise search in SAP Fiori are stored in the SAP HANA database. This app can be used to enable and disable the tracking of users' search results. Enabling this personalized search allows end users to carry out personalized searches based upon keywords and previous search history. The app offers the following extremely simple settings: **Disabled for all users, Enabled for all users, Enabled after user's approval (opt-in)**, and **Enabled until user's rejection (opt-out)**.

Create Electronic Documents Manually – Special Cases (F7437)

Application Type: Transactional

Automatic creation of electronic documents in SAP, especially billing documents, is central to many integrated processes. With this app, it is possible to review and create electronic documents manually in the event of automatic creation failures. Using the app, you can see a list of all source documents for which electronic processing and creation are applicable and not applicable, and those for which no electronic document currently exists. From the list, you can navigate to the source document or create electronic documents.

Custom Analytical Queries (F1572)

Application Type: Transactional

This app is very useful in that it can be used by analytics specialists to design queries based upon standard or custom core data services (CDS) views. The functionality exists to search for queries, create new queries, change existing queries, and create hierarchies. It is also possible to simulate the results of the query. In addition, it is possible to create a custom SAP Fiori tile based upon the query that has been created or changed. This app is a no-code solution for transforming raw SAP data into usable analytical queries for an organization.

Custom CDS Views (Version 2) (F1866A)

Application Type: Transactional

Administrators can build custom core data services (CDS) views using this app. Custom CDS views have several applications in organizational IT; for example, they can be consumed in analytical queries within the SAP Fiori launchpad, or they can be consumed via APIs in external analytical platforms such as those found in SAP Business Data Cloud, SAP Datasphere, and SAP Analytics Cloud. When amending custom CDS views, you can add fields from multiple sources, create custom fields for calculation purposes, and add custom filters.

Custom Fields (F1481)

Application Type: Transactional

SAP Fiori apps can be customized by implementation consultants using this app to add custom fields into reports, email templates, and form templates. Labels, descriptions, and possible values for the custom fields can be added. In addition, translations can be added, and fields can be changed and deleted and saved as drafts. Once the custom fields are ready, they can be published to the test system.

Custom Logic (F6957)

Application Type: Transactional

This app can be used to implement standard BAPIs that have been released by SAP. The app guides development users through the steps required to implement the BAPIs. These steps are as follows:

1. Specify the extension point from a list of standard SAP extension points.
2. Define the implementation attributes.
3. Create the implementation ID and description.
4. View and amend the code, then publish it.

Custom Logic Tracing (F3438)

Application Type: Transactional

This app can be used by administrators to trace the use of development objects in given processes. For example, the app can be used to determine the input and output values in certain variables, parameter changes, and the duration of the execution of the overall process. In addition, it is possible to view which exceptions are raised at which points in the process. Traces can be created and deleted using this app, and development objects can be traced to understand the sequence of determinations, validations, and actions of the development objects used.

11

Data Migration Status (F3280)

Application Type: Analytical

This app is extremely useful for data migration specialists to review the migration object processes (both live and in simulation) that have been carried out using the Migrate Your Data app (F3473). Using the app, it is possible to see a full overview in real time of the progress of all migrations. Selection filters are available with options such as migration period, migration project, object name, and record status. The objects are listed out as a result of the selection filters, with the number of imported records, number of records ready for import, number of failed records, total number of records, and the progress, presented in a color-coded chart to show success and failure rates. Multilevel drilldowns are available into each migration process first to show the list of individual records, then further to show the attributes of each record. The app uses situation handling for notifying key users of the progress and failures of the data migration processes.

Data Migration Status (Load Statistics) (F4110)

Application Type: Analytical

This app is extremely useful for data migration specialists to review the migration object processes (both live and in simulation) that have been carried out using the Migrate Your Data app (F3473). Using the app, it is possible to see a full overview of the load statistics in real time for the progress of all migrations. Selection filters are available using options such as migration period, migration project, object name, and record status. The objects are listed out as a result of the selection filters, showing the number of imported records, number of records ready for import, number of failed records, total number of records, and the progress, presented

in a color-coded chart to show success and failure rates. Multilevel drilldowns are available into each migration process first to show the list of individual records, then further to show the attributes of each record. The app uses situation handling for notifying key users of the progress and failures of the data migration processes.

Display Business Event Logs (F6021)

Application Type: Analytical

A business event can be triggered when data is created, changed, or deleted in SAP. The activation of business events is carried out in Customizing activities for the relevant SAP object. A normal business event would notify an interested party that an object has been changed. Using this app, it is possible to view all logged business events and their details and to navigate to the business event sequence. The data is retrieved using CDS view C_BusEventLogEventAnlys.

Display Business Events by Business Object (F6035)

Application Type: Analytical

A business event can be triggered when data is created, changed, or deleted in SAP. The activation of business events is carried out in Customizing activities for the relevant SAP object. A normal business event would notify an interested party that an object has been changed. Using this app, it is possible to view all logged business events by SAP object node and their details and to navigate to the business event sequence. It is also possible to view the processing variants and number of events for each business node, alongside the processing cycle time. The data is retrieved using CDS view C_BusEventLogObjectNodeAnlys.

Flexible Workflow Administration (F5343)

Application Type: Transactional

Business process specialists can use this app to monitor and review flexible workflows. Features of the app include the ability to display a list of all flexible workflows, cancel or resume multiple flexible workflows at one time, view the flexible workflow log, and display the steps of the flexible workflows. In addition, the details of each step can be viewed along with the log related to that step. It is also possible to refresh the allocated processors for a given step and forward a step that is ready to be processed to a new processor.

Import Consents from File (F3306)

Application Type: Transactional

As part of the consent administration suite of apps, it is necessary to link any personal data that has been collected to a predefined purpose, like a contract. Using this app, you can import consent records from a file. To import the file, you must specify the source file, select the import log detail level, and select the **Test Run** checkbox.

Tip

The file used for loading must be in JavaScript Object Notation (JSON) format. The fields relevant for a consent record can be found by viewing SAP Note 2607792.

Inbound Messages (F7317)

Application Type: Analytical

Automatic creation of electronic documents in SAP, especially billing documents, is central to many integrated processes. With this app, you can display a list of inbound messages for electronic documents. You also can process messages that have the **Ready for Processing** or **Processed with Error** status.

Intelligent Scenario Management (F4470)

Application Type: Transactional

This app is part of the Intelligent Scenario Lifecycle Management (ISLM) functionality. ISLM is a set of tools that allows you to perform operations using machine learning scenarios. The machine learning scenarios comprise of the use of trained models, deployment, and activation; this converts an intelligent scenario into an ABAP representation. There are embedded and side-by-side scenarios:

- In embedded scenarios, the SAP S/4HANA application is in the same technology stack as the machine learning provider. This typically utilizes SAP HANA Automated Predictive Library (APL) or SAP HANA Predictive Analysis Library (PAL).
- In side-by-side scenarios, the SAP S/4HANA application runs in a separate technology stack as the machine learning provider. Examples can include services found on the SAP Business Technology Platform (SAP BTP), such as Document Information Extraction or Data Attribute Recommendation.

Using this app, you can display a list of all the defined intelligent scenarios, train a specific scenario, and deploy a fully trained machine learning model. It is also possible to deactivate deployed versions that are no longer required.

Intelligent Scenarios (F4469)

Application Type: Transactional

This app is part of the Intelligent Scenario Lifecycle Management (ISLM) SAP functionality, similar to the Intelligent Scenario Management app

(F4470; see its entry for more information). This app is used to display and create custom developed intelligent scenarios, whereas the Intelligent Scenario Management app is used to manage the machine learning activities required for the efficient usage of the scenarios. The differences are as follows:

- **Intelligent Scenarios (F4469)**
  Allows you to create and display custom developed intelligent scenarios.

- **Intelligent Scenario Management (F4470)**
  Allows you to train the model of the intelligent scenario, deploy to a quality assurance system, and activate for a specific business domain.

Maintain Collaboration Users (F4911)

Application Type: Transactional

As part of the solution for business user management, users are represented as a business partner and a user assignment. Business users are employees, external resources, or collaboration users. A collaboration user is a person who is employed by an external company and has no active contract with the organization. However, this type of user collaborates with the organization via scenarios such as vendor-managed inventory. This app can be used to create, edit, archive, and export collaboration users. It is also possible to import a CSV file of collaboration users into the app.

Maintain Employees V2 (F2288A)

Application Type: Transactional

As part of the solution for business user management, users are represented as a business partner and a user assignment. Business users can be employees, external resources, or collaboration users. Employees are defined as persons employed by the organization. This app can be used to create, edit, archive, and export employees. It is also possible to import a CSV file of employees into the app.

This app replaces the deprecated Maintain Employees app (F2288) and has been modified to improve the user experience.

Transaction

The associated transaction in SAP GUI is Transaction BP (Business Partner).

Maintain External Resources (F3505A)

Application Type: Transactional

As part of the solution for business user management, users are represented as a business partner and a user assignment. Business users can be employees, external resources, or collaboration users. External resources are defined as external resources working for another company but in the same location as the organization, with a contract. This app can be used to create, edit, archive, and export external resources. It is also possible to import a CSV file of external resources into the app.

Manage Business Partner Master Data (F3163)

Application Type: Transactional

This app is used to create, change, display, copy, and search for business partner data. When creating a business partner, the **Person** or **Organization** option must be selected. This then opens a dialog box where you can enter more details quickly and efficiently. When using the **Copy** feature, all data values are copied from the source record to the new record.

Transaction Code

The associated transaction in SAP GUI is Transaction BP (Business Partner).

Manage Change Activities (F3405)

Application Type: Transactional

Change management is a central component within SAP S/4HANA in which you can manage and coordinate activities related to organizational change management. This app lets you see an overview of all change activities, such as tasks, approvals, notifications, or structure nodes. Selection filters can be defined for elements such as type, status, due date, main person responsible, role of person involved, and person involved. The results are shown in tabular format, with the ability to navigate to each activity to make changes where necessary.

Manage Change Requests (Management of Change) (F3406)

Application Type: Transactional

Change management is a central component within SAP S/4HANA in which you can manage and coordinate requests related to organizational change management. This app shows an overview of all change requests. Selection filters can be defined for elements such as type, status, role of person involved, and person involved. The results are shown in tabular format, with the ability to navigate to each change request to make changes where necessary.

Manage Consents (F3307)

Application Type: Transactional

As part of the consent administration suite of apps, it is necessary to link any personal data that has been collected to a predefined purpose, like a contract. This consent data can be imported in a file via the Import Consent from File app (F3306). The Manage Consents app (F3307) can be used to view and search for consent records that were loaded from such a file. Selection fields are available, including data subject ID type, consent ID, data subject ID, valid-to date, consent status, application template ID, form name, purpose name, data controller name, application name, jurisdiction, lifecycle status, third-party name and function, start of expiration, valid-from date, and consent method.

Manage Date Functions (Version 2) (F2595A)

Application Type: Transactional

This app can be used by analytics specialists to define date functions for use in reporting. To create a new date function, click the **Add** button in the **New Date Function** screen of the app. Each date function must be created relative to the base date, which can be defined as the current date or via a calculation based upon the current date. Examples of relative dates could be dates such as one month ago or two quarters from now. The combination of the base date and the relative date supply the date range, which can be used in SAP Fiori apps. It is also possible to edit and delete date functions. Date functions are transportable.

Manage Documents (F2733)

Application Type: Transactional

Document info records can be created by engineers in the SAP system. A document info record contains metadata about the document as well as the original content of the document itself. This app can be used to search, edit, copy, and create documents. Selection filters are available for editing status, document number, document type, document version, document part, document description, document status, and user. From the ensuing list, it is possible to download the original document.

> **Tip**
>
> There is a configuration setting available to restrict the uploading of files to one main file, based upon the document type. The IMG path for the setting is **Cross-Application Components • Document Management • Control Data • Define DMS Document Types**. The setting to change for the document type is to check the **One Main File per Document** checkbox.

Manage Electronic Documents (F4306)

Application Type: Transactional

Automatic creation of electronic documents in SAP, especially billing documents, is central to many integrated processes. With this app, you can create and send electronic documents to tax authorities (set up as business partners), navigate to the source documents, view statuses and process flows for electronic documents, and analyze errors in creation.

Navigation to related apps such as Manage Document and Reporting Compliance (F5218) and Check Electronic Document Consistency (F6680) is supported.

Manage Global Hierarchies (F2918)

Application Type: Transactional

This app offers users a centralized hub for the management of hierarchies, mainly in the finance area. However, additional hierarchies

such as customer hierarchies are also supported. Features such as creation, editing, copying, and deletion are all available. It is also possible to upload a hierarchy from a spreadsheet. Hierarchy assignments are also supported, and you can also use drag and drop to change assignments.

> **Tip**
>
> It is possible to import nodes from one hierarchy to another, so long as they are of the same hierarchy type. Note that if you do this, it is useful to select **Auto Update from Source**; that way, the nodes will automatically update when the referenced node is updated.

Manage KPIs and Reports (F2814)

Application Type: Transactional

This extremely useful app allows users to create their own KPIs and then produce a report that calls the data according to the KPIs. The report can then be embedded within an SAP Fiori tile, without any coding. The app is natively integrated with SAP Analytics Cloud. Features in the app include the following:

- **Groups**: This section is used to group together similar KPIs.
- **Key Performance Indicators (KPIs)**: This section is used to define business metrics.
- **Reports**: This section is used to configure reports based upon activated KPIs. It is also used to define how the report is visualized.
- **Stories**: This section is used to define stories and data when integrating with SAP Analytics Cloud.

Manage Launchpad Pages (F4512)

Application Type: Transactional

The SAP Fiori launchpad is built from spaces and pages, and this app is used by administrators to define the layout of the pages. This refers to defining which apps are available in which sections on the page, editing an existing page, copying an existing page, or deleting a page (non-SAP-standard pages only). The layout of the app is similar to the UI displayed when you configure your own personal SAP Fiori launchpad in that sections and tiles can be added to pages and converted to links and flat tiles. Tiles can also be dragged and dropped around the page easily.

Manage Launchpad Spaces (F4834)

Application Type: Transactional

The SAP Fiori launchpad is built from spaces and pages, and this app is used by administrators to define the layout of the spaces. Spaces are areas that bring together multiple pages for visualization in a unified user experience. With this app, administrators can create, copy, and delete spaces (non-SAP spaces only). The results are displayed in a table showing IDs, titles, how many roles the spaces are assigned to, and created-by and changed-by names. You can search in the list using free text or filter using any of the table fields.

Manage Output Items (F2279)

Application Type: Transactional

This app allows you to get a full overview of all outputs issued from SAP regardless of the channel used for the output. Selection filters are available such as application object ID, application object type, and status. From the resulting list, you can select multiple entries and use quick action options to duplicate the output items, retry existing items, set items to completed, and resend items. The status of each item is color-coded (red for error, orange for in preparation, and green for completed).

Manage Product Master (F1602)

Application Type: Transactional

This app is central to any organization's management of its master data related to products or articles. With this app, you can display, edit, copy, delete, and create product master records. The app is tailored toward a central master data function within the organization; all the usual features—such as attachment services, save as draft, classification attribute maintenance, product variants, and mass processing—are available.

Navigation to related apps in the product master function, such as Product (F2773), is supported.

Transaction Code

The associated transaction in SAP GUI is Transaction MM03 (Display Material).

Manage Responsibility Contexts (F4636)

Application Type: Transactional

Responsibility management in SAP refers to the process of aligning development object types or business frameworks to teams of users through a process of agent determination. For example, a business workflow or situation handling can be aligned via *responsibility definitions* (such as for a plant, sales organization, or purchasing group) to specific teams and then to specific agents. This assignment is carried out through the use of responsibility rules. SAP provides a list of standard team categories and responsibility rules for use in responsibility management. In this function, a *responsibility context* is a development object or a business framework; again, SAP delivers standard versions of responsibility contexts. With this app, you can define responsibility contexts and assign rules for agent determination.

11

Manage Responsibility Rules (F4637)

Application Type: Transactional

Responsibility management in SAP refers to the process of aligning development object types or business frameworks to teams of users, through a process of agent determination. For example, a business workflow or situation handling can be aligned via *responsibility definitions* (such as for a plant, sales organization, or purchasing group) to specific teams and then to specific agents. This assignment is carried out through the use of responsibility rules. SAP provides a list of standard team categories and responsibility rules for use in responsibility management. In this function, a *responsibility context* is a development object or a business framework; again, SAP delivers standard versions of responsibility contexts. With this app, you can create or copy responsibility rules for agent determination. You can also associate an implementation of the RSM_BADI_STATIC_RULE business add-in (BAdI) with a customized responsibility rule.

Manage Scheduling Worklists – Settlement Management (F5856)

Application Type: Transactional

Settlement management of condition contracts can be carried out via worklists by using this app. The scheduling worklists contain selection options for filters, such as editing status, scheduling worklist ID, scheduling worklist template, scheduling status, last run status, job execution date and time, and scheduling worklist run category. Once the list of scheduling worklists for settlement management documents is displayed, you can carry out the following activities:

- **Correct**: Reschedule scheduling worklists in the case of errors
- **Close**: Close the scheduling worklists
- **Revoke**: Revoke the scheduling worklist completely

- **Schedule**: Create a new schedule for existing scheduling worklists
- **Create**: Create new scheduling worklists for settlement management documents
- **Delete**: Delete scheduling worklists for settlement management documents

Manage Search Models (F3036)

Application Type: Transactional

The enterprise search functionality in SAP Fiori is one of the key innovations that helps differentiate the SAP Fiori UX from SAP GUI. Administrators can use this app to view and create copies of the standard search models delivered by SAP. The app is a low-code solution and allows administrators to view two types of enterprise search models: delivered by SAP, and custom. There is some limited functionality available for amending some search model fields in the **Delivered by SAP** area, but most changes are carried out in the **Custom** tab. In this tab, customized search models can be edited, deleted, and created from CDS views.

Manage Situation Objects (F5609)

Application Type: Transactional

Situation handling is a way to inform key users about business situations that arise and require some manual input. The notifications for key users appear in the notifications area of the SAP Fiori launchpad. Examples of situations include upcoming deadlines, such as expiration of quotations and contracts. Initially, situation handling delivered around 90 standard situations, and it was not possible to create customized situations. However, this has changed in recent years; this app now allows extensibility specialists to create customized business objects that enable situation handling. The business object itself connects CDS views, events, and actions in a single situation.

Manage Situation Scenarios (F5698)

Application Type: Transactional

Situation handling is a way to inform key users about business situations that arise and require some manual input. The notifications for key users appear in the notifications area of the SAP Fiori launchpad. Examples of situations include upcoming deadlines, such as expiration of quotations and contracts. Initially, situation handling delivered around 90 standard situations, and it was not possible to create customized situations. However, this has changed in recent years; this app now allows extensibility specialists to model customized situation scenarios. Each business object, as defined in the Manage Situation Objects app (F5609), is defined as an *anchor object*. To build up the situation scenario, the app is used to first select the appropriate anchor object and then map it to the trigger object and events.

The follow-up process is to define the layout of the situation in the My Situations – Extended app (F4537).

Tip

Situation scenarios can only be created, modified, or deleted if the Adaptation Transport Organizer (ATO) has been set up. This is an authorization procedure that is configured in SAP using Transaction S_ATO_SETUP (Configure ATO Setup).

In addition, situation scenarios cannot be created in production systems.

Manage Situation Types (F2947)

Application Type: Transactional

Situation handling is a way to inform key users about business situations that arise and require some manual input. The notifications for key users appear in the notifications area of the SAP Fiori launchpad. Examples of situations include

upcoming deadlines, such as expiration of quotations and contracts. This app is more aligned to business process configuration experts than it is to extensibility specialists, as the creation of the situation type is dependent upon existing situation templates. A *situation type* can be defined as a representation of a business situation that requires attention. From the templates provided, it is possible to do the following:

- Create situations from templates.
- Maintain the relevant conditions that trigger the situation.
- Maintain the output text regarding the situation, which appears in the notification for the relevant user.
- Assign recipient groups to situations.
- Enable situation monitoring; see the related Monitor Situations app (F3264).

Manage Situation Types – Extended (F5437)

Application Type: Transactional

Situation handling is a way to inform key users about business situations that arise and require some manual input. The notifications for key users appear in the notifications area of the SAP Fiori launchpad. Examples of situations include upcoming deadlines, such as expiration of quotations and contracts. This app differs from the Manage Situation Types app (F2947) in that you can create new situation templates and types rather than just use ones based on existing situation templates. These templates and types can be used in the management of extended situation handling.

Manage Subscription Product-Specific Data (F3560)

Application Type: Transactional

As part of the subscription order management suite in SAP S/4HANA, this app allows users to create and manage subscription product master

597

data. Unlike standard product master data, subscription product master data includes pricing configurations, discounts, surcharges, billing cycles, contract terms, and autorenewal indicators.

Navigation to numerous related apps, such as Manage Product Master (F1602) and Manage Prices – Sales (F4111), is supported.

Tip

A product is defined as a subscription product only if the **Subscription** (ID: **3**) product type is assigned.

Manage Team Hierarchies (F3077)

Application Type: Analytical

Responsibility management in SAP refers to the process of aligning development object types or business frameworks to teams of users through a process of agent determination. For example, a business workflow or situation handling can be aligned via *responsibility definitions* (such as for a plant, sales organization, or purchasing group) to specific teams and then to specific agents. Team members must be set up as business partners in SAP. This app can be used to create, edit, delete, and copy teams and to manage the hierarchies to which they are assigned. When creating a team, it is possible to give the team a global ID, which can be used across the system landscape—in quality assurance and production, for example. This ID is separate from the internal ID, which is system specific. Other functions of the app include creating team owners, defining responsibility definitions (based on, for example, organizational structure objects such as the plant), assigning functions to team members, creating hierarchical structures for team members, and viewing changes to teams.

Several selection options are available for filters when searching for teams, such as the editing status, name, description, status, type, category, responsibility definition, and team members.

Navigation to the related Manage Teams and Responsibilities (F2412) app is supported.

Tip

Enterprise search can be used from the SAP Fiori launchpad home screen to search for team members within teams. Simply select the **Teams** search criterion from the enterprise search list.

Manage Teams and Responsibilities (F2412)

Application Type: Transactional

Responsibility management in SAP refers to the process of aligning development object types or business frameworks to teams of users through a process of agent determination. For example, a business workflow or situation handling can be aligned via *responsibility definitions* (such as for a plant, sales organization, or purchasing group) to specific teams and then to specific agents. This app allows business process specialists to manage team members' functions to determine which users are responsible for workflows, business scenarios, and situation handling. Team members must be set up as SAP business partners. This app can be used to create, edit, delete, and copy teams. When creating a team, it is possible to give the team a global ID, which can be used across the system landscape—in quality assurance and production, for example. This ID is separate from the internal ID, which is system specific. Other functions of the app include creating team owners, defining responsibility definitions (based on, for example, organizational structure objects such as the plant), assigning functions to team members, and viewing changes to teams.

Several selection options are available for filters when searching for teams, such as editing status, name, description, status, type, category, responsibility definition, and team members.

Manage Workflow Scenarios (F3067)

Application Type: Transactional

Workflows consist of a series of *stages*, which can be defined as entities from a specific business process that are grouped together into a linear structured process. Business process specialists can use this app to define stages for use in the creation and maintenance of flexible workflows. Functions available in the app include viewing all available stages for a specific workflow scenario and displaying stages and details. You can also import and export workflow scenarios.

Manage Workflow Template for Change Records (F3402)

Application Type: Transactional

Administration users can use this app to define workflow templates for change records. With the app, process templates can be created as new templates or as copies of existing templates. In addition, you can edit a process template to add or remove steps and assign users or roles to steps.

Manage Workflow Templates (F2787)

Application Type: Transactional

Business process specialists can use this app to define templates. These templates can be reused in the workflow modeling environment or through APIs. In this app, you can view all available templates with filters available and define steps in the template that are relevant to the scenario.

Manage Workflows (F2190)

Application Type: Transactional

Business process specialists can use this app to define and manage approval workflows for business processes in the organization. The app can be used to view existing approval workflows, create new approval workflows, edit workflows, define the sequence of start conditions, and activate and deactivate workflows.

Mass Maintenance for Master Data (F2505)

Application Type: Transactional

This app is for mass maintenance of products and business partners only, including relationships in business partners. You can search for specific entries for business partners or products, then enter the new values in the **New** field on the screen. The app then displays information about the change, including number of errors, warnings, and changed records and affected tables. This information is also displayed as a bar chart.

11

Message Monitoring (F4516)

Application Type: Transactional, Analytical

Business process specialists can monitor specific interfaces by using this app. If an alert exists, you can click the **Confirm Alert** button to clear the alert. Note that alerts can be configured using the Alert Settings app (F7056). The following functions are available in the app:

- **Restart**: Allows you to restart a message for which the message status is **Error** or **In Process**
- **Cancel**: Allows you to cancel a message for which the message status is **Error** or **In Process**
- **Set Process Information**: Allows you to add an **In Process**, **Blocked**, or **Completed** status manually, with a corresponding comment
- **Settings**: Defines which attributes can be shown for each message

Message Monitoring for Integration Experts (F5169)

Application Type: Transactional

Specific interfaces can be monitored by integration experts using this app. The screen is organized into three optional UIs. The first is a tabular message overview, where each interface is listed as a row, with messages organized by status in a table. The second option is for message details, where a specific interface has already been selected and the collected messages are shown in a table with their details. The third option is used for mass error handling, where messages for previously selected interfaces are displayed with options to perform error handling for multiple messages at the same time.

Message Monitoring Overview (F4515)

Application Type: Transactional, Analytical

Configuration experts can monitor specific interfaces using this app. Each interface is represented by a tile in the overview screen, and clicking the tile will show the statuses and the number of messages. The dynamic data on the front of the tile will display the name and version of the interface, as well as if there is an active alert for that interface. It is possible to navigate to the Message Monitoring app (F4516) to clear the active alerts. The selection of the data can be refined using the filter options for time and date settings.

Migrate Your Data (F3473)

Application Type: Transactional

This app follows a long line of data migration tools and is the recommended function for migrating data to SAP S/4HANA. This app is especially useful in migrating data from SAP ERP to SAP S/4HANA systems. Migration templates are available for the creation of projects to migrate data. Simulation options are also available for the projects. The app uses situation handling functionality to inform you of the data migration progress.

Transaction Codes

This app is the natural replacement for the traditional data migration transactions in SAP GUI: Transaction LSMW (Legacy System Migration Workbench) and Transaction LTMC (Migration Cockpit).

Monitor Situations (F3264)

Application Type: Analytical

You can monitor standard object-based situations using this useful analytical list page. Selection options for filtering include activities by situation type ID, activities by year, and instances by status. The number of activities is tracked over a given time frame using a line graph, and the instance activities are shown in a tabular format underneath. The type of graph can, as with other analytical apps, be changed to suit your preferences. Navigation to the related Manage Situation Types app (F2947) is supported.

Monitor Situations – Extended (F6780)

Application Type: Analytical

You can monitor standard object-based and message-based situations using this useful analytical list page. Selection options for filtering include activities by situation type ID, activities by year, and instances by status. The number of activities is tracked over a given time frame using a line graph, and the instance activities are shown in a tabular format underneath. The type of graph can, as with other analytical apps, be changed to suit your preferences. Navigation to the related Manage Situation Types (F2947) app is supported.

My Inbox (F0862)

Application Type: Transactional

This app gives you access, on any device, to your SAP S/4HANA inbox. Standard and custom workflows generate messages that are sent to the app, with the option to select preconfigured tasks from within such a message. Completed tasks and workflows are pushed to the My Outbox app (F6511), where it is possible to view all completed and suspended tasks. This app needs to be configured by a system administrator.

To implement and start using this app, SAP recommends the following set of actions to begin with:

- Have at least one workflow in the backend system that defines the task types to be used.
- Ensure that all the authorizations required for the workflows are maintained.
- Configure connections between the SAP Gateway system and backend system or systems.
- Activate the Task Gateway Service.

Tip

For the full details (prerequisites, system landscape requirements, SAP Notes, implementation, and configuration tasks) of the My Inbox implementation steps, visit *http://s-prs.co/v612001*.

Transaction Code

The corresponding transaction in SAP GUI is Transaction SO01 (My Inbox).

My Outbox (F6511)

Application Type: Transactional

Standard and custom workflows generate messages that are sent to the My Inbox app (F0862), with the option to select preconfigured tasks from within each message. Completed tasks and workflows are pushed to the My Outbox app. In addition, suspended tasks can be resumed from within this app. This app needs to be configured by a system administrator.

My Situations (F4154)

Application Type: Transactional

Situation handling is a way to inform key users about business situations that arise and require some manual input. The notifications for key users appear in the notifications area of the SAP Fiori launchpad. Examples of such situations include upcoming deadlines, such as expiration of quotations and contracts. All of your object-based situations can be viewed in an overview table using this app. From the table, you can navigate to the SAP Fiori app relevant to the situation in order to resolve the issue. Statuses can be updated manually in the app by setting them to one of the following:

- **Resolved**: Set when the situation is completed
- **Obsolete**: Set for outdated situations
- **Invalid**: Set for situations created in error

My Situations – Extended (F4537)

Application Type: Transactional

Situation handling is a way to inform key users about business situations that arise and require some manual input. The notifications for key users appear in the notifications area of the SAP Fiori launchpad. Examples of such situations include upcoming deadlines, such as expiration of quotations and contracts. All of your object-based and message-based situations can be viewed in an overview table using this app. From the table, you can navigate to the SAP Fiori app relevant to the situation in order to resolve the issue. Statuses can be updated manually in the app by setting them to one of the following:

- **Resolved**: Set when the situation is completed
- **Obsolete**: Set for outdated situations
- **Invalid**: Set for situations created in error

Product (F2773)

Application Type: Fact Sheet

This fact sheet app allows you to navigate to a product overview page that shows contextual information in an easily digestible format. The information is organized into sections of data under tabs, including such data as purchasing info records, purchase orders, and sales orders. In addition, images can be viewed for each product. For retail-specific industries, information such as a list of sites is available. Navigation to the related apps is supported.

As with all factsheet apps, this app is accessed via other SAP Fiori apps, such as Manage Product Master (F1602), Find Maintenance Order (F2175), Manage Production Operations (F2335), Manage Production Orders (F2336), Procurement for Maintenance Planner (Purchase Order) (F2827), Analyze Delivery Performance – Shipped as Planned (F2878A), Procurement for Maintenance Planner (Purchase Requisition) (F3065), and Display Classified Products (International Trade) (F3789).

Product List (F6518)

Application Type: Analytical

This app allows you to display full lists of products based on an extensive array of selection options for filtering, including attributes such as plant, company code, product, product description, created by, and more. The resulting table of products can be displayed using the same columns as the filter options, and you can navigate directly to the Manage Product Master Data app (F1602). You can also view the inventory price summary in the selected currencies by clicking the **Inventory Price Total** tab on the screen.

Query Browser (F1068)

Application Type: Transactional

All queries that have been created using SAP Fiori functionality based on CDS views can be found, displayed, and tagged using this app. The only queries that will be displayed are those to which you already have access. From the list of queries displayed, you can drill down into the query itself or tag a query as a favorite. You can also personalize custom tags to add for each query.

Situation Handling Demo (F5697)

Application Type: Transactional

Situation handling is a way to inform key users about business situations that arise and require some manual input. The notifications for key users appear in the notifications area of the SAP Fiori launchpad. Examples of such situations include upcoming deadlines, such as expiration of quotations and contracts. This app allows you to become familiar with the concepts of situation handling through a ready-made scenario in the app. The scenario focuses on a booking portal for flight data. In the app, you can define and enable the situation type, then trigger situations to view the full flow. SAP includes the following predelivered situation templates:

- **Ecological Footprint (SITDEMO_ECOLOGICAL_ FOOTPRINT)**
  Used in situation scenario ID SITDEMO_ FLIGHT_PROFITABILITY. This situation checks the object's *eco index* (a calculation based on occupied seats versus luggage weight). If the

defined threshold is passed, then a situation is raised.

- **Sales Rate (SITDEMO_SALES_RATE)**
  Used in situation ID SITDEMO_FLIGHT_PROFIT-ABILITY. This situation checks the number of sales against the capacity for a given flight. If the defined threshold is not reached, then a situation is raised.

- **Booking Rate**
  Used in situation ID SITDEMO_FLIGHT_PROFIT-ABILITY. This situation checks overbooked flights. If the defined threshold is reached, then a situation is raised.

View Browser (F2170)

Application Type: Transactional

Analytics specialists can use this app to search for core data services (CDS) views that have been released by SAP or customized internally. Details of the CDS views are available in the app, such as category, view type, dimensions, measures, annotations, and the modeling pattern used. Search options are available via filters, such as description, application component, and tables. Key features of the app include the **Views** and **Variants** sections. In the **Views** section, you can view the number and type of CDS views available. The summary number displayed will change with the selection options you have chosen. In the **Variants** section, you can see the standard SAP-delivered variants, all views, favorite views, released views for cloud deployment, released views for key users, and virtual data model views available, as well as customized variants. Variants can be customized and marked as default using this app by selecting the **Manage** button. You can also use various search methods for finding CDS views, including searching based on annotations. It is possible to use the @ and : symbols to search for annotations in combination with the annotation value. For example, searching for "@VDM.VIEW-TYPE:#BASIC" will return all CDS views with the VDM.VIEWTYPE annotation and the #BASIC value.

It's also possible via alternative search to search for CDS views using tables, columns, annotations, and tags. The annotation search is different from the standard search in that the annotation must be selected from the list. To search for CDS views by tags, select the **Tags** option.

Chapter 12
Additional Resources

In this short addendum, we will recommend some additional SAP Fiori resources that we have found useful over the years. These resources include SAP-delivered information as well as independent sources that you can consume at your leisure. Inevitably, some of these resources come at a cost, but we have tried to focus on free resources where possible, with one eye firmly on quality and accuracy.

Although we are focused overwhelmingly on SAP Fiori, it has probably become clear now how interlinked SAP Fiori is with the remaining SAP ecosystem. As a result, we will also include here resources that will be useful in your overall SAP S/4HANA journey in SAP Fiori—for example, information on SAP Business Technology Platform (SAP BTP), which has been mentioned a few times in the book.

12.1   SAP Resources

SAP has myriad SAP Fiori resources at your disposal, many of them 100% free. These resources include introductions, blogs, videos, training manuals, development guidelines, and certification tracks. Let's look at some popular options:

- **SAP Fiori web design guidelines**
  A good place to start is with the SAP Fiori web design guidelines, which can be found at *https://experience.sap.com/fiori-design-web/*.

  Cast your mind back to the Preface of this book, in which we explored SAP Fiori briefly as a design system. All the guidelines for SAP Fiori can be found at this location, including how to deliver a consistent look and feel for SAP Fiori apps, how to use general patterns, the structure of and a guide to using SAP Fiori elements and floorplans, and even guidelines on how to embed Joule and AI into SAP Fiori.

  In addition, the website offers access to the *designer toolkit*, which is designed to be a one-stop-shop destination for all of SAP's design resources and services. This means it offers fonts, icons, *method cards* (which describe the most commonly used research methods), and the SAP design community, where it is possible to ask questions as well as post design ideas.

- **SAP Community**
  Over the years, SAP has had many incarnations of an SAP community, originating with the SAP Developer Network (SDN). Recently, these offerings have merged into a single web page: SAP Community, which can be found at *https://community.sap.com/*.

Think of this page as a kind of SAP central hub, where not only it is possible to search for solutions from other users, but it is also possible to navigate within the page to see SAP events, learn what's new, and find the latest blogs from members.

- **SAP Help Portal**
  SAP Help Portal has been the "go-to" place for information for SAP consultants for a long time. It can be found at *https://help.sap.com/docs/*. From SAP Help Portal, you can access information about the entire SAP portfolio of products.

- **SAP Fiori apps reference library**
  Specifically for SAP Fiori, the SAP Fiori apps reference library is a good resource, which can be found at *https://fioriappslibrary.hana.ondemand.com/sap/fix/externalViewer/*. Here you can learn anything about any app in SAP Fiori, in the public or private cloud. You can also build your own upgrade path to tell you how to identify what will change in your new release of SAP S/4HANA.

- **SAP Fiori roadmap**
  To dig a little deeper into SAP Fiori, you can see an excellent overview of the product and its roadmap at *https://storyboard.cfapps.eu10.hana.ondemand.com/#/story/sap_fiori_roadmap_2021q4*.

  The roadmap is a few years old now, and it has not yet been updated with changes in the last couple of years, but it's still very helpful.

- **SAP Discovery Center**
  SAP BTP is at the center of the SAP roadmap currently, including for SAP Fiori development and integrations, and most innovations are appearing in this space. To find out more about SAP BTP, SAP Discovery Center is an excellent resource, available at *https://discovery-center.cloud.sap/index.html*.

- **SAP for Me**
  It's useful to understand your own SAP system and landscape, including SAP Fiori. How can you look at all the products you have installed as well as all the SAP Notes you have logged, your licensing details, upcoming expiry dates, and the latest services and support on offer? The best resource for this is SAP for Me, which you can find at *https://me.sap.com/home*.

- **SAP newsletters**
  Now let's consider SAP news and events and how you can keep abreast of the latest in SAP technologies and solutions. For this information, there are SAP newsletters available (there are a *lot* of them, so pick wisely!). Each newsletter is aligned with a different area, such as product newsletters, industry newsletters, and so on. There is an SAP Fiori Development Newsletter released on a bi-monthly basis under the "Products" section. You can find these newsletters at *https://www.sap.com/dashboard/newsletters.html*.

- **SAP Learning Hub**
  SAP Learning Hub is a fantastic resource to upskill yourself on SAP Fiori for free. It's

available at *https://learning.sap.com/*. Searching for "SAP Fiori" on SAP Learning Hub as of the time of writing returns 464 results, so there is plenty of content to choose from—everything from a one-hour introductory course on SAP Fiori development to a 20-hour course on the basics of SAP Fiori. This resource is definitely recommended!

■ **SAP Certifications**
This is a part of SAP Learning Hub, but the URL ahead will navigate you directly to the certification page. If you go to *https://learning.sap.com/certifications?page=1* and search for "SAP Fiori", you can find two certification exams for SAP Fiori:

 – SAP Certified Associate – SAP Fiori Application Developer (C_FIORD)

 – SAP Certified Associate – SAP Fiori System Administration (C_FIOAD)

Each certification exam is three hours long, and the website points you to the preparatory courses you can study.

12.2   External Resources

SAP is such a large ecosystem that there are plenty of external resources at your disposal for additional learning and development. Here are some of our favorites:

■ **SAP PRESS**
The SAP PRESS library of SAP Fiori titles is always growing. For a full list of SAP Fiori books and E-Bites, as well as relevant titles about the broader SAP user interface landscape, visit *https://www.sap-press.com/programming/user-interfaces/*.

Also, you can check out online courses on offer from SAP PRESS at *https://www.sap-press.com/online-courses/*.

■ **Stack Overflow**
Stack Overflow has an SAP Fiori community (*https://stackoverflow.com/questions/tagged/sap-fiori*) in which you can post your SAP Fiori questions and receive help and tips from other members.

■ **YouTube**
We all know YouTube, the global leader in online videos. Search for "SAP Fiori", and you will find a lot of content. These are some of our favorite channels:

 – UI5 Community Network: *https://www.youtube.com/@ui5cn*

 – SAP Design: *https://www.youtube.com/@SAP_Design*

 – Secure SAP: *https://www.youtube.com/@SecureSAP*

 – SAP PRESS: *https://www.youtube.com/c/SAPPRESS_Official*

- **Udemy**
  You're probably already familiar with Udemy, which gives you access to user-managed courses on any topic. A quick search for SAP Fiori returns a huge number of results (*https://www.udemy.com/courses/search/?src=ukw&q=SAP+Fiori*), so there are plenty to choose from.

- **Michael Management**
  This website specializes in SAP training, and its courses, while more expensive than Udemy's, are of a high quality. Searching for "Fiori" will return dozens of courses (*https://www.michaelmanagement.com/browse?q=fiori*).

- **LinkedIn**
  No additional resources collection for SAP would be complete without a reference to LinkedIn, which can be a very rich source of information.

  The following are a few of our favorite pages and people on LinkedIn:

  - **SAP Fiori (page)**
    https://www.linkedin.com/products/sap-fiori--sap-fiori/
    A page created by LinkedIn itself to link to content and people who discuss SAP Fiori.

  - **SAP Fiori Group (group)**
    https://www.linkedin.com/groups/5161110/
    A community group for SAP Fiori users and consultants to share their SAP Fiori experiences.

  - **SAP Fiori (group)**
    https://www.linkedin.com/groups/6611596/
    A group focused on SAP Fiori and SAP UI5.

  - **Jocelyn Dart (people)**
    https://www.linkedin.com/in/jocelyndart/
    Jocelyn is a member of the SAP Asia and SAP ANZ Global Shared Services team, focused on user experience, and she regularly posts insightful information about SAP Fiori.

12.3   Conclusion

These resources should give you more information and understanding about your SAP Fiori learning journey. However, the most important learning tool is real-life practice, so our final callout to you is to get your hands dirty! SAP Fiori is a fantastic tool for your business and can offer real benefits in terms of productivity, predictive analytics, and revenue growth. It's up to you now, as an SAP Fiori evangelist, to spread the word. Good luck!

The Authors

Anand Seetharaju is a certified SAP S/4HANA financial and management accounting consultant. He holds active CPA and CMA licenses and has more than 25 years of experience in business process design, consulting, and accounting system implementation. He has led several SAP S/4HANA Finance and Central Finance transformation projects and has presented on those topics at SAP conferences. Additionally, he has extensive implementation experience with the SAP classic general ledger, new general ledger, and SAP S/4HANA general ledger, including conversion from the classic general ledger. Anand's experience includes global SAP finance and controlling implementations and rollouts with an emphasis on receivables management, revenue accounting and reporting, cash and liquidity management, profitability analysis, Project System, product costing, material ledger, capital asset management, leasing, global tax, and legal reporting.

Jon Simmonds has more than 23 years of SAP experience with countless end-to-end SAP implementation projects and system rollouts. Having worked in various industries including manufacturing, media, software, financial services, and life sciences, Jon has fulfilled many different roles, including sales and distribution consultant, SAP Business Warehouse (SAP BW) consultant, ABAPer, and architect.

Tinotenda F. Chiraudi is a trailblazer in digital transformation and SAP implementations. With an extensive background in manufacturing engineering and SAP consulting, he has successfully spearheaded large-scale SAP implementations, specializing in supply chain management and project management. His high-impact solutions have not only optimized business processes but have also inspired teams to embrace innovation, making him a trusted advisor for enterprises seeking impactful, future-proof digital strategies.

Index

I

Q

S

T

Y

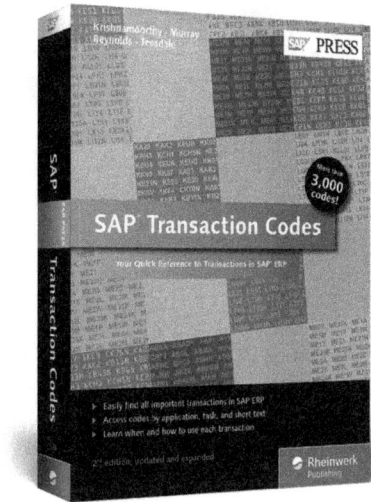

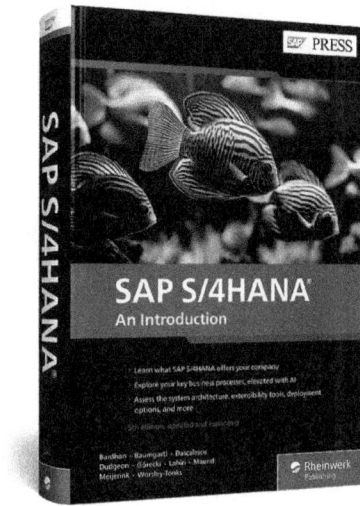

Bardhan, Baumgartl, Dascalescu, Dudgeon, Górecki, Lahiri, Maund, Meijerink, Worsley-Tonks

SAP S/4HANA

An Introduction

Whether you're making the move to SAP S/4HANA, or just curious about what the suite has to offer, this introductory guide is your starting point. Explore your business processes in SAP S/4HANA, now optimized by AI: finance, manufacturing, supply chain, sales, and more. Get insight into SAP S/4HANA's architecture and key features for reporting, sustainability, extensibility, and artificial intelligence. Build your business case, choose your adoption path, and get ready for your implementation!

744 pages, 5th edition, pub. 01/2025
E-Book: $74.99 | **Print:** $79.95 | **Bundle:** $89.99

www.sap-press.com/5973

Rheinwerk
Publishing